Springer Tracts in Advanced Robotics

Volume 162

The Springer Tracts in Advanced Robotics (STAR) publish new developments and advances in the fields of robotics research, rapidly and informally but with a high quality. The intent is to cover all the technical contents, applications, and multidisciplinary aspects of robotics, embedded in the fields of Mechanical Engineering, Computer Science, Electrical Engineering, Mechatronics, Control, and Life Sciences, as well as the methodologies behind them. Within the scope of the series are monographs, lecture notes, selected contributions from specialized conferences and workshops, as well as selected PhD theses.

Special offer: For all clients with a print standing order we offer free access to the electronic volumes of the Series published in the current year.

Indexed by SCOPUS, DBLP, EI Compendex, zbMATH, SCImago.

All books published in the series are submitted for consideration in Web of Science.

Giacinto Barresi · Sascha Wischniewski
Editors

Ergonomics in Robotics: Advances and Innovations

Volume 1

Springer

Editors
Giacinto Barresi
Bristol Robotics Laboratory, School of Engineering, CATE
University of the West of England
Bristol, UK

Sascha Wischniewski
Human Factors, Ergonomics Group
Federal Institute for Occupational Safety and Health (BAuA)
Dortmund, Germany

ISSN 1610-7438 ISSN 1610-742X (electronic)
Springer Tracts in Advanced Robotics
ISBN 978-3-032-19373-5 ISBN 978-3-032-19374-2 (eBook)
https://doi.org/10.1007/978-3-032-19374-2

This work was supported by German Federal Institute for Occupational Safety and Health and International Ergonomics Association.

This Springer imprint is published by the registered company Springer Nature Switzerland AG
The registered company address is: Gewerbestrasse 11, 6330 Cham, Switzerland

Series Editor's Foreword

At the dawn of the century's third decade, robotics is reaching an elevated level of maturity and continues to benefit from the advances and innovations in its enabling technologies. These all are contributing to an unprecedented effort to bringing robots to human environment in hospitals and homes, factories, and schools; in the field for robots fighting fires, making goods and products, picking fruits, and watering the farmland, saving time and lives. Robots today hold the promise for making a considerable impact in a wide range of real-world applications from industrial manufacturing to healthcare, transportation, and exploration of the deep space and sea. Tomorrow, robots will become pervasive and touch upon many aspects of modern life.

The *Springer Tracts in Advanced Robotics (STAR)* is devoted to bringing to the research community the latest advances in the robotics field on the basis of their significance and quality. Through a wide and timely dissemination of critical research developments in robotics, our objective with this series is to promote more exchanges and collaborations among the researchers in the community and contribute to further advancements in this rapidly growing field.

The volume edited by Giacinto Barresi and Sascha Wischniewski is focused on the important research challenge of human–robot interaction from an important aspect: that of ergonomics. The contents are organized in five parts from the fundamental issues to application areas into real-world scenarios, such as industrial, agricultural, social, and medical contexts. The role of the methodologies and technologies as enablers for ergonomics in robotics is also discussed. The human-centric perspective is highlighted, and the societal impact is pondered in terms of culture, trust, values, aging, and sustainability.

A nice blend of original and interesting results in the field of humans and robots, this volume is a great addition to the STAR series.

Naples, Italy
August 2026

Bruno Siciliano
STAR Editor

IEA Preface

As President and Past-President of the International Ergonomics and Human Factors Association (IEA), we are proud to introduce readers to Volume 1 of the first comprehensive book on 'Ergonomics in Robotics'. The IEA is a coalition of more than 62 federated and affiliated national societies of human factors and ergonomics (HFE), as well as 5 regional IEA networks and sustaining corporate and individual members. The purpose of the IEA is to work closely with our coalition to expand the application of HFE and to contribute to society through the improvement of quality of life for all. We therefore work with international partners such as the International Standards Organisation, the International Labour Organisation, the World Health Organisation, and a range of related disciplines to ensure that HFE is embedded in fundamental global initiatives. The official definition of HFE, adopted at the IEA Congress in 2000, is: "*the scientific discipline concerned with the understanding of interactions among humans and other elements of a system, and the profession that applies theory, principles, data, and methods to design in order to optimize human well-being and overall system performance.*" In essence, HFE is a systems discipline that combines our whole-of-person systemic understanding of people (i.e., their anatomical, physiological, biomechanical, psychological, and social dimensions) and their interactions with systems (i.e., environmental, technological, and sociological) to improve system performance while simultaneously improving the quality of life for all.

This book applies an HFE lens to the study of robotic system and systems that include robotics. In its simplest form, this book considers a wide range of human-robot interactions in order to understand what is required to create appropriate human-robotic systems. This means considering aspects such as the physical, cognitive, and organisational components of human-robotic system interactions. These are recognised by the IEA as the three core domains of HFE. In its more complex form, this book also considers the broader implications of human societies that include robotic systems. This means considering aspects such as the implications of human-robotic systems in the context of culture, values, trust, oversight and control, changes to social and community dynamics, and the impact on our environment. In doing so, this book examines the HFE tools and methods that can be used to understand and design human-robotic systems and considers the HFE implications in a range of different application domains including industrial, agricultural, health care, and social contexts.

For decades, human-robotic systems have been considered by much of society to be just science fiction. Yet, robotic systems have slowly, but surely, infiltrated almost all aspects of our society. Robotics are now ubiquitous with industrial production lines, robotics in farming, robotics in hospitality, robotics in the military, robotics for space and planetary exploration, and robotics to help us complete our domestic chores. In some parts of the world it is not unusually to see robots checking us into hotels, vacuuming our homes, mowing our lawns, serving us meals, and delivering parcels to rooms. Sometimes robotic systems are hidden from sight for all but those creating the systems, but more often we are seeing robotic systems becoming both more visible and more physically independent from their human operators.

The melding of artificial intelligence and autonomy with robotics may soon give us robotic systems that may replace many of our mundane, unsafe, and physically demanding tasks. However, these robotic systems will need to appropriately integrate with the needs and abilities of humans. There are instances where robotic systems threaten to take jobs away from people, where poorly designed robotic systems pose potential physical harm to human operators, and where human-robot teaming has not be taken into account. This book is therefore a timely reminder of the importance of embedding HFE into the design processes early to ensure that human-robotic systems are appropriately designed to be useful, efficient, effective, healthy, and safe for humanity now and well into the future.

Finally, we would like to congratulate the members of the IEA's Human Factors in Robotics Technical Committee for this brilliant initiative fully articulated with the IEA's strategic policies where one of the main foci is to promote HFE's state-of-art understanding in emerging technologies.

Prof. Andrew Thatcher
IEA President (2024–2027),
Psychology Department
University of the Witwatersrand
Johannesburg, South Africa

Prof. Jose Orlando Gomes
Immediate IEA Past-President,
Industrial Engineering Department
Federal University of Rio de Janeiro
Rio de Janeiro, Brazil

Introduction

While robotics keeps moving from science fiction to history, we notice how robots integrate (not always successfully) into workplaces, homes, and public spaces. Consequently, the need for making robots truly human-centered (usable and safe, first of all) is one of the main drives in this area. Accordingly, this book serves as an introduction to the interplay between robotics and a multi-faceted domain based on diversified methodologies that can help handling the complexity of human-robot interactions and systems by paying attention to the human factors: ergonomics.

Starting with this first volume, this book will propose an overview of approaches to the study of the interactions between human beings and other elements (robots, in this case) of a system. This will certainly provide bridges between roboticists and ergonomists, but we prefer thinking that it just introduces problems and methods in the diverse domains of human-robot systems inside and, in particular, outside the laboratories. This is why we start from foundational knowledge and know-how to reach a rich (surely not exhaustive) set of examples in real-world contexts (ergonomics should always target ecologically valid settings and tasks, similar to the real ones). Accordingly, this first volume is organised around five core themes, one per section, progressively moving towards examples of actual applications.

The first part will introduce the fundamentals in this domain, exploring basic concepts of human-robot interaction alongside definitions and approaches in the main three branches of the study of human factors—physical, cognitive, and organisational ergonomics—beyond any disciplinary limitations. Hopefully, this book will be adopted by teachers to introduce students of various disciplines (e.g., engineering, psychology, computer science, medicine): we strongly suggest them to keep the first section as a whole as a compulsory reading.

The second part discusses application areas into real-world scenarios where human-robot interaction occur. Industrial, agricultural, social, and medical contexts are presented to offer examples of the contributions offered by one or more branches of ergonomics to the design and development of human-centered robots.

The third part focuses on examples of technologies and methods that can work as enablers for ergonomics in robotics: digital human models, wearable systems, intuitive programming, shared control, and teleoperation are among the solutions that can multiply the applications of robotic devices to solve human problems.

The fourth part proposes examples of developments and opportunities to enhance the approaches mentioned above to make ergonomic robots, especially considering the adaptation to the user, considering their experience, preferences, and challenges, with special attention to the complexities of robot control.

The fifth part moves from the previous ones to ponder the societal perspectives on human-robot systems by considering factors like culture, trust, values, aging, and sustainability: aspects that must be prioritized in any effort to design and assess human-centered robots through the lenses of ergonomics.

Through the chapters in these parts, the book illustrates advances that strengthen and extend established work and innovations that create entirely new possibilities. Furthermore, we mean to highlight not only the diversity in human-centered robotics but also in our audience: researchers, designers, developers, entrepeneurs, policy-makers, stakeholders in any applied domains, as well as anyone curious about the ways robotics is transforming human life. Thus, it can be helpful to anyone interested in this heterogeneous topic, regardless (by virtue, actually) of their background: this book has been devised to welcome everyone willing to improve the human wellbeing and the human-robot systems performance.

This was surely a goal established since the start of this editorial journey. Thanks to the impulse of the International Ergonomics Association (IEA) and its Technical Committee on Human Factors in Robotics, and the Federal Institute for Occupational Safety in Germany (BAuA), we are glad to finally propose this book to start organising the knowledge and know-how on ergonomics in robotics. This will hopefully make us build (and use) smart machines tailored on human needs we cannot yet foresee.

Bristol, UK — Giacinto Barresi
giacinto.barresi@uwe.ac.uk

Dortmund, Germany — Sascha Wischniewski
wischniewski.sascha@baua.bund.de

Contents

Societal Perspectives

Fundamentals

Human–Robot Interaction and Ergonomics

Eva Heinold, Patricia Helen Rosen, and Sascha Wischniewski

Abstract Robots in the workplace are becoming a familiar sight. Hence, they have become the focus of debates and research on workplace ergonomics. While this emerging technology holds great promise to improve environments, this can only be achieved through the intentional application of ergonomic principles. Physical, cognitive and organizational ergonomics must be considered when talking about human-robot interaction (HRI) in a work environment. This chapter presents fundamentals on both human-robot interaction and ergonomics. This includes key topics such as potential roles of humans in HRI, interaction design and interfaces. Furthermore, it presents key domains of HRI and the emerging perspectives in research and application. It underscores the importance of an interdisciplinary approach in assessing the impact of robotics on humans and emphasizes the need for a human-centered design philosophy when developing innovative robotic systems.

Keywords Ergonomics · Robotics · Human-robot interaction · Automation · HRI

1 Introduction

Robots have become an integral part of many workplaces. Their deployment in manufacturing has become a familiar sight; however, their application in service roles, in healthcare, entertainment, transportation, the arts, and numerous other domains of modern life represents new frontiers. Historically, robots were confined to repetitive tasks following preset, unchanging instructions. This made them especially well-suited for tasks in certain sectors, such as manufacturing, while more complex, changing, or socially oriented tasks were out of their scope [1]. Consequently, for safety considerations, physical barriers were maintained between robots and humans.

E. Heinold · P. H. Rosen · S. Wischniewski (✉)
Human Factors, Ergonomics Group, Federal Institute for Occupational Safety and Health (BAuA), Dortmund, Germany
e-mail: wischniewski.sascha@baua.bund.de

G. Barresi and S. Wischniewski (eds.), *Ergonomics in Robotics: Advances and Innovations*, Springer Tracts in Advanced Robotics 162,
https://doi.org/10.1007/978-3-032-19374-2_1

In contrast, today's robotic systems have much more varied task portfolios. They can perform intricate and less structured tasks, even tasks that involve collaborating with humans directly, without physical separation [2]. This evolution has resulted in a new research field, focusing on human-robot interaction (HRI), which aims to find and develop strategies for designing and implementing robots within human-centered environments. HRI research aims to create a knowledge base to formulate principles that enable humans and robots to interact directly, safely, and efficiently, while supporting or enhancing human well-being and improving system performance [3].

Numerous aspects of HRI research intersect with fields such as psychology, anthropology, neuroscience, philosophy, industrial design, and ethics, establishing HRI as an interdisciplinary field. It is only when taking into account the multitude of perspectives that we are able to assess the use of robots for its impact on humans, and thereby are able to create technology with a human-centered approach.

When discussing HRI, it is necessary to define the types of robotic systems one is talking about. A distinction is often made between industrial and non-industrial applications. The international standard ISO 8373:2021 names two types of robots: industrial and service robots [4]. According to ISO 8373:2021, an industrial robot is an automatically controlled, reprogrammable, multi-purpose manipulator, programmable in three or more axes, which can be used either in a stationary or mobile capacity in industrial applications. Its movements or auxiliary functions can be changed without physical modifications. It is multipurpose; it can therefore be adapted to another application by physical changes. These physical changes are related to the mechanical structure or the control system. The axis describes the direction used to move the robot in a linear or rotary way [4]. Service robots are robotic systems that are not used in a fully automated environment and perform useful tasks for humans. A commercial service robot is used for commercial tasks and is typically operated by a trained operator. A personal service robot is used instead for a non-commercial task, such as cleaning a private space, and the primary user group is laypeople [4].

While this binary separation can be helpful in some contexts, it is limited in reflecting the increased complexity in which robotic systems can be used. The robotic guide by the Institute of Electrical and Electronics Engineers (IEEE) offers a more faceted classification for robots based on their application context. They make a distinction between aerospace robots, disaster robots, drones, teaching robots, entertainment robots, exoskeletons, industrial robots, medical robots, or autonomous vehicles [5]. These two approaches differ in their scope. While ISO 8373:2021 draws on the technological properties of the robot, IEEE focuses more on contextual factors, resulting in more fine-grained groups. Knowing what type of robot is used in a given context may help users understand the system and the type of HRI the technology is capable of.

1.1 The Human-Technology-Organization Framework

The distinction based on application context brings to the foreground that a technology and its use should be described and studied under the consideration of a larger context. A framework that is often used in human factors and ergonomics to contextualize a technology is the Human-Technology-Organization (HTO) framework (Fig. 1) [6]. It is used as a guide to understand and study the interactions between humans, technology, and the organization within which they work. The model is based on recognizing that these three factors do not exist independently from one another, but instead have relationships and interdependencies between them. The HTO framework therefore seeks to understand how these three domains influence each other in various contexts. It emphasizes that to design optimized systems and processes, it requires an in-depth understanding of how these three components interact. For instance, an organization might introduce a new technology (such as an interactive robotic system) to reduce physical strain on their workers. However, the success of an implementation depends on how well the technology aligns with the capabilities and limitations of the workers and how the organization adapts established workflows and working environments to integrate and accommodate the new technology. Each of the three foundational elements in the framework encompasses a variety of different facets that can interact with aspects of the other factors.

The human component of the HTO framework focuses on the individuals involved in the system. Considerations regarding the human component of a socio-technological system include their physical, cognitive, and emotional capabilities, as well as skills, knowledge, and limitations of any kind. Understanding these facets is

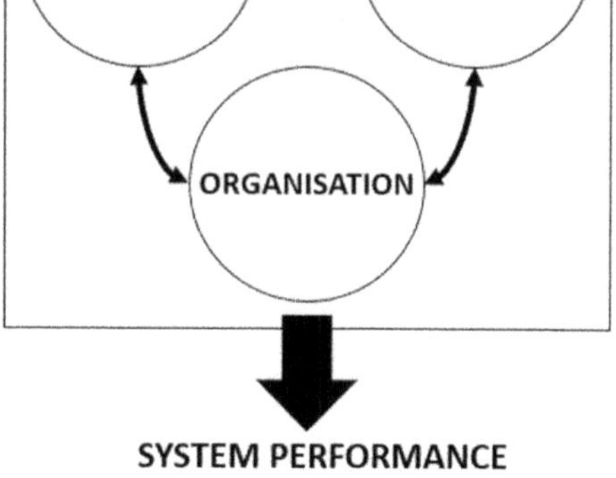

Fig. 1 The HTO framework illustrating how human, technology and organization effect system performance

crucial for designing technology and organizational processes that align with human needs, abilities, and limitations.

The technology component encompasses all tools, equipment, systems, and structures that workers use to perform their tasks. Examples include computers, software, and machinery, but increasingly, robotic systems and artificial intelligence to assist humans in their work. Technology should be designed with user-friendly interfaces, so that workers can easily learn and navigate the system. Furthermore, reliability, as well as adaptability, are important facets of human-centered technology design. Failures and errors should be minimized, while also allowing users to adapt the technology to their preferences or accommodate their needs. Overall, the design of the technology impacts how efficiently and effectively a task can be performed.

Lastly, the organizational component of the HTO-framework refers to the broader context within which human-technology interactions take place. Interaction here refers to reciprocal actions occurring between entities. This includes established organizational structures, internal policies, procedures, culture, internal and external communication, and management practices, such as providing necessary training and support for workers. These organizational factors determine how tasks are assigned, resources allocated, and how information flows within the company [6].

Researchers and practitioners use the HTO framework to analyze and optimize various systems, including, but not limited to, industrial processes and processes involving robotic systems [7]. The HTO framework emphasizes the dynamics and relationships between its three foundational components and that changes in one can ripple through the others. It highlights the necessity of a holistic perspective when it comes to system design and organizational management. By understanding these complexities, it becomes possible to create systems and work environments that are efficient, supportive of human well-being, and productive.

2 Types of Human-Robot Interaction

Apart from the different types of robots, distinctions can also be made regarding the various forms of interaction between humans and robots. These interactions are often categorized on a spectrum, beginning with no interaction between human and robot and ending in full collaboration. Onnasch and colleagues [8] present a taxonomy that can be used to classify different types of human-robot interaction. They differentiate between three basic forms, based on the degree of closeness between the two subjects: coexistence, cooperation, collaboration (see Fig. 2). Aaltonen and colleagues [9] refined these levels of collaboration further and added ‘No coexistence’ as an option.

No coexistence describes the complete physical separation of humans and robots. This may be achieved via fencing or other spatial manners. However, it is possible that humans and robots within this physical separation work on sub-tasks that contribute to a larger, shared goal. It has to be noted, that this would typically not be described as human-robot interaction.

Fig. 2 Types of human-robot Interaction based on [8]

Coexistence refers to scenarios where humans and robots share the same environment without direct coordination or interaction. In this type of interaction, humans and robots exist side by side, with each performing their own tasks independently. For instance, a transportation robot, which crosses the shop floor, while humans engage in other tasks unrelated to and unaffected by the robot. A challenge in coexistence scenarios lies in ensuring safe and efficient navigation and operation in the shared space. Robots must be designed to recognize and adapt to human presence and behavior, to avoid collisions and interference, thereby promoting a sense of safety for humans [8].

Cooperation involves interactions where humans and robots work together towards a common goal, but *their* tasks remain separate and parallel. This type of interaction is common in industrial settings, where robots assist humans by performing tasks that require, for example, extensive repetition, precision, or strength. A typical example of this is the heavy lifting of parts on an assembly line. In cooperative HRI scenarios, creating effective communication and coordination between humans and robots can pose a challenge. Interfaces should allow humans to convey their intentions clearly to the robot while also receiving feedback about the robot's next actions [8].

Collaboration represents the most complex form of interaction, where humans and robots engage in joint tasks in a shared space. Collaborative robots, sometimes referred to as *cobots*, are designed to work alongside humans. Their ability to work in close proximity to humans is often reflected in the quantity and complexity of their sensors. These robotic systems often possess enhanced abilities to perceive their surroundings, in order to create a safe environment for humans. As humans and robots share their tasks in collaborative scenarios, fluid communication is vital. Interfaces capable of real-time communication and dynamic task allocation can support this. Additionally, ensuring that robots respond to human actions in a safe and timely manner enhances the collaborative experience [8].

3 Ergonomics in Human-Robot Interaction

Human factors and ergonomics (HFE) focus on understanding the interactions of humans and other elements in a system. The goal is to optimize the interaction for both human well-being and system performance [10]. HFE is paramount in ensuring seamless and safe interactions between humans and robots. Given the diversity and complexity of HRI, it is crucial that all facets of ergonomics are thoroughly considered and applied. These facets encompass physical ergonomics, which aims to optimize strain and prevent injuries; cognitive ergonomics, which seeks to optimize mental workload and enhance decision-making; and organizational ergonomics, which involves the integration of robots into existing working environments [11] (see Fig. 3).

Physical ergonomics is relevant at virtually all levels of HRI. It supports user comfort and mitigates injury risks. It is based on the premise that physical compatibility between humans and robots, tasks, and environments is important for safe and effective interaction [12] as robots are a physical presence in all listed forms of HRI (coexistence, cooperation, and collaboration) [8]. Even a social robot, primarily tasked with collecting medical data through patient-to-robot conversations, must be evaluated for its physical ergonomic impact. Naturally, there are also robotic systems where their impact on physical ergonomics is among their primary intended effects (e.g., a manufacturing robot, which lifts heavy parts instead of the worker). This involves creating robots with appropriate sizes, shapes, and movements that allow for comfortable and safe interaction. In this context, physical ergonomics aims to prevent strain, discomfort, and injuries during interactions. It can also play an important role in the prevention of long-term strain injuries like musculoskeletal disorders. Ergonomically designed robot interfaces and controls can improve HRI, just like task automation through a robot may reduce physical strain and discomfort, support physical well-being and operational efficiency [13]. Equipping a robotic system with the necessary sensors and abilities to avoid collisions with workers is also vital for their impact on physical ergonomics during the interaction. Finally, when creating interactive robots, the system designers should also take into consideration how to reduce the chance of ergonomic strain arising from extended interactions with them.

Fig. 3 Facets of ergonomics based on [11]

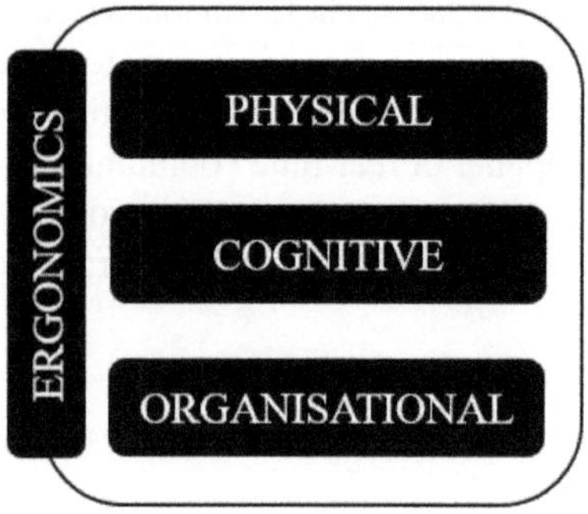

Cognitive ergonomics is central to designing interfaces that support human understanding and is becoming increasingly relevant as robots undertake more cognitive tasks. Optimizing mental workload and enhancing decision-making becomes vital for good HRI [14]. This opens up the challenges to streamline information exchange and optimize user interface designs to ensure seamless interaction between humans and robots. Robots should provide sufficient information that support decision-making without imposing additional cognitive burden by providing too little or too much information [11]. The goal is to minimize the cognitive effort needed to understand and interact with robots, particularly in complex or high-stress scenarios. Cognitive ergonomics may be affected during HRI through the human's perception of and attitude towards the robot. Trust, or more importantly, distrust in the technology is a vital factor for the quality and comfort of using a robot [15]. Furthermore, human-robot interaction may also trigger concerns in humans about their job security. The fear of job loss caused through automation, and by extension, robotic systems may influence the attitude towards the technology negatively [16]. Lastly, using a robot may also come with the need for additional and possibly continuous training for the workforce to learn how to operate the technology safely and efficiently. This may pose additional cognitive load on humans as a result of the robots' deployment. Hence, cognitive ergonomics relating to HRI needs to be considered beyond facets like user interface design and should take into consideration a wider range of influences.

Organizational ergonomics in HRI focusses on the management of robots within existing operational structures and the structural consequences these technologies have on the people working with them. Organizational ergonomics contributes to enhanced worker satisfaction, increased productivity, and reduced disruptions [17] both long- and short-term. The integration of robots into established workflows often necessitates organizational changes. Inadequate implementation planning can result in disruptions, employee resistance, and various inefficiencies. To achieve a seamless integration, considerations towards task allocation, role definitions, communication channels, training, and safety procedures should be made [11]. Organizational ergonomics focuses on change management strategies that create an open-minded reception of robots among the workforce. Worker engagement becomes vital in this context and can take the form of involving employees in the decision-making process, addressing general and safety concerns, and providing safe and open lines of communication. Effective organizational ergonomics promotes an optimal synergy between human and robotic capabilities, recognizing the interdependence of these three facets (physical, psychosocial and organizational).

The HTO framework [6] can be related to various branches of ergonomics, including *physical*, *cognitive*, and *organizational* ergonomics. Physical ergonomics aligns with the human component of the HTO framework by addressing physical strain, range of motion, and physical limitations. Incorporating ergonomic principles into technology and organizational processes can enhance long-term worker health and boost productivity [18]. Cognitive ergonomics aligns with the human component of the HTO framework by considering cognitive abilities. However, it is also reflected in any given technology through usability principles. User interface design, intuitive navigation, and logical workflows support cognitive processes. Clear communication and effective training programs are essential aspects of cognitive ergonomics within the organizational facet of the HTO framework. Organizational ergonomics considers human skills, knowledge, and limitations while assigning tasks and responsibilities. From a technology standpoint, it includes ensuring that the implemented technologies serve specific organizational needs and workflows. Lastly, organizational ergonomics directly relates to the organization component of the HTO framework. It focuses on adjusting previously established organizational structures, internal policies, procedures, culture, internal and external communication, as well as management practices as needed in order to support worker well-being and efficiency.

Looking beyond the fundamental categories of physical, cognitive and organizational ergonomics, one sees the field of *multidisciplinary ergonomics*. This encompasses both interdisciplinary and cross-disciplinary approaches, as it integrates different perspectives and contributions to human-robot interaction in order to achieve a more nuanced understanding of it. A multidisciplinary approach can include everything, from independent contributions to fully developed frameworks. The range of fields spans from neuroergonomics to extended reality (XR) as well AI enhanced ergonomic optimization and beyond. The fundamental goal of improving HRI in a human-centered way, remains at the core of this approach [19].

4 Potential Roles of Humans in Human-Robot Interaction

Humans can assume various roles in human-robot interaction, each of which defines their relationship with the robot. Their roles partially depend on the type of tasks that are automated or supported by the robot. From operators and supervisors to co-workers and bystanders, each role has unique responsibilities and implications for interaction dynamics. Understanding one's role in HRI can help for example in anticipating and preparing for potential trajectories and challenges in the interaction [2].

Robot Operators: Robot operators are responsible for controlling and managing robots' actions. This can take place by directly interacting with the robot, or via a remote system. A teleoperator is a subcategory of robot operator, that faces a unique set of challenges. Maintaining situational awareness and precise control over the robot, while managing interruptions, time delays or sensory limitations in remote environments, can impose additional mental strain on the operator [20]. Designing intuitive and responsive interfaces is crucial to optimize cognitive load and support efficient control in these situations.

Supervisors: Supervisors oversee robots' actions and intervene when necessary, ensuring safe and effective task completion. In environments like warehouses, where robots may coexist, cooperate or even collaborate with humans, supervisors are needed to ensure seamless coordination and intervene in case of anomalies or emergencies. An ergonomic challenge for supervisors lies in managing both the robot's activities and the team of human workers simultaneously. Ergonomic designs that offer clear and accessible information are essential for balancing the cognitive demands of overseeing multiple ongoing tasks and making decisions during running operations.

Co-Workers/Collaborators: The role of co-worker can exhibit a variety of HRI forms. They have a shared workspace, but the type of interaction ultimately is decided by the type of tasks both parties perform. Typically, collaborative robots, or cobots, are designed to work alongside humans, assisting in tasks that require precision or strength. The role of co-workers involves cooperating with the robot, communicating intentions, and to some extent adapting to the robot's capabilities. However, in an ergonomic workplace design the robot should be adapted to the human's capabilities and needs, rather than forcing the human to adapt to the machine's limitations. Co-workers also have the challenge of possible physical risks, as they are in direct proximity of the robotic system. Ergonomic interaction design must prioritize optimizing joint tasks to ensure that humans and robots can work together safely.

Bystanders: While a bystander is not intended to perform any form of interaction with a robotic system, this role still may arise. While in coexistence scenarios it is intended that an operator works in proximity to a robotic system, a bystander may enter such a workspace simply due to the nature of the robotic system's task. For instance, a robotic system designed to transport medicine through a hospital creates a scenario where patients and visitors are considered bystanders. So, while they do not have any responsibilities towards the robot, parameters need to be set to make sure all involved actors are safe and can perform their tasks as intended. The bystander needs to be able to assess possible next actions of the robot, a likely range of behaviors it can exhibit, and what behaviors might be caused due to their presence. One way to communicate this information to a bystander passively is through intentional system design [2].

As displayed in Table 1, these roles do not exist independently from the previously presented forms of human-robot interaction (HRI). Depending on the work environment, a person can hold several roles at the same time, if, for example, several robotic systems are present in the workplace. However, there are some combinations that typically do not exist. For the given examples, it is assumed that there is a single robot present.

If a robot operator directly controls the robotic system and its actions; coexistence is not possible. Teleoperators, who perform this task remotely, would experience collaboration in a special form, as they must not necessarily share a physical workspace with the robot they control. A supervisor typically does not share a task with the robotic system. Their tasks are independent from one another; hence they

Table 1 Types of HRI in relation to the human's role

	Coexistence	Cooperation	Collaboration
Robot (Tele-)Operators		†	†
Supervisors	†		
Co-Workers/Collaborators		†	†
Bystanders	†		

do not cooperate or collaborate. A co-worker or collaborator may, by definition, step into these two forms of human-robot interaction. Depending on the type of robot and task they are to perform, these workers either directly cooperate or collaborate with the robot. As the bystander role is defined by not sharing a higher goal with the robot, they may only experience the form of coexistence. However, they may transition in their role, depending on the situation. It may happen that a mobile robot asks bystanders for assistance in order to accomplish their tasks. An example of this would be asking a bystander to open a door for the robot [21].

5 Interaction Design and Interfaces

User interfaces provide a medium through which a user can interact with a system. They can take various forms and are fundamental for enabling human-technology interaction [22]. The design of these interfaces can enhance system usability but may also be constrained by external factors. Interface modalities refer to the different methods or modes of interaction between users and technology. Every interface modality may not be suitable for all working environments, task designs, or user requirements [23]. However, interface modalities may also help in increasing efficiency, safety, and enhancing inclusion in the workplace. As user interface design evolves with technology, there is no fixed number of modalities; rather, there are established types and ongoing developments.

Command-Line Interfaces (CLI) allow users to interact with the system by typing text-based commands. This necessitates some form of input device, like a keyboard or touchscreen. Today, they are predominantly utilized by developers and system administrators for tasks demanding precision and automation [24].

Graphical User Interfaces (GUI) allow users to interact directly by touching the screen, employing gestures such as tapping, swiping, and pinching. Commonly found in operating systems, web browsers, and most software applications, GUIs are the most widely used interface modality [24]. Input typically comes via additional technology, such as keyboards and a computer mouse.

Touchscreen Interfaces enable users to interact directly by touching the screen, employing gestures such as tapping, swiping, and pinching. They can use elements of GUI, like windows and icons; however, their unique input modalities distinguish them as a separate interface type [25]. Touchscreen interfaces have become widespread with the rise of smartphones and tablets.

These described types of interfaces require the user to be able to manually enter commands. Certain work environments, safety gear, or physical constraints may limit the suitability of these interface types. There are different interface modalities that can be considered to circumvent these limitations; however, their suitability for a task is also dependent on external factors. These are some examples:

Voice User Interfaces (VUI) enable users to interact with devices through voice commands using natural language processing technology. These interfaces have seen a rise in use due to the introduction of virtual assistants and other forms of smart home devices, however they also facilitate human-technology interaction that does not rely on manual input. This makes them well suited for environments where both hands are already occupied [26].

Gesture-Based Interfaces (GBI) allow users to interact with systems using hand movements, gestures, or body language. GBI uses advanced software to track relevant body parts and perform commands accordingly. First popularized in gaming consoles, today they find use in other types of environments, like hospitals, where factors of hygiene and maintaining sterility are highly important [27].

Augmented Reality (AR) Interfaces overlay digital information onto the real world, thereby allowing users to interact with virtual elements in their physical environment. AR applications can be found on smartphones, smart glasses, and AR headsets to facilitate a wide variety of tasks [28].

Brain-Computer Interfaces (BCI) are considered a recent addition to interface modalities. They allow for direct communication between the user's brain and technology. They find diverse applications, though they are frequently discussed in the context of assistive technology and healthcare [29].

Each interface modality has unique characteristics and applications. For system designers to choose the appropriate modality, they need to consider specific user and technology requirements and the context of use. Outside of selecting a suitable form of user interface for a task, technology, and environment, the execution of the user interface design itself is meaningful. Well-designed interfaces enhance user experience as users find it easy to accomplish tasks and interact with the system as a whole. This can streamline the work process, reducing the time and effort users need to complete tasks. It is also possible to decrease the learning curve via good user interface design and reduce errors. In order to create such an interface, system designers may consider incorporating facets from established standards like the interaction principles of the ISO 9241-110 [30].

6 Application Areas

There are areas of application where robots are revolutionizing various aspects of life, including various forms of HRI. The following section describes key domains of robotic application. They are characterized by widespread use of robots in the respective domain or the accumulated efforts and interests to use robots in this domain in the near future. Each domain receives unique benefits from using a robot, but overall, aiming to increase human well-being and efficiency.

Industrial robots are already a cornerstone of modern manufacturing. The use of robotic arms in automobile assembly lines is a well-known example. Typically, these robots perform repetitive tasks that require a high level of accuracy and may pose physical stress on workers, like welding or the assembly of heavy components. Robots help reduce physical strain on human workers, leading to improved worker well-being and prevention of musculoskeletal disorders (MSDs). Although non-coexistence, or the complete separation of humans and robots, is still the norm for many industrial applications, increasingly complex HRI workspaces are emerging [11].

The integration of robots in *agriculture* is changing the industry's landscape significantly. By automating repetitive and strenuous tasks such as fruit picking, weeding, harvesting, and sorting, robots reduce physical strain of farm workers, potentially decreasing the incidence of musculoskeletal disorders and occupational injuries [31]. Advanced robotic systems can also handle pesticide spraying, milking, and precision seeding, tasks traditionally associated with high physical demands and health risks. This shift enables workers to engage in more complex and less physically taxing activities, potentially enhancing job satisfaction and productivity. However, the transition to robotic automation requires addressing job displacement concerns and implementing retraining programs to ensure a harmonious balance between human and robotic contributions in agriculture.

The domain of *social robotics* focuses on robots designed to interact with humans in social contexts. This also implies that a social robot may be deployed in previously mentioned domains like the medical sector. Social robots typically offer a human-like presence, enhancing user comfort and natural communication, particularly in environments where human interaction is essential. Here, design considerations regarding facets of anthropomorphism and zoomorphism are vital. Social robots may also become the most frequently involved robots in coexistence, as they enter public spaces. Hence, social robots may need to be capable of interacting not only with professional users but also with laypeople [32].

The field of *medical robots* is currently highly researched and undergoing rapid changes. Specifically, surgical robots are already supporting procedures, enhancing precision, and minimizing invasiveness. Ergonomically, these robots offer surgeons dexterity and precision beyond human abilities, which translates to reduced tissue trauma and quicker recovery time for patients. However, these are not the only types of robots used or researched in the medical sector. Social robots interact with patients to document their medical history, and transport robots deliver medicine and equipment, subject to great developmental efforts [33].

In healthcare *assistive robots* form a subset of biomedical robotics. They have the potential to aid individuals with temporary or permanent disabilities, providing increased independence and mobility which can significantly improve a user's quality of life. Rehabilitation robots is another crucial area, which facilitates therapy and recovery for patients. Robotic prosthetic limbs may incorporate sensors and actuators to replicate natural movements, unlike their non-robotic predecessors. This can have profound significance to the user, as they may regain functionality and mobility to a previously impossible degree [34].

7 Human-Robot Interaction and Artificial Intelligence

Emerging technologies like Artificial Intelligence (AI) and Machine Learning (ML) impact HRI. AI-driven robots may become capable of adapting, learning, and anticipating human intentions with increasing accuracy. Machine Learning algorithms may soon equip robots with the capacity to not just perform tasks but evolve alongside human collaborators. However, innovation can also pose new challenges. A fundamental building block in HRI is trust in the system. Increasingly complex and capable systems may become less transparent and predictable for humans. Establishing and maintaining trust will be crucial for the future of HRI. This factors into more than just the technology's reliability and capability. Trust in robots can also be related to rising ethical considerations. AI-driven robots may face challenges regarding algorithmic bias and accountability [35]. The balance between human oversight and the delegation of decision-making to machines will also become increasingly important in the future. Privacy and data security are also likely to be central in future debates about human-robot interaction [36].

8 Technologies and Methods

The following section contains an overview of exemplary emerging approaches and methods regarding frontiers in human-robot interaction. While not an exhaustive list, it includes areas of interest which currently see strong research activity. This includes Digital Human Modeling (DHM), occupational wearable robotics and intuitive robot programming. The concept of shared control in assistive robotics is also examined as well as telepresence and teleoperation technologies.

Digital Human Modeling (DHM) involves creating detailed, computer-generated representations of human anatomy, physiology, and behavior to simulate interactions with robots. Using 3D bodyscans, motion capture, and biomechanical simulations DHM creates the opportunity for virtual representation and simulation of HRI. It enables robotic designers to assess ergonomic factors, predict user responses, and optimize robot design for enhanced safety, and efficiency, without direct human involvement [37]. In industrial settings, DHM helps ensure that robots and workspaces are tailored to human capabilities.

Occupational wearable robotics find applications in physically demanding jobs, providing support and assistance to workers. A possible example are exoskeletons used by construction workers to lift heavy loads [38]. These wearable devices augment the wearer's strength, enabling them to handle heavy objects with reduced effort. Being wearable, they can be used in mobile work situations where other forms of permanently installed robots are not viable. However, while they are designed to decrease physical strain on the wearer, any machinery worn close to the body may also introduce new ergonomic risks like pressure points.

Intuitive programming simplifies the complexity of operating robots by enabling users with varying technical expertise to command and control it. Opportunities in developing intuitive interfaces lie in a reduce hurdle to interact with the robotic system [39]. It may also reduce training time with a new robotic system. In order to achieve intuitive programming one can look towards natural language processing, gesture recognition, and possibly augmented reality. Ensuring reliability across different tasks and environments however, remains a challenge, requiring robust error handling, compatibility testing, and feedback mechanisms.

Shared control mechanisms in robotics involve the dynamic allocation of tasks between humans and robots, requiring clear division of responsibilities to support efficiency and safety [40]. Challenges lie in defining interaction modalities while balancing user preferences and system capabilities. Furthermore, the question of responsibility for errors must be addressed. Opportunities lie in adaptive interfaces that adjust control authority based on real-time situational changes and user feedback. Effective shared control systems may not only improve task completion but also optimize cognitive workload on human operators.

Telepresence and teleoperation in human-robot interaction enable users to remotely control robots and experience a sense of presence in distant locations. Telepresence technology leverages cameras, sensors, and communication networks to provide immersive, real-time visual and auditory feedback, allowing users to interact with remote environments as if they were physically present. Teleoperation involves direct manipulation of robots from a distance, facilitating tasks such as remote surgery, hazardous material handling, and space exploration. These technologies expand human capabilities by bridging physical distances and enabling precise, controlled actions in challenging or inaccessible settings [41]. As telepresence and teleoperation continue to evolve, they promise to revolutionize fields like healthcare, industry, and beyond by expanding the scope and efficiency of human-robot collaboration.

9 Developments and Opportunities

While the subject of HRI is in and of itself still new to many workplaces, there are trends emerging that present challenges and opportunities for the field. Designing robots that seamlessly integrate into human environments will become more complex as robotic abilities increase. Shared control and user preferences shape robot adaptation in asisting tasks and concepts such as errors and trust, roboethics, and sustainability in robotic systems shift into focus. Recent advancements and emerging possibilities for HRI have allowed robotic systems to become wider spread than ever before. Innovations such as artificial intelligence and newly developed sensor technologies enable robots to better perceive and adapt to human behavior. Challenges like ensuring safety during close interaction, improving communication and maintaining autonomy in HRI remain relevant through the recent changes [40]. However, if interaction design takes a human-centered approach these advancements may facilitate a more seamless and intuitive interaction.

In order to *quantify human-robot interaction* researchers and developers can fall back on a variety of different quantitative and qualitative methods. Understanding subjective factors such as trust, workload and satisfaction is equally important to the objective output of the interaction to measure the success of an interaction. By utilizing different measures, like self-report questionnaires, physiological monitoring or behavioral analysis, we can move towards a more seamless integration of robots into a wide variety of work environments [41].

Tailoring *robot behavior to user preferences* is essential for optimizing performance and user satisfaction when assisting tasks. Understanding user needs, capabilities, and preferences requires a robust user-centric design approaches that integrate feedback and presents customization options [42]. By accommodating individual preferences through adaptive interfaces and personalized interaction modes, robots can better support workers. Challenges include balancing personalization with universal usability standards and addressing varying cultural or situational preferences.

Anticipatory robots, equipped with advanced artificial intelligence and predictive analytics, hold the promise to foresee and respond to production needs before issues arise. These robots may one day optimize workflows by anticipating equipment failures, predicting maintenance needs, and adjusting operations in real-time to prevent disruptions. This proactive approach not only holds the potential to enhance productivity but also to reduce downtime and operational costs. Moreover, anticipatory robotics may adapt to dynamic manufacturing environments, learning from data to improve performance continuously [43].

Gestures offer another mode of communication in HRI. They enable intuitive exchange by leveraging non-verbal cues. This can elevate the interaction, especially under the constraint that not every work environment allows for other forms of information exchange (e.g. loud environments might interfere with verbal commands). The goal is to design effective and ergonomic gestures, which take into account human motor abilities [44].

Human-swarm interaction (HSI) focuses on facilitating effective interaction between humans and robot swarms. Robot swarms, inspired by biological systems, consist of simple agents working collectively to achieve complex tasks. They offer advantages like robustness, scalability, and flexibility, making them well suited for applications such as search and rescue, surveillance, and environmental monitoring [45]. However, the complexity of managing swarms poses challenges for human operators due to the large volume of information and the interdependencies among agents. Key here is to manage workload while maintaining sufficient situational awareness.

10 Societal Perspectives

Robotic systems have a complex relationship to society. It is important to address cultural and ethical implications for HRI. Cultural norms and personal attitudes can influence comfort and acceptance towards robotics systems [46]. This also includes acknowledging that cultural factors shape expectations and fears about robot integration. At the same time, it is likely that robotic systems will impact how society is shaped and help us tackle some of the burdens we´ll face in the future. An aging society can benefit greatly from robotic systems [47]. However, these benefits will only be accessible if ethical considerations are taken in the progress we make, which includes addressing biases and ensuring equitable access to the technology. Overall, it is highly important that we design robotic systems that align with societal needs, values and challenges.

How we interact with a technology is shaped by *cultural aspects*. Cultural norms, values, and societal expectations significantly affect how individuals perceive and engage with robots. For instance, cultures with high uncertainty avoidance may exhibit more apprehension towards robots, potentially heightening social anxiety during interactions [48]. Additionally, familiarity with technology, societal trust in automation, and cultural attitudes towards non-human agents can either mitigate or exacerbate these anxieties. Understanding these cultural nuances is essential for designing robots that are socially acceptable and effective across diverse populations.

Establishing *trust in robotic systems* involves transparency, reliability, and predictability in their behavior and their decision-making processes [49]. Challenges include addressing perceived risks associated with autonomous actions and overcoming psychological barriers to acceptance in human-robot interactions. One way to improve trust is by designing intuitive interfaces that provide clear feedback on system status and intentions [16]. Building trust requires robust safety protocols, testing for reliability, and adherence to ethical guidelines to mitigate concerns regarding privacy, security, and unintended consequences of HRI.

Roboethics encompasses ethical considerations arising from the development, deployment, and impact of robotic technologies on individuals and society. Roboethics navigates ethical dilemmas related to privacy, autonomy, accountability, and the distribution of benefits and risks associated with robotic automation [50]. Establishing ethical frameworks and guidelines that promote responsible innovation and address societal concerns is vital. Ideally this topic should be an interdisciplinary collaboration among engineers, ethicists, policymakers, and stakeholders to anticipate ethical implications, develop governance mechanisms, and uphold principles of fairness, transparency, and accountability in robotic design and implementation [51].

The global trend towards an *aging society* poses new challenges and opportunities for robotic systems. They can support the health, mobility and independence of older adults, while alleviating pressures on caregivers and the healthcare system [52]. Here it is important that they enhance the autonomy of elderly people, while optimising caregiver workloads, instead of increasing stress or failing to perform tasks to the level of satisfaction needed for the caregiver to benefit from the automation. Following the human-centric perspectives of Society 5.0 and Health 5.0 can help tackle these issues. Robots may offer physical assistance as well as social companions that reduce loneliness. When designing robotic systems for these purposes it is important to create intuitive interfaces and adaptive systems that improve accessibility and acceptance among elderly users.

Integrating *sustainability* principles in robotics involves optimizing energy efficiency, reducing environmental impact, and promoting responsible resource use throughout the lifecycle of robotic systems. Sustainability can be increased by integrating sustainable materials and promoting the recyclability of robotic components [53]. The goal is to reduce the carbon footprints, manage electronic waste, and ensure ethical sourcing of materials in manufacturing processes of robots. Furthermore, HFE focuses on designing robotic systems that are user-friendly and maintainable, reducing the need for frequent replacements and associated resource consumption. Ultimately, incorporating HFE considerations in sustainability efforts supports the development of eco-friendly robotic systems that support long-term environmental goals.

11 Conclusion

Robotic systems have the potential to fundamentally change the way we work. However, in order to reap the benefits that technological advancements provide, it is important to approach them with human factors in mind. It is important to analyze the interplay between human, robots and organizational structures, and how they are affected by and through HRI. Adaptive, user-centered design of intuitive interfaces and human-centered workplace design is vital for this transition. Frameworks like the HTO can help guide this process. Emerging trends in artificial intelligence, also pose challenges and opportunities to the field. Furthermore, considerations regarding the ethics of robotic use, and the nuances between different cultural sensitivities and accessibility must not be overlooked. A shared factor amongst all the emerging fields in HRI is, that an interdisciplinary approach will be key to creating a future where humans and robots can work together in a productive and ergonomic way.

References

1. Hägele, M., Nilsson, K., Pires, J.N., Bischoff, R.: Industrial robotics. In: Siciliano, B., Khatib, O. (eds.) Springer Handbook of Robotics. Springer Handbooks. Springer, Cham. (2016). https://doi.org/10.1007/978-3-319-32552-1_54
2. Onnasch, L., Roesler, E.A.: Taxonomy to structure and analyze human-robot interaction. Int. J. Soc. Robot. **13**, 833–849 (2021). https://doi.org/10.1007/s12369-020-00666-5
3. Bartneck, C., Belpaeme, T., Eyssel, F., Kanda, T., Keijsers, M., Šabanović, S.: Human-robot interaction: an introduction. Cambridge University Press (2020)

4. International Organization for Standardization (ISO).: Robots and robotic devices–vocabulary (ISO 8373: 2021) (2021)
5. Guizzo, E.: Types of robots (2023). https://robotsguide.com/learn/types-of-robots
6. Strohm, O., Ulich, E.: Unternehmen arbeitspsychologisch bewerten: Ein Mehr-Ebenen-Ansatz unter besonderer Berücksichtigung von Menschen, Technik und Organisation, pp. 9–10, 265–266 (1997) (vdf Hochschulverlag AG an der ETH Zürich, Zürich)
7. Kopp, T., Baumgartner, M., Kinkel, S.: Success factors for introducing industrial human-robot interaction in practice: an empirically driven framework. Int. J. Adv. Manuf. Technol. **112**, 685–704 (2021)
8. Onnasch, L., Maier, X., Jürgensohn. T.: Mensch-Roboter-Interaktion-Eine Taxonomie für alle Anwendungsfälle, p. 5. Bundesanstalt für Arbeitsschutz und Arbeitsmedizin, Dortmund (2016)
9. Aaltonen, I., Salmi, T., Marstio. I.: Refining levels of collaboration to support the design and evaluation of human-robot interaction in the manufacturing industry. Procedia CIRP **72**, 93–98 (2018). https://doi.org/10.1016/j.procir.2018.03.214
10. International Ergonomics & Human Factors Association (IEA).: What Is Ergonomics (HFE)? https://iea.cc/about/what-is-ergonomics/
11. Rosen, P.H., Heinold, E., Fries-Tersch, E., Wischniewski, S.: Advanced robotics, artificial intelligence and the automation of tasks: definitions, uses, policies and strategies and occupational safety and health. EU-OSHA (2022). https://doi.org/10.2802/681779
12. De Santis, A., Siciliano, B., De Luca, A., Bicchi, A.: An atlas of physical human-robot interaction. Mech. Mach. Theory **43**(3), 253–270 (2008)
13. Peeters, M.C., Plomp, J.: For better or for worse: the impact of workplace automation on work characteristics and employee well-being. In: Digital Transformation-Towards New Frontiers and Business Opportunities. IntechOpen (2022). https://doi.org/10.5772/intechopen.102980
14. Coronado, E., Kiyokawa, T., Ricardez, G.A.G., Ramirez-Alpizar, I.G., Venture, G., Yamanobe, N.: Evaluating quality in human-robot interaction: a systematic search and classification of performance and human-centered factors, measures and metrics towards an industry 5.0. J. Manuf. Syst. **63**, 392–410 (2022)
15. Hancock, P.A., Billings, D.R., Schaefer, K.E., Chen, J.Y., De Visser, E.J., Parasuraman, R.: A meta-analysis of factors affecting trust in human-robot interaction. Human Factors **53**(5), 517–527 (2011). https://doi.org/10.1177/0018720811417254
16. McClure, P.K.: "You're fired", says the robot: the rise of automation in the workplace, technophobes, and fears of unemployment. Soc. Sci. Comput. Rev. **36**(2), 139–156 (2018). https://doi.org/10.1177/0894439317698637
17. Carayon, P., Smith, M.J.: Work organization and ergonomics. Appl. Ergon. **31**(6), 649–662 (2000)
18. Dul, J., Neumann, W.P.: Ergonomics contributions to company strategies. Appl. Ergon. **40**(4), 745–752 (2009)
19. Chikwendu, O.C., Ezeanyim, O.C., Igbokwe, N.C.: Human-Robot interaction enhancement through ergonomics and human factors: future directions. Int. J. Eng. Res. Dev. **19**(6), 34–40 (2023)
20. Vidulich, M.A., Tsang, P.S.: Mental workload and situation awareness. Handbook of Human Factors and Ergonomics, 243–273 (2012)
21. Funk, M., Rosen, P.H., Wischniewski, S.: Human-centered HRI Design–the More Individual the Better?. Behavioural Patterns And Interaction Modelling For Personalised Human-Robot Interaction (2020)
22. Blair-Early, A., Zender, M.: User user interface design principles for interaction design. Des. Issues **24**(3), 85–107 (2008)
23. Oviatt, S., Cohen, P.R.: Definition and types of multimodal interface. In: The Paradigm Shift to Multimodality in Contemporary Computer Interfaces. Synthesis Lectures on Human-Centered Informatics. Springer, Cham (2015). https://doi.org/10.1007/978-3-031-02213-5_1
24. Stephenson, N.: In the Beginning was the Command Line, pp. 1–60. Avon Books, New York (1999)

25. Saffer, D.: Designing Gestural Interfaces: Touchscreens and Interactive Devices. O'Reilly Media, Inc. (2008)
26. Pearl, C.: Designing Voice User Interfaces: Principles of Conversational Experiences. O'Reilly Media, Inc. (2016)
27. Wachs, J.P., Stern, H.I., Edan, Y., Gillam, M., Handler, J., Feied, C., Smith, M.: A gesture-based tool for sterile browsing of radiology images. J. Am. Med. Inform. Assoc. **15**(3), 321–323 (2008)
28. Billinghurst, M., Grasset, R., Looser, J.: Designing augmented reality interfaces. ACM Siggraph Comput. Graph. **39**(1), 17–22 (2005). https://doi.org/10.1145/1057792.1057803
29. Kawala-Sterniuk, A., Browarska, N., Al-Bakri, A., Pelc, M., Zygarlicki, J., Sidikova, M., Gorzelanczyk, E.J.: Summary of over fifty years with brain-computer interfaces—a review. Brain Sci. **11**(1), 43 (2021)
30. International Organization for Standardization (ISO).: Ergonomics in Human-System-Interaction—Part 110: Interactionprinciples. (ISO 9241-110:2020-05) (2020)
31. Benos, L., Bechar, A., Bochtis, D.: Safety and ergonomics in human-robot interactive agricultural operations. Biosys. Eng. **200**, 55–72 (2020)
32. Ferrara, E., Varol, O., Davis, C., Menczer, F., Flammini, A.: The rise of social bots. Commun. ACM **59**(7), 96–104 (2016). https://doi.org/10.1145/2818717
33. Dupont, P.E., Nelson, B.J., Goldfarb, M., Hannaford, B., Menciassi, A., O'Malley, M.K., Yang, G.Z.: A decade retrospective of medical robotics research from 2010 to 2020. Sci. Robot. **6**(60), eabi8017 (2021). https://doi.org/10.1126/scirobotics.abi8017
34. Bogue, R.: Exoskeletons and robotic prosthetics: a review of recent developments. Ind. Robot. **36**(5), 421–427 (2009). https://doi.org/10.1108/01439910910980141
35. Rosen, P.H., Donoghue, R., Heinold, E., Moore, P., Niehaus, S., Wischniewski, S.: Cognitive automation: implications for occupational safety and health. EU-OSHA (2022). https://doi.org/10.2802/979082
36. Rosen, P.H., Heinold, E., Fries-Tersch, E., Wischniewski, S.: Advanced robotics and automation: implications for occupational safety and health. EU-OSHA (2022). https://doi.org/10.2802/789276
37. Malik, A.A., Brem, A.: Digital twins for collaborative robots: a case study in human-robot interaction. J Rob Comput-Integr Manuf (2021)
38. Zhu, Z., Dutta, A., Dai, F.: Exoskeletons for manual material handling—a review and implication for construction applications. Autom. Constr. **122**, 103493 (2021). https://doi.org/10.1016/j.autcon.2020.103493
39. Lentini, G., Grioli, G., Catalano, M.G., Bicchi, A.: Robot programming without coding. In: 2020 IEEE International Conference on Robotics and Automation (ICRA), pp. 7576–7582. IEEE (2020)
40. Gualtieri, L., Rauch, E., Vidoni, R.: Emerging research fields in safety and ergonomics in industrial collaborative robotics: a systematic literature review. Robot. Comput.-Integr. Manuf. **67**, 101998 (2021)
41. Abdulazeem, N., Hu, Y.: Human factors considerations for quantifiable human states in physical human-robot interaction: a literature review. Sensors **23**(17), 7381 (2023)
42. Mohebbi, A.: Human-robot interaction in rehabilitation and assistance: a review. Curr. Robot. Rep. **1**(3), 131–144 (2020)
43. Huang, C.M., Mutlu, B.: Anticipatory robot control for efficient human-robot collaboration. In: 2016 11th ACM/IEEE International Conference on Human-Robot Interaction (HRI), pp. 83–90. IEEE (2016)
44. Bartneck, C., Belpaeme, T., Eyssel, F., Kanda, T., Keijsers, M., Šabanović, S.: Human-Robot Interaction: An Introduction. Cambridge University Press (2024)
45. Din, A., Jabeen, M., Zia, K., Khalid, A., Saini, D.K.: Behavior-based swarm robotic search and rescue using fuzzy controller. Comput. Electr. Eng. **70**, 53–65 (2018S)
46. Bartneck, C., Suzuki, T., Kanda, T., Nomura, T.: The influence of people's culture and prior experiences with Aibo on their attitude towards robots. AI & Soc. **21**, 217–230 (2007)

47. Pino, M., Boulay, M., Jouen, F., Rigaud, A.S.: "Are we ready for robots that care for us?" Attitudes and opinions of older adults toward socially assistive robots. Front. Aging Neurosci. **7**, 141 (2015)
48. Kim, D., Kim, S.: A model for user acceptance of robot journalism: Influence of positive disconfirmation and uncertainty avoidance. Technol. Forecast. Soc. Chang. **163**, 120448 (2021)
49. Hancock, P.A., Billings, D.R., Schaefer, K.E., Chen, J.Y., De Visser, E.J., Parasuraman, R.: A meta-analysis of factors affecting trust in human-robot interaction. Hum. Factors **53**(5), 517–527 (2011)
50. Bartneck, C., Lütge, C., Wagner, A., Welsh, S.: An Introduction to Ethics in Robotics and AI, p. 117. Springer Nature (2021)
51. Veruggio, G., Operto, F.: Roboethics: a bottom-up interdisciplinary discourse in the field of applied ethics in robotics. In: Machine Ethics and Robot Ethics, pp. 79–85. Routledge, (2020)
52. Smarr, C.A., Mitzner, T.L., Beer, J.M., Prakash, A., Chen, T.L., Kemp, C.C., Rogers, W.A.: Domestic robots for older adults: attitudes, preferences, and potential. Int. J. Soc. Robot. **6**, 229–247 (2014)
53. Chourasia, S., Tyagi, A., Pandey, S.M., Walia, R.S., Murtaza, Q.: Sustainability of industry 6.0 in global perspective: benefits and challenges. Mapan **37**(2), 443–452 (2022)

Physical Ergonomics in Robotics

Pauline Maurice, Luka Peternel, Wansoo Kim, Simona Crea, Vineet Vashista, Giacinto Barresi, Alberto Ranavolo, Anna Bucchieri, Giulia Aurora Albanese, Michael Holmes, and Jacopo Zenzeri

Abstract With the emergence of embodied intelligent systems like collaborative robots (cobots), the landscape of physically demanding work is changing. Indeed, robotic devices can augment human strength, optimize performance, and minimize the physical demands associated with a job or a task. These achievements of human-robot interaction (HRI) surely match the objectives of physical ergonomics: making physical interactions between humans and other elements of a system more efficient, more gratifying, and safer. One of the key goals of physical ergonomics certainly is to prevent work-related musculoskeletal disorders (WMSDs). Therefore, a cobot must

---

P. Maurice, L. Peternel contributed equally to this work.

P. Maurice · A. Bucchieri
Université De Lorraine, CNRS, Inria, LORIA, Nancy, France

L. Peternel (✉)
Cognitive Robotics, Delft University of Technology, Delft, The Netherlands
e-mail: l.peternel@tudelft.nl

W. Kim
Department of Robotics, Hanyang University ERICA, Ansan, Korea

S. Crea
The BioRobotics Institute, Scuola Superiore Sant'Anna, Pisa, Italy

Department of Excellence in Robotics & AI, Scuola Superiore Sant'Anna, Pisa, Italy

V. Vashista
Human-Centered Robotics Lab, Indian Institute of Technology Gandhinagar, Gandhinagar, India

G. Barresi
Bristol Robotics Laboratory, School of Engineering, CATE, University of the West of England, Bristol, UK

A. Ranavolo
Department of Occupational and Environmental Medicine, Epidemiology and Hygiene, INAIL, Rome, Italy

G. A. Albanese · J. Zenzeri
ReWing S.r.l., Milan, Italy

M. Holmes
Department of Kinesiology, Brock University, St. Catharines, Canada

G. Barresi and S. Wischniewski (eds.), *Ergonomics in Robotics: Advances and Innovations*, Springer Tracts in Advanced Robotics 162,
https://doi.org/10.1007/978-3-032-19374-2_2

have insight into the effect of physical stress on the human body. One of the most common ways to provide that is to use the established ergonomics look-up tables that indicate which working configurations and conditions may cause WMSDs based on experience. Another approach is to use musculoskeletal models to get real-time estimates of key internal body parameters, such as joint torques, muscle forces, and muscle fatigue. A cobotic or, in general, robotic system can then use these ergonomic indexes and variables to guide a human co-worker to more favourable configurations and conditions. In this chapter, we examine physical human-robot interactions (pHRIs) to have an introductory overview of possible scenarios and constraints. We then look into the fundamentals of physical ergonomics such as anthropometry, biomechanics, and environmental factors, including important ergonomic assessment tools and innovative indicators developed in robotics. Subsequently, concrete examples of methods that address physical ergonomics with collaborative robotics are examined in terms of design and control methods.

Keywords Robotics · Physical human-robot interaction · Ergonomics · Biomechanics · Human modeling · Work-related musculoskeletal disorders · Robotic assessment

1 Introduction

Robots are becoming indispensable assistants for various daily tasks in both industrial and household environments. The research efforts on human-robot interaction design and assessment aim to propose novel ways how to efficiently and safely integrate robots into our environments to perform goal-directed human-robot interactions (HRIs) [1]. Robots and humans have complementary strengths. Traditionally, robots tend to have superior strength, speed, and precision, while humans show higher flexibility across perceptual and cognitive processes [2]. The design of ergonomic physical human-robot interaction paradigms aims to find impactful solutions to best combine these strengths in a way to optimise the human-robot teamwork. Specifically, it focuses on the direct and indirect physical effects of the robot morphology and activity on the human well-being and the system performance.

This chapter will focus specifically on physical ergonomics, which investigates and designs interactions according to the anatomical, anthropometric, physiological, and biomechanical characteristics as they relate to physical activity [3]. Beyond traditional basic studies of physical ergonomics on the user posture while sitting at a desk or the ease of holding and moving a tool, it is evident that understanding the intricate relationship between job demands and worker capacity is essential, especially on how these factors evolve over time and contribute to work-related musculoskeletal disorders (WMSDs) [4]. WMSDs are a set of degenerative and inflammatory diseases, with high incidence and prevalence, that impact the muscles, cartilage, spinal discs, tendons, ligaments, peripheral nerves, and joints. WMSDs are associated with manual activities, such as manual material handling (MMH), performed by workers with significant, repetitive, or prolonged physical efforts. Any occupational task that involves lifting, lowering, carrying, pushing, pulling, grasping, or otherwise moving or restraining an object with the application of human force is considered an MMH activity. Most of these WMSDs do not manifest suddenly but rather develop gradually from repetitive microtrauma that accumulates over years. Such insidious onset of WMSDs makes early injury detection challenging, often necessitating a reactive approach in implementing ergonomic solutions. For this reason, ergonomists and occupational health and safety technicians have the aim of designing targeted (pro-active) ergonomic interventions for prevention. Considering how vast this field can be, this chapter will introduce the reader to the main aspects of physical ergonomics in human-robot interaction (HRI) by dedicating special attention to WMSDs as sample target of human factors research and human-centered design in robotics, especially in cobotics.

WMSDs remain a pressing issue and are often the most reported reason why people miss work. WMSDs are associated with high direct and indirect financial costs (to both employees and employers) and their consequences on health are far-reaching. Occupational biomechanics research has evolved our understanding of how physical work requirements (including forces, repetition and postures) contribute to WMSDs [5, 6]. Yet, over the past few decades, the global prevalence of WMSDs has not declined to the extent many would expect [7–10], given vast research in the fields of work disability, health, safety and ergonomics. Systematic reviews suggest that many participatory ergonomics programs and multifaceted workplace interventions only have a moderate effect on reducing workplace injury [11]. Whatever the reason may be for this–some of which may include poor implementation, policy or a research failure–there is exciting potential for our field moving forward. As we look to the future of work, characterised by increasing automation and robotic systems, there emerges a unique opportunity for physical ergonomics to have a significant impact on the prevention of WMSDs. Specifically, the integration of physical ergonomics with robotics represents a promising and emerging area of research.

There is a growing trend of providing humans with access to robotic devices that can augment strength, optimize performance, and reduce the physical demands associated with various jobs and occupational tasks [1]. While it may be apparent that wearable devices and collaborative robots (cobots), when optimized, can reduce physical work demands, robotics can also be harnessed to simulate and evaluate physical job demands. This capability opens doors to quantitative assessments of human capacity, potentially offering biomarkers and predictive (rather than reactive) solutions for workplace changes. In addition, robotics can evaluate strength and sensorimotor function following injury to help guide the return-to-work process. With a shift towards increased automation and devices designed to reduce physical demands in modern workplaces, integrating robotics into the entire ergonomics process takes on critical importance.

The first section of the chapter examines two key aspects of physical HRI: types of robots and interactions. These define the basic framework and constraints we start with when addressing physical ergonomics in specific human-robot collaboration scenarios. The second section provides some key fundamentals of physical ergonomics: anthropometry, biomechanics, environmental factors, ergonomic assessment tools and innovative indicators developed with the use of robotics. These can serve as building blocks to analyse and address ergonomics with an interactive robotic system. The third section looks at various concrete examples of how ergonomics can be addressed with different human-robot collaboration systems. We categorized the scenarios into robot design and robot control.

2 Physical Human-Robot Interactions

When risk factors for WMSDs are detected in a workplace, the first approach is usually to try and act on the work organization, either temporal or spatial, to reduce the exposure to biomechanical risk factors. It is also recommended to train workers for best practice, especially in terms of posture, for instance through verbal, written or visual instructions. Yet, many cases remain where the WMSDs risk factors cannot be reduced enough through traditional ergonomic interventions. In such situations, robots that can work in physical interaction with workers can be used as tools to influence the working conditions and improve ergonomics. There are different scenarios of physical HRI (pHRI). In this section, we start to explore and categorize the most common pHRI scenarios by types of robots (e.g., robotic arms, mobile robots, exoskeletons) and types of interactions (e.g., coexistence, cooperation, collaboration).

2.1 Types of Robots

The most common robots in pHRI are cobotic arms. These devices are designed to work alongside workers, assisting in tasks that require strength, or repetitive motions in co-carrying, handover, co-manipulation, etc. (a more detailed description of such tasks is given in the next sections). Mobile robots are autonomous systems that navigate and operate within the workspace, where the task requires mobility. A growing trend within this category is the development of mobile cobots, which combine the mobility of autonomous robots with collaborative features. The integration of mobile cobots into the workspace enhances flexibility and reconfigurability [12].

Another major type of robot is the exoskeleton, which is a wearable robot attached to the human limbs or body and performs coupled movements with humans. The role of occupational exoskeletons is to augment the existing physical capabilities of humans. Exoskeletons can be either passive or active [13]. Passive exoskeletons are not actuated and support workers with passive elements such as springs. In this case, to influence ergonomics, the optimization has to be done in the stage of mechanical design of the exoskeleton so that it supports in the desired way. On the other hand, active exoskeletons are actuated by motors and can support human movements actively during the task execution. This gives more dynamic possibilities to optimize ergonomics online. Exoskeletons can also be divided depending on the body part they assist: back [14], upper-limb (for manipulation tasks) [15], or lower-limb (for locomotion tasks) [16]. Note that cobots are the main focus of this chapter and the peculiarities of exoskeletons themselves are more extensively covered in the chapters on occupational wearable robotics and rehabilitative, assistive, and prosthetic robotics.

Small wearable devices such as vibrotactile devices have also been designed to monitor and assist workers in real-time. These devices can provide feedback on posture and muscle activity, to help avoid awkward postures by guiding workers on how to perform the tasks in an ergonomic way [17].

Finally, a peculiar case of interaction is what is called physical Human Aerial Robot Interaction (pHARI). Aerial robots widely known as drones are used in many applications such as aerial imaging, surveillance, mapping, payload transportation, agriculture etc. Furthermore, recent developments in mechanical design and robust control modalities have enabled aerial robots to physically interact with humans in performing physical tasks. The researchers have explored this pHARI paradigm in applications such as collaborative aerial transportation [18, 19], guiding a differently abled person [20], package delivery [21] etc. In such physical interaction tasks with the aerial robots, ergonomics plays a pivotal role in enabling effective, safe, and smooth interaction.

2.2 *Types of Interactions*

Here we examine more in detail the types of HRI that can physically affect (directly or indirectly) the human being with a set of representative tasks. We categorize them based on the relative proximity and frequency of interactions: coexistence where humans and robots share the same environment without direct interaction, cooperation where humans and robots interact and work together towards a common goal with their tasks separate and parallel, and collaboration where humans and robots engage in joint tasks in a shared space. For an alternative or a more extensive analysis of the taxonomy in HRI, one can refer to literature (e.g., [22]).

A coexistence is where the robot works next to the human without barriers, but they both have their own tasks, which can themselves involve physical work. The aim is usually to avoid physical interaction in these scenarios and is often treated as a threat to safety. The vast majority of robots are used in this way today in real industrial settings. In this scenario, safety is often the main focus as the robot is typically not torque-controlled and since there is no physical interaction the robot cannot be directly used as a tool to help the human physically to improve ergonomics related to the tasks (i.e., through physical guidance, etc.). On the other hand, the physical separation of humans and robots in this context can be considered an intervention of ergonomics in terms of improvement of user safety according to the hazardous presence of the robot. Thus, it can be categorised as a solution of physical ergonomics in the setting where the human-robot system is located.

In cooperation, the human and the robot both work on the same general task, but each has their own subtask to carry out. For instance, they work on the same object, but sequentially and the physical interaction may happen only for a short time. A representative task is object handover, where the object has to pass from the robot to the human, or vice-versa, at a specific location and time [23]. While spatial and temporal coordination is essential, transitional interaction forces are also very important. Another aspect of pHRI is the scheduling of different actions, which puts more focus on temporal coordination. The sequence and timing of required actions to complete a collaborative task can be adjusted to optimize physical ergonomics [24].

Collaboration is when the human and the robot both work on the same object simultaneously, with different roles. In this scenario, the human and the robot are often continuously physically coupled through a common object. This is often the case in tasks such as carrying, where a heavy bulky object needs to be moved from one location to another [25]. For instance, the human-robot system is adjusted through the mechanical application of forces pushing or pulling the operator in different directions for improving their posture. In co-manipulation, while physical interaction can be continuous, the human and the robot are not necessarily physically coupled through a common object. In this case, the robot might be holding an object while the human is holding a tool or a machine to perform some actions on the held object [26].

Teleoperation is a concept where humans can interact with a remote robot at a distance. We categorize it as a special case of collaboration where the interaction happens through interfaces that enable the human operator to influence the actions of the robot and also perceive what is happening at the remote site. A direct pHRI happens when the interface used is a haptic device, i.e., another type of robot, which can render force feedback for the human. Such teleoperation is classified as bilateral since the information flow between the human operator and the remote robot is bidirectional. With such communication established, there are possibilities to adjust the robot or haptic device in a way to account for physical ergonomics [12].

3 Fundamentals of Physical Ergonomics

When humans are physically interacting with a robot, the first and major concern is about the safety of the human operator, since there are no physical barriers between the human and the robot. Safety in pHRI has been addressed in several ways: by limiting the overall robot speed or power [27] (e.g., ISO 15066 standard [28]), by adapting the robot trajectory to avoid collision [29], or by monitoring the needed deceleration to stop the robot before collision [30]. Once safety is ensured, the next major challenge is physical ergonomics. Physical ergonomics is a crucial aspect of workplace design, focusing on the physical interaction between workers and their tasks, tools, and environment [31].

Despite efforts to improve working conditions, many jobs still require workers to perform physically demanding tasks, which can lead to musculoskeletal disorders and other health issues [32]. Among working adults, WMSDs are the single largest category of work-related illness, the primary cause of global morbidity and work disability, and amount to almost 60% of all work-related injuries [33]. When reporting WMSDs, the specific body areas most impacted vary by region and occupational sector, with links to the job-specific manual materials handling demands. However, low back and upper extremity disorders are at the forefront of most prevalence statistics and estimates of financial burden. 37% of all low back pain cases can be attributed to occupation [34] and it is estimated that upper extremity WMSDs cost close to 2% of the European Union's gross national product [35]). While the financial consequences of WMSDs are well established, the true employee (i.e., human) burdens are more difficult to quantify.

To address this, several important aspects have to be considered, such as anthropometry, biomechanics, environmental factors, and key indexes and parameters. This section provides an overview of each of these aspects.

3.1 Anthropometry

Anthropometry is the study of the physical characteristics of the human body, encompassing aspects such as height, limb length, shape, body mass, etc. These characteristics are not uniform; they vary widely based on factors like gender, age, ethnicity, and occupation, which in turn influence physical attributes and capacities, and hence the risk of developing WMSDs [36]. Anthropometric data is essential in analyzing and quantifying the physical state of individuals, providing valuable parameters for examining how forces and torques interact across body segments. Such data is crucial for tools like biomechanical models, which simulate the body's internal states to assess factors like posture, movement, and physical stress on the musculoskeletal system. In ergonomics and workplace safety, anthropometric data enables the scaling and personalization of these models to fit specific individuals, allowing for precise analyses of physical demands and load distribution in various tasks. This level of customization is particularly important when designing solutions for a diverse workforce, as it helps tailor interventions to the physical attributes of each worker, thereby enhancing both safety and productivity.

3.2 Biomechanics

WMSDs are caused by repetitive loading of body tissues, associated to forces in the muscles and the joints, which are therefore relevant variables for physical ergonomics. Since it is impossible to directly measure muscle forces and joint torques in the human body, such states can be estimated using biomechanical models. In particular, for physical ergonomics, the most important type of biomechanical model is a musculoskeletal one which faithfully models the human body's skeletal system and muscle-tendon units, the interaction between them, as well as accounting for forces coming from the external environment which can vary in complexity, but often replicates the muscle and skeletal parameters of humans. Such models can be scaled by anthropometric data and driven by kinematics (motion) and/or kinetics (external forces, muscle activity) to estimate muscle forces and/or joint loads.

Models can be categorized based on their complexity in terms of whether they include only the skeletal system, the muscular system, or both [37]. The choice of using different levels of complexity depends on the application and computational requirements/capabilities. For example, full body, three-dimensional musculoskeletal models have a very high computation complexity and can be difficult to employ for real-time state estimation. On the other hand, a simplified skeletal model can be computationally very fast at the expense of the accuracy of state estimation.

Some examples of musculoskeletal models are OpenSim [38], AnyBody [39], Mujoco [40], SCONE [41], and MyoSim [42]. These models typically have a dedicated community developing them or they are developed for commercial purposes by companies. Simplified skeletal models are based on rigid-body dynamics and are typically developed for a specific application or research, such as in [25]. Furthermore, some ergonomics analysis tools are also integrated with biomechanical models, such as ScaleFit [43].

3.3 Environmental Factors

There are many environmental factors to be considered in the general field of ergonomics. However, since this chapter is focused on physical ergonomics in HRI, we will limit our analysis to some of the key factors related to the sub-field. These factors refer to task-related requirements and constraints, such as external interaction forces acting between the environment and the human body, human body motion needed to execute the task, workspace limitations, and robot capabilities and limitations.

External forces can be measured using force/torque sensors. These can be mounted on or integrated into tools, working equipment, or on the robot. Some robots are torque-controlled, which means they have integrated torque sensors in each joint [26, 44]. Using the geometric robot Jacobian matrix, these torques can be transformed into end-effector forces where the interaction with human/task is typically happening. Standard industrial robots can be equipped with an additional multi-axis force/torque sensor that is mounted between the robot tool/gripper and the final link [30, 45]. Ground reaction forces (GRF) are interaction forces of specific interest for ergonomics, since combined with human kinematic state, they allow for the estimation of internal (joint) forces via inverse dynamics computation. While they are traditionally measured with force plates, wearable sensors such as sensorized insoles have been developed to facilitate GRF measurements outside of the lab.

Human posture and environmental constraints (e.g., objects in the workspace, etc.) can be measured using external devices such as RGB depth cameras [46, 47], and optical [48, 49] or inertial [50–52] motion capture systems. In the case of wearable robots like exoskeletons, the human posture can be measured directly by internal sensors such as encoders [15, 53]. The advantage of depth cameras is that they are easy to set up and maintain, however, the accuracy of human posture estimation cannot compete compared to motion capture systems. Optical motion capture systems typically require multiple infrared cameras placed around the workspace to measure the markers placed on the human body. Setting markers on the human body can be a time-consuming process, while occlusion can happen in the workspace. While these systems are the gold standard for in-lab measurement, they can hardly be transferred for regular use in real-world settings. Inertial motion capture systems are based on wearable inertial measurement units (IMUs), containing three-axis accelerometers and three-axis gyroscopes, that estimate position by integrating the measured accelerations and thus do not require external cameras. IMU-based motion capture systems are more portable than infrared-based optical motion capture, and can be more easily used outside. Nevertheless, IMUs are prone to drift and can be more bulky than markers. The advantage of robot-mounted sensors like encoders is their simplicity, however, they can only be applied for direct human posture estimation in wearable robots. Recently, markerless motion capture based on RGB cameras has made significant progress, supported by deep learning techniques to reliably detect skeletons, and is being increasingly used for non-invasive recordings of human movements.

3.4 Ergonomic Assessment Tools and Variables

In physical ergonomics, several key indexes and parameters are used to assess the risk associated with specific tasks and guide the design of ergonomics interventions. We divide the indexes into two categories. One major category is based on harnessing practical experiences and observations from workplaces that can tell us what kind of conditions caused WMSDs in the past. Such experiences and observations have been transformed into look-up tables such as RULA [54], REBA [55], NIOSH [56], OCRA [57], OWAS [58], etc. For example, for each approximate human body posture look-up table provides a score that indicates how risky that posture is in terms of physical ergonomics. The key advantages of these metrics are their simplicity and the fact that they are based on actual experiences and observations from workplaces by using a camera and predefined templates to take handwritten notes. As such, these metrics directly provide a level of risks with respect to WMSDs.

One disadvantage of look-up tables or preset posture binning categories, like RULA and REBA is that they are discrete and do not continuously cover all states. Thus, we cannot know what the actual score is between two discrete steps. This is a significant challenge in pHRI, where ergonomic scores are often used in a numerical optimization, to optimize the assistance provided by the robot. However discrete scores are not well-suited for numerical optimization, so they have to be interpolated.

Attempts were made to make RULA and REBA continuous so that they can be differentiable for the calculation [48]. However, this too is based on interpolation and does not guarantee that in-between values are accurate. Another key disadvantage of look-up tables is that they only give an overview score and cannot tell us what is actually happening in specific parts of the human body. Additionally, these traditional approaches were developed from epidemiological studies conducted without considering robotic assistance. This creates a potential mismatch when robots are used to support human movements. This is typically the case with exoskeletons that assist in supporting the weight of the user's body segments (e.g. arms in overhead work or torso in bent postures): for the same posture, the muscular effort is reduced, hence postural scores such as RULA may not apply anymore. Thus, relying on traditional scores might misrepresent the ergonomic conditions in pHRI scenarios.

Alternative to these discrete approaches is the estimation of human state via biomechanical modelling (e.g. estimation of joint torque or muscle force). This allows for the real-time analysis of various variables that reflect the physical demands placed on specific parts of the human body during activities. This approach allows for dynamic adjustment of tasks to optimize assistance and mitigate risk through external supports (e.g. feedback interfaces, HRIs, etc.). Furthermore, unlike in ergonomic look-up tables, robot dynamics can be integrated into biomechanical models to account for the influences of pHRI. In this context, Plantard et al. [59] used an inverse-dynamics approach to calculate joint torques during work-related tasks. This measure can aid in predicting the body areas most affected by loads, which can subsequently inform a robot on how to optimally assist the user. Kim et al. [25] demonstrated this by developing an online optimization technique that adjusts the trajectories executed by the robot based on estimated joint torques, enabling the human to work in postures that reduce joint loading throughout the task. Another method, proposed by Peternel et al. [44], involves modifying task configurations based on estimated muscular fatigue. Since prolonged loading of joints (either static or dynamic) contributes to fatigue, Lorenzini et al. [51] incorporated a whole-body overloading fatigue model into a human-robot collaboration framework, where the robot advises the user on optimal timing for posture adjustments. Lastly, Fortini et al. [60] pointed out how compressive forces generated between adjacent body segments could serve as an additional real-time index for the assessment of ergonomics. Such a hypothesis was first tested on simulation and then validated on a small pool of participants, showing the correlation between compressive forces and increased muscle activation throughout the task.

However, a drawback of these biomechanical quantities is that they provide relative information, therefore they may not be sufficient for identifying non-ergonomic conditions in absolute terms. Unlike tools like ergonomic look-up tables, such as RULA, REBA, etc., biomechanical quantities do not provide explicit levels of risk and are not based on years of experience. Therefore, while these biomechanical quantities are valuable for pHRI applications, particularly in optimizing the behaviour of the robot online, they may have limitations when evaluating a pHRI system with respect to ergonomics.

Lately, another category has been emerging which is based on data-driven approaches. Unlike real-time modeling, this method involves collecting ergonomics datasets. Maurice et al. [50] proposed methods on how an ergonomic dataset can be obtained. Using this dataset one can employ machine learning approaches to create surrogate ergonomic models. For example, data from the AnyBody [26] and OpenSim [44] biomechanical models were used to design machine learning models to enable real-time computation of biomechanical quantities that would otherwise be too costly to compute online from a full biomechanical model. The approach in [52] learned the ergonomic model for walking using learning by demonstration and then employed online using model predictive control. In [61], a data-driven approach is used for automatic ergonomics evaluation based on human body activity recognition. In [62], Artificial Intelligence (AI) algorithms are used for on-line estimation of the biomechanical risk level. Such AI models can be based on kinematic, kinetic, and even muscular data (e.g. using bipolar (sEMG) or High Density (HDsEMG) surface electromyography) coming from wearable sensors. Artificial Neural Networks (ANNs) have shown good performance in classifying the biomechanical risk in lifting heavy loads activities [63, 64], even with extremely simple architectures. Ergonomic databases, such as the Human Data Corpus [65] that contain measurements with wearable sensors during the execution of MMH activities, are then valuable to train and evaluate ANN models. The instrument-based biomechanical risk assessment tools offer multiple advantages:

1. overcome some limitations of the traditional methods listed in the international ergonomics standards of the ISO 11228 series [66];
2. accurately and precisely estimate the biomechanical risk in real-time [67];
3. inform workers, with appropriate acoustic, visual and vibrotactile feedback stimuli, of their physical effort during the execution of MMH activities;
4. optimize the control of cobotic technologies that assist the worker [68];
5. estimate the risk, otherwise not classifiable with traditional methods, even in activities performed with the aid of wearable assistive devices such as passive and active exoskeletons [69].

Finally, an important aspect is also to visualise computational data and metrics from the ergonomic models. For example, the values of joint torques can be visualised using colours on a simulated human body [70]. A more complex way of visualisation is to label the shared human-robot workspace. In the work of Figueredo et al. [23] the workspace was labelled with different colours based on the suitability of a specific pose. In Peternel et al. [49], different approaches to labelling the workspace were explored, including thresholding of metrics that results in a binary map, and a weighted sum of metrics that results in a continuous map. Lastly, in Vianello et al. [71] latent ergonomic maps were proposed to address the redundancy in projecting the indexes of many degrees of freedom in the human body into the workspace.

3.5 *HRIs as Sources of Information for Physical Ergonomics Assessment*

As highlighted in the previous section, when assessing jobs that involve multiple individual tasks, practitioners frequently face the challenge of selecting and combining assessment tools. This process may involve subjective interpretation or be limited to assessing posture and force demands at a single point in time. Therefore, there is a growing need in fields like HRI for holistic approaches that seamlessly integrate these tools and offer tailored assessments of job demands, taking into account individual physical capabilities. Interestingly, mechatronic devices not only offer assistance to the humans in their activities, but they can also become sources of information on the ergonomics of HRIs (especially in terms of the user conditions). Indeed, they can provide mechanical biomarkers as quantitative indexes of the individual conditions based on any metrics derived from the mechanical activity of the robotic devices interacting with a user. A similar perspective has been explored in rehabilitation robotics about neuromechanical biomarkers that unveil information on the conditions of the patients [72].

In general, we can exploit the fact that robotic devices can provide precise control over force/torque and motion, enabling us to establish and design innovative indicators and indexes for understanding specific work demands. Additionally, they enable the acquisition of repeatable, valid, and accurate measurements of sensorimotor function. We will now present some relevant examples of innovative robotic assessment quantitative indexes within the framework of well-known time-varying work-related exposures (e.g., force, repetition, and posture).

Regarding the assessment of the mechanical impedance of specific human joints, researchers have developed different robot-based methods designed to compute mechanical impedance models allowing extraction of objective and quantitative measurements of joint mechanical properties (including stiffness, viscosity, and spasticity) [73], which are fundamental to comprehending the intricacies of healthy joint functions and could serve as valuable biomarkers for predicting injuries. Another aspect to be treated is the development of indicators related to strength demands, and fatigue. In particular, while acute injuries resulting from task demands that exceed tissue tolerance are often readily identifiable, the detection of submaximal loading and cumulative trauma associated with injuries presents a greater challenge. Utilizing robotics, neuromuscular fatigue models can investigate the impact of muscle fatigue [74]. Recognizing the variances in central and peripheral fatigue manifestations based on sex and task, different dynamic fatigue protocols using robots have been proposed [75] in order to mimic workplace movements. Notably, these dynamic fatigue protocols have been found to impair performance and to have a more enduring impact on stiffness and proprioception [76], compared to isometric fatigue. In professions such as aviation and surgery, where motor control and sense of position and movement are of utmost importance, any compromise in joint accuracy and proprioception can give rise to circumstances with life-or-death implications.

How posture influences the human performance is another aspect that can be unveiled with the integration of robotic tools. Coupling surface electromyography and functional protocols administered through robots, taking into account different postures, can predict joint loading [77]. Most prior studies on muscle activity and demands have focused on isolated joint motions. Using robotic devices, the knowledge can be expanded to complex dynamic movements [78]. The last relevant aspect is the quantitative assessment of proprioception: the sense responsible for the self-perception of position, motion and orientation of the body or body segments [79]. Using models similar to the ones mentioned above it is possible to investigate the impact of postures and movement direction on joint proprioception [80]. These studies demonstrated that different postural conditions alter proprioceptive acuity suggesting that certain joint postures may be more important for actions requiring high accuracy.

4 Physical Ergonomics for pHRI

There are several ways one can approach addressing physical ergonomics in pHRI. Two major categories are based on whether we use design principles to make an ergonomics setup or we use real-time control for the robot to adjust its behavior on-line. The design approach can be separated into two sub-categories: appropriate mechanical design of the setup (spatial considerations), and optimal scheduling of tasks (temporal considerations). The control approach can be separated based on how we use the ergonomic metrics: preventive control which attempts to prevent injuries from happening in the future (e.g., reducing joint torque to prevent injury), and reactive control which reacts to something that has already occurred (e.g., address the fact that muscles become fatigued). Similar metric categorisation was proposed in a recent study on shared control of collaborative force production based on ergonomic metrics [81].

4.1 Design: Mechanical Aspects

Designing a pHRI robot, or workcell, with ergonomics in mind comes to selecting the different geometric, kinematic, and inertial elements of the robot structure, such that the resulting motion of the human interacting with this robot respects ergonomics guidelines. Design choices that impact the ergonomics are at all levels of the design, e.g. kinematic structure, actuator choices, and human-robot mechanical interface. But it also includes the design of the pHRI workcell (for non-wearable robots), i.e. the positioning of the robot with respect to the human and to the task at hand, and of teleoperation workcells [82].

Digital human simulation is a powerful tool to guide the design process. The digital model of the human is animated performing the pHRI task, with different degrees of detail in the motion features and biomechanical modeling. The ergonomics of the motion features can thereby be tested without the need for a physical mock-up, facilitating and speeding up the iterative design process. This is crucial for co-design of pHRI robots, which is gaining interest. At the level of the workcell design, ergonomics is considered in the spatial organization of the environment (for temporal organization, see the section on Scheduling). Among the standard digital human modeling tools for workstation design (e.g. Delmia, Ema, Jack, IMMA) [83], some offer the possibility to simulate simple features of a pHRI workcell. Design choices can thereby be validated, by ensuring that they match standard ergonomics guidelines, such as calculating RULA scores. For example the case study of Ruiz Castro et al. [84] demonstrates how software like IPS IMMA can support the design of collaborative workstations, particularly by simulating safety-related environments. While basic collaborations are simulated, the software also supports more complex interactions, like simultaneous object handling by a human and a robot, which can help the design of various workstation arrangements early in the development phase. The software provides general ergonomic and efficiency insights, though detailed time and ergonomic accuracy might be limited compared to real scenarios. For instance, the ergonomic assessment is affected by fixed object paths that might not optimize human movement.

While this is a traditional approach in workplace design, research work has proposed to go one step beyond by not only validating but also optimizing the design parameters of pHRI applications. This is possible by coupling the digital human-based evaluation with optimization algorithms to automatically identify well-performing solutions. For instance, Maurice et al. [85] explored which robot parameters most impact the physical demands on the user, aiming to optimize these parameters based on key ergonomic indicators to create an efficient robot design. Their method, based on a genetic algorithm approach, was applied to optimizing robot morphology for assisting with a manual task, and the resulting robot designs demonstrated the effectiveness of this approach. Building on this approach, Sartore et al. [86] emphasize the importance of ergonomics in designing control architectures for effective human-humanoid collaborations. Often, ergonomic considerations are neglected during the robot design phase, leading to suboptimal outcomes in pHRI tasks. Sartore et al. address this by proposing a novel methodology that optimizes humanoid robot designs with a focus on ergonomic factors linked to multi-agent interactions. Their bilevel optimization approach utilizes a genetic algorithm to generate robot designs by selecting link and motor specifications, followed by nonlinear optimization to evaluate ergonomic indexes during collaborative tasks, such as lifting payloads with varying human partners and weights. This method demonstrated significant improvements in ergonomic outcomes, particularly in energy expenditure, across static and dynamic motion scenarios.

Optimizing the design with respect to ergonomics has also been proposed for wearable robots. Zhou et al. [53] propose a design optimization approach for exoskeletons that integrates simulations of both the exoskeleton and a human body model. This human-centered approach tackles the challenges of physical human–exoskeleton interactions by modeling and simulating the mechanics involved for both systems.

This enables designers to effectively analyze and assess exoskeleton functionality. The authors developed a simulation platform that combines a musculoskeletal human body model with an exoskeleton model, allowing for detailed evaluations of assistive designs. In their study, they simulated an exoskeleton to assist individuals with brachial plexus injuries, testing two types of passive, gravity-compensating exoskeletons to identify optimal spring stiffness values. The optimization results validate the approach, highlighting its potential for improving exoskeleton design. A similar approach was proposed for the electromechanical design and development of a passive spinal exoskeleton for low back prevention [14].

Due to the tight interaction between the exoskeleton and the human, exoskeleton design uses detailed biomechanical models, which often include musculoskeletal modeling [53], and aims at optimizing exoskeleton design by addressing challenges such as misalignment, variability in anatomy, and the need for compliant control. Detailed biomechanical models help in simulating and analyzing these factors that lead to a design for minimizing the undesirable interaction force and moments [87]. Furthermore, such optimization approaches are often employed to handle the design objectives, such as minimizing weight while maximizing ergonomic benefit [88]. By considering such criteria, the developer can design exoskeletons that not only augment human capabilities but also promote long-term musculoskeletal health.

Another important design aspect for exoskeleton is the problem of kinematic compatibility, namely, the alignment between the human and robot joint axes. Indeed, kinematic misalignment induces reactive loads on the human musculoskeletal structure. Misalignments can happen for various reasons, e.g. initial wearing of the robot, migration during use, and physiological movements of the human joint rotation axis. Therefore, self-alignment mechanisms or misalignment compensation strategies should be considered in the kinematic design of robots to avoid detrimental effects in the transfer of forces/torques [87]. Similarly, a mechanical robot design that includes compliant actuators, such as series elastic actuators (instead of traditional rigid actuators), can provide more ergonomic HRI [89, 90].

4.2 Design: Scheduling Aspects

The question of scheduling with respect to ergonomics in pHRI applications arises in applications where several tasks can be performed either by the human or by the robot, i.e. they both have the capabilities to carry out these tasks. However, because the human and the robot have different physical abilities, the allocation of the tasks to one or the other impacts the overall task execution. The objective is then to distribute the task allocation in a way that optimizes the task execution (time, performance) but also the physical state of the agents, i.e. the ergonomics for the human with the goal of reducing the accumulation of fatigue. Task allocation includes both which agent performs which task, but also in which order the different tasks are executed.

In pHRI, scheduling has been addressed in two ways: static and dynamic. In static task/role allocation, the scheduling is done offline before the pHRI activity starts [91].

For each task, the allocation algorithm estimates, thanks to a model of the human, the ergonomics if the human were to perform the task. Ergonomic scores, such as REBA [47] or the strain index [91] are computed. Then, tasks can be allocated to the robot if the ergonomic scores exceed a given threshold. The different allocations are compared, and the one resulting in the best compromise between cumulative ergonomic scores and overall execution time is selected.

Scheduling can also be applied in a dynamic way, i.e. online, with task allocation based on the current physical or physiological state of the human. In such a case, the state of the human is monitored throughout the execution of the activity, and cumulative strain metrics, such as fatigue accumulation, are considered [24]. The decision of which agent will perform the next task, or which is the next task that should preferably be performed, is then re-evaluated regularly. El Makrini et al. [47] proposed a mixed approach in which both cumulative and non-cumulative metrics were used. Specifically, the task allocation considers both a threshold on the REBA score and a cumulative workload index. The framework was tested and validated on a gearbox assembly case scenario while employing an end-effector robot.

4.3 Control: Preventive Aspects

Historically, the first HRI robots introduced to improve physical ergonomics at work were called "Intelligent Assist Devices", and provided weight compensation for the manipulation of heavy loads [92]. With these devices, there was no measurement of ergonomics to design or adapt the action of the robot. However, by supporting the weight, and sometimes full inertia, of the manipulated object, these robots allowed the human operator to move the object around with reduced effort. Weight compensation is a mode that is generally present in the commercially available control modes of modern robots, where the human intent to move the load is detected by direct or indirect measurement of interaction force.

Most occupational exoskeletons belong to a similar category since they often aim at providing supporting torques to reduce human effort, especially by supporting the weight of human body segments and/or of a manipulated load. Yet, due to the tighter interaction between the human and the device with exoskeletons compared to non-wearable cobots, seamless exoskeleton assistance calls even more for human motor intention detection and prediction. When the task involves efforts that are not due to load manipulation, some cobots provide strength amplification [93]: the force applied by the robot on the object or tool is an amplified image of the force applied by the human on the robot. This is close to teleoperation, with related stability questions, but the human, robot and object are co-located.

More recently, several research works have been proposed to directly account for ergonomics in the control of the cobot, i.e. optimize the assistance of the robot with respect to ergonomics. The ergonomics of the motion is estimated with the metrics described in Sect. 3.4, either beforehand or online, and the movement or assistance of the robot is adjusted accordingly. The approach proposed in [45] is an extension of weight compensation or strength amplification, where the assistance provided by the robot is adapted online based on the posture of the human. The posture-dependent human force capacity at the hand (maximum Cartesian force exertion) is estimated using a biomechanical model, and the force assistance provided by the robot is adjusted, such that the force the human has to exert is adjusted based on the current force capacity [94]. Monitoring the human effort and adjusting the robot assistance accordingly has also been proposed in prosthetics: in [52] the assistance provided by an ankle prosthesis is adapted to minimize the human knee torque, using Bayesian primitives to predict the future torques.

Other works propose to use the robot to guide the human towards a neutral posture, by modifying the robot end-effector pose. In [46], local deviations from an optimal posture given by RULA or REBA postural guidelines are used as an error signal in the robot control law. A resembling approach using the deviation from a reference human posture as an error signal has been applied to prosthesis control, to reduce compensatory movements [95]. The methods in [48, 96] also rely on RULA/REBA postural scores, but conduct a whole-body postural optimization. They compute an optimal human posture for the task at hand according to RULA/REBA scores, thereby determining the optimal human hand position, and adapting the robot pose accordingly. In [97], the human posture is also optimized with respect to standard ergonomic scores, but these works aim at planning the future robot poses, instead of only reactively adjusting the robot motion. Similar approaches based on the RULA score have been used for teleoperation applications, where the master haptic device is used to guide the human towards a more ergonomic posture [98].

Optimization-based approaches have also been applied using biomechanical measurements instead of ergonomic scores such as RULA. Kim et al. [25] computed the human posture that minimizes the joint torques (or overloading joint torques, when the focus is on the effect of the manipulated load), and adapted the robot end-effector pose accordingly. Similar approaches aiming to optimize posture have been proposed with different optimization criteria, such as joint compressive forces [60], human hand manipulability [99], or muscle forces [23, 26]. In Bestick et al. [100], the authors introduce an approach to prevent ill-adapted postures in the future: in a hand-over application, the pose of the robot is adjusted to guide the human towards a posture that will facilitate their next action. All these approaches require the online use of a biomechanical model, with different levels of complexity (e.g. muscular modeling) depending on the considered biomechanical metrics.

It is also worth noting that all approaches based on postural optimization usually assume that the human is going to adopt the optimal posture that has been computed. Yet, due to kinematic redundancy, prescribing the robot end-effector, and hence the human hand, the pose does not fully constrain the whole-body posture. The human may still adopt a posture different from the one computed. Van der Spaa et al. [97] addressed this challenge following an inverse-kinematics based approach. The authors first tuned a biomechanical model for pose estimation on motion capture data. Then, they coupled the prediction output with an informed graph search algorithm for optimal human-robot interaction plans with improved ergonomics. The user case study involved the use of a 32 degrees-of-freedom bimanual mobile robot as a partner and showed reduced ergonomic cost with respect to a baseline planner. Another example is the study of Yaacoub et al. [101], in which a planning under uncertainty framework is presented to reduce fatigue in human-robot co-manipulation. The problem was modeled using Partially Observable Markov Decision Process, and a digital human simulation was employed to predict the fatigue cost induced by a diversity of plausible whole-body postures that a human may adopt in reaction to the available robot actions (end-effector pose). Based on these tools, the optimal robot policy was defined considering the probability of each human postural reaction and associated ergonomic consequences. The approach was validated under different conditions and showed how their approach outperforms other policies that do not take into account fatigue.

It should be noted that all approaches based on online adaptation of the robot to the human state have so far remained at the research stage, and have not yet been deployed for real-world applications. Thus, the deployment of online adaptation methods remains one of the future challenges.

4.4 Control: Reactive Aspects

While the majority of control methods for ergonomic pHRI involve a preventive approach, there are also methods that work in a reactive manner. In general, the reactive approach involves a robot making an action when an index/metric is integrated over time and reaches a certain critical threshold. A chief index/metric in such an approach is fatigue.

The method in Peternel et al. [44] monitored the muscle fatigue during a force production task and the robot reacted when a certain muscle group demonstrated signs. The robot changes the configuration of the task to change the direction of the force production to involve a different group of muscles allowing the previous muscles to rest. To estimate the muscle forces in real-time, data from an OpenSim musculoskeletal model was used in combination with fatigue models calibrated for individual muscles. A similar approach was also implemented in teleoperation, where the haptic device was temporarily decoupled to move the operator to a more comfortable configuration when fatigue was detected [102]. In Lorenzini et al. [51], overloading fatigue was introduced to account for human factors in a spray painting task. The proposed method monitored the cumulative effect of the overloading torque induced on the joints over time. Thus, the fatigue was estimated on a joint level instead of a muscle level and accounted only for the joint torque of the load and not the body dynamics.

The work of Shafti et al. [46] used a reactive robot control during a welding task to correct posture when the RULA score deviated from certain threshold zone. While RULA would normally be classified as a preventive index/metric, the way it is acted upon in this approach is more reactive.

Following a similar idea of inducing changes in posture through time to distribute the biomechanical load across different joints, [103] added random changes in the robot end-effector pose, while [101] proposed a probabilistic planning approach to minimize long term physical fatigue. All these works aim to reduce the repetitiveness of gestures, which is a known risk factor for WMSDs. Indeed, movement variability has been suggested to be beneficial in reducing WMSD risks [104].

The study [18] introduces an innovative pHARI paradigm for collaborative aerial transportation of a cylindrical payload. In this setup, one end of the payload is lifted by a human operator, while the other end is supported by a quadcopter using a cable. The quadcopter's control framework is designed to provide a natural experience for the human operator, akin to having another human assist in transporting the payload. In this scenario, the human operator lifts and maneuvers the payload along a path, while the quadcopter aids by stabilizing it. This collaborative framework enables individuals without prior knowledge of flying aerial robots to successfully participate in such tasks. This collaboration between humans and aerial robots not only enhances task efficiency but also enables the use of aerial systems in dynamic outdoor environments.

4.5 A Note on Physical-Cognitive Factors

Another aspect to consider is related to the internal structure of the field of ergonomics. Indeed, the different branches of the research in human factors should work in synergy to show impactful advances: for instance, physical and cognitive ergonomics can mutually reinforce in fields like HRI. Together, they can ponder holistically how physical and cognitive factors interact dynamically [105], especially during tasks requiring a person to monitor and handle the behaviour of an autonomous or semi-autonomous machine. The process of HRI design should certainly recognize that using a certain device, like an exoskeleton, can reduce the physical demand for a user while adding a degree of mental demand for controlling the robot and paying attention to the pace of its movements on the individual limbs [106]. An intuitive control interface can make the physical assistance offered by the robot feel seamless, increasing its efficiency. The evaluation of pHRI should also consider this point [107]: assessment tasks should be designed to evaluate both physical and cognitive demands in order to balance them through a subsequent re-design of the system - even using neuroergonomic methods [108], investigating neurocognitive processes underlying the user experience and performance. Literature further specifies the potential of a sub-field labeled as physical neuroergonomics, focusing on neural factors related to the execution of physical tasks. Interestingly, this field blend motor, cognitive, and affective aspects and their inter-relationships in connection to physical ergonomics [109]. Further solutions - including physical-cognitive approaches to user training, multimodal feedback, customization, and personalization - constitute keys for making pHRIs align with both the physical and mental capabilities of the users, enhancing their well-being, safety, and performance.

5 Conclusion

When addressing physical ergonomics in HRI systems, it is important to take into account the type of robots involved/available and the type of possible interactions for a given application. These define the basic framework and constraints for the application. In the next step, we can consider different tools to assess anthropometry, biomechanics, and environmental factors, and select the proper ergonomic indexes and metrics that fit within the framework and constraints of the specific application. Finally, these considerations will come into play when designing the robot mechanism, control, and specifics of robot spatial/temporal involvement.

As outlined and discussed in this chapter, the research considering physical ergonomics in the HRI process has boomed in the past ten years. There are many applications, ranging from different types of robots and tasks to different ergonomics assessment approaches and various robot control methods. While substantial first steps have been made, there are still many open challenges and opportunities for future research and application.

In terms of research, the quality and quantity of assessment indexes and metrics could be improved. On the robot control side, new ways of integrating and using assessment indexes and metrics could open possibilities for the development of new robot guidance control algorithms. Perhaps the largest amount of effort is required on the implementation and application side, as developing technology beyond the lab environment and onto the industry level demands the methods and algorithms to be intuitive, economical, robust, reliable, and fast. We hope to see the emergence of R&D teams in established companies, alongside innovative start-ups, fully focusing on physical ergonomics in HRI to bridge this middle gap between research and industry.

Moreover, achieving practical advances in physical ergonomics within HRI systems will require interdisciplinary collaboration across different domains, even different branches of ergonomics itself (motor and mental activities should be pondered together in most HRI investigations and assessments). Such interdisciplinary efforts should aim to introduce and validate experimental techniques to design, implement, and assess HRIs for updating the existing guidelines. This path should be followed by establishing partnerships between academic institutions, users associations, industrial stakeholders, and policy-makers to facilitate the development of regulatory frameworks, encouraging the widespread adoption of physically ergonomic HRI systems in workplaces worldwide. These initiatives will ultimately drive the creation of safer, more efficient, and human-centered robotic systems, with profound implications for users' health, productivity, and satisfaction.

References

1. Arash, A., Andrea M.Z., Ivaldi, S., Albu-Schäffer, A., Kosuge, K., Khatib, O.: Progress and prospects of the human-robot collaboration. Auton. Robot. **42**, 957–975 (2018)
2. Paul, M.: Fitts: human engineering for an effective air-navigation and traffic-control system (1951)
3. Maury, A.N., Jaap H van D.: Physical ergonomics. Mater. Mech. Des. **762** (2006)
4. Vesselinka, D.N., Iskra, S.M.: Work-related musculoskeletal disorders (wmsds): risk factors, diagnosis and prevention. Scripta Scientifica Salutis Publicae **4**, 15–21 (2018)
5. Stuart, M.M.: Evolving ergonomics? Ergonomics **52**(1), 80–86 (2009)
6. Don, B.: Chaffin: the evolving role of biomechanics in prevention of overexertion injuries. Ergonomics **52**(1), 3–14 (2009)
7. Silverstein, B., Viikari-Juntura, E., Kalat, J.: Use of a prevention index to identify industries at high risk for work-related musculoskeletal disorders of the neck, back, and upper extremity in washington state, 1990–1998. Am. J. Ind. Med. **41**(3), 149–169 (2002)
8. Tim, M., Charles, D., Kenta-Bibi, E., Weber, J., Diva, U., Warren, N., Grey, M.: Trends in work-related musculoskeletal disorder reports by year, type, and industrial sector: a capture-recapture analysis. Am. J. Ind. Med. **48**(1):40–49 (2005)
9. Vargas-Prada, S.: Work-related musculoskeletal disorders: are they truly declining? (2015)
10. Robert, B.D., Brian, D.L., Ming-Lun, L., Edward, F.K.: Trends in work-related musculoskeletal disorders from the 2002 to 2014 general social survey, quality of work life supplement. J. Occup. Environ. Med. **62**(8), 595–610 (2020)
11. Sundstrup, E., Seeberg, K.G.V., Bengtsen, E., Andersen, L.L.: A systematic review of workplace interventions to rehabilitate musculoskeletal disorders among employees with physical demanding work. J. Occup. Rehabil. **30**(4), 588–612 (2020)
12. Yuqiang, W., Balatti, P., Lorenzini, M., Zhao, F., Kim, W., Ajoudani, A.: A teleoperation interface for loco-manipulation control of mobile collaborative robotic assistant. IEEE Robot. Autom. Lett. **4**(4), 3593–3600 (2019)
13. Jan, B., Matteo, L., Federico, T., Tom, V., Domen, N., Nejc Šarabon, Barkan U., Luka, P., Diego T., Jan, F.V.: Challenges and solutions for application and wider adoption of wearable robots. Wearable Technol. **2**, e14 (2021)
14. Babič, J., Petrič, T., Mombaur, K., Kingma, I., Bornmann, J., González-Vargas, J., Baltrusch, S., Šarabon, N., Houdijk, H.: Spexor: design and development of passive spinal exoskeletal robot for low back pain prevention and vocational reintegration. SN Appl. Sci. **1**, 1–5 (2019)
15. Petrič, T., Peternel, L., Morimoto, J., Babič, J.: Assistive arm-exoskeleton control based on human muscular manipulability. Front. Neurorobot. **13**, 30 (2019)
16. Wansoo, K., Seunghoon, L., Minsung, K., Jungsoo, H., Changsoo, H.: Energy-efficient gait pattern generation of the powered robotic exoskeleton using dme. In: 2010 IEEE/RSJ International Conference on Intelligent Robots and Systems, pp. 2475–2480. IEEE (2010)
17. Wansoo, K., Virginia, R.G., Juan, M.G., Marta, L., Arash, A.: A directional vibrotactile feedback interface for ergonomic postural adjustment. IEEE Trans. Haptics **15**(1), 200–211 (2021)
18. Prajapati, P., Vashista, V.: Aerial physical human robot interaction for payload transportation. IEEE Robot. Autom. Lett. **8**(8), 4903–4910 (2023)
19. Li, G., Liu, X., Loianno, G.: Safety-aware human-robot collaborative transportation and manipulation with multiple mavs (2023)
20. Tognon, M., Alami, R., Siciliano, B.: Physical human-robot interaction with a tethered aerial vehicle: application to a force-based human guiding problem. IEEE Trans. Rob. **37**(3), 723–734 (2021)
21. Geonwoo, P., Hyungeun, P., Park, W., Lee, D., Kim, M., Lee, S.J.: The palletrone cart: human-robot interaction-based aerial cargo transportation. IEEE Robot. Autom. Lett. **9**(8), 6999–7006 (2024)
22. Onnasch, L., Roesler, E.: A taxonomy to structure and analyze human-robot interaction. Int. J. Soc. Robot. **13**(4), 833–849 (2021)

23. Luis, F.C., Figueredo, R.C., Aguiar, L., Chen, S., Chakrabarty, Mehmet, R., Dogar, A.G.: Cohn: human comfortability: integrating ergonomics and muscular-informed metrics for manipulability analysis during human-robot collaboration. IEEE Robot. Autom. Lett. **6**(2), 351–358 (2020)
24. Costanza, M., Anna, B., Andrea, M.Z., Paolo, R.: A dynamic task allocation strategy to mitigate the human physical fatigue in collaborative robotics. IEEE Robot. Autom. Lett. **7**(2), 2178–2185 (2022)
25. Kim, W., Lee, J., Peternel, L., Tsagarakis, N., Ajoudani, A.: Anticipatory robot assistance for the prevention of human static joint overloading in human-robot collaboration. IEEE Robot. Autom. Lett. **3**(1), 68–75 (2017)
26. Marin, A.G., Shourijeh, M.S., Galibarov, P.E., Damsgaard, M., Lars, F., Stulp, F.: Optimizing contextual ergonomics models in human-robot interaction. In: 2018 IEEE/RSJ International Conference on Intelligent Robots and Systems (IROS), pp. 1–9. IEEE (2018)
27. Mansfeld, N., Hamad, M., Becker, M., Marin, A.G., Haddadin, S.: Safety map: a unified representation for biomechanics impact data and robot instantaneous dynamic properties. IEEE Robot. Autom. Lett. **3**(3), 1880–1887 (2018)
28. International Organization for Standardization.: Robots and robotic devices - collaborative robots (iso/ts 15066:2016). Standard iso/ts 15066. International Organization for Standardization, (2016)
29. Lasota, P.A., Rossano, G.F., Shah, J.A.: Toward safe close-proximity human-robot interaction with standard industrial robots. In: 2014 IEEE International Conference on Automation Science and Engineering (CASE), pp. 339–344. IEEE (2014)
30. Wang, Q., Liu, D., Marc, G., Carmichael, S., Aldini, C.-T.: Lin: computational model of robot trust in human co-worker for physical human-robot collaboration. IEEE Robot. Autom. Lett. **7**(2), 3146–3153 (2022)
31. Salvendy, G. (ed.): Handbook of human factors and ergonomics. John Wiley & Sons, Oxford, England, UK (2012)
32. Krishnan, K.S., Raju, G., Shawkataly, O.: Prevalence of work-related musculoskeletal disorders: psychological and physical risk factors. Int. J. Environ. Res. Pub. health **18**(17), 9361 (2021)
33. Van Eerd, D., Munhall, C., Irvin, E., Rempel, D., Brewer, S., van der Beek, A.J., Dennerlein, J.T., Tullar, J., Skivington, K., Pinion, C., et al.: Effectiveness of workplace interventions in the prevention of upper extremity musculoskeletal disorders and symptoms: an update of the evidence. Occup. Eviron. Med. **73**(1), 62–70 (2016)
34. Punnett, L., Prüss-Ütün, A., Nelson, D.I., Fingerhut, M.A., Leigh, J., Tak, S., Phillips, S.: Estimating the global burden of low back pain attributable to combined occupational exposures. Am. J. Ind. Med. **48**(6), 459–469 (2005)
35. De Kok, J., Vroonhof, P., Snijders, J., Roullis, G., Clarke, M., Peereboom, K., van Dorst, P., Isusi, I.: Work-related musculoskeletal disorders: prevalence, costs and demographics in the eu. Eur. Agency Saf. Health work **1** (2019)
36. Viviani, C., Arezes, P.M., Braganca, S., Molenbroek, J., Dianat, I., Castellucci, H.I.: Accuracy, precision and reliability in anthropometric surveys for ergonomics purposes in adult working populations: a literature review. Int. J. Ind. Ergonomics **65**, 1–16 (2018)
37. Fang, C., Peternel, L., Seth, A., Sartori, M., Mombaur, K., Yoshida, E.: Human modeling in physical human-robot interaction: a brief survey. IEEE Robot. Autom. Lett. (2023)
38. Seth, A., Hicks, J.L., Uchida, T.K., Habib, A., Dembia, C.L., Dunne, J.J., Ong, C.F., DeMers, M.S., Rajagopal, A., Millard, M., et al.: Opensim: simulating musculoskeletal dynamics and neuromuscular control to study human and animal movement. PLoS Comput. Biol. **14**(7), e1006223 (2018)
39. Damsgaard, M., Rasmussen, J., Christensen, S.T., Surma, E., De Zee, M.: Analysis of musculoskeletal systems in the anybody modeling system. Simul. Model. Pract. Theory **14**(8), 1100–1111 (2006)
40. Todorov, E., Erez, T., Tassa, Y.: Mujoco: a physics engine for model-based control. In: 2012 IEEE/RSJ International Conference on Intelligent Robots and Systems, pp. 5026–5033. IEEE (2012)

41. Geijtenbeek, T.: Scone: open source software for predictive simulation of biological motion. J. Open Source Softw. **4**(38), 1421 (2019)
42. Wang, H., Caggiano, V., Durandau, G., Sartori, M., Kumar, V.: Myosim: fast and physiologically realistic mujoco models for musculoskeletal and exoskeletal studies. In: 2022 International conference on Robotics and Automation (ICRA), pp. 8104–8111. IEEE, 2022
43. ScaleFit. Scalefit - digital transformation in manufacturing
44. Peternel, L., Fang, C., Tsagarakis, N., Ajoudani, A.: A selective muscle fatigue management approach to ergonomic human-robot co-manipulation. Robot. Comput. Integr. Manuf. **58**, 69–79 (2019)
45. Carmichael, M.G., Aldini, S., Khonasty, R., Tran, A., Reeks, C., Liu, D., Waldron, K.J., Dissanayake, G.: The anbot: an intelligent robotic co-worker for industrial abrasive blasting. In: 2019 IEEE/RSJ International Conference on Intelligent Robots and Systems (IROS), pp. 8026–8033. IEEE (2019)
46. Shafti, A., Ataka, A., Lazpita, B.U., Shiva, A., Wurdemann, H.A., Althoefer, K.: Real-time robot-assisted ergonomics. In: 2019 International Conference on Robotics and Automation (ICRA), pp. 1975–1981. IEEE (2019)
47. El Makrini, I., Merckaert, K., De Winter, J., Lefeber, D., Vanderborght, B.: Task allocation for improved ergonomics in human-robot collaborative assembly. Interact. Stud. **20**(1), 102–133 (2019)
48. Busch, B., Maeda, G., Mollard, Y., Demangeat, M., Lopes, M.: Postural optimization for an ergonomic human-robot interaction. In: 2017 IEEE/RSJ International Conference on Intelligent Robots and Systems (IROS), pp. 2778–2785. IEEE (2017)
49. Peternel, L., Tofte Schøn, D., Fang, C.: Binary and hybrid work-condition maps for interactive exploration of ergonomic human arm postures. Front. Neurorobot. **14**, 590241 (2021)
50. Maurice, P., Malaisé, A., Amiot, C., Paris, N., Richard, G.-J., Rochel, O., Ivaldi, S.: Human movement and ergonomics: an industry-oriented dataset for collaborative robotics. Int. J. Robot. Res. **38**(14), 1529–1537 (2019)
51. Lorenzini, M., Kim, W., De Momi, E., Ajoudani, A.: A new overloading fatigue model for ergonomic risk assessment with application to human-robot collaboration. In: 2019 International Conference on Robotics and Automation (ICRA), pp. 1962–1968. IEEE (2019)
52. Clark, G., Amor, H.B.: Learning ergonomic control in human-robot symbiotic walking. IEEE Trans. Rob. **39**(1), 327–342 (2022)
53. Zhou, L., Li, Y., Bai, S.: A human-centered design optimization approach for robotic exoskeletons through biomechanical simulation. Robot. Auton. Syst. **91**, 337–347 (2017)
54. McAtamney, L., Corlett, E.N.: Rula: a survey method for the investigation of work-related upper limb disorders. Appl. Ergonomics **24**(2), 91–99 (1993)
55. Hignett, S., McAtamney, L.: Rapid entire body assessment (reba). Appl. Ergon. **31**(2), 201–205 (2000)
56. Work practices guide for manual lifting. US Department of Health and Human Services. Health Div. Biomed. Behav. Sci. (1981)
57. Colombini, D.: An observational method for classifying exposure to repetitive movements of the upper limbs. Ergon. **41**(9), 1261–1289 (1998)
58. Graham, B., Scott, N.R.: Lambe: working practices in a perchery system, using the ovako working posture analysing system (owas). Appl. Ergon. **27**(4), 281–284 (1996)
59. Plantard, P., Muller, A., Pontonnier, C., Dumont, G., Hubert, P.H., Shum, F.: Multon: inverse dynamics based on occlusion-resistant kinect data: is it usable for ergonomics? Int. J. Ind. Ergon. **61**, 71–80 (2017)
60. Fortini, L., Lorenzini, M., Kim, W., De Momi, E., Ajoudani, A.: A real-time tool for human ergonomics assessment based on joint compressive forces. In: 2020 29th IEEE International Conference on Robot and Human Interactive Communication (RO-MAN), pp. 1164–1170 (2020)
61. Malaisé, A., Maurice, P., Colas, F., Ivaldi, S.: Activity recognition for ergonomics assessment of industrial tasks with automatic feature selection. IEEE Robot. Autom. Lett. **4**(2), 1132–1139 (2019)

62. Ranavolo, A., Draicchio, F., Varrecchia, T., Silvetti, A., Iavicoli, S.: Wearable monitoring devices for biomechanical risk assessment at work: current status and future challenges-a systematic review. Int. J. Environ. Res. Public Health 15(9), (2018)
63. Varrecchia, T., De Marchis, C., Draicchio, F., Schmid, M., Conforto, S., Ranavolo, A.: Lifting activity assessment using kinematic features and neural networks. Appl. Sci. **10**(6), 1989 (2020)
64. Varrecchia, T., De Marchis, C., Rinaldi, M., Draicchio, F., Serrao, M., Schmid, M., Conforto, S., Ranavolo, A.: Lifting activity assessment using surface electromyographic features and neural networks. Int. J. Ind. Ergon. **66**, 1–9 (2018)
65. The human data corpus. https://humandatacorpus.org/. Accessed on 23 July 2026
66. Ranavolo, A., Ajoudani, A., Cherubini, A., Bianchi, M., Fritzsche, L., Iavicoli, S., Sartori, M., Silvetti, A., Vanderborght, B., Varrecchia, T., Draicchio, F.: The sensor-based biomechanical risk assessment at the base of the need for revising of standards for human ergonomics. Sensors **20**(20), 5750 (2020)
67. Ranavolo, A., Ajoudani, A., Chini, G., Lorenzini, M., Varrecchia, T.: Adaptive lifting index (ali) for real-time instrumental biomechanical risk assessment: concepts, mathematics, and first experimental results. Sensors **24**(5), 1474 (2024)
68. Ajoudani, A., Albrecht, P., Bianchi, M., Cherubini, A., Del Ferraro S., Fraisse, P., Fritzsche, L., Garabini, M., Ranavolo, A., Rosen, H.P., Sartori, M., Tsagarakis, N., Vanderborght, B., Wischniewski, S.: Smart collaborative systems for enabling flexible and ergonomic work practices [industry activities]. IEEE Robot. Autom. Mag. **27**(2), 169–176 (2020)
69. Sørensen, K., Fox, F., Hayot, C., Ajoudani, A., Mobedi, E., Lorenzini, M., Ranavolo, A., Chini, G., Silvetti, A., Varrecchia, T., Ringelberg, A., Draicchio, F., Rodriguez Cianca, D., Torricelli, D., Turcksin, T.: Cwa 17938: guideline for introducing and implementing real-time instrumental-based tools for biomechanical risk assessment (2023) https://www.cencenelec.eu/media/CEN-CENELEC/CWAs/RI/2023/cwa17938_2023.pdf
70. Lorenzini, M., Kim, W., De Momi, E., Ajoudani, A.: A real-time graphic interface for the monitoring of the human joint overloadings with application to assistive exoskeletons. In: Wearable Robotics: Challenges and Trends: Proceedings of the 4th International Symposium on Wearable Robotics, WeRob2018, October 16–20, 2018, Pisa, Italy 3, pp. 281–285. Springer (2019)
71. Vianello, L., Gomes, W., Stulp, F., Aubry, A., Maurice, P., Ivaldi, S.: Latent ergonomics maps: Real-time visualization of estimated ergonomics of human movements. Sensors **22**(11), 3981 (2022)
72. Garro, F., Chiappalone, M., Buccelli, S., De Michieli, L., Semprini, M.: Neuromechanical biomarkers for robotic neurorehabilitation. Front. Neurorobot. **15**, 742163 (2021)
73. Falzarano, V., Holmes, M.W. R., Masia, L., Morasso, P., Zenzeri, J.: Evaluating viscoelastic properties of the wrist joint during external perturbations: influence of velocity, grip, and handedness. Front. Hum. Neurosci. 15 (2021)
74. Forman, G.N., Sonne, M.W., Kociolek, A.M., Gabriel, D.A., Holmes, M.W.R.: Influence of muscle fatigue on motor task performance of the hand and wrist: a systematic review. Hum. Mov. Sci. **81**, 102912 (2022)
75. Mugnosso, M., Marini, F., Holmes, M., Morasso, P., Zenzeri, J.: Muscle fatigue assessment during robot-mediated movements. J. Neuroeng. Rehabil. **15**(119), (2018)
76. Albanese, G.A., Falzarano, V., Holmes, M.W. R., Morasso, P., Zenzeri, J.: A dynamic submaximal fatigue protocol alters wrist biomechanical properties and proprioception. Front. Hum. Neurosci. **16**, (2022)
77. Kumar, R.I., Forman, G.N., Forman, D.A., Mugnosso, M., Zenzeri, J., Button, D.C., Holmes, M.W.R.: Dynamic wrist flexion and extension fatigue induced via submaximal contractions similarly impairs hand tracking accuracy in young adult males and females. Front. Sports Act. Living **2** (2020)
78. Forman, D.A., Forman, G.N., Mugnosso, M., Zenzeri, J., Murphy, B., Holmes, M.W.R.: Sustained isometric wrist flexion and extension maximal voluntary contractions similarly impair hand-tracking accuracy in young adults using a wrist robot. Front. Sports Act. Living **2** (2020)

79. Proske, U., Gandevia, S.C.: The proprioceptive senses: their roles in signaling body shape, body position and movement, and muscle force. Physiol. Rev. **92**(4), 1651–1697 (2012)
80. Albanese, G.A., Holmes, M.W.R., Marini, F., Morasso, P., Zenzeri, J.: Wrist position sense in two dimensions: between-hand symmetry and anisotropic accuracy across the space. Front. Hum. Neurosci. **15** (2021)
81. Andrés, Á.G., Beckers, N., Abbink, D.A., Luka, P.: Arbitration of authority in physical human-robot collaboration with combined preventive and reactive fatigue management. In: 2022 9th IEEE RAS/EMBS International Conference for Biomedical Robotics and Biomechatronics (BioRob), pp. 1–6. IEEE (2022)
82. Gholami, S., Lorenzini, M., De Momi, E., Ajoudani, A.: Quantitative physical ergonomics assessment of teleoperation interfaces. IEEE Trans. Hum. Mach. Syst. **52**(2), 169–180 (2022)
83. Scataglini, S., Paul, G. (eds.): DHM and Posturography. Academic Press, Cambridge, Massachusetts, USA (2019)
84. Castro, P.R., Mahdavian, N., Brolin, E., Högberg, E., Hanson, L.: Ips imma for designing human-robot collaboration workstations. In: 5th International Digital Human Modeling Symposium, Bonn, Germany, June 26–28, 2017, pp. 263–273. Federal Institute for Occupational Safety and Health (2017)
85. Maurice, P., Padois, V., Measson, Y., Bidaud, P.: Human-oriented design of collaborative robots. Int. J. Ind. Ergon. **57**, 88–102 (2017)
86. Sartore, C., Rapetti, L., Bergonti, F., Dafarra, S., Traversaro, S., Pucci, D.: Codesign of humanoid robots for ergonomic collaboration with multiple humans via genetic algorithms and nonlinear optimization. In: 2023 IEEE-RAS 22nd International Conference on Humanoid Robots (Humanoids), pp. 1–8. IEEE (2023)
87. Näf, M.B., Junius, K., Rossini, M., Rodriguez-Guerrero, C., Vanderborght, B., Lefeber, D.: Misalignment compensation for full human-exoskeleton kinematic compatibility: state of the art and evaluation. Appl. Mech. Rev. **70**(5), 050802 (2018)
88. Gopura, R.A.R.C., Bandara, D.S.V., Kiguchi, K., Mann, G.K.: Developments in hardware systems of active upper-limb exoskeleton robots: a review. Robot Auton. Syst. **75**, 203–220 (2016)
89. Livolsi, C., Conti, R., Guanziroli, E., Friðriksson, Þ, Ásgeir Alexandersson, K., Kristjánsson, A., Esquenazi, R., Molino Lova, D., Romo, F.: Giovacchini: An impairment-specific hip exoskeleton assistance for gait training in subjects with acquired brain injury: a feasibility study. Sci. Rep. **12**(1), 19343 (2022)
90. Vanderborght, B., Albu-Schäffer, A., Bicchi, A., Burdet, E., Caldwell, D.G., Carloni, R., Catalano, M., Eiberger, O., Friedl, W., Ganesh, G., et al.: Variable impedance actuators: a review. Robot. Auton. Syst. **61**(12), 1601–1614 (2013)
91. Pearce, M., Mutlu, B., Shah, J., Radwin, R.: Optimizing makespan and ergonomics in integrating collaborative robots into manufacturing processes. IEEE Trans. Autom. Sci. Eng. **15**(4), 1772–1784 (2018)
92. Colgate, J.E., Peshkin, M., Klostermeyer, S.H.: Intelligent assist devices in industrial applications: a review. In: Proceedings 2003 IEEE/RSJ International Conference on Intelligent Robots and Systems (IROS 2003)(Cat. No. 03CH37453), vol. 3, pp. 2516–2521. IEEE (2003)
93. Maurice, P., Measson, Y., Padois, V., Bidaud, P.: Assessment of physical exposure to musculoskeletal risks in collaborative robotics using dynamic simulation. In: 19th CISM-IFToMM Symposium on Robot Design, Dynamics, and Control-RoManSy, pp. 325–332. Springer (2012)
94. Skuric, A., Padois, V., Rezzoug, N., Daney, D.: On-line feasible wrench polytope evaluation based on human musculoskeletal models: an iterative convex hull method. IEEE Robot. Autom. Lett. **7**(2), 5206–5213 (2022)
95. Legrand, M., Jarrassé, N., Richer, F., Morel, G.: A closed-loop and ergonomic control for prosthetic wrist rotation. In: 2020 IEEE International Conference on Robotics and Automation (ICRA), pp. 2763–2769. IEEE (2020)
96. Gomes, W., Maurice, P., Dalin, E., Mouret, J.-B., Ivaldi, S.: Multi-objective trajectory optimization to improve ergonomics in human motion. IEEE Robot. Autom. Lett. **7**(1), 342–349 (2021)

97. van der Spaa, L., Gienger, M., Bates, T., Kober, Jens.: Predicting and optimizing ergonomics in physical human-robot cooperation tasks. In: 2020 IEEE International Conference on Robotics and Automation (ICRA), pp. 1799–1805. IEEE (2020)
98. Rahal, R., Matarese, G., Gabiccini, M., Artoni, A., Prattichizzo, D., Giordano, P.R., Pacchierotti, C.: Caring about the human operator: haptic shared control for enhanced user comfort in robotic telemanipulation. IEEE Trans. Haptics **13**(1), 197–203 (2020)
99. Yao, K., Billard, A.: An inverse optimization approach to understand human acquisition of kinematic coordination in bimanual fine manipulation tasks. Biol. Cybern. **114**(1), 63–82 (2020)
100. Bestick, A., Bajcsy, R., Dragan, A.D.: Implicitly assisting humans to choose good grasps in robot to human handovers. In: 2016 International Symposium on Experimental Robotics, pp. 341–354. Springer (2017)
101. Yaacoub, A., Thomas, V., Colas, F., Maurice, P.: A probabilistic model for cobot decision making to mitigate human fatigue in repetitive co-manipulation tasks. IEEE Robot. Autom. Lett. (2023)
102. Peternel, L., Fang, C., Laghi, M., Bicchi, A., Tsagarakis, N., Ajoudani, A.: Human arm posture optimisation in bilateral teleoperation through interface reconfiguration. In: 2020 8th IEEE RAS/EMBS International Conference for Biomedical Robotics and Biomechatronics (BioRob), pp. 1102–1108. IEEE (2020)
103. Zolotas, M., Luo, R., Bazzi, S., Saha, D., Mabulu, K., Kloeckl, K., Padır, T.: Imposing motion variability for ergonomic human-robot collaboration. IISE Trans. Occup. Ergon. Hum. Factors **12**(1–2), 123–134 (2024)
104. Srinivasan, D., Mathiassen, S.E.: Motor variability in occupational health and performance. Clin. Biomech. **27**(10), 979–993 (2012)
105. Ranjana, K.: Mehta: integrating physical and cognitive ergonomics (2016)
106. Bequette, B., Norton, A., Jones, E., Stirling, L.: Physical and cognitive load effects due to a powered lower-body exoskeleton. Hum. Factors **62**(3), 411–423 (2020)
107. Lee, G.I., Lee, M.R., Clanton, T., Sutton, E., Park, A.E., Marohn, M.R.: Comparative assessment of physical and cognitive ergonomics associated with robotic and traditional laparoscopic surgeries. Surg. Endosc. **28**, 456–465 (2014)
108. Mehta, R.K., Parasuraman, R.: Neuroergonomics: a review of applications to physical and cognitive work. Front. Hum. Neurosci. **7**, 889 (2013)
109. Karwowski, W., Siemionow, W., Gielo-Perczak, K.: Physical neuroergonomics: the human brain in control of physical work activities. Theor. Issues Ergon. Sci. **4**(1–2), 175–199 (2003)

Cognitive Ergonomics in Robotics

Patricia Helen Rosen, Eva Heinold, Giacinto Barresi, and Sascha Wischniewski

Abstract Work tasks performed by humans within any work system are the connecting elements between humans and other system elements such as deployed technologies. Fundamental to any task performance are mental processes of those individuals acting within any socio-technical system. Perception, attention, learning, memory and reasoning are just examples of these processes. Within human factors, they are typically studied to evaluate, for instance, mental workload and situation awareness of technology users, whose errors can dramatically impact the safety and the performance of the system. Accordingly, this chapter presents concepts from cognitive ergonomics. These concepts should be considered within Human–Robot Interaction (HRI) studies and applications in order to assess and improve the design of robots which can interact with humans in terms of technology acceptance, usability, accessibility, and, overall, user experience. This chapter describes different effects on humans related to the interaction with robotic systems. Furthermore, specific HRI design elements and recommendations will be outlined. The chapter closes with an overview on quantitative and qualitative methods proposed to evaluate and to assess different aspects of HRI.

Keywords Human factors · Cognition · HRI · Interaction quality · Human effects · Human-centered design

P. H. Rosen (✉) · E. Heinold · S. Wischniewski
Human Factors, Ergonomics Group, Federal Institute for Occupational Safety and Health (BAuA), Dortmund, Germany
e-mail: rosen.patricia@baua.bund.de

G. Barresi
Bristol Robotics Laboratory, School of Engineering, CATE, University of the West of England, Bristol, UK

G. Barresi and S. Wischniewski (eds.), *Ergonomics in Robotics: Advances and Innovations*, Springer Tracts in Advanced Robotics 162,
https://doi.org/10.1007/978-3-032-19374-2_3

1 Introduction

Every human action is accompanied by mental activities. Even the most physical working tasks cannot be performed without mental processes. For this reason, it might be too short-sighted to only address physical aspects like anthropometric or biomechanical metrics in relation to the design of a workplace. Cognitive ergonomics, a subset of ergonomics or human factors, includes psychological aspects of how work affects the human mind and how the human mind affects work [1]. Cognitive ergonomics is strongly linked to a number of different fields of psychology which themselves show some overlapping areas: cognitive psychology, automation psychology, occupational psychology and social psychology. However, a distinct aspect of cognitive ergonomics is to focus on mental processes in real contexts, intertwining psychological research with human-centered design. The main aim is not a laboratory-based understanding of our mental processes, but rather to describe how work and workplaces affect the human mind and vice versa [1]. Within cognitive ergonomics, we can find work-system design elements which can be associated with a number of consequences for the individual human, a team or even a whole organization. Relevant design elements can, among others, relate to the interface design, task design, design of automation or the modelling of users and systems; their relevance and effects may be very technology specific and the focus of this chapter lies on relevant design elements for human–robot interaction. As described by Hollnagel, a particular concern within cognitive ergonomics is the occurrence of situations which in consequence lead to negative outcomes or effects of work [1]. This results in the following agenda for cognitive ergonomics:

- identification or prediction of potential problems;
- description of the conditions which may either cause problems or significantly affect the development of situations; and
- provision of means to avoid or weaken negative impact situations [1].

Reducing negative effects of work on humans is a key concern of cognitive ergonomics. Ensuring that work systems do not harm workers physically nor mentally is the minimum requirement that must be met. However, a truly human-centered design of workplaces contributes to the development of a person's personality and health [2]. This is also in line with a more holistic understanding of health as defined by the World Health Organization (WHO), which states that health is not merely the absence of disease or infirmity but a state of complete physical, mental and social well-being [3]. Thus, health can be seen more as a continuum with factors that support different aspects of health, including the workplace [4].

2 Fundamental Principles of Mental Processes

The basic principles of mental processes revolve around understanding and enhancing various brain functions that play a crucial role in human performance and behavior. These functions include perception, attention, memory, decision-making, and affective states.

2.1 Perception

Perception, a key mental process, entails integrating and interpreting sensory stimuli from the environment. It relies on interpreting visual, auditory, and other sensory cues [5]. Sensation and perception have a very long research history within psychology and thus substantial knowledge on each sensory occurrence. Of great importance, especially in relation to technology design, are visual and auditory perception as many direct technology interaction forms use visual or auditory cues. Furthermore, tactile and haptic perception play an important role for human-technology interaction design [6]. This is especially relevant for teleoperated systems. Perception enables humans to act and react in complex environments and furthermore supports the construction of mental representations of surrounding environments. Cognitive ergonomics emphasizes (also through approaches like psychophysics) the design of systems that take into account how individuals interpret information through their senses. Perception is linked to cognitive load and attentional resources, with the human mind having limited capacity for both, each of which will be described in more detail later in this chapter. Thus, technology and work systems design should optimize cognitive load, guide attention effectively, and prioritize relevant information.

2.2 Attention and Executive Functions

Attention directs selective focus on specific stimuli while filtering out others. In cognitive ergonomics, managing attentional resources is crucial to optimize performance. Strategic allocation of attention through interfaces that guide focus and reduce distractions is essential to avoid depleting users' cognitive resources. Various models have emerged in the history of attention research, ranging from simplified to complex, describing characteristics of attention. With his research Kahneman has laid relevant foundations in the study of attention [5]. A model which has gained high levels of prominence in human factors research is Wickens' multiple resource model [7]. Wickens closely examines attentional resources, identifying three key dimensions: stages of processing, processing codes, and input/output modalities [7, 8]. In the updated model, Wickens adds a fourth dimension: visual channels, distinguishing between focal and ambient vision [8]. Modern work environments dominated by displays and interactive systems highlight the significance of ergonomic interface design. Ergonomic interfaces facilitate focusing on specific stimuli while filtering out irrelevant information, potentially optimising cognitive load. Excessive cognitive load can result in errors if attention spans are depleted. In the workplace, multitasking and task switching can strain attention. Meanwhile, sustained attention, also referred to as a source of cognitive stress [9], should be avoided. Inattentional blindness and change blindness are phenomena where individuals fail to notice unexpected stimuli or changes in visual, auditory or tactile scenes, respectively, due to a lack of cognitive resources or the activation of attention diverting cues.

2.3 *Learning and Memory*

Theories, models and research in relation to human memory address the encoding and retrieval of information, recognition and recall of stimuli and information, memory capacity, as well as a broad differentiation between distinct memory types, often referred to as the sensory memory, working memory and long-term memory. An influential model introduced by Broadbent laid the foundation for many multi-storage memory models [10]. Understanding how information is encoded, processed, stored, and retrieved is essential for creating human-centered systems. Long-term memory typically stores information for extended periods, while working memory handles temporary storage and manipulation during tasks. Working memory, however, is not infinite and can be overloaded, especially with dense information presentation. Recognition involves identifying information from options, while recall requires retrieving information without cues. In work settings, both recognition and recall are common, so task design should include recognition-based cues and clear retrieval paths for efficient recall. Context cues and consistent navigation can help with effective memory recall [11].

2.4 *Decision-Making*

Decision-making is a cognitive process involving the selection of optimal courses of action. One can choose between at least two options aiming to reach a desired consequence and to avoid unwanted consequences. In an optimal situation, this process leads to a decision which then leads to an action. The nature of decision making can vary greatly, some are routine and repeated frequently, others take place over a longer time span and even occur under severe time pressure [12].

For cognitive ergonomics, the focus here lies in designing interfaces that facilitate informed decision-making through the presentation of relevant information in a comprehensible manner. This includes considerations about cognitive biases, information presentation strategies, and the integration of decision support tools when appropriate. Interfaces should allow users the autonomy to make decisions based on their expertise and preferences. Balancing decision support with user autonomy ensures that a person feels in control of the decision-making process and can help create a sense of empowerment and engagement.

2.5 *Affective States and Emotions*

Affective states, emotions and moods are an essential part of the human mind and behavior. Research distinguishes between the concepts of affective states, emotions and moods, although quite often all terms are used interchangeably. Affect is considered as the broadest concept including emotions and moods. In relevant definitions the main difference is seen in the aspect that affect and emotions are referred to as shorter, momentary states with a clearer component of appraisal while mood is considered as a more diffuse, prolonged mental state [13]. As stated by the Affective Events Theory (AET), emotions can be the consequence of workplace related events and design elements [14]. By using human-centered design principles and elements one can create interfaces and working environments that elicit positive emotional responses, mitigate stressors, and support emotional well-being. Research has also shown that emotional states influence attention, memory, and decision-making [15]. Understanding these dynamics helps create user experiences that align with users' emotional needs and goals. However, people may differ in their emotional responses based on their cultural or demographic background and context clues.

3 HRI Specific Design Elements

New work environments can challenge cognitive abilities such as perception, attention, and decision-making. To mitigate undesired, effects cognitive and sensory aids must be provided. The quality and effectiveness of these aids are closely linked to design elements such as robot morphology, task design, and how well a system incorporates fundamental interaction principles. This section outlines specific elements of HRI that should be thoughtfully considered in the technology design process and during organizational implementation in working systems. In principle, these considerations apply to any robotic application, albeit with varying degrees of potential impact.

3.1 Interaction Design Principles

One guiding aspect when designing robotic systems are the basic principles of interaction as stated in the international standard ISO 9241-110:2020 (en) Ergonomics of human-system interaction—Part 110: Interaction principles [16]. Formerly known as "dialogue principles", they were developed for software systems. In the most recent updated version however, the scope has been broadened to any interactive system and the analysis, design and evaluation of such. The overall aim of these design recommendations is to enhance system usability and a smooth user experience. The seven design principles "Suitability for the user's tasks", "Self-descriptiveness", "Conformity with user expectations", "Learnability", "Controllability", "Use error robustness" and "User engagement" form a basis of general design recommendations for technology including robots. They can either refer to software implemented functions and behavior as well as hardware related features and communication channels. The ISO standard contains detailed design suggestions for each principle and furthermore states that these principles should not be considered independently from each other. The following section provides a short description of each principle.

Suitability for the user's tasks: A system supports the user in efficient and effective task completion. This involves providing appropriate system functionalities, making them comprehensible to users and minimizing unnecessary information and interactions.

Self-descriptiveness: The system provides clear and appropriate information describing the capabilities of the system to the user wherever necessary.

Conformity with user expectations: The behavior of the interactive system is predictable and aligns with the user's experience in the given domain and general system conventions, both in presentation and operation.

Learnability: Interactive systems should facilitate users' discovery of functions through exploration ("trial and error"), reducing the learning curve. Help functions like configuration wizards, context-sensitive help, or manuals can assist users in this process.

Controllability: The system allows the user to maintain control over interface and interactions, including the speed, sequence, and customization of user-system interaction.

Use error robustness: The system supports the user in avoiding errors, tolerates user mistakes in the case of recognizable errors, and supports the user in error correction.

User engagement: The system presents functions and information in an inviting and motivating manner, thus encouraging a continuous interaction with the system.

3.2 Robot Morphology

The concept of embodiment, particularly in anthropomorphic and zoomorphic robotic design, has received significant attention in scientific literature. Robot morphology features can act as cues that influence human expectations and trigger various mental processes and reactions, including behavior and attitudes towards robotic systems, both intended and unintended [17]. Considering consequences of the specific robot morphology carefully, plays a crucial role in creating human-centered workplaces and protecting health and well-being of individuals as design choices may impact user related outcomes significantly.

Anthropomorphism: Anthropomorphism refers to the extent to which a person perceives a robotic system as human-like. This can be based on intentional outward design choices (e.g., giving the robot eyes), but anthropomorphism also includes robotic behavior, and attribution of human-like features despite them not being intentionally human in design [18]. Anthropomorphic features in robotic systems can lead to increased willingness to interact and familiarity with a technology [19]. It can also improve interpretability of actions and intentions of the robotic system [20]. However, anthropomorphic design features can raise expectations and beliefs about the abilities of the robot among users and observers [20]. If those are not met, this in return may lead to irritation or frustration. Preferences regarding anthropomorphic features may also depend on the work environment in which the technology is to be deployed. In industrial settings, lower degrees of anthropomorphism are generally preferred while in social settings higher degrees of anthropomorphism are valued [21]. Additionally, a humanoid appearance can also lead to attributing human-like cognitive or emotional abilities to robots [22, 23].

Zoomorphism: Zoomorphism, like anthropomorphism, refers to the extent to which the robotic design can be derived from animals or nature, or the extent to which a person projects animal-like characteristics onto the robot. Zoomorphic robotic systems often use specific features or behaviors of animals to better fulfil their tasks [24]. Zoomorphism can also trigger emotional reactions and expectations in users [25]. In social situations, certain zoomorphic design elements can evoke positive emotional reactions by using familiar, positively charged animal characteristics [25]. However, if the zoomorphic characteristics are perceived as inappropriate, they can also contribute to the rejection of the robotic system [25]. There are initial studies examining the attribution of mental states to zoomorphic robotic systems [26, 27], indicating similar tendencies as towards anthropomorphic robotic systems.

Functional Design: A functional or technical robot design is more task driven and prioritizes efficiency and task performance over human- or animal-like design features [23]. These robots typically have more abstract forms and might be optimized for specific tasks like assembly or transportation. Industrial robots are often examples of system with a functional design morphology. Functional robot design might lead to the challenge of creating an adequate interaction design as often only a limited number of interaction channels are available.

3.3 Task Design

In social-technical system theories and models, the working task serves as a link between human, technical, and organizational components. Task design and characteristics are vital in the context of task automation, as automation involves redesigning working tasks within a work system. The allocation of specific functions within a work system between humans and one or more robotic systems leads directly to the need to design the assigned work task (or work content) allocated to the human. An important aspect of task design is the level of job control assigned to human workers. Job control, encompassing decision latitude (or autonomy), timing, and method control, has a significant history in occupational psychology. Scientific literature extensively describes the positive effects of job control on workers' well-being, motivation, satisfaction, and mental health, particularly in balancing high job demands, as for example stated in the Job Demand-Resources Model [28] or the Job Strain Model [29, 30]. Overall, the level of job control or autonomy within a given task can be considered a strong determinant for job quality and well-being [31]. Especially in areas where working tasks are highly standardized and the level of job control is naturally lower, timing and method control are influential factors on workers' mental health, motivation and satisfaction [32]. Using robotic systems may increase the risk of reducing job control even further. A tight and non-flexible coupling of the human tasks to robotic actions could lower task performance, flexibility and increase a machine-determined work rate. These factors can lead to various negative psychosocial consequences like emotional exhaustion, nervousness or irritability, an overall poorer mental health and lower intrinsic job satisfaction [32, 33]. Task design activities in relation to HRI working tasks should therefore include the following design recommendations for timing, method control and decision latitude [34]:

Task design recommendations regarding timing control: Avoidance of robot-determined work rates by creating opportunities…

- …to interrupt the work process, e.g., by starting the robot itself;
- …to initiate the process at one's own pace, e.g., by adjusting the robot's work speed;
- …to add buffers to the working system.

Task design recommendations regarding method control: The use of the robot should enable the human…

- … to realize greater variance in task procedures, i.e., allowing more than one way to perform an assembly task, for example.
- …to change task sequences and procedures, i.e., a form of task rotation without actually moving to another workstation.

Task design recommendations regarding decision latitude: Where possible, the use of the robotic system should allow the human…

- … to decide on the sequence of product flows.

- …to be able to perform quality checks independently.
- …to correct potential errors in the quality control process of a product independently.

3.4 Implementation Strategy and Introduction Process

As organizations increasingly integrate new technologies into the workplace to stay competitive, a systematic approach is essential, particularly when implementing robotic systems. These systems may offer unprecedented levels of freedom and autonomy, necessitating organizational adjustments compared to older technology. Moreover, it's important to acknowledge that employees may hold various beliefs, opinions, expectations, or fears regarding organizational changes, especially concerning robotic technology. Addressing these concerns is crucial for successful implementation [35]. Especially when people are lacking first-hand experience, they are forced to rely on their social representations or mental images and are more likely to develop negative attitudes [36]. Clear and straightforward communication methods, along with worker participation approaches, can offer employees insights into the reasons, methods, and timing of upcoming alterations in their workplace and routines. This serves dual purposes: it informs workers about task modifications, policies, and other organizational changes they'll need to adapt to, while also bridging any perceived gaps between implementers and affected individuals. Empirical studies suggest that establishing formal communication channels within organizations initiating change activities reduces uncertainty and enhances commitment [37]. Transparent and unambiguous communication has also been shown to encourage supportive behavior from workers during periods of change [37]. Furthermore, the combination of transparent communication with additional factors such as a sense of meaning and responsibility, and involving users and stakeholders in decision-making can lead to improved safety, psychological well-being, and performance [38].

Alongside pre-implementation communication efforts, it is vital to redefine the workers' tasks accordingly. Creating a use case and analyzing specific needs, possibly with an on-site team, human factors specialists, and ergonomics experts, is recommended. Risk assessment, particularly, is crucial for a detailed analysis of the working situation. Additionally, implementing risk management protocols, employee training, feedback mechanisms, and maintenance procedures can facilitate ongoing evaluation and encourage workforce contributions for improvement.

4 Effects of HRI Design on Humans

The following section describes effects commonly associated with the design, use and implementation of robotic systems. It will elaborate on different concepts related to the use of automation technologies in general, which also apply to the interaction of humans with interactive robots in particular.

4.1 Mental Workload

Mental workload is one of the most invoked and relevant concepts in human factors and cognitive ergonomics. Many attempts have been made in defining this concept, each reflecting slightly different aspects. A very general definition has been introduced by Parasuraman et al., who define mental workload as.

> The relation between the function relating the mental resources demanded by a task and those resources available to be supplied by the human operator" (Parasuraman, Sheridan & Wickens, 2008 pp. 145–146) [39].

Due to the importance of the concept of mental workload, it has found its way into an international standard, the ISO 10075-1:2017(en) Ergonomic principles related to mental workload—Part 1: General issues and concepts, terms and definitions. The standard outlines the terminology with regards to mental workload, including mental stress and mental strain. Furthermore, it focusses on short- and long-term effects when under mental stress or strain, as well as outcomes related to this. In the ISO standard mental workload is considered as an umbrella term referring to all concepts addressed in the standard.

In order to understand mental workload more intuitively, one can evoke a more illustrative analogy to physical workload. How successful a person can perform a physical task is determined by two major factors: the fitness of the person, and the demand of the tasks. Frist, one outlines the exogenous task demands (such as task difficulty, task priority, situational contingencies) as well as the endogenous supply of attentional or processing resources, which are needed to support cognitive functions such as information processing including perceiving, memory updates, planning, decision making, and response processing [40]. Finally, when it comes to mental workload, one needs to take the individuals level of expertise and experience into consideration as well [41]. Young et al. outline mental workload as a multidimensional construct. It is defined by task characteristics (e.g. demands, complexity), intraindividual factors of the operator (e.g. skill, attention) and environmental factors (e.g. unforeseen distractors, risks) [42].

Considering all these factors, an operator can, broadly speaking, end up in one of three states during a working task: in a state of overload or underload, if the workload is suboptimal [43], or in a state of optimal workload, where performance is typically best [9]. Overload may occur during excessive multitasking, or when the task demands exceed that person's skill level. This can result in decreased performance, higher error rate or exhaustion. Mental underload however is characterized by an acute lack of engagement with the task at hand [42]. It may occur during monotonous and repetitive tasks and also result from a mismatch of task and skill. However, rather than demanding a skill that the person does not sufficiently possess or is currently able to display due to a lack of resources, underload may occur when too little resources are needed for successful task engagement. Hence, it will also lead to performance degradation [42].

To adjust mental workload, one can leverage advanced technologies such as interactive robotics, tailored to individual tasks and personal needs. Rauch et al. highlight that new work environments will challenge cognitive abilities more than previous physical tasks, emphasizing the goal of advanced technologies to reduce cognitive burden and enhance productivity [44]. While the specific technology used depends on circumstances, there are shared traits among them. They may automate parts of tasks heavy on limited resources like working memory and attention. Additionally, they should display relevant information to the user to prevent information overload and ensure important feedback isn't missed.

As task automation through advanced technology becomes more prevalent, concerns arise regarding two potential negative outcomes: a reduction or overload in mental workload and an increase in human error [45]. Critical effects associated with automation efforts have already been raised in the 1980's. Under the title "Ironies of Automation" Bainbridge describes how the automation of industrial processes may actually expand problems for the human than eliminating those, as complexity might be increased and operators lose the ability to handle critical situations [46]. That the increasing complexity of technology poses a challenge in terms of mental workload remains relevant, even in recent research [47, 48]. However, the relationship between task automation, mental workload, and error rate is complex and cannot be simplified to a linear correlation.

4.2 Situation Awareness (SA)

Knowing 'what is going on around you' presents an informal way of describing the concept of situation awareness. Although there is no universally agreed definition, it is often described in terms of three facets: the perception of elements in the environment, comprehension of the current situation, and projection of likely changes and events [49]. According to Endsley situation awareness is:

> The perception of the elements in the environment within a volume of time and space, the comprehension of their meaning, and the projection of their status in the near future. (Endsley, 1990, p. 3) [49].

In recent definitions it has been expanded to reflect the complexity of situation awareness (SA) in socio-technical systems more as:

> Activated knowledge for a specific task within a system…[and] the use of appropriate knowledge (held by individuals, captured by devices, etc.) which relates to the state of the environment and the changes as the situation develops. (Stanton et al., 2006, p. 6) [50].

Situation awareness also typically includes factors such as environmental (physical location or display arrangement and size etc.) and group aspects (communication, use of collaboration tools) that affect an individual's situation awareness [51]. Shared situation awareness also includes the degree of communality of team members' individual SA about the current situation [51]. In situations, where the other agent in a team is a technological system of any kind, even a robot, how it communicates and displays its knowledge on a given situation is a highly important factor in interface design [52].

Situation awareness also stands in relation to automation bias [53] and complacency (see Chap. 4.3 on Errors). Especially with increasing levels of automation the risk arises that a person presumes a passive role instead of an active role and thereby reduces their vigilance towards monitoring the situation [49]. This shift, from active engagement to passive reliance, can reduce situation awareness and undermine the human operator's ability to detect and respond to critical cues effectively.

4.3 Errors

Neither humans, nor machines are infallible. As much as one strives to create automation that is reliable and robust to user error, the chance of an error occurring remains. A crucial part of risk assessment and safety analysis of different kinds of applications, especially in high risk environments such as nuclear plants, is the human reliability analysis which is a systematic approach to identify and reduce the probability of human error [54]. Influential work on the classification of human error have been developed by Reason [55] and Rasmussen [56]. Next to this classification, one important distinction can be made between human error (e.g., due to a false input, or mishandling of the machine) or system error [57]. In consequence, this can lead to an error of omission which refers to a mistake that occurs when someone (either human or machine) fails to perform an action that is required or expected of them. Alternatively, an error of commission refers to a mistake that occurs when someone performs an action incorrectly or inappropriately. Hence, one type of error is characterized by a lack of action (omission), while another involves incorrect action (commission). In human-technology interactions with high risk, determining which type of error is more favorable for long-term interaction is nuanced. Both false alarms and missed alarms can undermine trust in the technology [58], with implications discussed later in this chapter. During HRI overreliance on technology may lead to errors of omission, as the system may be left unsupervised or its capabilities overestimated. Conversely, overcompliance involves unnecessary steps taken to manage the technology [59].

Automation complacency: An error that may arise from increased automation is automation complacency. Parasuraman and Manzey characterized it as a poorer detection of system malfunction under automation compared with manual control [60]. While automation complacency can theoretically occur in any kind of human-technology interaction, the phenomenon is more frequently observed during multi-tasking between automated and non-automated tasks. When operators are required to split their attention, and end up relying on the automation to the point that malfunctions either go unnoticed, or errors are not anticipated, the error can be attributed to automation complacency. It can be described as a failure in a person's attention allocation strategy, as the operator primarily pays attention to the manual (non-automated) task, while neglecting supervising the automated task [60]. The phenomenon can occur either consciously or subconsciously; however, it might be influenced by the level of trust placed in the specific technology. The relationship between trust in automation and complacency is complex [57]. Task complexity, operator workload and prior experience with technology as well as the reliability of this specific technology overtime can change the degree to which an individual shows automation complacency [57]. Additionally, studies indicate that neither expertise nor training mitigate the occurrence of it [60].

Automation bias: Automation bias is a significant phenomenon in the realm of automation-related error sources, stemming from excessive trust in a specific technology to the extent that it overrides human judgment. Individuals may rely on the outputs or recommendations of automated systems without critically assessing or verifying them [60], leading to avoidable errors, inappropriate decision-making, overlooking critical information or causing harm to the machine or humans nearby. This bias is intricately tied to trust in automation but can also be influenced by factors such as perceived accuracy, cognitive workload, time pressure, system transparency, and task complexity [53, 61]. Particularly noteworthy is the relationship between automation bias and verification complexity, with research indicating its prevalence in situations where the system's choice involves high verification complexity and is not limited to multitasking environments but can also occur in single-task scenarios [61]. Although reducing unfavorable levels of cognitive load and time pressure may mitigate automation bias, completely eliminating this phenomenon remains undetermined. It may persist even after additional manual training, intentional exposure to automation failure, and explicit training on automation bias [62–64].

4.4 Trust

Trust in automation, regardless of the specific automation technology, level of automation or task, is an important factor in human–machine interaction and often determines the use of automation technologies [65]. The complex nature of trust is shown in its multiple definitions in different disciplines, categorizing trust as a belief, attitude, intention or behavior. Trust between humans can be defined as an individual's willingness to depend on another party because of that party's characteristics [66]. That successful collaboration is influenced by trust between the cooperating parties is well-known in occupational psychology concerning human–human interaction

[67]. However, trust also plays a role when it is no longer human–human interaction but extends to human–machine interaction, including human–robot interaction [68]. According to Lee and See, trust can be understood as the believe that a technology, such as interactive robotics, will help achieve one's goals in uncertain and vulnerable situations [69].

Having an adequate level of trust in a system encourages its appropriate use [65, 70]. However, excessive trust can lead to automation complacency, while insufficient trust may result in neglect of the technology [70]. Trust levels need to match the actual capabilities of the system, known as calibrated trust [69]. Trust is closely tied to the reliability of automation technology or robotic systems [70]. When trust is misjudged, it can lead to misuse (over- or under-relying on the robot), disuse (stop using the robot all together), or abuse (use the robot for purposes other than as designed) of the technology [70]. Various studies have explored factors influencing trust in robotic systems. These factors include human-related aspects like user satisfaction and comfort with robots, as well as robot-related factors such as performance and attributes like anthropomorphism. Task difficulty and the reliability of the robot also impact trust significantly [68, 70]. Proximity of the robot to humans, experience with the robot, and anthropomorphism may influence trust positively. Literature shows that the closer the location of the robot to the human, the higher the degree of trust [70]. Furthermore, robot personality such as positive facial expressions, empathy, likeability and sociability, show positive relations with trust. Overall robot-related factors have a stronger impact on trust than human-related factors [70].

Trust is crucial in occupational safety and health (OSH) as it affects the use of systems and worker safety. It is essential to balance trust-enhancing aspects with potential drawbacks that could affect task completion or safety [70]. Hence, the calibration of trust, considering actual system capabilities, becomes crucial in evaluating the balance between excessive trust and distrust [69].

4.5 Acceptance

An important outcome to consider when humans interact with technology is their acceptance of the technology. A large body of research has evolved around this topic. An important model in this area is the Technology Acceptance Model (TAM) which is based on the influential psycho-social Theory of Reasoned Action [71]. The TAM is a theoretical framework developed to understand and predict how users adopt and accept new information technology [72]. While it was developed in the early 1980´s, it still finds use today, as it focusses on principles that outlive technological change. The TAM itself has been the subject to a number of evolutions and updates, for example the development of the updated model UTAUT (Unified Theory of Acceptance and Use of Technology) by Venkatsh, Morris and Davis [73]. The original TAM suggests that user acceptance of a technology depends on two primary factors: perceived usefulness (PU) and perceived ease of use (PEOU). Perceived usefulness (PU) refers to the degree to which a person believes that using a specific technology will enhance their performance and productivity. If individuals perceive a technology as beneficial and relevant to their tasks or goals, they are more likely to accept and use it. Perceived ease of use (PEOU) relates to the user's belief in how much effort (or lack thereof) it is to use a technology. Users are more likely to accept a technology if they find it easy to understand, learn, and operate as a whole.

Both of these factors have continuously and repeatedly been found to be significant factors when it comes to technology acceptance and use intention for a wide variety of technologies [72, 74, 75], including robotic systems [76]. Beyond PU and PEOU, the model also takes the effect of additional variables into account. Training, support, and social influence, can also impact users' perceptions of a technology, and thereby the degree to which the technology is accepted. The extended model by Venkatesh and Davis considers two additional factors of influence: social influence and cognitive instrumental processes [77]. Social influence consists of subjective norms, voluntariness and image, while cognitive instrumental processes cover job relevance, output quality, result demonstrability, and perceived ease of use. These different aspects are described in the following:

Subjective norm: An individual's perception that other individuals, who the person values or thinks highly of, could perform the task in question. This concept aligns with the theory of reasoned action (TRA), highlighting how social expectations and comparison influence our behavior and the significance of perceived social pressures in influencing behavior [77].

Voluntariness: This describes how much choice a person has over adopting a new technology. An individual's willingness to accept a technology is higher when they perceive the decision to adopt it as a voluntary one, rather than a mandatory one [77].

Image: This refers to what benefits to one's outer perception the use of the technology brings to an individual. If using the technology is perceived as a way to enhance one's status, or to align oneself with desirables trades associated with the technology, it is more likely that people will accept and adopt it. This aspect highlights social influence on technology adoption, which is entirely detached from technology related factors like PEOU and PU.

Job relevance: Considering that technology typically is used to perform some sort of task or job, this facet describes the extent to which it is perceived as suitable for said job. If a technology is perceived as both applicable and relevant to a specific task, users are more inclined to accept and adopt it.

Output quality: This facet also relates to the technology's suitability for the task; however, it is specifically addressing the output, not like PEOU the experience of the process when using a technology. This factor is based on the user's evaluation of the technology's performance and capability in delivering outputs of an expected quality.

Result demonstrability: The model states that producing a perceivable result will directly influence how useful the system is perceived. This does not mean that the result must be physically tangible, but rather that the user is able to perceive and evaluate the output of the technology.

4.6 Smoothness of Interaction

Creating seamless interactions is a key goal of ergonomic design for interactive systems, encompassing various design and interaction features. This seamless experience offers numerous benefits applicable to human–robot interaction as well. It can enhance efficiency by optimizing mental workload and minimizing interruptions during tasks. A smooth transition between human and robotic tasks supports workflow and may increase productivity, while also improving the overall user experience. Operators perceive smooth interactions as more intuitive and user-friendly [78]. Furthermore, an uninterrupted workflow reduces the source of errors. Clear, understandable interfaces, intuitive design, and consistent feedback mechanisms help operators navigate seamlessly, minimizing the risk of incorrect inputs [79]. Non-smooth interactions between human and technology can lead to increased unfavorable mental workload. Operators may find it difficult to understand and navigate unoptimized interfaces. Hence, frustration, mental fatigue, and reduced performance can result. This may also negatively affect satisfaction, and trust in the technology, decreasing the willingness to use it in the future [80]. These effects highlight the importance of creating intuitive, understandable systems for HRI. That also includes considering what types of input and output modalities are best suited for a specific task and environment.

4.7 Motivation

Work motivation is considered an important factor in organizations and human factors research. A number of influential theories strive to explain motivated behavior in work situation. For a comprehensive overview see for example Luczak et al. [81]. Most motivation researchers fall back on the psychological concepts of needs and goals. Motivation is considered goal oriented; it serves to meet a need or to achieve a goal. In a work context, motivation and satisfaction are closely linked, however they are two distinct concepts. While motivation describes the disposition to act goal oriented in a work process, job satisfaction is the outcome of performance-based rewards [81]. One can also distinguish between intrinsic and extrinsic motivation. Intrinsic motivation refers to the internal drive, while extrinsic motivation involves external factors such as rewards or performance feedback. Both concepts come into play when assessing worker motivation. Understanding and predicting human behavior and reactions in terms of motivation and job satisfaction are important factors for organizations. Motivation research shows that motivation theories and models have been applied to a large number of workplace related factors like worker involvement and participation, work and task design, or working time—each to some extend serving as factors relevant to motivation [81].

Motivation is crucial in HRI and influenced by various factors. For instance, the introduction process of interactive robots can enhance employee involvement and participation, impacting motivation. Task design changes after introducing robotic systems can also affect motivation and satisfaction. Direct interaction with robots can shape motivation, with feedback from the system influencing willingness to use it. User engagement, as per ISO 9241-110, includes aspects of robotic design that can impact motivation. As robots become more complex, personalization becomes possible, adapting to users' needs and preferences applying specific algorithms or software. Individualization has been an important factor for all types of software (as it is also part of the ISO 9241-110). However, the issue of data privacy must not be ignored when implementing such systems.

4.8 Well-Being

Well-being is a complex and multidimensional phenomenon that contains physical, mental and social dimensions, and is certainly linked to a number of work-related factors. Yet in relation to well-being HRI can also play a vital role as is may be considered as a enduring long-term effect of human-cetered workplace design.

Well-being can be described as the combination of the experience of positive emotions like happiness, a physical state of being unharmed and not overexerted, and the ability to exercise control over your own life [82]. Well-being is often included in conversations about mental health, to the point that it is a fundamental building block of the World Health Organizations definition of positive mental health: "*a state of well-being in which the individual realizes his or her own abilities, can cope with the normal stresses of life, can work productively and fruitfully, and is able to make a contribution to his or her community*" [3].

Well-being has been linked to success at professional levels [83]. Workers with higher subjective well-being show greater productivity in the workplace, are more effective at acquiring new skills, exhibit higher creativity, and engage and support their colleagues more [84]. The positive impact of well-being goes so far, that it can be used as a possible predictor for job performance, job satisfaction, engagement, commitment, burnout, and work stress [83, 85, 86]. Regarding job performance, the relationship to well-being might be bidirectional. Having higher well-being is crucial for effective performance at work, and successful performance, in turn, contributes to overall well-being [87]. Higher well-being is linked to a number of better outcomes regarding physical health, resilience during illness, and recovery [88, 89]. However, this relationship is also bidirectional, as good physical health contributes to experience a higher quality of life [88].

4.9 Emotional Responses

As mentioned earlier in this chapter, affect and emotions are a crucial part of mental processes and a constant companion of everyday lives. Robotic systems may be connected to the topic of emotions during HRI in more than one way affecting and shaping human emotional responses. While they do not experience emotions, robotic systems may be capable of displaying emotional responses, or eliciting emotions in the person interacting with them. Robotic systems operating in a social environment sometimes have the inbuilt ability to recognize human emotions and display emotional responses themselves. This can affect user experience and performance. Robots designed to recognize and respond to human emotions can enhance engagement and satisfaction levels [90, 91]. Robotic systems may display emotions via voice modulation (e.g., they may have a voice databank with "happy" or "sad" versions of the same response line), through visual cues (e.g., showing a happy or sad expression on a display), or (dependent on their degree of anthropomorphism) through gestures (e.g., lowering their head). Humans have been shown to be able to interpret robotic expressions in great nuance [90]. Emotional displays in HRI should be context-sensitive. System developers must weigh whether adding emotional responses enhances interaction or distracts the operator. Recognizing the interaction partner's emotional state is crucial for social robots, but accurately interpreting human emotions is technically challenging and raises privacy and surveillance concerns. Feeling connected to the system might be valuable in domains such as healthcare and education. However, robotic systems may evoke various emotional responses beyond social interaction, such as the uncanny valley effect in anthropomorphic robots [92]. Other emotions include for example the fear of job loss in work environments.

Fear of job loss: Being worried to lose one's job to progressing automation is nothing unique to robotic systems. The impact of uncertainty of one's employment is a well-documented emotional response negatively affecting mental well-being [93–95]. The stress associated with job insecurity can affect a person's physical and mental health, but may also be detrimental to their job satisfaction [93], commitment and effort [94]. Individuals in job positions that are particularly likely to be automated tend to be more fearful toward robots in the workplace. This tendency is stronger in countries with unfavorable economic conditions and where employees have less protection [96]. There is emerging evidence that this can elicit a negative attitude towards robotic systems which is linked to maladaptive workplace behavior [97] and lower job satisfaction [98].

5 Measuring HRI Elements and Potential Outcomes

Understanding and measuring different constructs within cognitive ergonomics is essential for analyzing, evaluating, and ultimately improving HRI. The following section aims to provide an overview of different data collection methods and specific tools for assessing HRI-specific design elements and associated effects on humans. We strive to offer a comprehensive overview of methods and tools, encompassing both quantitative and qualitative data collection approaches. Additionally, we present a selected toolkit for evaluating specific design elements and outcomes. However, this collection is not exhaustive, and other tools can also be applied. This chapter serves as an initial entry point into the diverse field of data collection methods for assessing cognitive ergonomics and human factors in HRI. It's important to note that each presented data collection method should be carefully considered, as each comes with its own constraints and limitations, requiring deeper engagement with the specific method or tool as well as a careful consideration of the application context. Deeper insights into human factors methods are for example presented in the book of Stanton et al. [52].

5.1 Quantitative Data Collection Methods

Quantitative data collection in human factors research involves systematically gathering and analyzing numerical data to explore patterns, relationships, and trends within populations or specific groups of interest. Quantitative approaches prioritize the measurement and quantification of observable phenomena. Researchers utilize techniques such as surveys, experiments, observations, and psychometric assessments to collect data in a structured and standardized manner. These methods often employ predetermined scales, questionnaires, or instruments designed to assess specific constructs, behaviors, attitudes, or outcomes. Data may also be collected using psychophysiological methods and techniques, for example such as heart rate variability, electrodermal activity, electroencephalography (EEG) or eye-tracking, each of which requires specific equipment. These methods are also especially used in the field of neuroergonomics [99] by focusing on the neural mechanisms—essentially the "neural human factors", related to the central and the peripheral nervous system—underlying human cognition and performance during interactions with other system elements, including robots. Through neuroscientific methods to monitor neural and neurocognitive processes in ecologically valid conditions, neuroergonomics can be employed to evaluate HRIs or to enable live neuroadaptive functions, making the machine respond dynamically to human commands and needs. Through the methods of neuroergonomics it is possible to assess factors like mental workload, for instance, in fields like rehabilitation robotics [100] and cobotics [101]. An example of the aforementioned neuroadaptive systems is based on the detection of the Error-related EEG Potential (ErrP), informing the robot when the person notices that its actions are going against their expectations (thus, guiding the robot learning) [102]. Such promising solutions, anyway, are mainly explored in research settings, at present. To bring neuroscientific methods outside the laboratory, portable, particularly wearable, devices are they key solutions [103].

Overall, these approaches (even the most experimental ones bridging laboratory research with the specific requirements of robot design for the real world) enable researchers to quantify variables, test hypotheses, and statistically analyze relationships between different factors. Statistical methods such as correlation analysis, regression analysis, and inferential statistics can be applied to draw conclusions about associations, effects, and generalizability of findings. Even when statistical analysis isn't feasible, descriptive quantitative data can serve as anchors for evaluation purposes and further design activities. The focus of this sections lies on describing relevant scales for subjective, quantitative measures in relation to HRI.

5.1.1 Interaction Principles and Usability Assessment

Interaction principles (IsoMetrics): As already described in this chapter, the ISO 9241-110 standard defines seven interaction principles for designing interactive systems: "Suitability for the user´s tasks", "Self-descriptiveness", "Conformity with user expectations", "Learnability", "Controllability", "Use error robustness" and "User engagement". The interaction principles have through history been repeatedly used as a foundation for usability assessment tools. One instrument which has been originally developed for software applications and is based on the former version of the standard, is the IsoMetrics questionnaire [104]. A long version of the questionnaire consists of 75 items in total. Statements are rated on a five-point Likert scale from 1 "strongly disagree" to 5 "strongly agree." The long version of the questionnaire targets experts and is recommended for proactive use in the development process. Experts are encouraged to provide specific examples while completing it. In contrast, a short version aims at end users and does not ask about specific application situations. The IsoMetrics has also been found suitable for assessing human–robot interaction and been validated for this use [105, 106].

Usability: **The System Usability Scale (SUS)** is a questionnaire designed to evaluate the subjective perception of usability in technology [107]. It consists of ten items that measure overall usability based on various aspects. The ten items collectively cover aspects related to ease of use, learnability, efficiency, interface quality, and overall user satisfaction. Respondents rate their agreement or disagreement with statements about the system's usability on a five-point Likert scale. The SUS provides a quick way to assess usability, offering a single score ranging from 0 to 100, where higher scores indicate better perceived usability. It's valued for its simplicity, versatility, and ability to provide actionable insights for improving system usability, making it a popular tool in user experience research and evaluation that is used to assess numerus technologies [108].

5.1.2 Job Control

Within occupational psychology there are a number of well-established questionnaires dedicated to measure one or more facets of the construct of job control as one of the major working task characteristics.

The Job Content Questionnaire (JCQ) is a widely used instrument to assess psychosocial aspects of work environments and their impact on employee health and well-being [30]. The questionnaire was developed based on the influential Job Demand-Control (JDC) model by Karasek [29]. The JCQ is a tool designed to measure the constructs proposed by the JDC Model, specifically job demands (physical and psychological effort, workload, time pressure) and job control (decision latitude, autonomy). It focuses on measuring components of job stress, but particularly the balance between job demands and job control as worker experiences, it contains the following sub-scales:

Decision Latitude: This scale assesses the level of control and autonomy workers have over their work, including both. skill variety and decision authority they experience during their work.

Psychological Job Demands: This measures the intensity and psychological load of work-related demands, such as time pressure, workload, and conflicting priorities.

Social Support: This scale evaluates the perceived support employees receive from colleagues and supervisors in the workplace.

Physical Job Demands: This component assesses physical workload, such as the amount of physical effort and demands in the work environment.

Job Insecurity: This facet describes the general perceived level of job insecurity, as well as how much a worker experiences skill obsolescence.

In its full version the JCQ contains 49 items that are answered on a four-point scale from "strongly disagree" to "strongly agree". The questionnaire has been used to assess work with different kinds of technologies, including robotic systems [109, 110].

The Job Diagnostic Survey (JDS) assesses core task characteristics in a specific job. The complete version also allows to analyze their impact on satisfaction, motivation and performance [111]. However, the most common versions of the JDS focus on the five core dimensions (skill variety, task identity, task significance, autonomy, feedback) of which the dimension skill variety and autonomy are directly related to concept of job control. The five dimensions are answered on a seven-point Likert scale (from "very inaccurate" to "very accurate") and are described as followed:

Skill Variety: This factor describes the extent to which a job requires a range of different skills and abilities in order to complete tasks.

Task Identity: This refers to the degree to which a job involves completing a whole and identifiable piece of work which contributes to a sense of accomplishment and responsibility.

Task Significance: The impact and importance of the job's outcomes on others or the organization, influencing the perceived meaningfulness of the work.

Autonomy: This factor relates to the level of independence and discretion a worker has in making decisions and exercising control over their work processes.

Feedback: The extent to which a worker receives clear and direct information about their performance and the outcomes of their work.

While created almost half a decade ago, the JDS still finds use today, also extending into the world of HRI [112].

5.1.3 Mental Workload

In order to assess mental workload in a specific working situation, the NASA Task Load Index (NASA-TLX) has gained remarkable reception among human factors researcher. The NASA-TLX is a subjective assessment tool designed to evaluate perceived workload across various tasks or systems [113]. It measures mental workload, providing insights into the cognitive, physical, and temporal demands experienced by a person while performing a task. The questionnaire consists of six subscales, each represented by one item:

Mental Demand: Assesses the perceived level of mental effort, complexity, and cognitive workload required.
Physical Demand: Focuses on the physical exertion or effort required to perform the task.
Temporal Demand: Evaluates the perceived time pressure or urgency associated with completing the task within a given time frame.
Performance: Assesses self-perceived task performance and how well the task was completed.
Effort: Measures the overall level of effort exerted in order to complete the task.
Frustration: Reflects the degree of insecurity, dissatisfaction, or emotional stress experienced while performing the task.

Each item is rated on a 21-point bipolar scale with the poles "very low" and "very high" based on a person's perceived workload during or after a specific task or activity. In addition to the individual dimensions, the additive combination of the individual scales allows to calculate an overall mental workload value. This tool has been widely used to assess workload in a variety of technologies and remains a staple of human factors research [114]. However, one has to notice that this scale does not capture positive mental states, according to the conceptualization of an optimal state of mental load.

5.1.4 Human Robot Trust Scales

Trust is a significant consequence and influencing factor in human actions regarding HRI. It was demonstrated that trust is a multifaceted construct, with a non-linear relation to other. Trust expresses itself in issue of under- and over-trust. However, evaluating trust in HRI is crucial, as it directly impacts effectiveness, safety, and acceptance. This importance has led to an increased number of scales measuring various aspects of trust in recent years. The measures presented here represent a selection of available tools for quantifying this complex concept in different ways. It is essential to choose a measure that suits one's needs and technology to effectively study human–robot trust.

Trust Perception Scale-HRI: The Trust Perception Scale HRI [115] is designed to measure trust perceptions over time, across domains, by individuals in all the major roles of HRI and across differing levels of system autonomy and intelligence.

The scale is designed as a pre-post interaction measure used to assess changes in trust specific to HRI, while remaining context and domain independent. The scale contains 40 items which participants are asked to rate on a scale from 0 to 100%. Forty attributes of a robotic system are listed, following the question "What % of the time will this robot…". This is then cumulated into an overall percentage of trust towards the robot.

Multi-Dimensional Measure of Trust (MDMT): This questionnaire captures trust in human–robot or human-automated agent interaction in form of two superordinate factors (Performance trust, Moral trust) that each break into two subscales (Reliable and Capable within Performance, and Sincere and Ethical within Moral). It is designed to be context and robot-type independent. The questionnaire contains 16 items and users evaluate the robot in question on an 8-point discrete rating scale from 0 "not at all" to 7 "very" [116].

Trust in Industrial Human–Robot Collaboration: Charalambous, Fletcher, and Webb developed a questionnaire to measure trust in HRI in an industrial context [117]. The questionnaire consists of three subscales (Robot's Motion and Pick-up Speed, Safe Cooperation, Robot and Gripper Reliability) and a total of ten items. The items mainly focus on whether the user feels safe during interaction with the industrial robot, whether the robot appears reliable, and how the appearance and movement of the robot affect the user. They are designed to be evaluated on a 5-point discrete rating scale from "strongly disagree" to "strongly agree".

Human–Robot Interaction Trust Scale: The Human–Robot interaction trust scale encompasses eight dimensions that are each assessed via four items, designated one of four attributes: reliable, dependable, accessible und consistent. This scale focusses noticeably on the technological characteristics of the robotic system, as the eight dimensions that are rated are User Interface, Sensor Data, Navigation Capabilities, Signal/Bandwidth, End Effectors, Remote Information Processing, Level of Automation and Type of control. The HRI trust scale uses 32 items that are to be scored on a 7-point Likert rating scale ranging from "strongly disagree" to "strongly agree"[118].

5.2 *Qualitative Data Collection Methods*

While quantitative methods focus on collecting and analyzing numerical data, qualitative data collection involves systematically gathering and analyzing non-numerical data to explore the complexities, meanings, and contexts of human experiences, behaviors, and interactions. The emphasis here is not on statistical analysis but on the depth, richness, and subjective interpretations that humans bring to any situation. Researchers utilize various techniques such as interviews, focus groups, participant observations, and content analysis to collect qualitative data. These methods typically allow participants to express their experiences, perspectives, beliefs, or values more freely and naturally than questionnaire-based assessments.

Interviews: Interviews are face-to-face or remote conversations between a researcher and one or more participants. They can be structured, semi-structured, or unstructured, offering flexibility in exploring experiences, opinions, attitudes, and behaviors. Open-ended questions are commonly used to elicit detailed responses and insights into participants' thoughts and feelings. Interview recordings are often transcribed for analysis, which can involve content analysis or quantitative methods like tallying the frequency of specific topics mentioned.

Workshops or Focus Groups: Workshops involve interactive sessions, conducted either in person or remotely, with a group of participants facilitated by a researcher. These sessions emphasize group discussions, brainstorming, and collaboration. Participants are often selected to represent diverse expertise and perspectives. Various techniques and media are utilized during workshops to facilitate participants in sharing their perspectives, ideas, opinions, and experiences on specific topics. Focus groups, a subtype of workshops, consist of individuals sharing crucial characteristics relevant to the research questions, such as professional background, geographic location, or attendance at a specific event. Despite these shared characteristics, participants in focus groups may still exhibit differences in beliefs, attitudes, and opinions. It is important to be mindful of group behavior effects when using this method.

Participant Observation: This method entails researchers immersing themselves in participants' natural settings to observe and document specific behaviors, interactions, or social dynamics. It provides firsthand insight into context-specific behaviors and social phenomena, which are challenging to quantify otherwise. However, it is important to note that participant observation may lead to behavior modification due to being observed.

Diary Method: During data collection using the diary method, individuals maintain personal diaries or journals to document their thoughts, experiences, activities, and emotions over a set period. This enables researchers to gather data over an extended timeframe, offering detailed insights into participants' daily lives, perspectives, and experiences.

Content Analysis: A content analysis systematically examines texts, documents, or audiovisual materials to discern patterns, themes, or meanings. It aids in grasping the content, context, and underlying messages in the materials, applicable to various qualitative methods' outputs. It involves coding and categorization, breaking down data into smaller units aligned with research themes. Categories may be developed for the analysis or derived from established psychological models.

The presented section offered a brief overview of different qualitative methods for assessing HRI. No qualitative method is specifically designed for, or better suited to assess HRI; the decision on which method to use must be based on the research question and contextual factors such as availability of participants or organizational constraints.

6 Conclusion

Any human performance is deeply rooted in mental functions—such as perception, attention, learning, memory, and reasoning—that process information for directing our actions, especially in the complexities characterizing the current socio-technical systems. Within the domain of human factors, these mental processes are often investigated (through approaches mainly derived from cognitive psychology and neuroscience) to understand key aspects like mental workload and situational awareness, which for example may lead to negative health consequences or to errors due to a mismatch between our capabilities and the task demands. Such errors can generate hazardous situations for humans interacting with robots despite the effort to design machines that should support us: without knowing how the human mind "works", a designer would ignore how to align the robot to the user.

This chapter has explored foundational concepts of cognitive ergonomics, emphasizing their importance in HRI. By integrating these concepts in the know-how for human-centered robot design and assessment, we can improve technology acceptance, usability, accessibility, and overall user experience. Additionally, the chapter has discussed design recommendations specific to HRI, and concluded with an overview of quantitative and qualitative methods presenting options for evaluating diverse aspects of HRI.

References

1. Hollnagel, E.: Cognitive ergonomics: it's all in the mind. Ergonomics **40**(10), 1170–1182 (1997). https://doi.org/10.1080/001401397187685
2. Tegtmeier, P., Weber, C., Sommer, S., Tisch, A., Wischniewski, S.: Criteria and guidelines for human-centered work design in a digitally transformed world of work: findings from a formal consensus process. Int. J. Environ. Res. Public Health **19**(23) (2022)
3. W. H. Organization.: Constitution. Geneva. https://www.who.int/about/accountability/governance/constitution
4. Antonovsky, A.: The salutogenic model as a theory to guide health promotion. Health Promot. Int. **11**(1), 11–18 (1996)
5. Kahneman, D.: Attention and Effort. Prentice-Hall (1973)
6. van Jan, B.: The multi-dimensional nature of encoding tactile and haptic interactions: from psychophysics to design guidelines. In: Proceedings of the Human Factors and Ergonomics Society Annual Meeting, vol 50, no. 5, pp. 685–688. SAGE Publications Sage CA, Los Angeles, CA (2006)
7. Wickens, C.D.: Multiple resources and performance prediction. Theor. Issues Ergon. Sci. **3**(2), 159–177 (2002)
8. Wickens, C.D.: Multiple resources and mental workload. Hum. Factors **50**(3), 449–455 (2008)
9. Hancock, P.A., Warm, J.S.: A dynamic model of stress and sustained attention. Human Factors **31**(5), 519–537 (1989), https://doi.org/10.1177/001872088903100503
10. Broadbent, D.E.: Perception and Communication. Pergamon, London (1958)
11. Chun, M.M.: Contextual cueing of visual attention. Trends Cogn. Sci. **4**(5), 170–178 (2000)
12. Salvendy, G.: Handbook of Human Factors and Ergonomics. John Wiley & Sons (2012)
13. Arnold, M.B.: Emotion and Personality. Columbia University Press, New York (1960)

14. Weiss, H.M., Cropanzano, R.: Affective events theory. Res. Organiz. Behav. **18**(1), 1–74 (1996)
15. LeBlanc, V.R., McConnell, M.M., Monteiro, S.D.: Predictable chaos: a review of the effects of emotions on attention, memory and decision making. Adv. Health Sci. Educ. **20**, 265–282 (2015)
16. Ergonomics of human-system interaction Part, 110, I. O. f. Standardization (2020)
17. Roesler, E., Manzey, D., Onnasch, L.: A meta-analysis on the effectiveness of anthropomorphism in human-robot interaction. Sci. Robot. **6**(58) (2021)
18. Bartneck, C., Kulić, D., Croft, E., Zoghbi, S.: Measurement instruments for the anthropomorphism, animacy, likeability, perceived intelligence, and perceived safety of robots. Int. J. Soc. Robot. **1**, 71–81 (2009). https://doi.org/10.1007/s12369-008-0001-3
19. Anzalone, S.M., Boucenna, S., Ivaldi, S., Chetouani, M.: Evaluating the engagement with social robots. Int. J. Soc. Robot. **7**, 465–478 (2015)
20. Złotowski, J., Proudfoot, D., Yogeeswaran, K., Bartneck, C.: Anthropomorphism: opportunities and challenges in human–robot interaction. Int. J. Soc. Robot. **7**, 347–360 (2015)
21. Roesler, E., Naendrup-Poell, L., Manzey, D., Onnasch, L.: Why context matters: the influence of application domain on preferred degree of anthropomorphism and gender attribution in human–robot interaction. Int. J. Soc. Robot. **14**(5), 1155–1166 (2022)
22. Marchetti, A., Manzi, F., Itakura, S., Massaro, D.: Theory of mind and humanoid robots from a lifespan perspective. Zeitschrift für Psychologie **2** (2018). https://doi.org/10.1027/2151-2604/a000326
23. Thellman, S., de Graaf, M., Ziemke, T.: Mental state attribution to robots: a systematic review of conceptions, methods, and findings. ACM Trans. Human-Robot Inter. (THRI) **11**(4), 1–51 (2022)
24. Onnasch, L., Roesler, E.: A taxonomy to structure and analyze human–robot interaction. Int. J. Soc. Robot. **13**(4), 833–849 (2021)
25. Löffler, D., Dörrenbächer, J., Hassenzahl, M.: The uncanny valley effect in zoomorphic robots: the U-shaped relation between animal likeness and likeability. In: ACM/IEEE International Conference on Human-Robot Interaction IEEE, ACM/IEEE, pp. 261–270 (2020)
26. Eyssel, F.A., Hegel, F.: (S)he's got the look: gender-stereotyping of social robots. J. Appl. Soc. Psychol. **42**(2), 213–223 (2012)
27. Keijsers, M., Bartneck, C., & Eyssel, F.: What's to bullying a bot? Correlates between chatbot humanlikeness and abuse. Interact. Stud. **22**(1), 55–80 (2021)
28. Bakker, A.B., Demerouti, E.: The job demands-resources model: state of the art. J. Manage. Psychol. **20**(7), 743–757 (2007). https://doi.org/10.1108/02683940710733115
29. Karasek, R.A.: Job demands, job decision latitude, and mental strain: implications for job design. Adm. Sci. Q. **24**(2), 285–308 (1979). https://doi.org/10.2307/239249
30. Karasek, R.A., Brisson, C., Kawakami, N., Houtman, I., Bongers, P., Amick, B.: The job content questionnaire (JCQ): an instrument for internationally comparative assessments of psychosocial job characteristics. J. Occupat. Health Psychol. **3**(4), 322 (1998)
31. Van der Doef, M., Maes, S.: The job demand-control (-support) model and psychological well-being: a review of 20 years of empirical research. Work & Stress **13**(3), 87–114 (1999)
32. Rosen, P.H., Wischniewski, S. Rosen, P.H., Wischniewski, S.: Scoping review on job control and occupational health in the manufacturing context. Int. J. Adv. Manuf. Technol. **102**(5), 2285–2296 (2019). https://doi.org/10.1007/s00170-018-03271-z
33. Robelski, S., Wischniewski, S.: Human-machine interaction and health at work: a scoping review. Int. J. Human Fact. Ergonom. **5**(2), 93–110 (2018). https://doi.org/10.1504/IJHFE.2018.092226
34. Rosen, P.H., Wischniewski, S.: Task design in human-robot-interaction scenarios—challenges from a human factors perspective. In: International Conference on Human Factors and Systems Interaction, The Westin Bonaventure Hotel, Los Angeles, California, USA 2018: Springer International Publishing, in Advances in Human Factors and Systems Interaction: Proceedings of the AHFE, pp. 71–82 (2017)

35. Naneva, S., Sarda Gou, M., Webb, T.L., Prescott, T.J.: A systematic review of attitudes, anxiety, acceptance, and trust towards social robots. Int. J. Soc. Robot. **12**(6), 1179–1201 (2020)
36. Savela, N., Turja, T., Oksanen, A.: Social acceptance of robots in different occupational fields: a systematic literature review. Int. J. Soc. Robot. **10**(4), 493–502 (2018)
37. Bordia, P., Hobman, E., Gallois, C.: Uncertainty during organizational change: types, consequences and management strategies. J. Business Psychol. **18**(4), 507–532 (2004)
38. Fraboni, F., Brendel, H., Pietrantoni, L.: Evaluating organizational guidelines for enhancing psychological well-being, safety, and performance in technology integration. Sustainability **15**(10), 8113 (2023)
39. Parasuraman, R., Sheridan, T.B., Wickens, C.D.: Situation awareness, mental workload, and trust in automation: viable, empirically supported cognitive engineering constructs. J. Cognit. Eng. Decision Making. **2**(2), 140–160 (2008)
40. Salvendy, G., Karwowski, W.: Handbook of Human Factors and Ergonomics. John Wiley & Sons (2021)
41. Just, M.A., Carpenter, P.A., Miyake, A.: Neuroindices of cognitive workload: neuroimaging, pupillometric and event-related potential studies of brain work. Theor. Issues Ergonom. Sci. **4**(1), 56–88 (2003)
42. Young, M.S., Brookhuis, K.A., Wickens, C.D., Hancock, P.A.: State of science: mental workload in ergonomics. Ergonomics **58**(1), 1–17 (2015)
43. Brookhuis, K.A., de Waard, D.: Assessment of drivers' workload: performance, subjective and physiological indices. In: Desmond, P.A.H.P.A. (ed.) Stress, Workload and Fatigue Lawrence Erlbaum, Mahwah, NJ (2000)
44. Rauch, E., Linder, C., Dallasega, P.: Anthropocentric perspective of production before and within Industry 4.0. Comput. Ind. Eng. (2020)
45. Moray, N.: Monitoring behavior and supervisory control. In: Boff, L.K.K.R., Thomas, J.P. (ed.) Handbook of Perception and Performance, vol. 2, pp. 40-1–40-51. Wiley, New York (1986)
46. Bainbridge, L: Ironies of automation. In: Analysis, Design and Evaluation of Man–Machine Systems, pp 129–135. Elsevier (1983)
47. Marchand, C., De Graaf, J.B., Jarrassé, N.: Measuring mental workload in assistive wearable devices: a review. J. NeuroEng. Rehabil. **18**(1), 1–15 (2021)
48. Hossain, G., Yeasin, M.: Cognitive load based adaptive assistive technology design for reconfigured mobile android phone. In: Mobile Computing, Applications, and Services: Third International Conference, Heidelberg, MobiCASE, Ed., Springer, pp. 374–380 (2012)
49. Endsley, M.R.: Situation awareness global assessment technique (SAGAT). In: Proceedings of the National Aerospace and Electronics Conference (NAECON), New York, NY, IEEE, pp. 789–795 (1988)
50. Stanton, N.A., Stewart, R., Harris, D., Houghton, R.J., Baber, C., McMaster, R., Salmon, P., Hoyle, G., Walker, G., Young, M.S., Linsell, M., Dymott, R., Green, D.; Distributed situation awareness in dynamic systems: theoretical development and application of an ergonomics methodology. Ergonomics **49**(12–13), 1288–1311 (2006). https://doi.org/10.1080/00140130600612762
51. Kulyk, O., van der Veer, G., van Dijk, B.: Situational awareness support to enhance teamwork in collaborative environments. In: European Conference on Cognitive Ergonomics: The Ergonomics of Cool Interaction Enschede, The Netherlands, pp. 1–5 (2008)
52. Stanton, N.A., Salmon, P.M., Rafferty, L.A., Walker, G.H., Baber, C., Jenkins, D.P.: Human Factors Methods: A Practical Guide for Engineering and Design. CRC Press (2017)
53. Goddard, K., Roudsari, A., Wyatt, J.C.: Automation bias: a systematic review of frequency, effect mediators, and mitigators. J. Am. Med. InFormat. Assoc. **19**(1), 121–127 (2012)
54. Ciani, L., Guidi, G., Patrizi, G.: Human reliability in railway engineering: literature review and bibliometric analysis of the last two decades. Saf. Sci. **151**, 105755 (2022)
55. Reason, J.: Human Error. Cambridge university press (1990)

56. Rasmussen, J.: Human errors. a taxonomy for describing human malfunction in industrial installations. J. Occup. Accid. **4**(2–4), 311–333 (1982)
57. Wickens, C.D., Clegg, B.A., Vieane, A.Z., Sebok, A.L.: Complacency and automation bias in the use of imperfect automation. Human Factors **57**(5), 728–739 (2015)
58. Dixon, S.R., Wickens, C.D., McCarley, J.S.: On the independence of compliance and reliance: are automation false alarms worse than misses?. Human Factors **49**, 564–572 (2007)
59. Dixon, S.R., Wickens, C.D.: Automation reliability in unmanned aerial vehicle flight control: a reliance compliance model of automation dependence in high workload. Human Factors **48,** 474–486 (2006)
60. Parasuraman, R., Manzey, D.H.: Complacency and bias in human use of automation: an attentional integration. Human Factors **52**(3), 381–410 (2010)
61. Lyell, D., Coiera, E.: Automation bias and verification complexity: a systematic review. J. Am. Med. Inf. Assoc. **24**(2), 423–431 (2017)
62. Singh, A.L., Tiwari, T., Singh, I.L.: Effects of automation reliability and training on automation-induced complacency and perceived mental workload. J. Indian Acad. Appl. Psychol. **35**(2), 9–22 (2009)
63. Bahner, J.E., Hüper, A.D., Manzey, D.: Misuse of automated decision aids: complacency, automation bias and the impact of training experience. Int. J. Human-Comput. Stud. **66**(9), 688–699 (2008)
64. Mosier, K.L., Skitka, L.J., Dunbar, M., McDonnell, L.: Aircrews and automation bias: the advantages of teamwork? Int. J. Aviation Psychol. **11**(1), 1–14 (2001)
65. Parasuraman, R., Riley, V.: Humans and automation: use, misuse, disuse, abuse. Human Factors **39**(2), 230–253 (1997)
66. Rousseau, D.M., Sitkin, S.B., Burt, R.S., Camerer, C.: Not so different after all: a cross-discipline view of trust. Acad. Manage. Rev. **23**(3), 393–404 (1998)
67. Costa, A.C., Roe, R.A., Taillieu, T.: Trust within teams: the relation with performance effectiveness. European J. Work Organizat. Psychol. **10**(3), 225–244 (2001)
68. Hancock, P.A., Billings, D.R., Schaefer, K.E., Chen, J.Y., De Visser, E.J., Parasuraman, R.: A meta-analysis of factors affecting trust in human-robot interaction. Human Factors **53**(5), 517–527 (2011)
69. Lee, J.D., See, K.A.: Trust in automation: designing for appropriate reliance. Human Factors **46**(1), 50–80 (2004)
70. Hancock, P.A., Kessler, T.T., Kaplan, A.D., Brill, J.C., Szalma, J.L.: Evolving trust in robots: specification through sequential and comparative meta-analyses. Human Factors **63**(7), 1196–1229 (2021)
71. Ajzen, I., Fishbein, M.: Understanding attitudes and predicting social behavior. Prentice-Hall (1980)
72. Davis, F.D.: Technology acceptance model: TAM. Information Seeking Behavior and Technology Adoption, pp. 205–219 (1989)
73. Venkatesh, V., Morris, M.G., Davis, G.B., Davis, F.D.: User acceptance of information technology: toward a unified view. MIS Quarterly 425–478 (2003)
74. Ramayah, T., Ignatius, J.: mpact of perceived usefulness, perceived ease of use and perceived enjoyment on intention to shop online. CFAI J. Syst. Manag. (IJSM) **3**(3), 36–51 (2005)
75. Raza, S.A., Umer, A., Shah, N.; New determinants of ease of use and perceived usefulness for mobile banking adoption. Int. J. Electron. Custom. Relation. Manage. **11**(1), 44–65 (2017)
76. Liu, Y., Liao, S.: Would humans want to work side-by-side with autonomous robots: the effect of robot autonomy on perceived usefulness, ease of use and desire for contact. In: International Conference on Control and Intelligent Robotics: IEEE-RAS (2021)
77. Venkatesh, V., Davis, F.D.: A theoretical extension of the technology acceptance model: four longitudinal field studies. Manage. Sci. **46**(2), 186–204 (2000)
78. Kondo, Y., Takemura, K., Takamatsu, J., Ogasawara, T.: Smooth human-robot interaction by interruptible gesture planning. In: International Conference on Advanced Intelligent Mechatronics. IEEE/ASME, Montreal, QC,Canada, pp. 213–218 (2010)

79. Zahabi, M., Kaber, D.B., Swangnetr, M.: Usability and safety in electronic medical records interface design: a review of recent literature and guideline formulation. Human Factors. **57**(5), 805–834 (2015)
80. Babamiri, M., Heidarimoghadam, R., Ghasemi, F., Tapak, L., Mortezapour, A.: Insights into the relationship between usability and willingness to use a robot in the future workplaces: Studying the mediating role of trust and the moderating roles of age and STARA. PloS One **17**(6) (2022)
81. Luczak, H., Kabel, T., Licht, T.: Task design and motivation. In: Handbook of Human Factors and Ergonomics, pp. 397–440 (2012)
82. Huppert, F.A.: Psychological well-being: evidence regarding its causes and consequences. Appl. Psychol. Health Well Being **1**(2), 137–164 (2009)
83. Wright, T.A., Cropanzano, R.: Psychological well-being and job satisfaction as predictors of job performance. J. Occupat. Health Psychol. **5**(1), 84 (2000)
84. Huppert, F.A., So, T.T.: Flourishing across Europe: application of a new conceptual framework for defining well-being. Soc. Indicat. Res. **110**(3), 837–861 (2013)
85. Ayala, C.: Four well-being-performance patterns: personal and organizational antecedents and age as moderator. RODERIC [Published doctoral dissertation, Universitat de Valencia, Valencia, Spain] (2017) http://hdl.handle.net/10550/6131
86. Sonnentag, S.: Dynamics of well-being. Annu. Rev. Organ. Psych. Organ. Behav. **2**(1), 261–293 (2015). https://doi.org/10.1146/annurev-orgpsych032414-111347
87. Gutierrez, O.I., Polo, J.D., Zambrano, M.J., Molina, D.C.: Meta-analysis and scientific mapping of well-being and job performance. Spanish J. Psychol. **23** (2020)
88. Hernandez, R., Bassett, S.M., Boughton, S.W., Schuette, S.A., Shiu, E.W., Moskowitz, J.T.: Psychological well-being and physical health: Associations, mechanisms, and future directions. Emot. Rev. **10**(1), 18–29 (2018)
89. Diener, E.: New findings and future directions for subjective well-being research. Am. Psychol. **67**(8), 590 (2012)
90. Breazeal, C.: Emotion and sociable humanoid robots. Int. J. Hum Comput Stud. **59**(1), 119–155 (2003)
91. Gorny, P.M., Woodard, R.C.,. U.O.E.A. Karlsruhe Institute of Technology.: Don't Fear the Robots: Automatability and Job Satisfaction (2020)
92. Koschate, M., Potter, R., Bremner, P., Levine, M.: "Overcoming the uncanny valley: displays of emotions reduce the uncanniness of humanlike robots. In: International Conference on Human-Robot Interaction (HRI) ACM/IEEE, IEEE pp. 359–366 (2016)
93. De Witte, H., Vander Elst, T., De Cuyper, N.: Job insecurity, health and well-being. In: Sustainable Working Lives: Managing Work Transitions and Health Throughout the Life Course, pp. 109–128 (2015)
94. Ferrie, J.E.: Is job insecurity harmful to health? J. R. Soc. Med. **94**(2), 71–76 (2001)
95. Brenner, M.H., Mooney, A.: Unemployment and health in the context of economic change. Soc. Sci. Med. **17**(16): 1125–1138 (1983)
96. Dekker, F., Salomons, A., Waal, J.V.D.: Fear of robots at work: the role of economic self-interest. Socio-Economic Rev. **15**(3), 539–562 (2017)
97. Yam, K.C., Tang, P.M., Jackson, J.C., Su, R., Gray, K.: The rise of robots increases job insecurity and maladaptive workplace behaviors: multimethod evidence. J. Appl. Psychol. **108**(5) (2023)
98. Schwabe, H., Castellacci, F.: Automation, workers' skills and job satisfaction. PLOS ONE **15**(11) (2020)
99. Parasuraman, R., Rizzo, M.: Introduction to neuroergonomics. Neuroergonomics: The Brain at Work, pp. 3–12 (2007)
100. Zhu, Y., Johnson, C., Chang, S.-H. J., Mehta, R.K.: Neuroergonomics metrics to evaluate Exoskeleton based Gait Rehabilitation. In: 2020 IEEE International Conference on Systems, Man, and Cybernetics (SMC), IEEE, pp. 3447–3453 (2020)
101. Caiazzo, C., Savkovic, M., Pusica, M., Milojevic, D., Leva, M.C., Djapan, M.: Development of a neuroergonomic assessment for the evaluation of mental workload in an industrial human-robot interaction assembly task: a comparative case study. Machines **11**(11), 995 (2023)

102. Kim, S.K., Kirchner, E.A., Schloßmüller, L., Kirchner, F.: Errors in human-robot interactions and their effects on robot learning. Front. Robot. AI **7**, 558531 (2020)
103. Brandt-Rauf, P.W., Ayaz, H.: Occupational Health and Neuroergonomics: The Future of Wearable Neurotechnologies at the Workplace, vol. 66, pp. 456–460.LWW (2024)
104. Gediga, H., Hamborg, K.C., Düntsch, I.: The isometrics usability inventory: an operationalization of ISO 9241-10. Behav. Inf. Technol. **18**(3), 151–164 (1999)
105. Rosen, P.H., Sommer, S., Wischniwski, S.: Evaluation of human-robot interaction quality: a toolkit for workplace design. In: Ergonomics in Design, Design for All, Activity Theories for Work Analysis and Design, Affective Design IEA, Ed., vol. 20. Springer International Publishing, in Proceedings of the 20th Congress of the International Ergonomics Association (IEA 2018) Volume VII, pp. 1649–1662 (2019)
106. Heinold, E., Rosen, P.H., Wischniewski, S.: Usability questionnaire for robotic systems based on the ISO 9241–110. IEEE Robotics and Automation Letters **10**(3), 2231–2238 (2025). https://doi.org/10.1109/LRA.2025.3526557
107. Brooke, J.: SUS: A quick and dirty usability scale. In: Jordan P.W., Weerdmeester B.A., McClelland I.L. (ed.). Usability Evaluation in Industry. Taylor & Francis, London (1996)
108. Vlachogianni, P., Tselios, N.: Perceived usability evaluation of educational technology using the system usability scale (SUS). A systematic review. J. Res. Technol. Educ. **54**(3), 1–74 (2022)
109. d'Errico, A., Punnett, L., Gold, J.E., Gore, R.: JCQ scale reliability and responsiveness to changes in manufacturing process. Am. J. Ind. Med. **51**(2), 138–147 (2008)
110. Tran, T.A., Péntek, M., Motahari-Nezhad, H., Abonyi, J., Kovács, L., Gulácsi, L., Ruppert, T.: Heart rate variability measurement to assess acute work-content-related stress of workers in industrial manufacturing environment—a systematic scoping review. IEEE Trans. Syst. Man, Cybern. Syst. (2023)
111. Hackman, J.R., Oldham, G.R.: Job diagnostic survey. J. Appl. Psychol. (1974)
112. Sadeghian, S., Hassenzahl, M.: The" artificial" colleague: evaluation of work satisfaction in collaboration with non-human coworkers. In: 27th International Conference on Intelligent User Interfaces, pp. 27–35 (2022)
113. Hart, S.G., Staveland, L.E.: Development of NASA-TLX (Task Load Index): results of empirical and theoretical research. Adv. Psychol. **52**, 139–183 (1988)
114. Hart, S.G.: NASA-task load index (NASA-TLX); 20 years later. In: Proceedings of the Human Factors and Ergonomics Society Annual Meeting, vol. 50, pp. 904–908.:Sage publications, Los Angeles, CA (2006)
115. Schaefer, K.E.: Measuring trust in human robot interactions: development of the "trust perception scale-HRI. In: Robust Intelligence and Trust in Autonomous Systems Boston, pp. 191–218. Springer US, MA (2016)
116. Malle, B.F., Ullman, D.: A multidimensional conception and measure of human-robot trust. Trust in Human-Robot Interaction, pp. 3–25 (2021)
117. Charalambous, G., Fletcher, S., Webb, P.: The development of a scale to evaluate trust in industrial human-robot collaboration. Int. J. Soc. Robot. **8**, 193–209 (2016)
118. Yagoda, R.E., Gillan, D.J.: You want me to trust a ROBOT? The development of a human–robot interaction trust scale. Int. J. Soc. Robot. **4**, 235–248 (2012)

Organizational Ergonomics in Robotics

Vivek Kant, Francesco Rea, and Fabrizio Bracco

Abstract The chapter introduces the organizational dimension of ergonomics for robotics. Specifically, it introduces the foundations of organizational ergonomics and sociotechnical principles for effective work systems design and development. It then highlights key challenges involved in issues related to work, activities, and quality of life in robotic setups. In this realm, attitudes towards robots and automation play a major role in the adoption of new technologies and change management in organizations. These broad-scale changes are important to be recognized as they shape human–robot interaction in various work settings, ranging from smart manufacturing to defence. The chapter then addresses specific aspects of teamwork and the design of high-performance teams. These topics include team performance and well-being, human function allocation, shared situation awareness, and job roles, amongst other such topics that will be explored later in the book by subsequent chapters. This chapter supplements the previous chapters on physical and cognitive ergonomics, but it goes beyond these aspects to highlight the organizational basis required to develop robotics holistically.

Keywords Robots · Human robot interaction · Sociotechnical systems · Work systems design · Organizational human factors

All authors contributed equally.

V. Kant (✉)
Indian Institute of Technology Kanpur, Kanpur, India
e-mail: vkant@iitk.ac.in

F. Rea
Istituto Italiano di Tecnologia, Genova, Italy

F. Bracco
Department of Education Science, University of Genova, Genova, Italy

G. Barresi and S. Wischniewski (eds.), *Ergonomics in Robotics: Advances and Innovations*, Springer Tracts in Advanced Robotics 162,
https://doi.org/10.1007/978-3-032-19374-2_4

1 Introduction

The aim of this chapter is to introduce the foundations of organizational ergonomics for robotics. Ergonomics as a discipline is concerned with the well-being, safety, and productivity of workers in a variety of work situations. This requires that workers should be considered from a holistic viewpoint—in terms of "factors of the human", "factors affecting the human", along with systemic human factors that go beyond individuals [1]. The factors of the human are generally addressed in terms of physical/physiological ergonomics as well as cognitive ergonomics. The external factors affecting work performance could belong to broader team, institutional, regulatory, and governance aspects that shape everyday work performance in multiple ways.

In order to address the notion of organizational ergonomics holistically, this current chapter complements the earlier chapters on physical and cognitive ergonomics for robotics. In addition, this chapter introduces key themes that are addressed later in the book.

The rest of this chapter is divided into nine sections. Section 2 introduces why there is a need for organizational ergonomics in the area of robotics. Section 3 introduces the foundations of addressing organizational ergonomics in the form of sociotechnical systems design and insights for modern-day work systems in terms of jobs, roles, and quality of life. Sections 4, 5, and 6 address issues of collaboration in human robot setups. These include challenges related to trust, social formation, and human function allocation. Section 7 presents issues related to job autonomy, task characteristics, and work challenges that could be useful for characterizing robotics. Sections 8 and 9 present quality of life with robotics and attitudes towards mechanization, cyberization, and automation. The chapter concludes with Sect. 10, discussing the steps enabling the transitions to robotic workspaces of the future and the challenges thereof.

2 Why Do We Need an Organizational Dimension of Human Factors and Ergonomics (HFE) for Robotics?

Human Factors and Ergonomics (HFE) is the study and design of work and work systems. In the past century, major changes in manufacturing, such as the assembly line, put forth novel challenges in the manner work was conducted. Due to these widespread changes, jobs and occupations have changed in society in the last century. One major inclusion in modern-day workplaces is the growth of automation in a variety of forms [2]. These include the automation of several activities that would have been done earlier as part of manual work. In addition, the growth of digital forms of control resulting in newer control systems design and different means of engagement with technologies in the workplace, such as manufacturing setups. A form of automation is seen in the development of robotics and inclusion in highly predictable systems, such as assembly lines.

In these cases, the robots perform low-level, highly regular tasks, in turn changing the role of the human to that of a supervisory controller, who attends to the disruptions to work more than the continuous monitoring of the constant ongoing work. Thus, these changes in the nature of work have resulted in considerations of factors above and beyond the body-based and mind-based dimensions of work. In addition, with the use of robotics in uncertain environments and complex sectors, such as military, disasters, and emergency response, there is a need for addressing the concerns that these new work scenarios present for ergonomics considerations. Therefore, organizational ergonomics, which moves beyond physical and cognitive ergonomics, is important in several modern-day work settings involving robots.

The current technological shift of Industry 4.0 is characterized by the centrality of cutting-edge technology for the next industrial revolution. In this technological trend, human well-being is claimed to be a core value [3]. Robotics is one of the main drivers of this revolution, and strategic planning is needed for its effective introduction in the workplace. This strategic design should go beyond the patent advantages of robotics in terms of functionalities (production time, accuracy, process optimization) and should consider the whole socio-technical system: workers, technology, and organizational aspects in their mutual interaction [4].

If workers' well-being is really a value of the new industrial revolution, organizational conditions will contribute to the creation of meaningful jobs. In other words, we have to consider organizational structures, policies, processes and risk assessment actions in light of broader issues of regulation and governance. A few authors [e.g. 5] have highlighted that the main factors that contribute to work meaningfulness and describe how robots could be a threat or an opportunity, depending on how they are designed and introduced in the workplace. People feel that their job is rich in meaning if they feel a purpose in what they do, and robots should not impoverish the quality of their tasks. People value social relationships, and robots should not isolate workers [2, 6]. People need to feel competent, and robots should be perceived as an opportunity for professional growth. People need to protect their self-esteem and need recognition, and collaborating with robots should be seen as a positive professional skill. Finally, people need autonomy; therefore, the tasks assigned to robots should not be perceived as a reduction of competencies but rather as a transfer of duties for the sake of higher and more engaging activities.

The organizational conditions should therefore be aimed at a few goals. First, the robot should be introduced, and its activities should be designed to let workers perceive the real advantage of its implementation. The robots should be seen as an aid for the workers, not as a better-performing substitute. The task allocation should give the robot the execution and the people the high level tasks that requires human intervention; otherwise, workers could feel as observed and judged by the machine. The task design should allow workers to give priority to their preferences, and the robots should adapt to them [6].

The discussion above highlights that various issues are involved in organizational ergonomics. It has grown out of multiple roots in its current inclusion of the discipline of ergonomics. It is found under the labels of "organizational design and management", sociotechnical systems, along with risk and safety management in complex systems. Despite these categorical labels, we find a few prominent approaches which will be helpful for understanding basic concepts in organizational ergonomics which researchers and practitioners can use to develop applications, services and products that support human robot interaction.

3 Background and Principles of Organizational HFE for Robotics

HFE researchers have continued to regularly address the roles and activities of humans in various technological setups in the past century, beginning with the assembly lines. As the century progressed, new modes of interaction between people and technologies were developed. These included increasing the productivity of factory workers, wage administration, and social benefits programs for workers. The role of an industrial-organizational psychologist (and sociologists, in general) sprang up to address the newer jobs and roles created for workers and managers. Thus, the nature of work was being transformed, and so were new roles for the HFE practitioner coming into being [7, 8].

During the first half of the last century, many themes started to develop, which have now become commonplace in organizational ergonomics as a discipline. These include advanced mechanization and automation of several tasks which were earlier conducted by humans. By the mid-1900s, there was a concomitant growth of electrical systems and control systems engineering, along with systems and information sciences. The development of industrial robots is one outgrowth of the convergence of these various technological trends of the mid-twentieth century. Not only were these industrial robots bringing about changes in various tasks, but they could also be programmed effectively to handle a variety of tasks. The result is a trend of an increase in automation and robotization on the shop floor.

This increase also resulted in a significant shift from when robots were first seen as tools to now being seen as social actors in the workplace. The radical increase in the case of robotization in the second half of the past century resulted in a significant increase in the public perception of threats to jobs and livelihood, which continues until now. This enhanced automation brings about a certain "irony"—the role of the human is extended rather than terminated [9–11]. Thus, there is a greater need for the organizational dimension of the human and social organizations of groups and teams under the label of organizational ergonomics to be developed for future human–robot work systems.

3.1 The Socio-Technical Direction of Organizational Ergonomics and Its Implications

While robots were being developed mid-century, a concomitant shift was occurring in the realm of HFE in Britain. One notable emphasis was on the improvement of the nature of work and working life in the industry by researchers from the Tavistock Institute [12, 13]. The Tavistock Institute was involved in a number of studies on working life. A few prominent studies in the coal mining sector provided a major direction to the subsequent work in the area of sociotechnical systems. In these studies, the focus was on the "longwall" coal mining method, in which the task of coal mining was reduced to a single person. Further, the control and coordination of the coal mining process were externalized and supervisory in nature, resulting in affecting the overall working morale. The researchers emphasized the social and organizational dimension of the work resulting in the growth of a new concept of "socio-technical". The usage of this new term emphasized that the nature of work and activity in technological sectors needs to be comprehended widely in terms of its social dimension. The social-technical approach from the Tavistock institute grew beyond its focus area of coal mining and was used in a variety of sectors and geographic locations, including textile mill operations in Ahmedabad, India [14, 15].

Researchers in the area of socio-technical systems also drew a number of insights from social and organizational sciences. These include concepts related to the formation of groups, working in teams, trust between humans and machines, job roles, morale, and locus of control, among many other aspects of work design that appear in mainstream models and methods of socio-technical researchers. A prominent theory of work, jobs, and tasks was put forward in 1975 by Richard Hackman and Greg Oldham [16] in the form of Job Characteristics Theory (JCT). JCT has continued to provide a sustainable viewpoint among a number of variables that provide a consolidated approach for addressing issues related to job autonomy, task characteristics, nature of work in groups and teams, along with attitudes.

The JCT theory consists of three key categories pertaining to (1) core job dimensions, (2) critical psychological states, and (3) personal and work outcomes. The core job dimensions include,

(a) skill variety involving a range of skills and activities related to the job.
(b) task variety that reflects the degree to which workers can complete them to the full.
(c) task significance that addresses the impact that workers feel their jobs brings to the organization.
(d) autonomy that involves the individual choice and freedom in decision-making they have while executing the job.

While these above elements together create the core job dimensions, they have an effect on critical psychological states and in turn to the organizational processes and functioning. These critical psychological states comprise experienced meaningfulness in work; experienced responsibility, which enables people to feel responsible for their work; and the extent to which people have knowledge of the actual results of the work activity. Since the core job dimensions affect critical psychological states, these, in turn, affect work outcomes. These outcomes are measured in terms of aspects including regular turnover at work, low levels of absenteeism on the job, work satisfaction and enhanced work performance along with internal motivation. These variables directly affect the functioning of the organization and impact the organization in terms of revenue. Thus, these everyday activities of workers have a long-term effect on an organization's longevity and productivity. Along with these outcomes, several variables, such as individual growth and context satisfaction, are also involved in the overall functioning of the work system performance.

In terms of our current organizational ergonomics approaches in human–robot shared workspaces, the key issues JCT will help us in deciding how jobs will change with the inclusion of robots. Further, how jobs can be designed and redesigned in order to consider all the variables of skills, task, autonomy, while including the critical psychological states, along with personal and work outcomes [17–19]. In addition, for successful human robot workspaces, the challenge is to use these theories to develop organizational structures, policies, and strategies to enable change management.

The JCT theory is one of the most prominent approaches that capture a number of important variables that together create work outcomes in relation to the job as well as the critical psychological states of the workers. These elements are of vital importance when considering co-workers in job scenarios that involve robots as tools or as partners. JCT and other related theories are one of the many that can be used to address organizational ergonomics. The field of organizational ergonomics needs to be systematically developed as more robots and automation get introduced in the workplace and become part of the fabric of next-generation infrastructure. Many concepts in organizational science are of vital importance in the field of organizational ergonomics for robotics. These include concepts related to trust, group dynamics, organizational structure and processes, subjective indicators of work with robots from the viewpoint of productivity and well-being, amongst many others. These variables will be revisited in the subsequent sections.

3.2 The Macroergonomics Direction of Organizational Ergonomics and Its Implications

A second direction of emphasis on organizations was developed in the 1980s by the researchers studying workers in manufacturing systems as well as considering the future of work more broadly [20–23]. In addition, during this time, Japanese manufacturing challenged Americans, especially in the automotive sector. A number of human factors researchers recognized that while traditional body-based ergonomics was important, extra-personal macro variables such as cultural aspects of workers and their ways of comprehending and acting in work systems were also necessary, resulting in an emphasis on *macro- dimensions* as opposed to the earlier micro-focus of physical ergonomics. A significant aspect of macroergonomics approaches complement the broader organizational ergonomics approaches in terms of basic principles and insights and thus should be considered as complementary. While macroergonomics also derives from the Tavistock-based sociotechnical systems approach, it has also been developed significantly in the past three decades.

Specifically, macroergonomics aims towards a participatory approach, where designers, workers, and end-users are co-participants in the creation and outcome of the work systems. While the participative work/design philosophy has several roots, in participative ergonomics, the key emphasis is on the development of understanding of the work system from the bottom up while ensuring the understanding of the work systems from the top down. Based on this mutually inclusive approach towards work systems, macroergonomics researchers have evolved a 10-step Macroergonomic Analysis and Design (MEAD) methodology [23–27]. This methodology listed below involves the work system in terms of unit operations, designing jobs and responsibilities, as well as accounting for variances. This 10-step methodology [24, 25] is helpful in terms of providing a structured approach towards designing work systems involving human–robot collaborations in its totality by identifying basic challenges in existing work spaces.

1. Scanning the environmental and organizational design sub-system
2. Defining production system type and setting performance expectations
3. Defining unit operations and work process
4. Identifying variances
5. Creating the variance matrix
6. Creating the key variance control table and role network
7. Performing function allocation and joint design
8. Understanding roles and responsibilities perceptions
9. Designing/redesigning support sub-systems and interfaces
10. Implementing, iterating, and improving

The research by macroergonomists has been promulgated further by setting up special interest work groups such as organizational design and management (ODAM) in the Human Factors and Ergonomics Society. In addition, there have been studies in macroergonomics and work systems design led by researchers such as Imada, Hendrick, and many others in a variety of application sectors [e.g. 21, 23, 25, 27]. The key idea from macroergonomics that is helpful for organizational robotics is that in order to setup a work system that involves human and non-human agents, it is essential to take into account the various aspects of the organization in terms of the performance expectations, unit operations, and work processes, as well as understanding the roles and responsibilities to account for changes at the systemic level.

Since the work of Cherns [28, 29] on the sociotechnical dimension of work, there has been a renewed emphasis on the *design dimension* of sociotechnical work. In its most comprehensive form, Clegg [30, also 31] has presented a structured set of principles that address sociotechnical systems design. Clegg's principles involve three categories: meta-principles, content principles, and process principles. The meta-principles address the worldview of design, whereas the content principles and the process principles describe the operational and subject matter-based dimensions of design. By incorporating the design dimension, Clegg [30] has connected the sociotechnical aspect to coherent design aspects and will be central to our discussion of robotics in the subsequent sections. Here we briefly recount Clegg's design principles. [30]

Meta-principles
1. Design is systemic.
2. Values and mindsets are central to design.
3. Design involves making choices.
4. Design should reflect the needs of the business, its users and their managers.
5. Design is an extended social process.
6. Design is socially shaped.
7. Design is contingent.

Content principles
8. Core processes should be integrated.
9. Design entails multiple task allocations between and amongst humans and machines.
10. System components should be congruent.
11. Systems should be simple in design and make problems visible.
12. Problems should be controlled at source.
13. The means of undertaking tasks should be flexibly specified.

Process principles
14. Design practice is itself a sociotechnical system.
15. Systems and their design should be owned by their managers and users.
16. Evaluation is an essential aspect of design.
17. Design involves multidisciplinary education.
18. Resources and support are required for design.
19. System design involves political processes.

These three categories of principles provide the human–robot interaction (HRI) designers and engineers human-centered approaches from an organizational viewpoint to guide them through their design and development activities. Therefore, if these sociotechnical principles are considered, they will lead to more acceptable and productive work systems for human–robot joint operations.

3.3 The HRI Direction of Organizational Ergonomics and Its Implications

Along with Tavistock approaches and macroergonomics, social psychology and social dynamics of human–human interaction have an impact on the artificial agent design. This trend is quite prominent in the field of HRI and is of vital importance to the study of group formation and team functioning [32–34]. Researchers from HRI rely on a number of social psychological insights that support the interpretation of how human partners react to the robot's social and physical presence. Studies on how robots and humans co-exist are fundamental to a better understanding of the level of acceptability of an HRI session and are detailed in the later chapters of this book. Within the heterogeneous and homogeneous group, it is important to understand intentionality and team dynamics and assess how all the parts intend to work towards shared or distinct objectives.

Along with the social dimension above, the goal of the technical HRI research community is to develop efficient algorithms capable of improving the interaction between humans and artificial agents. However, once the interaction mechanism is designed (both in relation to the algorithms and the system configuration), it is important for the research community to assess how human factors play relevant roles in the interaction through the observational process. There are three kinds of studies (or a combination of these) present in the research literature [35] that are typically referenced in various chapters in the rest of the book:

(1) System Design: Direct applications of HRI principles to create a system; e.g., [36],
(2) Observational: Studies which focus on different aspects of HRI from the viewpoint of analysis; e.g., [37],
(3) Algorithmic Design; The focus is on algorithms, computational models, and frameworks e.g., [38].

3.4 *The Safety and Large-Scale Systemic Direction of Organizational Ergonomics and Its Implications*

Along with the above three directions, a fourth direction that can be found in HFE was the shift towards the cognitive and organizational basis of human technology interaction. This change was brought about by the growth of large-scale complex systems post-1950s, ranging from nuclear power, defence, and civil aviation. As these systems grew, a crucial aspect came forward in the form of how to design for humans who were a part of these systems[39]. This problem was not simply of designing technology but also recognizing that humans are an integral part of the system under consideration. As a result, these large-scale systems, with people and technologies co-acting together, are best considered as sociotechnical systems[40, 41]. Here, organizational aspects of structures, policies, processes, and risk assessment actions are adopted and considered, bearing broader issues of regulation and governance in mind. This also includes risk assessment and the introduction of robotics for next-generation workspaces.

During the late 1970s, there was a prominent emphasis on designing for the human and human–machine interaction after the Three Mile Island disaster. Along with this, the Tenerife airport disaster occurring in the same era resulted in a shift towards understanding the cognitive and organizational dimensions of complex sociotechnical systems[42–46]. The new trends in ergonomics in these complex systems is to understand how to make them resilient. A related direction of thought includes how designers can support the situation awareness of the workers, as well as other cognitive capabilities to enable successful systemic outcomes.

In addition, a crucial insight that this direction of complex systems brings for the robotics sector is that these systems function at several levels of abstraction at various timescales[47, 48]. Therefore, at the level of operations, workers need to be addressed. Reciprocally, there is also the need to cater to the functioning of individuals and teams. At the management level, there is a focus on addressing the role of managers. Similarly, at higher levels, different strategic stakeholders, such as the government and other political regulatory agencies, play a role in the overall functioning of the system. Therefore, a multi-stakeholder approach with equal weight on rules, procedures, and strategies should be provided for comprehending the design, functioning, and operations of sociotechnical systems.

All four directions (3.1–3.4) of sociotechnical systems provide a broad basis that can be incorporated into the existing robotics research. For example, the direction from Tavistock studies and macroergonomics point towards a social dimension that involves the functioning of teams of humans and robots as well as differentiated job roles and job characteristics that enable a successful design of human–robot work situations. In turn, the view of sociotechnical systems from complex systems emphasizes human capabilities and limitations and their design thereof, which includes supporting the human-in-the-loop, and situation awareness of operators and teams, among others. In the next section, the steps towards collaboration and the social formation of groups are addressed in human–robot work contexts.

4 Collaboration in HRI: Groups, Teams, and Social Formations

An analysis of the types of interactions in organizational ergonomics is paramount. In order to classify the typologies of interaction, the concept of interaction is defined as the mutual relationship where information is exchanged between one or more individuals and one or more artificial agents in the environment [35, 49–51]. In this specific analysis, there are three entities involved in the interaction that are congruent with the classification in terms of an object (robot), a subject (worker), and a context (environment and organizational stance). In the analysis of different typologies of interaction ranging from dyads to swarms, we should also clarify how the number of interactants changes the modality of the interaction.

In dyadic systems, we find a pairing between a robot and a human. In contrast, a triadic interaction in HRI refers to a situation where three agents interact. These triad could consist of various permutations and combinations of robots and humans—2 robots, 1 human, or 1 robot, 2 humans [35, 49]. Furthermore, when more than three entities are involved the notion of teams or groups are used frequently to refer to these multi-agent HRI systems [50, 51]. Examples of these include human operators controlling multiple robots in factories, or even robots assisting humans in tasks in industry 4.0 contexts. Several studies have reviewed the state of the art in the field of HRI and collaboration e.g. [52–54].

Groups and teams play an important role in organizational ergonomics [55]. An aspect that requires attention in interaction for organizational ergonomics is the role of artificial agents in team performance. In this context, we are not simply considering the typologies of interactions where the artificial agent is exclusively interacting with another human individual (dyadic interaction), or the triadic interaction in all its forms is left for other and further discussion. Here, we would like to focus on the interaction of the robot with a group of individuals and the dynamics that spark the definition of a team [50, 51, 56].

In the last three decades, research in the field of HRI has mainly focused on dyadic scenarios (i.e., one human interacting with one robot). Although, this framework may not be completely suitable for everyday life since humans run most of their daily activities in groups composed of more than two people, such as teams, families, and classrooms. People normally consider groups as composed of two or more people; note that, in this work, with the word "group", we refer to a gathering of three or more human or robotic individuals. In order to integrate robots into humans' everyday life, HRI studies concerning one-robot-to-many-humans and many-to-many interactions need to be conducted.

Group dynamics cannot be fully understood by studying only individuals. A study [57] found that members' individual cognitive abilities are not associated with group intelligence; instead, it is correlated with a number of social factors. These include the social sensitivity of the group members, the presence of females in proportion to the males in their group, in addition to the turn-taking in conversations. This suggests that to understand group dynamics is necessary to observe groups as a whole [50, 51, 56]. In our case, the focus is to understand the relationship between robots and humans as part of the larger groups.

The ICE framework was recently proposed by Abrams et al. [58] to evaluate group dynamics in HRI. ICE is an acronym for In-group identification, Cohesion, and Entitativity. Each of these terms refers to a specific subjective indicator of the group status from a social and task point of view.

(1) Entitativity: This concept is defined as "the degree of having the nature of an entity" [59]—denotes to which extent a social unity is perceived as a group from an external point of view. In order to quantify entitativity, outsiders rate the unity of a group. It is an assessment on a group level: it does not take into account how individuals inside the group are perceived, but just the assembly as a whole. These studies [60] suggest that entitativity depends on a combination of variables. The most entitativity-correlated factors resulted in being perceived interaction between group members, the presence of common goals, in addition to the common outcomes, the similarity in appearance of the group members, as well as, speaking language or behavior, and the importance of the group. Further,

the researchers identified four categories of entitativity: intimacy groups (e.g., families), task groups (e.g., a relay race team), loose associations (e.g., people traveling on the same railway carriage), and social categories (e.g., "women").

(2) Cohesion: Intimacy groups, such as task groups with strong common goals, were linked to a high sense of "groupness", while loose associations were associated with a lower level. Social categories and other task groups were attributed with medium-scale ratings. Cohesion measures how a group is unified and coherent from an insider's point of view. Findings [61] suggest that cohesion is correlated with performance, especially in small groups. In particular, the relation seems to be bidirectional: cohesion causes a group to have better performance, and, at the same time, good performance (for example, reaching a goal successfully) makes group members feel more cohesive [61]. Cohesion is a feeling that evolves with time, along with group development, which comprises five different stages [62]:

   (i) forming (an initial orientation phase);
   (ii) storming (characterized by conflicts within the group);
   (iii) norming (definition of rules and roles, exchange of information, especially in this phase, group cohesion is developed);
   (iv) performing (emergence of solutions to the task);
   (v) adjourning (parting of the group). According to the previous observations, in addition to the norming phase, cohesion might also increase after the performing one.

3. In-group identification: This concept expresses the sense of belongingness of an individual to a group. A study identified five items measuring in-group identification: solidarity (i.e., the grade of commitment to the group), satisfaction (i.e., to what extent being part of the group evokes positive emotions in a member), centrality (i.e., how important is the group in individuals' lives), individual self-stereotyping (i.e., how much a member feels to have in common with the other individuals in the group), in-group homogeneity (i.e., to what extent a member feels that the other individuals in the group are similar).

ICE theory can play a vital role in the design of HRI, both for single-agent HRI as well as multi-agent HRI [57]. Specifically, the design of the interaction in multi-agent HRI requires particular attention since the number of mutual exchanges of information between the parts increases with the number of participants [63–65]. In several cases, multi-agent systems such as swarms require an understanding of the dynamics of the swarm itself as well as the humans associated with them. These include people in teams with robots of varying capabilities, resulting in multi-agent HRI with challenges of functional allocation, team cohesion, and parameters related to group identification, and balance between treating robots as tools or as agents.

Further, the very nature of the human and the robot (degree of automation capabilities) makes the interaction challenging. For example, based on the current technology, conformity and in-group identification cannot really be attributed to the robot partners but only to the human partners. The robot cannot conform to the group (lack of sentience), but only work according to the rules set up by the designers and implementers, who in turn may or may not be members of that particular group or understand its dynamics.

The dimension of the number of actors in the interaction can be extended to specific situations where the number of artificial agents is relevant: swarms of agents. Human–Swarm Interaction (HSI) is, in fact, a particular case of multi-agent systems where few humans (frequently limited to one) interact with a group of robots. These robots coordinate or even self-organize to present an outwards display of singular entity. Kolling et al. [66] present a survey of Human–Swarm interaction systems and discuss how the design of the interaction changes when swarms of robots interact with the human partner.

In the multi-agent HRI literature, we should also consider the opposite situation where a limited number of artificial agents (sometimes one) interact with a greater number of human partners. For example, Sebo et al. [51] present a review that focuses on robots' interactions with a small group or team of people and explores the attributes of such systems which includes the overall behavior of the group (sometimes also related to complex social dynamics; also [67]). A further excursion is needed into the dimensions of social formation in HRI.

5 Trust and Conformity in Robotic Work Setups

Social formation in HRI can be comprehended in several dimensions, such as trust, compliance, and cohesion, amongst others. In the context of the interaction between the workers and the artificial agents, it is fundamental to recognize and highlight the potential issues, such as trust towards the agents, social expectations and job dimensions of robotics-oriented workspaces, and other related challenges in the next-generation workforce.

A key direction also involves how robots are accepted in the workforce. As stated by Meissner et al. [68], workers' acceptance of robots depends on three main factors: the workers' attitudes, the robots' characteristics, and the organizational conditions that facilitate the effective and satisfactory interaction of humans and robots. These conditions could be observed in aspects such as organizational culture, quality of relationships, implementation process, and job design.

Organizational culture refers to explicit and implicit assumptions that lead the organizational life, which should be oriented towards fostering workers' empowerment and well-being, openness to change and innovation, willingness to learn, trust, and free communication. Quality of relationships refers to the open and fair relationship dynamics at every level, enabling workers to express suggestions, needs, and concerns to the executives; the vertical and horizontal relationships contribute to the feeling of a community, a network that connects and protects the people in the organization. When workers do not trust executives, they expect their needs and points of view will not be taken into account, and they are afraid their job is threatened by robots.

In addition, if they doubt the executives' skills in the implementation of robots at the organizational level, their resistance towards robots will be higher. concerning the quality of relationships with colleagues, supporting teams could be beneficial in uncertain times, especially when the colleagues have more experience in the cooperation with robots and can help those who are less skilled or have less self-confidence. Another social effect in this respect is the peer pressure concerning the attitude towards automation, highly respected workers can actually drive the team's attitude towards robots.

The implementation process refers to the strategies for the effective introduction of robots in the workplace, how the technological changes are communicated, to whom, how the training is aimed at fostering workers' growth and self-efficacy, reducing the fear of job insecurity. Concerning information and communication, it is very important for workers to be informed about the process of introduction and implementation of robotic technology, taking into account their perceptions, managing misconceptions, treating them as peers. This should be taken care of in particular in case of workers that are stereotypically considered less skilled with technology, like women. Linked to this aspect is participation: workers want to be asked for their opinions, since they are experts of the assembly line and may not trust managers in the implementation of robots in the productive process. Finally, the implementation process should also provide workers with training and support, since they may need help in case of malfunctions or operation problems, especially when the implementation is not well planned.

Job design refers to what kind of tasks are assigned to robots, and which are left to humans, the level of autonomy of humans, the redesign of tasks to protect workers' self-efficacy and well-being, the extent to which robots and automation are organizational resources for people, and not just a demand, the level of job crafting allowed for workers, giving them a feeling of agency and organizational commitment [69, 70].

A key factor that influences the social formation between artificial agents and human individuals is transparency; e.g. robot's transparency has a direct influence on the trust in the relationship [73–75]. Individuals trust a robot more that is able to explain its actions as compared to the ones that do not [76]. Therefore, comprehending the notion of trust is a key aspect of supporting meaningful HRI. It is also necessary to understand which other factors are linked to trust in the relationship as well as how these factors change longitudinally during the course of the development of the relationship. HRI researchers, by studying these aspects will be able to assess and predict risks such as overtrust between humans and robots, exploitation of trust and its perceived uses and misuse as an important variable characterizing HRI in a number of domains such as healthcare, home care [77], or education [78].

Apart from the different connotations given to the term "trust", trust is a fundamental component of human–human interaction and is naturally transferred in HRI contexts. Robots that are already integrated into our society and into the organizational process are often perceived as tools or as part of the machinery with limited agencies. However, in novel operative contexts, the artificial agent and robotic platform gain agency and intentionality and consequently become partners in activities of daily living [73]. In the context of new emerging social formation, there is a need to assess which factors influence trust. The individual‘s attitudes (subject), the environmental factors (context), and robot characteristics (object), such as performance, equally affect the development of trust [68].

The artificial agent‘s efficiency and how, in general, the system improves reliability [79] as a consequence of artificial agent adoption is important in establishing their partners’ trust [80]. In the context of organization HFE, the risk that trust can become overtrust must be considered, as demonstrated by the following insights. In recent research, authors have demonstrated that participants might trust and comply with the robot’s requests independently of their level of correctness or the level of transparency of the robot [81]. Evidence from the literature has demonstrated how the trust towards artificial agents is not altered even in case of the robot malfunctions [82, 83]. Obviously, this aspect is more than relevant, especially in contexts where technology is far from perfect. The sign of overtrust can be defined, as stated in [71], as a poor calibration between the person’s trust and the system’s capabilities.

It has been demonstrated that robots can act as effective persuasive agents and enable humans to change their ideas and conform to the robot’s suggestions [84]. The potential risk for the organization adopting this technology is not mainly in the overtrust that it can be exploited, but it is in overtrust that disables the power of individuals to make autonomous decisions about the ergonomics and safety in their activities.

It has been found that humans tend to conform easily to the robot‘s suggestions. Conformity is defined by Cialdini et al. [85] as “the act of changing one’s behavior to match the responses of others”. Thus, conformity is an attitude associated with the dimension of social formation. In fact, Asch [86] demonstrated the existence of a specific phenomenon of conformity elicited in the presence of groups of people. In a perceptual task in which participants had to match the length of a line with another one, choosing between three options, all the participants chose the wrong answer biased by participant actors. Specifically, the subject conformed to the wrong decision of the others, even if they disagreed with them. Asch’s study was replicated with robots to verify if mixed groups of humans and robots have a comparable influence on group decision-making as only-human groups since the dimension of the typology of group members impacts the in-group out-group paradigm. Different results were found depending on the age of the participants (children or adults) and the composition of the mixed group (number of robots and humans). More recently, Vollmer et al. [87] replicated Asch’s experiment with a human subject and three robot peers. Their results show that children between 7 to 9 years old conformed to the robots, while adults did not. Furthermore, Qin et al. [88], in a similar experiment with adults, using a different group composition: a human subject, two human confederates, and a robot confederate, found evidence that adults are subject to conformity when the social robot represents a minority in the group.

The hypothesis behind such a result is that this effect reflects the fact that humans perceive the group as a more familiar social environment, and robots’ presence is more acceptable when the robotic agents are outnumbered by humans (belonging to another societal group, most likely). Hertz et al. [89] further investigated how the respective number of artificial agents and humans in a mixed group influenced social conformity during a social task (i.e., recognizing the emotions of human faces) and an analytical one (i.e., arithmetic operations). The experiment result shows how higher rates of conformity during the analytical task: only in the social task was there found higher conformity with the group opinion as the group size increased. The rationale is that the same effect in the analytical task could have been balanced by the fact that humans, in general, see artificial devices as more proficient than human subjects in computational tasks. Similar assumptions are frequently found in similar situations where the artificial agent is required to perform computational tasks or tasks that are more accessible for computers. This is, in fact, a condition also found in different industrial scenarios.

Similar to previous studies, Salomon et al. [90] performed conformity experiments with a human subject and three robots, like Vollmer. The results of their study showed that participants conformed with the robots for 1/3 of the critical trials. In addition, they investigated trust and discovered that, at the beginning of the experiment, participants conformed to robots at a similar rate to Asch's subjects. Many participants, however, stopped following the robot's choices after a wrong answer was given. This suggests there is a decrease in robots' trust they make a mistake which contrasts with some other shreds of evidence unless we start to consider the interactions prolonged in time. The same effect was not detected with human actors in other studies.

In previous studies, where individuals were involved in demanding tasks with the support of robotic partners suggesting when to act and how, the authors noticed that after a brief period of assessment lasting a few minutes, the participants started to trust the robot‘s suggestion without double checking the suggestions [91]. Since this overtrust process occurs in a few minutes, what would happen for a longer period of interaction for events that are critical for the safety and ergonomics for the individual is a major concern.

In their meta-analysis of factors affecting trust in HRI, Hancock et al. stated that team collaboration aspects influence humans' perceived trust in robots [72]. In particular, in-group membership (i.e., the extent to which a robot is perceived as part of the group), sharing the same culture, and the capacity to easily communicate seem to play a significant role that can eventually be introduced in the new generation of artificial agents that consolidate trust on built rapport. This aspect is of key importance as there is the need to understand the task characteristics and, more broadly, job characteristics which are of crucial concern to support shared situation awareness in various HRI contexts.

6 Shared Situation Awareness and Job Roles

A primary challenge in work systems design is the shared situation awareness between robots and humans working together. Situation awareness (SA) is a determining factor for the effective and safe management of production processes [92].

In cases of poor SA, the operators do not have a clear representation of the state and dynamics of the operational context and may make mistakes.

In more recent years, so-called "situation awareness demons" have been identified [93], i.e., mechanisms and conditions that can intervene in compromising SA at one or more levels [94].

From an analysis of accidents that occurred in an industrial robotics context, Camblor et al. [95] found that the main demons involved were relative to errant mental models and out-of-the-loop conditions. Therefore, it seems necessary to improve the conditions under which robots help operators to understand the situation and its evolution. These aspects are relevant for organizational ergonomics, since job design, task allocation, proper training and communication could help workers in developing proper mental models and monitor their SA. In order to support these results, we mention the study by Gombolay et al. [96], in which they observed that the SA of operators decreased as the autonomy of the robot increased. The authors suggest that if the type of work requires task regulation and operators must intervene to regain control, and if the costs of a poor SA are too high, it would be advisable to organize the activity to leave operators priority in decision-making. Conversely, if the type of activity does not require human intervention and if the risks due to a poor SA are not high, then it is preferable to leave the decision-making authority to robots for the sake of production efficiency.

Similarly, Endsley [97] provides a series of recommendations to manage the automation conundrum: give space to automation only if necessary and mainly to carry out routine tasks, leaving operators tasks that require high-level cognitive functions; design person-centered tasks, giving operators degrees of control and helping them develop coherent and comprehensive representations of the situation, and providing support for projecting contextual data into the future.

One of the criticisms of Endsley's model is that it concerns only the cognitive processes of the single operator, but the typical operating context of industry 4.0 sees the interaction between multiple people and robots in environments regulated by procedures and organizational dynamics. Therefore, different versions of the original concept of SA have been proposed, broadening the perspective more and more. For example, Berx et al. [98] propose an interesting reflection applied to collaborative robots. In addition to "individual SA", the "team SA" concept has been proposed, which concerns the level at which each one of the members of the group possesses the necessary SA to carry out their task.

A further step is towards the "shared SA", which refers to the level of overlap of the SAs of the various team members. However, to design operational contexts of collaboration between operators and robots that are effective and safe, the authors speak of "compatible SA", which is based on the consideration that the SA of the various actors (operators and robots), must be neither similar nor have elements in common, as long as it is compatible. The roles, tasks, and ways of working of the various actors are different, and the representations of the situation are different, but it is necessary to design the interaction so that the global SA is a distributed phenomenon, an emergent property. Towards this end, the authors suggest guiding system design according to the principles of observability, predictability, and directability [99]. In the case of robotics, a prominent challenge is to ensure that proper situation awareness is maintained in human–robot work settings. This can be in cases of teleoperation and other related cases, where along with shared SA, compatible SA will also become necessary. In addition, the SA demons, such as attentional tunnelling, memory traps, and displaced salience, amongst others, will play a prominent role. In addition, macroergonomics principles and MEAD methodology (Sect. 3, this chapter) will ensure that SA activities are supported throughout the work system. Therefore, broader issues beyond immediate tasks must be explored in relation to human robot work setups.

7 Job Autonomy, Task Characteristics, and Work Challenges

As workspaces in many countries are moving towards human–robot interaction based workspaces in Industry 4.0, there is a need to characterize how human–robot interactions will be shaped in automated and semi-automated workspaces. Even in Industry 4.0, where many processes may take place completely autonomously, people are still needed to perform certain tasks. These people must be kept within the loop to effectively solve problems, react flexibly to workspace challenges and contextually understand processes.

Effective distribution of tasks in the relationship between people and robots allows for a combination of the advantages of humans and robots. At the human's end we achieve creativity, flexibility, and problem-solving ability. In contrast, at the robot end, we receive efficiency, accuracy, and speed of execution. The autonomy of the operators is complementary to that of the robot, and for the autonomy of the robot, we can refer to the definition provided by Beer, Fisk and Rogers [100]: "The extent to which a robot can sense the environment, plan based on that environment, and act upon that environment, with the intent of reaching some goal (either given to or created by the robot) without external control."

In HFE, researchers attempted to find the formula for the optimal balance between automation and human contribution. From the first studies of Fitts [102] up to the most recent contributions of Michalos et al. [103], we note the attempt to divide the activity into subtasks, and according to their nature, they can be assigned to the person or machine, to ensure precision, efficiency, and safety. However, it should be remembered that people can feel more motivated and have greater job satisfaction if they feel they are doing rich, stimulating, meaningful work (see JCT Sect. 3). For example, Gombolay et al. [104] found that operators prefer tasks that allow them to preserve their autonomy in decision-making and management processes. Similarly, Tausch and Kluge [105] report that when workers have the autonomy to decide the distribution of tasks between themselves and robots, they report greater satisfaction with both the task and the results of the process.

Cardoso et al. [106] report how workers can show positive attitudes towards collaborative robots since they allow a reduction in workload, but the adequate distribution of tasks and the distribution of autonomy between person and machine is strongly conditioned by the trust that the operator places in the robot. If trust is excessive or poor, it is possible that the distribution of autonomy is unbalanced between the parties [107]. The impact of task automation on workers' well-being was assessed by Peters and Plomb [108], who found that the use of robotic automation was negatively correlated with the perception of autonomy and variety of tasks, as well as work engagement.

An interesting contribution that collects and systematizes the most relevant factors for human–robot collaboration was conducted by Baltrusch et al. [6], who identify four characteristics of work that are decisive for satisfaction, worker well-being, and effective integration with robots: workload, fluency in collaboration, trust and acceptance, and satisfaction. The workload is reduced if robots have predictable behaviors and provide information that helps to understand and anticipate events. The workload must be well distributed between person and robot because both overload and underload can lead to errors.

Workload increases when the operating speed of the robot is perceived as too high, when it is not known in advance, when the operator is too close to the machine, and when its movements are performed in straight trajectories. These factors can create discomfort, surprise, and fear, together with cognitive workload. In order to address these issues, robots should provide visual input for the operators to inform them about the robot's parameters, and repetitive and predictable movements could help reducing workload. In turn, functional workload levels can enable the operators to develop appropriate situational awareness.

The fluidity of collaboration generates satisfaction in workers, so it would be useful to develop forms of dynamic adaptation of the robot to the behavior of the operator, as well as systems to keep the operator informed about the objectives and status of the robot. One possible way to adapt the robot to the operator's preferred collaboration strategy is to use the same strategies that are commonly used to effectively train human teams on human–robot teams. Iteratively switching roles with one another, team members discover a shared strategy for completing a cooperative task. Nikolaidis et al. [101] have demonstrated that when working with a real robot, this kind of virtual robot training results in a statistically significant improvement in team fluency. This suggests that methods that are effective in enhancing human–human interaction can also enhance human–robot interaction and should be taken into account when attempting to achieve a synergistic partnership. This tactic may also help determine the unique preferences of the operators and enable the robot to be modified to suit their requirements.

Trust in the robot is associated with the feeling of control over the situation, safety in carrying out activities, and reliability of automation. It also depends on robot design features, such as movement strategies, physical appearance of the robot, and safety characteristics (e.g., sensors to detect potential collisions). Finally, aspects such as the acceptance of robots and job satisfaction must be considered because workers may show resistance due to the perception of insecurity, unreliability of the system, rigidity of the robot, as well as the belief of being replaced by robots in their work.

Robots should be implemented as real cooperators with humans, as a tool that eases their work, and not as a constraint or something they must adapt their behavior to. Flexibility is an important feature to foster the feeling of cooperation centered on workers' needs and working styles. Again, providing inputs for operators to understand the system's state and predicting and controlling its behavior are important for situation awareness and, eventually, for well-being and satisfaction.

Robot acceptance can also depend on its physical and social appearance. Humanoid features could help workers to feel safe when operating with robots, but robots' social behavior is also important. If a robot is perceived as having a socially threatening role (e.g., a quality inspector), workers will probably have more concerns operating with them. Since previous experiences with robots are relevant for their acceptance, organizations should plan processes of acquaintanceship between robots and workers, for example starting with playful interactions, videos and immersive training. All these aspects of immediate work performance should be considered from a broader vantage point of how robots and humans can co-exist in workspaces and how we can transition towards such workspaces in the future.

8 Organizational Ergonomics and Quality of Life with Robots

The introduction of technology in the workplace aims to optimize processes, increase productivity, and improve results. The well-being of workers is a goal that has emerged only in more recent times since technology is not only a resource but can also become a threat if poorly integrated with humans [70]. There is evidence in the literature that the acceleration of work rhythms favored by technology can have effects on worker stress, workload, fatigue, and burnout [109, 110]. This is specifically important in the case of robotics, where robots can work around the clock in the same manner, imposing a steady rhythm on the human workers.

In Industry 4.0, despite the increasing autonomy of many processes, human involvement remains necessary for specific tasks. It is crucial to keep humans engaged to comprehend processes, adapt to changes, and handle problems effectively. By appropriately assigning tasks between humans and robots, we can harness the strengths of human creativity, flexibility, and problem-solving alongside the efficiency, accuracy, and speed of execution offered by robots.

The design of tools, procedures, tasks, policies, and organisational developments is the core challenge to ensure a properly human-centred digital transition in the workplace. Throughout the life cycle of a manufacturing system, the technological, organisational, and human aspects will undergo continuous evolution. Therefore, it is essential to establish principles and guidelines to effectively manage the co-evolution of robotic technology and human factors. Robots can become a job resource or a demand, depending on how they are designed and introduced in the workplace.

The well-being and mental health of workers can be compromised by industrial robotics if technology reduces resources and increases the demands perceived by workers. Poor work design, an increase in task-related demands, a decline in workers' autonomy, and an imbalance between costs and benefits in work have a high price on health. As well described by the Job Demands-Resources model [111], if the demands of the work exceed the resources (organizational and personal) of the worker, burnout takes over (also JCT Sect. 2). In addition, JCT describes the need for comprehending various job characteristics that support job morale and meaningfulness of activities promoting well-being.

Resources can be work control, support to carry out tasks effectively, adequate feedback, and learning opportunities. Technology can also be a resource and not a threat and can contribute to workers' mental health. In order to accomplish this, robots should manage routine tasks, leaving the most creative, flexible, and meaningful ones to workers [99]. The challenge for those who design and manage work contexts characterized by a high interaction between humans and robots is to consider how workers must have new digital skills and robots should have forms of artificial intelligence that require a new approach to work organization in terms of performance, job qualifications, skills requirements and their impact on human capital [112–114].

Turja et al. [115] investigated whether robotization had an impact on the quality of the working life of workers employed in Finnish companies. The results show that job satisfaction in robotic contexts is generally lower than in non-robotic ones. Work satisfaction and the perception of performing meaningful work are associated with varied tasks, which are reduced in robotized work contexts. The workers interviewed report positive opinions towards robotization in cases where diversity in work activities is maintained. Variety can be guaranteed if changes in collaboration between people are allowed, ensuring a balance between routine tasks and more challenging activities.

Mechanical and repetitive work, alongside a robot, reduces levels of well-being, satisfaction, and richness of meaning at work, with effects on commitment and motivation. Conversely, technologies, if well integrated, can represent a form of positive challenge, an opportunity for growth and professional learning. Similar results were obtained by Welfare et al. [116] following interviews with assembly-line workers. Workers especially appreciate activities that promote socialization, movement, autonomy, the possibility of solving problems, the variety of tasks, and manual skills. Conversely, the repetitiveness of tasks, time pressure, and conditions that put safety at risk are experienced as threats. Therefore, it is imperative that robots should be introduced in work contexts while taking into consideration the broader trends of mechanization, cyberization, and automation of advanced robotic workspaces.

9 Attitudes Toward Mechanization, Cyberization, and automation in Advanced Robotic Workspaces

The actual behavior of workers in their interaction with robots is strongly dependent on their attitude toward robots. According to Meissner et al. [68], it is possible to frame robot acceptance with three components (see Table 1):

- Factors related to the object, i.e., the robot characteristics
- Factors related to the subject, i.e., the workers' characteristics
- Factors related to the context, i.e., the organizational level and workplace setting.

In the existing literature, experimental results are sometimes contradictory, given the multifaceted nature of acceptance. Elprama et al. [117] suggest that workers perceive robots as a threat because they could eventually take their jobs. At the same time, robots are perceived as a resource to decrease workload. Another ambiguous issue is human–robot similarity. According to You and Robert [118], a high similarity between humans and robots promoted trust, but Latikka et al. [119] reported that workers have a more positive attitude towards robots if perceived as equipment than as coworkers, especially if they did not have previous experiences of interaction with robots and feel little self-efficacy with technology. Further, a robot's autonomy could trigger a fear of loss of control and a threat to autonomy. In addition, meaningful collaboration with robots requires that robots are transparent, i.e., that they clearly communicate what the system is doing and why.

According to Hampel et al. [also 120, 121], attitudes toward technology are related to workers' evaluation and satisfaction with their current work. In addition, workers that do not perceive robots as an enrichment may feel high technology-based job insecurity. The authors suggest designing robot implementation considering these resistance factors, addressing workers' needs and fears.

Table 1 Factors related to acceptance of robots

Object-related factors	Subject-related factors	Context-related factors
– Safety compliance – Trustworthy design – Process reliability – Adaptive automation to workers' behavior and needs – Controllable and autonomous workflow	– Perception of job security – Knowledge of technology – Experience – Self-efficacy – Work motivation – Trust in the system – Interest in robots – Personality traits – Moral values	– Organizational openness to change – Technology implementation (communication, participation, support…) – Relationship between workers and executives – Supportive colleagues – Risk management strategies – Organizational processes of inclusion – Regulatory principles for advanced robotics workspaces

As stated by Paluch et al. [121], the willingness to collaborate with a robot is a multi-stage process based on adoption-related perceptions. The authors ground their model on the idea that robots' acceptance depends on the evolution of coping mechanisms adopted by workers in their interaction with them. They outline four types of groups, which can be characterized by the following attitudes:

- The embracers: they cooperate and support the introduction of robots, aiming at the development of skills based on the effective interaction between humans and robots.
- The supporters: they value the benefits of technology just for its effects on reducing the workload and work repetitiveness, but they want to preserve their control over the operations and critical decisions.
- The resisters: they refuse to work with robots and complain about them to their superiors; however, they already have some degree of experience in interaction with technology.
- The saboteurs: they do not have previous experience with technology and are actively engaged in opposing robots' introduction in the workplace.

A strategic introduction of robots in the workplace should therefore be based on accurate knowledge of the workers' attitudes, possibly beginning the implementation of changes with those more open to robotization, with the slow and steady inclusion of the other groups. It should also take into account how organizational structures, policies, processes, and risk assessment strategies can be adopted for safer and healthier human–robot workspaces.

Jacob and colleagues [122] provide an extensive review of literature concerning the acceptance of collaborative robots and frame the results according to the Unified Theory of Acceptance and Use of Technology (UTAUT) model, proposed by Venkatesh et al. [123]. According to this model, the level of acceptance of using a certain technology depends on four main factors: performance expectancy, effort expectancy, social influence and facilitating conditions. These factors are moderated by personal characteristics like age, experience, gender, voluntariness of use, and they contribute to the formation of the behavioral intention, which precedes the actual use. Framing the literature about robots' acceptance with the UTAUT model we can summarize the following results:

- *performance expectancy*: workers can accept to operate with cobots if they believe the system will increase the productivity, shortening production times and/or promoting the quality of the output. This can be achieved through an effective introduction to cobots, by means of training and guidance, and a careful task planning for the optimal distribution of activities between cobots and workers; to enhance acceptance and, therefore, productivity the work and resources allocation should be performed by the workers and not by the robot.
- *effort expectancy*: it is associated to the ease of use of the technology and is strictly related to ergonomic design of the tools but also to the capacity of the cobots to dynamically adapt their performance to the operator, detecting conditions of fatigue and workload. Another aspect that could affect the perception of

effort concerns the controlling or programming of the cobot, tasks that should be designed according to a user-centered approach. Since the introduction of cobots is shifting the effort from physical to cognitive, proper training and support are of paramount importance.

- *perceived occupational safety*: cobots safety should not only be granted by the compliance to objective standards and regulations, it is important to invest also on perceived safety, the subjective appraisal of workers which depends on the personal evaluation of the level of safety of the operations, especially concerning the risk of collision. Distance and speed of activities should be monitored, and the system should enable the workers to understand and anticipate its movements.
- *legal and privacy factors*: the increasing numbers of sensors and the recording of many performance data could lead to some concerns about privacy, data protection and security. It is important to develop strong legal frameworks to increase the perception of security and promote acceptance.
- *psychosocial factors*: they encompass psychological aspects and abilities that are influenced by social contexts. When working with robots in the workplace, for instance, these factors may have an impact on humans. Demands that are either too high or too low, stress, disengagement, fear, etc. can have a negative impact on acceptance of collaboration. Again, work experience, personal preferences, qualification, and motivation could affect cobots' acceptance. Social aspects are relevant as well, for instance, team leaders are essential in the adoption of cobots because they train and motivate staff members to use cobots; balanced team structure, comprehensible task distribution, maintenance of interpersonal relationships are also important. In addition, by experiencing human discomfort, robots can learn how to be compassionate coworkers. This way, it is possible to counteract factors that reduce acceptance, like stress, perceived performance pressure, and perceived performance limits.

In general, we can conclude that the organization should invest in the involvement of workers in the implementation, programming, management of cobots, because their agency could increase the job crafting factors that foster well-being and productivity.

10 Conclusion

The aim of the chapter was to introduce the organizational dimension of HFE in robotics. The chapter first presented the need for addressing the organizational dimension of HFE for robotics. It then highlighted the background literature on HFE in terms of sociotechnical approaches as well as HRI, macroergonomics, and work systems design. The chapter delved into issues related to social formation, trust, and transparency, among others. The chapter also developed the issues related to job autonomy, task characteristics, and work challenges and their implications in human robot interactions in modern workspaces. The chapter highlighted that the robotics workspaces need to be considered in terms of the quality of life of humans and their attitudes toward mechanization, cyberization, and automation. These broader issues will enable the successful design of HRI and transitions to robotics workspaces in the future. Many of these themes are developed further in detail in multiple chapters throughout the book. Based on the issues raised in the book chapter, the authors derive some considerations to guide the introduction of robots in future contexts:

1. Robots should not replace workers but improve their working conditions, for example, in strenuous tasks and time management.
2. Take into account how organizational structures, policies, processes, and risk assessment actions can be adopted while including robots in the workforce.
3. Distribute tasks accurately, checking with workers which tasks are preferred and without adopting predefined approaches.
4. Introduce robots to increase workers' skills so that they are perceived as opportunities for empowerment.
5. Use robots to ensure workers' health and safety, both physically (e.g., handling loads) and psychologically (e.g., facilitating relationships, enriching work, etc.).
6. Plan human robot task allocation for systemic safety, productivity, and worker well-being so as to successfully transition to future robotics workspaces.

References

1. Shorrock, S.: Four Kinds of "Human Factors, 1. The Human Factor, https://humanisticsystems.com/2017/08/11/four-kinds-of-human-factors-1-the-human-factor/
2. Lorenzini, M., Lagomarsino, M., Fortini, L., Gholami, S., Ajoudani, A.: Ergonomic human-robot collaboration in industry: a review. Front. Robot. AI. **9**, 262 (2023)
3. Schwab, K.: The fourth industrial revolution. Currency (2017)
4. Bracco, F., Bruzzone, A.A., Carpanzano, E.: Transfactory: Towards a New Technology-Human Manufacturing Co-evolution Framework. Presented at the International Conference on System-Integrated Intelligence (2022)
5. Smids, J., Nyholm, S., Berkers, H.: Robots in the workplace: a threat to—or opportunity for—meaningful work? Philosophy Technol. **33**, 503–522 (2020)
6. Baltrusch, S.J., Krause, F., de Vries, A.W., van Dijk, W., de Looze, M.P.: What about the human in human robot collaboration? A literature review on HRC's effects on aspects of job quality. Ergonomics **65**, 719–740 (2022)
7. Kuorinka, I. (ed).: History of the International Ergonomics Association: The first quarter of a century. IEA Press (2000)
8. Helander, M.G.: Forty years of IEA: some reflections on the evolution of ergonomics. Ergonomics **40**, 952–961 (1997)
9. Bainbridge, L.: Ironies of automation. Automatica **19**, 775–779 (1983)
10. Romero, D., Bernus, P., Noran, O., Stahre, J., Fast-Berglund, Å.: The Operator 4.0: Human Cyber-Physical Systems & Adaptive Automation Towards Human-Automation Symbiosis Work Systems. Presented at the, Cham (2016)
11. Baxter, G., Rooksby, J., Wang, Y., Khajeh-Hosseini, A.: The ironies of automation: still going strong at 30, 65–71 (2012)
12. Dicks, H.V.: Fifty Years of the Tavistock Clinic (Psychology Revivals). Routledge (2014)
13. Trist, E., Murray, H., Trist, B.: The Social Engagement of Social Science, a Tavistock Anthology, vol. 2: The Socio-Technical Perspective. University of Pennsylvania Press (1993)

14. Rice, A.K.: Productivity and Social Organization, The Ahmedabad Experiment: technical Innovation. Work Organization and Management, Tavistock, London (1958)
15. Miller, E.J.: Socio-technical systems in weaving, 1953–1970: A follow-up study. Human Relations. **28**, 349–386 (1975)
16. Hackman, J.R., Oldham, G.R.: The Job Diagnostic Survey: An Instrument for the Diagnosis of Jobs and the Evaluation of Job Redesign Projects. Yale University, Department of Administrative Sciences, New Haven (1974)
17. Parker, S.K., Grote, G.: Automation, algorithms, and beyond: why work design matters more than ever in a digital world. Appl. Psychol. **71**, 1171–1204 (2022)
18. Hampel, N., Sassenberg, K., Scholl, A., Reichenbach, M.: Introducing digital technologies in the factory: determinants of blue-collar workers' attitudes towards new robotic tools. Behav. Inf. Technol. **41**, 2973–2987 (2022)
19. Johansson, B., Engberg, S., Lindgren, I., Sördal, T.: Impacts of RPA on employees' skill variety. Presented at the AMCIS Proceedings (2020)
20. Nagamachi, M., Imada, A.S.: A macroergonomic approach for improving safety and work design. Presented at the Proceedings of the Human Factors Society Annual Meeting (1992)
21. Imada, A.S.: Macroergonomics: integrating human factors and organizational behavior approaches to organizational design and management. Presented at the Proceedings of the Human Factors Society Annual Meeting (1987)
22. Hendrick, H.W.: Ergonomics in organizational design and management. Ergonomics **34**, 743–756 (1991)
23. Hendrick, H.W., Kleiner, B.: Macroergonomics: Theory, Methods, and Applications. CRC Press (2002)
24. Kleiner, B.M.: Macroergonomic analysis and design (MEAD) of work system processes. Presented at the Proceedings of the Human Factors and Ergonomics Society Annual Meeting (2002)
25. Kleiner, B.M.: Macroergonomics: analysis and design of work systems. Appl. Ergon. **37**, 81–89 (2006)
26. Kleiner, B.M., Hettinger, L.J., DeJoy, D.M., Huang, Y.-H., Love, P.E.: Sociotechnical attributes of safe and unsafe work systems. Ergonomics **58**, 635–649 (2015)
27. Kleiner, B.M.: Macroergonomics: work system analysis and design. Hum. Factors **50**, 461–467 (2008)
28. Cherns, A.: Principles of sociotechnical design revisted. Human Relations. **40**, 153–161 (1987)
29. Cherns, A.: The principles of sociotechnical design. Human Relations. **29**, 783–792 (1976)
30. Clegg, C.W.: Sociotechnical principles for system design. Appl. Ergon. **31**, 463–477 (2000)
31. Waterson, P.E., Older Gray, M.T., Clegg, C.W.: A sociotechnical method for designing work systems. Hum. Factors **44**, 376–391 (2002)
32. Gasteiger, N., Hellou, M., Ahn, H.S.: Factors for personalization and localization to optimize human–robot interaction: a literature review. Int. J. Soc. Robot. **15**, 689–701 (2023)
33. Kim, S.: Working with robots: human resource development considerations in human–robot interaction. Hum. Resour. Dev. Rev. **21**, 48–74 (2022)
34. Winkle, K., Lagerstedt, E., Torre, I., Offenwanger, A.: 15 Years of (Who) man robot interaction: reviewing the H in human-robot interaction. ACM Trans. Human-Robot Inter. **12**, 1–28 (2023)
35. Dahiya, A., Aroyo, A.M., Dautenhahn, K., Smith, S.L.: A survey of multi-agent human-robot interaction systems. Robot. Auton. Syst. **161**, 104335 (2023)
36. Karami, H., Darvish, K., Mastrogiovanni, F.: A task allocation approach for human-robot collaboration in product defects inspection scenarios. Presented at the 2020 29th IEEE International Conference on Robot and Human Interactive Communication (RO-MAN) (2020)
37. Podevijn, G., O'grady, R., Mathews, N., Gilles, A., Fantini-Hauwel, C., Dorigo, M.: Investigating the effect of increasing robot group sizes on the human psychophysiological state in the context of human–swarm interaction. Swarm Intelligence. **10**, 193–210 (2016)

38. Dahiya, A., Akbarzadeh, N., Mahajan, A., Smith, S.L.: Scalable operator allocation for multi-robot assistance: a restless bandit approach. IEEE Trans. Control of Netw. Syst. **9**, 1397–1408 (2022)
39. Hollnagel, E., Woods, D.D.: Cognitive systems engineering: new wine in new bottles. Int. J. Man Mach. Stud. **18**, 583–600 (1983)
40. Carayon, P.: Human factors of complex sociotechnical systems. Appl. Ergon. **37**, 525–535 (2006)
41. Carayon, P., Smith, M.J.: Work organization and ergonomics. Appl. Ergon. **31**, 649–662 (2000)
42. Rasmussen, J., Pejtersen, A.M., Goodstein, L.P.: Cognitive systems engineering. John Wiley & Sons, New York (1994)
43. Woods, D.D.: Cognitive technologies: the design of joint human-machine cognitive systems. AI Mag. **6**, 86 (1985)
44. Woods, D.D., Hollnagel, E.: Joint Cognitive Systems: Foundations of Cognitive Systems Engineering. CRC Press (2005)
45. Hollnagel, E., Woods, D.D.: Joint Cognitive Systems: Foundations of Cognitive Systems Engineering. CRC press (2005)
46. Woods, D.D.: Four concepts for resilience and the implications for the future of resilience engineering. Special Issue Resil. Eng. **141**, 5–9 (2015)
47. Sawaragi, T., Horiguchi, Y., Hirose, T.: Design of Productive Socio-Technical Systems by Human-System Co-Creation for Super-Smart Society. Presented at the IFAC World Congress, Berlin (2020)
48. Sawaragi, T.: Design of resilient socio-technical systems by human-system co-creation. Artificial Life and Robotics (2020)
49. Gvirsman, O., Koren, Y., Norman, T., Gordon, G.: Patricc: A platform for triadic interaction with changeable characters. Presented at the Proceedings of the 2020 ACM/IEEE International Conference on Human-robot Interaction (2020)
50. Jung, M.F., Šabanović, S., Eyssel, F., Fraune, M.: Robots in groups and teams. Presented at the Companion of the 2017 ACM Conference on Computer Supported Cooperative Work and Social Computing (2017)
51. Sebo, S., Stoll, B., Scassellati, B., Jung, M.F.: Robots in groups and teams: a literature review. Proc. ACM Human-Comput. Inter. **4**, 1–36 (2020)
52. Selvaggio, M., Cognetti, M., Nikolaidis, S., Ivaldi, S., Siciliano, B.: Autonomy in physical human-robot interaction: a brief survey. IEEE Robot. Autom. Lett. **6**, 7989–7996 (2021)
53. Bartneck, C., Belpaeme, T., Eyssel, F., Kanda, T., Keijsers, M., Šabanović, S.: Human-Robot Interaction: An Introduction. Cambridge University Press (2020)
54. Sheridan, T.B.: Human–robot interaction: status and challenges. Hum. Factors **58**, 525–532 (2016)
55. Kozlowski, S.W.: Enhancing the effectiveness of work groups and teams: a reflection. Perspect. Psychol. Sci. **13**, 205–212 (2018)
56. Correia, F., Mascarenhas, S., Prada, R., Melo, F.S., Paiva, A.: Group-based emotions in teams of humans and robots. Presented at the Proceedings of the 2018 ACM/IEEE International Conference on Human-Robot Interaction (2018)
57. La Barbera, F., Capone, V.: Five dimensions of European identity: a contribution to the Italian adaptation and validation of the in-group identification scale. Eur. J. Psychol. **12**, 288 (2016)
58. Abrams, A.M., Pütten, der, A.M.R.-V.: I–C–E Framework: concepts for group dynamics research in human-robot interaction: revisiting theory from social psychology on Ingroup Identification (I), Cohesion (C) and Entitativity (E). Int. J. Soc. Rob. **12**, 1213–1229 (2020)
59. Campbell, D.T.: Common fate, similarity, and other indices of the status of aggregates of persons as social entities. Behav. Sci. **3**, 14 (1958)
60. Lickel, B., Hamilton, D.L., Wieczorkowska, G., Lewis, A., Sherman, S.J., Uhles, A.N.: Varieties of groups and the perception of group entitativity. J. Pers. Soc. Psychol. **78**, 223 (2000)

61. Sidorenkov, A.V., Borokhovski, E.F.: The role of cohesion and productivity norms in performance and social effectiveness of work groups and informal subgroups. Behav. Sci. **13**, 361 (2023)
62. Tuckman, B.W., Jensen, M.A.C.: Stages of small-group development revisited. Group Organiz. Stud. **2**, 419–427 (1977)
63. Wang, X.V., Kemény, Z., Váncza, J., Wang, L.: Human–robot collaborative assembly in cyber-physical production: classification framework and implementation. CIRP Ann. **66**, 5–8 (2017)
64. Yanco, H.A., Drury, J.: Classifying human-robot interaction: an updated taxonomy. Presented at the 2004 IEEE International Conference on Systems, Man and Cybernetics (IEEE Cat. No. 04CH37583) (2004)
65. Thomaz, A., Hoffman, G., Cakmak, M.: Computational human-robot interaction. Found. Trends® Robot. **4**, 105–223 (2016)
66. Kolling, A., Walker, P., Chakraborty, N., Sycara, K., Lewis, M.: Human interaction with robot swarms: a survey. IEEE Trans. Human-Mach. Syst. **46**, 9–26 (2015)
67. Oliveira, R., Arriaga, P., Paiva, A.: Human-robot interaction in groups: methodological and research practices. Multimodal Technol. Interact. **5**, 59 (2021)
68. Meissner, A., Trübswetter, A., Conti-Kufner, A.S., Schmidtler, J.: Friend or foe? Understanding assembly workers' acceptance of human-robot collaboration. J. Hum.-Robot Interact. **10**, (2020)
69. Demerouti, E., Bakker, A.B.: Job demands-resources theory in times of crises: new propositions. Organizat. Psychol. Rev. **13**(3), 209–236 (2022)
70. Demerouti, E.: Turn digitalization and automation to a job resource. Appl. Psychol. **71**, 1205–1209 (2022)
71. Lee, J.D., See, K.A.: Trust in automation: designing for appropriate reliance. Hum. Factors **46**, 50–80 (2004)
72. Hancock, P.A., Billings, D.R., Schaefer, K.E.: Can you trust your robot? Ergonom. Des. **19**, 24–29 (2011)
73. Sanders, T., Oleson, K.E., Billings, D.R., Chen, J.Y.C., Hancock, P.A.: A model of human-robot trust: theoretical model development. Proc. Human Factors Ergonom. Soc. Annu. Meet. **55**, 1432–1436 (2011)
74. Dzindolet, M.T., Peterson, S.A., Pomranky, R.A., Pierce, L.G., Beck, H.P.: The role of trust in automation reliance. Int. J. Hum.-Comput. Stud. **58**, 697–718 (2003)
75. Ososky, S., Sanders, T., Jentsch, F., Hancock, P., Chen, J.Y.C.: Determinants of system transparency and its influence on trust in and reliance on unmanned robotic systems. Presented at the Proc.SPIE June 3 (2014)
76. Wang, L., Gao, R., Váncza, J., Krüger, J., Wang, X.V., Makris, S., Chryssolouris, G.: Symbiotic human-robot collaborative assembly. CIRP Ann. **68**, 701–726 (2019)
77. Robinson, S., Pearson, J., Jones, M., Joshi, A., Ahire, S.: Better together: disaggregating mobile services for emergent users. **44**, 1–44:13 (2017)
78. Basoeki, F., Dalla Libera, F., Menegatti, E., Moro, M.: Robots in education: new trends and challenges from the Japanese market. Themes Sci. Technol. Educ. **6**, 51–62 (2013)
79. Muir, B.M.: Trust between humans and machines, and the design of decision aids. Int. J. Man Mach. Stud. **27**, 527–539 (1987)
80. Young, J.E., Hawkins, R., Sharlin, E., Igarashi, T.: Toward acceptable domestic robots: applying insights from social psychology. Int. J. Soc. Robot. **1**, 95–108 (2009)
81. Salem, M., Lakatos, G., Amirabdollahian, F., Dautenhahn, K.: Would you trust a (faulty) robot? Effects of error, task type and personality on human-robot cooperation and trust. Presented at the Proceedings of the Tenth Annual ACM/IEEE International Conference on Human-Robot Interaction (2015)
82. Robinette, P., Li, W., Allen, R., Howard, A.M., Wagner, A.R.: Overtrust of robots in emergency evacuation scenarios. Presented at the 2016 11th ACM/IEEE International Conference on Human-Robot Interaction (HRI) (2016)
83. Parasuraman, R., Riley, V.: Humans and automation: use, misuse, disuse. Abuse. Hum Factors. **39**, 230–253 (1997)

84. Gaudiello, I., Zibetti, E., Lefort, S., Chetouani, M., Ivaldi, S.: Trust as indicator of robot functional and social acceptance. An experimental study on user conformation to iCub answers. Comput. Human Behav. **61**, 633–655 (2016)
85. Cialdini, R.B., Goldstein, N.J.: Social influence: compliance and conformity. Annu. Rev. Psychol. **55**, 591–621 (2004)
86. Asch, S.E.: Studies of independence and conformity: I. A minority of one against a unanimous majority. Psychol. Monographs: General Appl. **70**, 1 (1956)
87. Vollmer, A.-L., Read, R., Trippas, D., Belpaeme, T.: Children conform, adults resist: a robot group induced peer pressure on normative social conformity. Sci. Robot. **3**, eaat7111 (2018)
88. Qin, X., Chen, C., Yam, K.C., Cao, L., Li, W., Guan, J., Zhao, P., Dong, X., Lin, Y.: Adults still can't resist: a social robot can induce normative conformity. Comput. Hum. Behav. **127**, 107041 (2022)
89. Hertz, N., Shaw, T., de Visser, E.J., Wiese, E.: Mixing it up: how mixed groups of humans and machines modulate conformity. J. Cognit. Eng. Decision Making. **13**, 242–257 (2019)
90. Salomons, N., Van Der Linden, M., Strohkorb Sebo, S., Scassellati, B.: Humans conform to robots: Disambiguating trust, truth, and conformity. Presented at the Proceedings of the 2018 ACM/IEEE International Conference on Human-Robot Interaction (2018)
91. Aoki, M., Rea, F., Jirak, D., Sandini, G., Yanagi, T., Takamatsu, A., Bouet, S., Yamamura, T.: On the influence of social robots in cognitive multitasking. Int. J. Humanoid Rob. **19**, 2250022 (2022)
92. Endsley, M.R.: Toward a Theory of situation awareness in dynamic systems. Hum. Factors **37**, 32–64 (1995)
93. Endsley, M.R.: Designing for Situation Awareness: An Approach to User-Centered Design. CRC press (2016)
94. Stratmann, T.C., Boll, S.: Demon Hunt—The Role of Endsley's Demons of Situation Awareness in Maritime Accidents. Presented at the, Cham (2016)
95. Camblor, B., Salotti, J.-M., Fage, C., Daney, D.: Degraded situation awareness in a robotic workspace: accident report analysis. Theor. Issues Ergon. Sci. **23**, 60–79 (2022)
96. Gombolay, M., Bair, A., Huang, C., Shah, J.: Computational design of mixed-initiative human–robot teaming that considers human factors: situational awareness, workload, and workflow preferences. Int. J. Robot. Res. **36**, 597–617 (2017)
97. Endsley, M.R.: From here to autonomy: lessons learned from human–automation research. Hum. Factors **59**, 5–27 (2017)
98. Berx, N., Adriaensen, A., Decré, W., Pintelon, L.: Individual situation awareness to be set free from the collaborative robot. Presented at the Proceedings of the 31st European Safety and Reliability Conference (ESREL 2021) (2021)
99. Johnson, M., Bradshaw, J.M., Feltovich, P.J., Jonker, C.M., Van Riemsdijk, M.B., Sierhuis, M.: Coactive design: designing support for interdependence in joint activity. J. Human-Robot Interact. **3**, 43–69 (2014)
100. Beer, J.M., Fisk, A.D., Rogers, W.A.: Toward a framework for levels of robot autonomy in human-robot interaction. J. Human-Robot Interact. **3**, 74 (2014)
101. Nikolaidis, S., Lasota, P., Ramakrishnan, R., Shah, J.: Improved human–robot team performance through cross-training, an approach inspired by human team training practices. Int. J. Robot. Res. **34**(14), 1711–1730 (2015)
102. Fitts, P.M. (ed).: Human engineering for an effective air-navigation and traffic-control system. National Research Council, Division of Anthropology and Psychology, Committee on Aviation Psychology (1951)
103. Michalos, G., Spiliotopoulos, J., Makris, S., Chryssolouris, G.: A method for planning human robot shared tasks. CIRP J. Manuf. Sci. Technol. **22**, 76–90 (2018)
104. Gombolay, M.C., Gutierrez, R.A., Clarke, S.G., Sturla, G.F., Shah, J.A.: Decision-making authority, team efficiency and human worker satisfaction in mixed human–robot teams. Auton. Robot. **39**, 293–312 (2015)
105. Tausch, A., Kluge, A.: The best task allocation process is to decide on one's own: effects of the allocation agent in human–robot interaction on perceived work characteristics and satisfaction. Cogn. Technol. Work **24**, 39–55 (2022)

106. Cardoso, A., Colim, A., Bicho, E., Braga, A.C., Menozzi, M., Arezes, P.: Ergonomics and human factors as a requirement to implement safer collaborative robotic workstations: a literature review. Safety. **7**, 71 (2021)
107. Sadrfaridpour, B., Saeidi, H., Wang, Y.: An integrated framework for human-robot collaborative assembly in hybrid manufacturing cells. Presented at the 2016 IEEE International Conference on Automation Science and Engineering (CASE) (2016)
108. Peeters, M.C., Plomp, J.: For better or for worse: the impact of workplace automation on work characteristics and employee well-being. In: Digital Transformation-Towards New Frontiers and Business Opportunities. IntechOpen (2022)
109. Maier, C., Laumer, S., Eckhardt, A.: Information technology as daily stressor: pinning down the causes of burnout. J. Bus. Econ. **85**, 349–387 (2015)
110. Murray, W.C., Rostis, A.: Who's running the machine? A theoretical exploration of work stress and burnout of technologically tethered workers. J. Individ. Employm. Rights **12**, 249–263 (2007)
111. Schaufeli, W.B., Bakker, A.B., Van Rhenen, W.: How changes in job demands and resources predict burnout, work engagement, and sickness absenteeism. J. Organiz. Behav. **30**, 893–917 (2009)
112. Dixon, J., Hong, B., Wu, L.: The robot revolution: managerial and employment consequences for firms. Manage. Sci. **67**, 5586–5605 (2021)
113. Fletcher, S.R., Johnson, T.L., Larreina, J.: Putting people and robots together in manufacturing: are we ready? Robotics and Well-Being, 135–147 (2019)
114. Sadeghian, S., Hassenzahl, M.: The" artificial" colleague: evaluation of work satisfaction in collaboration with non-human coworkers. Presented at the 27th International Conference on Intelligent User Interfaces (2022)
115. Turja, T., Särkikoski, T., Koistinen, P., Krutova, O., Melin, H.: Job well robotized!–Maintaining task diversity and well-being in managing technological changes. European Manage. J. (2022)
116. Welfare, K.S., Hallowell, M.R., Shah, J.A., Riek, L.D.: Consider the human work experience when integrating robotics in the workplace. Presented at the 2019 14th ACM/IEEE International Conference on Human-Robot Interaction (HRI) (2019)
117. Elprama, S.A., Jewell, C.I., Jacobs, A., Makrini, El, I., Vanderborght, B.: Attitudes of factory workers towards industrial and collaborative robots. Presented at the Proceedings of the Companion of the 2017 ACM/IEEE International Conference on Human-Robot Interaction (2017)
118. You, S., Robert, L.P., Jr: Human-robot similarity and willingness to work with a robotic coworker. Presented at the Proceedings of the 2018 ACM/IEEE International Conference on Human-Robot Interaction (2018)
119. Latikka, R., Savela, N., Koivula, A., Oksanen, A.: Attitudes toward robots as equipment and coworkers and the impact of robot autonomy level. Int. J. Soc. Robot. **13**, 1747–1759 (2021)
120. Hampel, N., Sassenberg, K.: Needs-oriented communication results in positive attitudes towards robotic technologies among blue-collar workers perceiving low job demands. Comput. Human Behav. Rep. **3**, 100086 (2021)
121. Paluch, S., Tuzovic, S., Holz, H.F., Kies, A., Jörling, M.: "My colleague is a robot"–exploring frontline employees' willingness to work with collaborative service robots. J. Serv. Manag. **33**, 363–388 (2022)
122. Jacob F, Grosse EH, Morana S, König CJ.: Picking with a robot colleague: A systematic literature review and evaluation of technology acceptance in human–robot collaborative warehouses. Comput. Ind. Eng. **1**(180), 109262 (2023)
123. Venkatesh, V., Thong, J.Y.L., Xu, X.: Consumer acceptance and use of information technology: extending the unified theory of acceptance and use of technology. MIS Q. **36**(1), 157–178 (2012)

Ergonomics, Robotics, and Disciplinary Integration

Giacinto Barresi, Marta Bertolaso, Tiago H. Falk, Claudio Castellini, Patricia H. Rosen, and Sascha Wischniewski

Abstract Ergonomics is inherently based on disciplinary integration: a compulsory condition to lead the systemic understanding and improvement of real-world complexities like the ones of Human-Robot Interactions (HRIs). Accordingly, this chapter ponders the contributions of diverse disciplines to the study of human factors alongside their synergies with the multi-faceted area of robotics. Key examples are presented, considering problems requiring approaches like bio-inspired soft robotics, value-sensitive design, interactions and affective features, and neuroergonomics in HRI. The implications about education on HRI ergonomics are also pondered. Overall, this chapter remarks how diverse backgrounds can lead to contributions to the human-centered design and assessment of robots if the experts, regardless of disciplinary boundaries, just focus on learning novel perspectives and methods to solve the problems occurring in Human-Robot Systems.

Keywords Human-robot interaction · Human-robot systems · Interdisciplinarity

---

G. Barresi (✉)
Bristol Robotics Laboratory, School of Engineering, CATE, University of the West of England, Bristol, UK
e-mail: giacinto.barresi@uwe.ac.uk

M. Bertolaso
Research Unit of Philosophy of Science and Human Development, Università Campus Bio-Medico di Roma, Rome, Italy

T. H. Falk
INRS-EMT, University of Québec, Montreal, QC, Canada

INRS-UQO Joint Research Unit on Cybersecurity and Digital Trust, Gatineau, QC, Canada

C. Castellini
Department of Artificial Intelligence in Biomedical Engineering, Friedrich-Alexander Universität Erlangen-Nürnberg, Erlangen, Germany

Institute of Robotics and Mechatronics, German Aerospace Center (DLR), Oberpfaffenhofen, Germany

P. H. Rosen · S. Wischniewski
Human Factors, Ergonomics Group, Federal Institute for Occupational Safety and Health (BAuA), Dortmund, Germany

G. Barresi and S. Wischniewski (eds.), *Ergonomics in Robotics: Advances and Innovations*, Springer Tracts in Advanced Robotics 162,
https://doi.org/10.1007/978-3-032-19374-2_5

1 Introduction

The philosopher Karl Popper [1] argued that we *study problems, not disciplines*. Problems surely are independent from the disciplines offering perspectives and methodologies. This is definitely true in the complex domain of ergonomics, especially when it faces the challenges posed by human-centered robotics, another area growing across diverse disciplines. This chapter will discuss how systemic understanding and disciplinary integration can contribute to investigating and improving *Human-Robot Interactions* (HRIs). Following them means being open to learn from novel methodologies to achieve integrated solutions to real problems beyond a single expertise.

We will highlight why ergonomics and robotics can (separately and synergistically) grow when dealing with HRI in real contexts (and in teamworks) and how they actually do so by exceeding their disciplinary boundaries. This observation leads to discuss not only the challenges and the opportunities of the methodological integration as an initial pragmatic step (because each expert typically grew in one disciplinary field). Moreover, while introducing a reflection on the peculiar epistemic viewpoint of ergonomics, it also widens our reflection on what such methodological integration might look like.

The remainder of this chapter is organized as follows. The second section presents the epistemic aspects of the disciplinary integrations. The third section discusses the complexities in systems involving humans and robots, leading to the need of integrating disciplines to handle them. It also considers the perspectives of the three main branches of ergonomics (physical, cognitive, and organizational ergonomics) in terms of disciplinary integration. The fourth section presents key examples of problems requiring the application of ergonomics in robotics beyond disciplinary boundaries, focusing on bio-inspired principles, values, interactions and affective features, and neural processes in HRI. These are not only examples of the multi-faceted problems and solutions in this domain: they demonstrate its disruptive potential for innovation. Finally, the fifth section introduces topics related to the education on ergonomics in robotics to encourage any teachers and students to shape paths of disciplinary integration from their own distinctive fields.

These topics constitute the premise to obtain valid and helpful results from the integration of diverse disciplines by focusing on the problem before choosing the methods to overcome and manage it.

2 When Ergonomics Meets Robotics: Epistemic Issues for a Sustainable Design

No specific degree can include all knowledge and know-how to handle the complexities of the entities involved in *Human-Robot Systems* (HRSs): hybrid frameworks where humans and robots perform tasks by affecting each other through direct or indirect interactions.

On the one hand we have the human beings with their behaviours, shaped by biological, mental, social, and cultural factors. Medicine, psychology, and sociology are just examples of the disciplinary areas that are required to dare understanding a person as a whole. On the other hand, we have the robots: a heterogeneous set of devices functionally devised to directly or indirectly help people. While disciplines like engineering and computer science work on their development, their design should certainly be human-centered. This approach becomes even more crucial and interesting, from a scientific and technological point of view, when adding the ergonomic perspective to the scenario. Considering, in fact, HRI in specific dynamic contexts, not only requires involving additional disciplinary areas (e.g., management in socio-technical settings), but also poses a question about the most adequate way to frame the problem and the best way to address it.

Let's therefore disentangle the different levels at which an ergonomic HRI's goal brings human factors into the picture and into the design process. In order to do so, we start spelling out some epistemic peculiarities that the ergonomic perspective introduce. If engineering, in fact, aims at solving functional problems, ergonomics asks to do so for the human wellbeing and in a given environment that demands specific performances and tasks. That is, the goal to be fulfilled by the HRI is no further merely instrumental but relational. Moreover, the emphasis on the contextual needs brings about a change of perspective on the very notion of a problem. In any living system, in fact, a peculiar instrumental activity holds, which is a one-many relationship, i.e. living systems use the same tool to fulfil different tasks when required by different contexts. But there is also a many-one dynamic, by which living systems fulfil the same functional goal adopting different strategies. This means, for example, that human beings, when challenged by different functional tasks, might use the same tool (a hand, a movement, etc.) to fulfil them, but also that they might adopt different strategies to fulfil a specific goal. These evidences add an epistemic viewpoint to the multiplicity of biological, medical, psychological or social factors above-mentioned for any ergonomic HRI challenge, that asks reframing the very notion of problem when looking for sustainable solutions in these areas of inquiry and technological development. The kind of dynamics, in fact, that are object of inquire do not admit a mere instrumental, unique, best functional solution. By nature, they admit (and require) different solutions that might be equally effective and even efficient depending on contextual factors (in which, in fact, enter all the biomedical, social, etc. factors too).

When therefore thinking of the design processes of ergonomic HRI processes, the human factor's relevance is threefold. It enters in the very definition and identification of the real-world-problem (to distinguish it from mere functional problems that admit mechanistic outcomes, unique optimized solutions, and one-to-one causal relationships) to be solved (epistemic level) while constraining the possible relevant solutions to a given problem. It enters in the appreciation of the socio-cultural factors that might shape the preferences and expectation of the user (socio-cultural level). It finally enters in the need of combining and blending different disciplinary perspectives in order to move from the design to the implementation of a given ergonomic HRI solution (procedural level). Consequently, optimizing every HRS through the lens of a single discipline could never be feasible, if we want to solve real-world-problems [2, 3]. Combining different fields of knowledge and expertise is requested by the very nature of the 'problems' that ergonomic HRIs aim to solve. Therefore, since each expertise is typically achieved by a person through the study of a discipline, a pragmatic first step should be the integration of diverse disciplinary approaches. Thus, we must clarify the debated topic[1,2,3] of their integration [4–6]. The following labels constitute examples of such relationships, often expressed in terms of collaboration in teams of experts.

- *Multidisciplinarity* refers to the separate contributions of different disciplines-typically represented by different experts-to solve a problem. This can be often the first step in teamwork, before an advanced integration, expected as an outcome.
- *Crossdisciplinarity* represents a first type of integration, based on observing a problem in a discipline from the perspective of another one (e.g., translating psychological concepts on emotional language in computational ones).
- *Interdisciplinarity* focuses on the synthesis of concepts and methods from multiple disciplines to obtain novel perspectives, frameworks, and insights (e.g., engineers and psychologists working on the cognitive architecture of an assistive robot).
- *Transdisciplinarity* defines a further integration, possibly based on co-creative efforts to achieve global views on a problem and its solution (e.g., neuroprosthetics can emerge beyond methodological distinctions in neuroscience and neuroengineering).

Recognising these terms is quite important when a project or a task is proposed or assessed: it can unveil the real level of integration, detecting a mismatch between actual and planned activities. This also helps avoiding ambiguities, clarifying intermediate steps and goals, establishing shared ground and standards, facilitating the intra-team and inter-team communication. This can foster efficient coordination, training initiatives, integration sessions, co-design, and a mutual respect between teammates with diverse backgrounds [7, 8]. Both ergonomics and robotics work through these approaches, especially in terms of interdisciplinary integration [9]. However, they can also proceed sequentially (a project can start as a multidisciplinary activity, ending as an interdisciplinary one). The experts in a team can keep their own spaces (as in multidisciplinary and crossdisciplinary cases), focusing on the outcomes related to their own disciplinary areas although learning novel concepts from their colleagues. Nevertheless, these learnings can progress to obtain interdisciplinary or transdisciplinary results that can generate novel approaches (even disciplines) beyond any specific solutions devised by the team for a task. Each level can also help to expand the collaborations through its peculiar features: crossdisciplinary

[1] https://usher.ed.ac.uk/advanced-care-research-centre/acrc-internal/multidisciplinary-or-interdisciplinary-what-to-use.

[2] https://www.arj.no/2012/03/12/disciplinarities-2.

[3] https://www.archpsych.co.uk/post/disciplinarities-definitions.

approaches, for instance, can be helpful to ease the communication between scientists and policy-makers, while a transdisciplinary view could also be considered the final goal of their collaboration to effectively engage the public too. However, transdisciplinarity still needs to be truly achieved [10] and pragmatic approaches should probably aim more to interdisciplinarity by exploring the connections between disciplines brought by different experts facing the same problem (the premise of this chapter).

3 Humans and Robots

The study of HRSs reflects the complexities of users, machines, and their interactions.

First, we have the users: *humans* with anatomo-physiological features that did not change radically from the paleolithic age. Although the most advanced technologies can bring us beyond our limitations, our bodies and minds are optimised for activities like hunting and gathering in the savannah. This is why, for instance, the same heuristic processes that helped humans making decision quickly tend to oversimplify our choices in the technologically complex present days (becoming systematic distorsions, biases, in our mental functions) [11, 12]. Accordingly, we always need to design our interactions with any technologies by being aware of our evolutionary heritage.

In addition to this general mismatch between humans and complex technologies, each person can be different by various features, including cultural and professional contexts. Impairments can play a crucial role, especially if we consider (see Fig. 1) how they can be situational (e.g., carrying a bag, paying attention to the movements of an exoskeleton), temporary (e.g., wearing a plaster cast), and permanent (e.g., amputation or agenesis of a limb) impairments [13]. The first type of impairment can even be due to the usage of technologies with sub-optimal ergonomics (a device can help us to perform a task but it can create additional burden on some other aspects) [14].

Understanding how solving the problems in one category can help people in the other two would be a real step-change in human-centered design for inclusion-across accessibility (making a system usable regardless of impairments due to diverse characteristics) and universal design (making a service or an experience accessible for anyone).

Overall, each branch of the study of human factors brings and blends approaches and techniques derived from diverse disciplines, also generating original perspectives and methodologies, enabling ergonomics to bridge diverse domains. Below, we list a (certainly non exhaustive) set of examples of disciplines contributing to (and working in synergy with) each one of the three branches of ergonomics, paying special attention to HRIs.

Physical Ergonomics [15] can work with disciplines like Anthropometry (by applying data on human body dimensions to tailor the range of robot motion according to diverse physical profiles), Biomechanics (investigating the human body

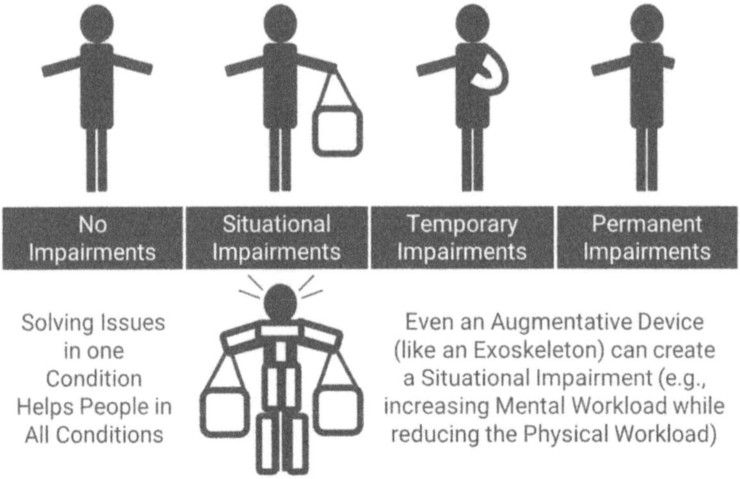

Fig. 1 Different users can have different problems but solving them in a category can help finding a solution in others based on the Inclusive Design toolkit (https://inclusive.microsoft.design/articles/inclusive-101-guidebook) of Microsoft, mentioned by [13]

mechanics to reduce the risk of physical strain and injury during physical HRIs), Physiology (to analyse how physical HRIs can affect human performance and fatigue in diverse contexts, both clinical and industrial), Mechanical Design and Engineering (by designing robotic components that align with human physical sizes and capabilities, considering both mobility and comfort).

Cognitive Ergonomics [16] can work with disciplines like Experimental Psychology (studying our mental processes through methods adopted from its different branches, from psychophysics to decision-making), Cognitive Neuroscience (focusing on the neurocognitive processes underlying the user behaviour for optimising the devices in terms of workload, for instance), User Experience (UX) and Usability (caring for aspects evaluated while assessing and designing robotic systems, including their user interfaces), Computer Science and Engineering (developing user interfaces and intelligent systems to adjust the robot autonomy to paradigms like the shared control).

Organizational Ergonomics [17] can work with disciplines like Organizational Psychology, analysing and improving teamwork and policies supporting and optimizing HRIs and HRSs in organizational contexts), Occupational Medicine (by understanding how a certain HRI design can affect the workers' health and well-being, for diagnosis and prevention), Management (improving the impact of the introduction of robots interacting with the workers in terms of individual well-being and system outcome), Sociology (studying socio-technical systems to understand their structure and development with implications for the human society as a whole).

As already mentioned, these (and other) disciplines can work in synergy with ergonomics (and the aforementioned main branches, despite many cross-branch sub-fields exist: e.g., healthcare ergonomics) [18] or they can provide it with methodologies to devise (e.g., industrial design to adjust the product to the UX) or assess (e.g., ethnographic methods for the videoanalysis of user behaviours) the HRIs [19].

These synergies can also produce fertile dialectics for enriching each discipline with methodologies derived or refined from ergonomics and related human-centered practices as well (e.g., using ecologically valid settings, similar to real contexts, in laboratory) [20]. This surely helps the journey of understanding the human being in the digital age while proposing solutions to care for their criticalities [21]. In the light of the epistemic issues mentioned in section two, caring means supporting and accompaning the human being in an adaptive and technologically mediated process to a given environment. This process admits and requires a plurality of solutions.

We must therefore consider how there can be overlaps in the disciplinary contributions to different branches of ergonomics: for instance, computer science can improve the physical ergonomics, especially in terms of safety, of an industrial robot through the assistance offered by Artificial Intelligence (AI) solutions [22]. In all these contexts, ergonomics can further advance through disciplines like ecological psychology (studying the interdependence of human and environment), control engineering (focusing on design and control of systems), systems engineering (integrating and managing complex systems), and cybernetics (investigating control, feedback, and information processing in systems through transdisciplinary views), each one dedicated to specific relationships in complex systems without losing attention to their features [23–26]. Interestingly, ergonomics itself can be considered as a *systems science*, as in the definition provided by the International Ergonomics Association (IEA) with its reference to the system and its elements, including the human beings[4]. The debate on such a point of view can surely improve the study of human factors as a whole, especially considering how ergonomics should go beyond the improvement of performance and well-being by means of systems design: it should predict how work in the system could change [27], first of all, and it should also ponder global effects beyond the anthropocentric perspective [28]. Facing such a challenge requires a deep understanding of the problem space of human factors, with metrics of the system complexity that help optimizing the interventions of ergonomics [29]. Without the integration of disciplines catching different perspectives on problems in systems (with their multiplicity, dynamism, and uncertainty), this would be impossible.

In such a view of the complex systems addressed by ergonomics, we can now introduce *robots*. These (relatively) smart machines may have different degrees of autonomy, shapes, and functions. Even before pondering HRI, we should mention how robotics itself [3] can show diverse examples of disciplinary integration. Disciplines like mechanical engineering and electronics (and their blend in mechatronics), computer science, computer engineering, and material science are just examples of the expertise domains required to build a robot. They can progress through multidisciplinary and crossdisciplinary efforts, even through the contribution of behavioural and social sciences in several cases of HRI-for instance, to visually program social robots [30]. They can also move towards interdisciplinary and transdisciplinary inte-

[4] https://iea.cc/about/what-is-ergonomics.

grations, generating fields that transcend the initial expertise of each specialist working on them. This is the case of robotic biomimicry [31]-which requires a synthesis of knowledge on biological and robotic systems to integrate living and mechatronic ones in terms of sustainability (a trait that is not directly shared by other branches of bio-inspired robotics).

Specifically, we focus here on robots designed to interact with us while we bring our specific skills, limitations, needs, and goals in a specific context. Different types of robots are made to meet the peculiarities of different types of users, possibly with features for personalization (automated adaptation to the user) or customization (adjustment guided by the user).

Accordingly, robots may resemble, for instance: handheld tools (like a robotic spoon mitigating the tremor in the hand of the user) [32], wearings (like exoskeletons to assist or rehabilitate the control of lower limbs) [33], limbs (like prosthetic legs) [34], mediums (like telepresence robots) [35], assistants (like robots in manufacturing) [36], or companions (like affective robots for the social stimulation of the elderly) [37]. According to the type of interactions that are required, each robot will be designed and assessed through methodologies typically adopted by different disciplines, as showcased in Fig. 2.

Every example presented above helps introducing the need of moving beyond disciplinary boundaries, especially when we must establish a high degree of human-robot integration-as extension of the human-technology integration or human-computer integration, occurring when the boundary between humans and technology tends to become blurred as users and machines work jointly and share agency to execute tasks [38]. That is, the complexity of the systems addressed by ergonomics eventually simplifies and levels where correlation maximizes in the realization of a functional task. In this context we can refer to the seamless condition of a user that is focusing on the ongoing activity without paying attention to the (experientially

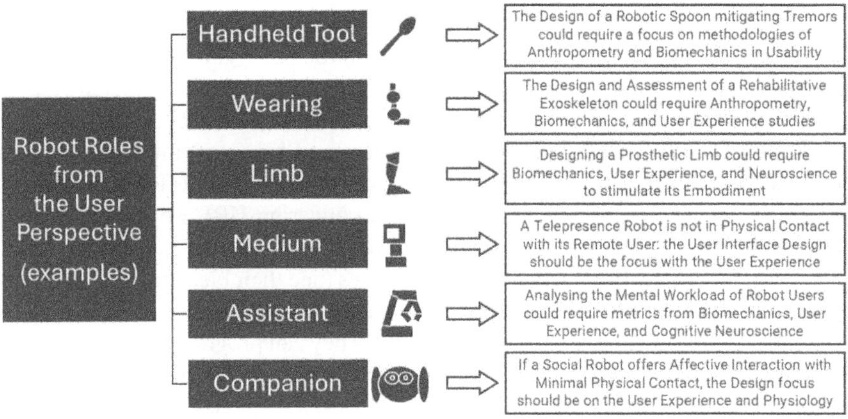

Fig. 2 Examples of robot roles from the perspective of the users, and disciplinary contributions to the design of their interactions

integrated) devices enabling its execution, as in case of the embodiment of wearable and prosthetic devices [39]. These examples of human-robot merging also highlight the need of integrating ergonomics and robotics. This can happen in two ways.

- *Ergonomics can help Robotics* by offering methods (also shared with diverse disciplinary domains) for improving the compatibility of smart machines with the people they are interacting with;
- *Robotics can help Ergonomics* by offering smart machines to make certain tasks, environments, and systems more ergonomic (the quality leading to improved human well-being and system performance).

The integration between these two fields can also work by considering the connections between the disciplines contributing to their growth: e.g., a cross-disciplinary collaborations between experts in linguistics and computer science, or the inter/transdisciplinary approaches of computational linguistics, can be surely advantageous to power the development of versatile Large Language Models for conversational agents embodied in robots [40]. This highlights even more the need to focus on a problem and the opportunity to get the most appropriate methodologies regardless of the initial expertise in the specialist or the team approaching this task, leading to systemic perspectives and learnings that must be human-centered. About the latter term, we have already referred to the *Human-Centered Design* [41] (simultaneously an intermediate goal of ergonomics and a related discipline) in this chapter: making a system compatible with human skills, limitations, and goals in a certain context, paying special attention to individual needs and experience.

This approach surely requires heterogeneous methodologies to understand the user needs and lead the processes of design and assessment with the involvement of diverse stakeholders. Consequently, it can offer a series of tools (closer to the operative aspects of creating and planning solutions than to the scientific investigation of physical, cognitive, and organizational factors related to the human being) to ignite and assist the processes of ergonomics to design and assess HRIs. By adopting this design perspective, we must define the problem we are trying to face (and the solutions to find) in human-centered terms, prioritizing human needs, experiences, and abilities we want to reach. Afterwards, we must select and employ a set of methodologies and criteria (derived from diverse disciplinary integrations) to, respectively, plan interventions, based on ergonomics, and establish our objectives to obtain an "ergonomically successful" HRI (improving the individual well-being and the system performance). The reader can consider this example as an introductory proposal, obviously.

The general vision expressed so far in this chapter requires integrating perspectives that converge in the multi-faceted field of ergonomics in robotics. However, the literature already presents several journeys of disciplinary integration: they can be pondered as sources of inspiration to move the next steps in this domain. Accordingly, the next section will introduce examples of problems and solutions beyond disciplinary boundaries in HRI.

4 Problems and Solutions Beyond Disciplines: Some Examples

The previous section pointed at the complexities in humans, robots, and HRSs, requiring the integration of disciplines for solving problems related to HRI. This section will present examples of domains where ergonomics and other disciplines can express different roles for improving the systems involving humans and robots (providing additional concepts that can suggest further explorations in this multi-faceted area):

- *soft robotics* may be based on bio-inspired approaches that can improve HRIs;
- *value-sensitive design* coordinates disciplines for the dignity of robot users;
- *interactive and affective features* can make robots more compatible with the user;
- *neuroergonomics* connects laboratory studies and real-world interventions.

4.1 *Bio-Inspired Soft Robotics*

As already discussed, robotics can also merge with different domains. For instance, by blending with life sciences, it can generate interdisciplinary (possibly transdisciplinary) areas like biorobotics, which is literally based on the integration of robotic and biological systems. It is especially interesting to consider the implications of bio-inspired robotics [42], when robots are designed by imitating living systems (as in biomimetics) also for enhancing or replacing them (as in bionics) [43]. Accordingly, robotic actuators can imitate the tentacles of an octopus or the trunk of an elephant (based on principles like morphological computation) [44], groups of robots can coordinate like swarms (versatile yet challenging to control by a user-thus, generating additional needs of ergonomics) [45], and neuromorphic technologies can provide robots with animal-like sensory systems (as in neurorobotics) [46].

These solutions disclose a groundbreaking potential, solving our problems by employing solutions already “developed” by the nature. This is especially interesting because it highlights how a specialized field of robotics, integrating knowledge and know-how from life sciences (alongside material science, mechatronics, mechanical design, and more), can offer novel solutions to improve the ergonomics of complex devices like rehabilitative, assistive, and augmentative exoskeletons by practically generating an all-new sub-set of solutions like the *soft exosuits* [47, 48]: wearable robots constituted by soft, compliant components that improve wearability, versatility, comfort, safety, and acceptance. Achieving these devices is possible exclusively by integrating different technologies, devised through different types of expertise, according to the user needs as unifying principles. These examples surely describe the high potential of getting inspiration from nature to make wearable robots more ergonomic at physical level. However, soft biorobotics, and soft robotics in general, may also ease social HRIs with embodied agents by eliciting the user’s feelings of bonding and safety in physical interaction. However, these observations need further investigations, especially considering the exploration of their consequences on the

HRI tasks [49, 50]. For instance, we should consider how a certain design could trigger uncomfortable sensations, introducing value-sensitive issues described in the next sub-section.

4.2 Values as Human Factors

Values-what a group of people consider important in their life-are surely among the most relevant aspects of human experience and drives of our behaviours. They can be principles that motivate our actions and a constant reference for appraisal and self-appraisal. They might also be epistemic values that typically refer to scientific successful processes, such as what counts in a given field of knowledge as relevant explanation and deep understanding. Accordingly, values become a crucial aspect in human-centered design, specifically in the domain of *Value-Sensitive Design (VSD)* [51], which literally accounts for human values in the design process. VSD is inherently tied to iterative and participatory design: iterative, because it requires a cyclical evaluation of each version of a product or a service to revise its features; participatory, because it engages the end-users and, possibly, all stakeholders-anyone affected by the product or the service that is the object of the design process.

These can be considered features of collaborative design (Co-Design), shared with the larger perspective of Human-Centric Design [52] with its attention to the societal impact of the design processes and results.

In general, VSD addresses ethical issues through a tripartite approach [53] encompassing, as highlighted by Fig. 3:

- *Conceptual Studies* defining the problem (e.g., what terms like "informed" and "consent" mean in the "informed consent" concept) according to disciplines like law and philosophy (in particular ethics);
- *Empirical Studies* enacting empirical (possibly systematic, thus experimental) investigations on how a problem occurs in real-world, according to methodologies from domains like psychology and ergonomics;
- *Technical Studies* proposing solutions to the problem observed empirically according to the concepts initially established-such solutions are devised through design and engineering, for instance.

Such an approach can be adopted in different contexts, including, for instance, the privacy issues in smart environments and the approach of privacy-by-design, demonstrating the opportunity of embedding design principles to maintain the individual dignity in the technology itself [54]. In this perspective, values (like autonomy and inclusion) become human factors that ergonomics should consider in multiple ways. Another example is offered by fair, diversity-sensitive language solutions for HRI [55].

Although the tripartite approach described above can sound multidisciplinary (or cross-disciplinary, since concepts can be translated in relationships between variables

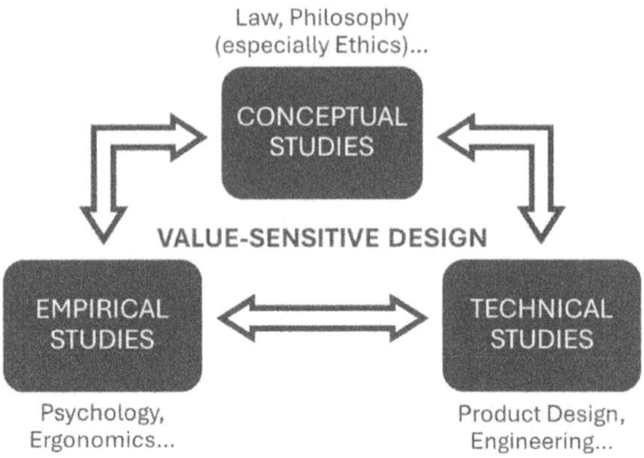

Fig. 3 The three types of studies involved in VSD (with examples of disciplines contributing with their methodologies). Ergonomics can contribute to the empirical studies (as in the picture) but can also be informed by the VSD process, based on [53]

in experiments which describe interactions that must be re-designed), the debate on VSD methodologies is quite intense (e.g., the process not necessarily starts with the conceptual studies) [56], and different types of disciplinary integrations should be pondered according to the role, for instance, of ergonomics. Ergonomics can be part of the VSD process (in terms of empirical studies, as in Fig. 3, or even in technical studies, when it works in synergy with product design and diverse areas of engineering), including a positive UX as a value, for example. However, it can be also informed by VSD to proceed in the analysis, design, and re-design of assets like robots and processes like their interactions with humans.

Nevertheless, even co-design could move hazardously beyond the problems that we decide to face and the solutions we resolve to test in ethically sensitive contexts. This is quite obvious if we ponder how robots like autonomous cars may be programmed to priorities the human safety with criteria that, in complex scenarios, could be advantageous for the passenger or the pedestrians. Accordingly, moral decision-making [57] is currently investigated in this context, highlighting how a social dilemma could (and should) be challenging and non-trivial for human beings before intentionally scripting biases in smart machines. On the other hand, the deployment of such advances also require ethically critical achievements like the trust of the users (even before achieving technology acceptance) which is one of the goals of implementing explainable AI (XAI, informing the user about the processes leading to their output) solutions [58].

Overall, robots influencing or persuading people constitute criticalities that cannot be ignored. Robot nudgers, for instance, can steer users toward the expected behaviours through stimuli (nudges) [59] designed according to those human biases that have been mentioned in the second section of this chapter as examples of the natural constraints in our mind. By understanding these biases, even robots can be designed to nudge individuals toward certain choices (typically positive decisions for their health and wealth) without restricting their freedom. However, robot nudgers [60] can show crucial issues: first of all, they can adjust (customize) the transparency granted to a specific user; secondly, they can monitor the impact of any potential combination of nudges and transparency features on the individual decisions. Such situations require the involvement of ethicists at the early stages of robot nudgers' design.

In general, any HRIs not only can be devised to influence the user but is always designed to provide them with certain abilities or experiences, even emotions. The next sub-section will discuss this aspect, encompassing interactions and affective processes.

4.3 Interactions and Affective Processes

The idea of "designing" a subjective event can be considered controversial: people often refer to the concept of design as the arrangement of patterns in objects or environments. However, we can design interactions between (and the systems involving) humans and artifacts (like robots) by adjusting the configuration or the activity of the second in order to influence the behaviour of the first (as in the interaction design). We can also design a UX by implementing features that stimulate the users or influence their behaviour in order to elicit a specific subjective event. In both cases, we are manipulating the probability that a certain observable series of events can trigger an expected experience and behaviour in a person (according to the relationships we discovered or assumed between such events). Accordingly, affordances (what a setting offers, how an object invites for a certain action) and constraints (features channeling the user activity on a tool, for instance) define the relationships and the interactions planned by the designer (and the ergonomist) [61–63].

Clearly, all these aspects require to integrate concepts of psychology and cognitive science (especially on perception, motor cognition, action-perception coupling, and enactive cognition) that directly inform cognitive ergonomics. An example can be how the concept of stimulus–response compatibility (based on the matching between the properties of perception and action) can be employed to improve the "naturalness" of interactions (making them seamless and intuitive) in (cross-disciplinary) terms of control-feedback compatibility in a reality-based user interface [64–67].

Furthermore, some features can be based on the activity itself of the machine, for instance to engage the user, as in recent studies [68, 69] employing the approach of the radical constructivism (which defines how the knowledge is actively built by the learner while the cognition organizes the experience) to robot learning. In such studies, the robot (a myoelectric prosthetic hand) can guide the user data collection (literally "asking" for more trials, when needed) in order to build computational models enabling the control of the machine itself through an actual human-robot co-adaptive training. The integration of theoretical aspects of radical constructivism from psychology in HRI is a clear example of the fertile synergies occurring in cognitive ergonomics.

Similarly, certain features can lead the user to experience a certain emotional or motivational change and we implement such features in our device, including an affective robot. This is a general view on the highly interdisciplinary concept of affective design, requiring knowledge and know-how from psychology even through its relationships with fields like affective ergonomics, as in the Kansei framework, for choosing which features can trigger a certain mood in the user, for instance [70–72]. However, we must also consider how these features are not only defined before the HRI itself as static ones: they can also be dynamically adjustable to the current conditions of the person. Affective skills (understanding the emotional states of the user or expressing emotional patterns to elicit responses in the user) are crucial for social robots [73]. In this case, we need features of affective computing, based on the collection and analysis of multimodal (physiological, visual, acoustic) data that can inform the machine on the emotional and motivational state of the user in order to "orient" it as expected through specific (tactile, visual, acoustic) stimulations. These adaptive (also co-adaptive) processes require a systemic approach (considering the HRS) where interdisciplinary and transidisciplinary integrations become necessary to use a set of heterogeneous methodologies able to catch the subjective meaning of different human signals and machine stimulations and translate it (in a crossdisciplinary manner) in the design of the robot and of its actions [74–77]. Both static and dynamic features for affective interaction must be shaped according to principles and guidelines established in the literature of affective research with its integrated disciplines.

Obviously, the design of the robot actions and reactions is especially critical, and it leads to merge the approach of diverse interactive features, including the affective ones, possibly extending them through the opportunities offered by novel paradigms that create further bridges between human and machine. For instance, XAI can be especially helpful for providing the user with explainable and transparent feedback, establishing a trustworthy HRI, through multimodal interactions [78].

Importantly, all approaches based on the design of HRIs for eliciting certain (positive) experiences in the user should always be updated to the most recent state-of-the-art on sensors detecting the user conditions in real-world settings, exactly as in the case of neuroergonomics, the topic of the next sub-section.

4.4 Neural Factors: Neuroergonomics

Neuroergonomics studies human factors and proposes human-centered solutions by adopting methods of the diverse fields of neuroscience. By bringing such approaches (based, for instance, on recording physiological and motor data through wearable devices) outside laboratory settings, neuroergonomics can study the nervous system at work, in real contexts or ecologically valid environments populated by complex devices like robots, exploring HRIs through innovative methods [79–82]. Nowadays, this is especially true if we consider the opportunities offered by portable and wearable devices that can enable ubiquitous data collection processes (despite limitations in reliability and signal quality). By avoiding oversimplified settings in laboratories (after exploratory tests that can occur even through bulky systems for neuroimaging) and (possibly) targeting "in the wild" contexts, studies in neuroergonomics can engage the expected neural processes of the subjects (making the latter compliant to the procedure itself), without necessarily losing the compulsory degree of experimental control on the proposed tasks. This way, both the subjective meaning of the activity and the statistical significance of the test results can be maintained, offering conclusions that can improve the corpus of knowledge and know-how in the neuroscience domain, closing the loop by providing neuroergonomics with novel methodologies. Figure 4 depicts this with three examples of interventions in neuroergonomics [83–85]:

- the *assessment* of human-machine interactions through physio-motor indices;
- the *training* of a user by means of procedures like biofeedback and neurofeedback;
- the *adaptation* of the machine activity to the current conditions of the user.

The second example could fall in the third category, although it focuses more on improving the skills of the user in simulated trainings than to adjust the functions of the machine to the user conditions regardless of their initial skills. Interestingly, they can converge in processes of co-adaptation [86].

We may consider neuroergonomics as an interdisciplinary field, blending approaches of neuroscience and ergonomics to obtain a third domain, or a crossdisciplinary one, translating the methods from a discipline to the other (being each one highly diversified, anyway). However, we must also ponder how neuroergonomics can be specified with the same attributes of the main branches of ergonomics [81, 87, 88] – physical neuroergonomics, cognitive neuroergonomics, organizational neuroergonomics – just by intertwining them with neuroscientific methods that can further differentiate this discipline across the Central Nervous System and the Peripheral Nervous System (e.g., peripheral neuroergonomics is based on peripheral biosignals) [89].

Furthermore, we also need to highlight how neuroergonomics can be considered as part of the field of neuroscience just when we highlight its peculiarities [81], like the focus on human-centered design and assessment, alongside a clear attention towards technology transfer and, possibly, market access.

This type of complex dialectics is destined to grow with the advances in wearable devices able to collect the user data in the real world, enabling breakthroughs in our capability to model systems involving humans and robots, for instance, according to the diverse methodologies offered by disciplines like neuroscience, at the convergence of diverse domains encompassing the complexity of human experience and action.

However, we must recollect how this chapter introduction remarked that understanding such a complexity in a pragmatic way requires strategies of disciplinary integration because every expert typically grew in one disciplinary field. Furthermore, the first statement of the second section highlighted how no specific academic degree can include all knowledge and know-how to handle the complexity in the HRSs. Thus, how could we (as teachers and students) facilitate such a learning for handling the real-world problems in HRIs?

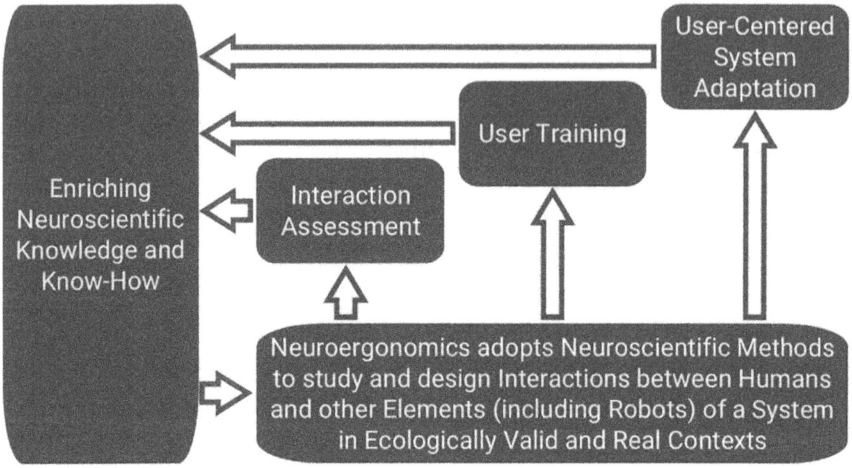

Fig. 4 Examples of the relationships between neuroscience and neuroergonomics

5 Education Beyond Disciplinary Boundaries

Every disciplinary integration approach described in the previous sections may be considered as a practical response to real-world problems. Each expert in teams working in such domains typically bring the knowledge and know-how from their own fields. However, when we move across disciplines we must also ponder the risk to face the expert's dilemma [90]: the interdisciplinary growth of a specialist can lead to a loss in expertise and, consequently, credibility and acceptability among the peers in the initially main field. We could also add that entering a new field can show barriers in terms of credibility and acceptability, obviously. However, the nature of ergonomics is inherently beyond disciplinary boundaries, especially when it targets a multi-faceted area like HRI: both are areas of convergence for diverse types of expertise. This means that ergonomics in robotics has the potential for becoming a fertile and safe area to nurture the new generations of HRI experts. They would be focused on real-world problems with their epistemic peculiar status instead of academic constraints, feeling the right to bring the richness of their backgrounds in a journey to make innovative solutions emerge-also filling methodological gaps [91]. This should work as a preliminary advice about making people with different expertise meet in the same room to collaborate. This effort can lead to innovations flowing towards a technology transfer driven by a market access vision based on the effort of responding to unmet needs of the users, achieving extra-academic impact.

The first unmet need we should consider is to make the diverse teammates' background compatible. Education on ergonomics in robotics surely plays a key role in this effort. Ergonomics-and each one of its branches and sub-branches-can be an academic subject in diverse types of degrees, from psychology to architecture, from medicine to Human-Computer Interaction (HCI). In each context, it can be characterised differently-we must not forget the use of the "human factors engineering" term. On the other hand, HRI can be a subject in degrees on robotics or HCI, while robotics (not necessarily including HRI) can be a course in several engineering programs, from computer engineering to mechatronics. Each one of these subjects typically needs initial lectures to bridge the background of the students (in the second half of their undergraduate period, or in their graduate one) with the new discipline (e.g., students in architecture could need concepts of cognitive psychology at the start of classes on cognitive ergonomics, while students in psychology could need fundamentals of design within a course with the same title) and to give them adeguate heuristic and epistemological tools to navigate the complexity of the system to be addressed. Therefore, when we talk about ergonomics in robotics, we expect that this should be part of the widest courses on ergonomics, or it can become a specific subject in programs dedicated to HRI. The literature offers several publications on such topics (from education to practice), such as [92–94]. Any classes involving learners from diverse domains should move through perspectives of disciplinary integration in order to make each student bring their own expertise (still in development) as a value within the new field that they are going to explore. The new teaching must make sense in their storytelling, and this can happen just if they focus on real-world problems that constitute the reason for the effort to learn new perspectives and methods (and to collaborate, today or in future, with people from different backgrounds). Enhancing HRIs through ergonomics can be the scope of engineers, and studying HRIs for investigating how the human behaves in certain contexts can be the goal of cognitive scientists. Interestingly, each one should learn from the domain of the other to accomplish their goals, possibly enlarging their perspectives when the phenomena tends to blend in novel ways.

For instance, teachers of ergonomics should make engineering students (including the ones in programs for Continuing Professional Development, CPD) aware of (for instance) the need for practices of user engagement and co-design, the value of high-level descriptions of a system to integrate the lower-level ones, the differences between quantitative and qualitative methods and between objective and subjectives sources of information. This would anticipate the design, development, and test of any machine. However, teaching ergonomics applied to HRI in psychology programs would require similar topics but the implementative options could be limited, and the focus should be on the design and assessment in behavioural studies without the development of complex devices or simulations. A synergistic approach could be helpful. Indeed, for instance, collaborative sessions of human-centered design and assessment can make students from diverse backgrounds care for cross-disciplinary efforts to establish a clear communication across different experts within a team.

Such a process can find an effective implementation in fields like computational ergonomics [95] in HRI, related to digital models representing systems involving humans and robots. The cross-disciplinary effort to translate the models from a discipline to another to create an interactive simulation of a specific HRS could be an impactful scenario to train all HRI ergonomists by asking them to contribute according to their own specialty. These practical strategies (like creating a game-like simulation, often helpful for outreach and public engagement too) can be quite successful in executive courses, targeting professionals like managers that can contribute to The dialectics between academic and extra-academic contexts. Each extra-academic expert can bring novel terms and trends during the discussion of case studies and exercises.

However, sometimes terms do not match. For instance, the recent revision of ISO 10218:2025 [96, 97] exemplifies the epistemological tension between academic and extra-academic discourses: indeed, it emphasised how collaboration is a property of the application, not an intrinsic characteristic of the robot itself-weakening the use of terms like "collaborative robot" and "cobot". Similar concerns may arise with words like "intelligence" or "agent", which anthropomorphise machines or suggest ontological parity between humans and robots (as in "human-robot teaming" too), obscuring fundamental asymmetries in agency and responsibility. A rigorous argumentation is therefore essential for facing such issues (potentially hindering the disciplinary integration), as Floridi does in [98] to propose the AI as agency without intelligence (making the term agent legitimate for an artificial entity).

Consistent argumentations are necessary for neologisms too: Hybrid Intelligence (HI), for instance, is justified as a framework emphasising synergy through collaboration and augmentation while preserving the distinct role of the human being [99]. Another example of neologism is constituted by the Phygital (physical and digital) Twins for HRI proposed in [100, 101] with different meanings (showing the need for an agreement).

Overall, all terminological proposals must emerge from a dialectical co-design process involving all stakeholders-e.g., policy makers, trade unions, health and safety institutions-for allowing definitions to evolve in an operative and collectively refined manner (always rooted in the real world). However, we expect that words like "cobotics" will probably persist in the scholarly lexicon: in such a case, the term must be considered as an academic entry instead of one from international standards.

A final note on education should also highlight how recent reports [102] identify workforce training-not technology itself-as the primary barrier to robotics adoption. This underscores the need for educational interventions that prepare not only operators but also managers and all stakeholders for forthcoming technological transitions, especially to overcome biases on the effects of robotics on employment. Indeed, countries like Germany and South Korea, which lead global robotics adoption, maintain among the lowest unemployment rates, suggesting that automation complements rather than displaces labour when accompanied by appropriate skills policies [103]. Promoting through educational exercises the synergies between human and machine can thus be an effective strategy for fostering acceptance and facilitating the introduction of robotic devices designed to assist-not replace-workers. The barrier constituted by the perception of machines "stealing" jobs remains strong, despite the emergence of robots explicitly designed to be safe for humans and requiring human versatility to operate effectively [104].

All educational proposals in this section express the need for a deep knowledge of how HRIs can be positively established with a focus on real-world problems:

we cannot achieve this without designing (by involving all actors in socio-technical systems) our way to study such problems.

6 Conclusion

This chapter introduced and highlighted the importance of transcending the disciplinary boundaries in the study and application of ergonomics in robotics. Such an attitude should certainly be oriented towards problems. However, since every expert grows in a specific field, a first step to approach problems in human-centered robotics must certainly rely upon disciplinary integration (which inherently characterize both ergonomics and robotics). This is why we described examples of search for helpful methodologies from any disciplinary fields to face the real-world challenges of the mentioned problems. Accordingly, the main goal of this chapter is to stimulate the mindset of the reader towards the critical analysis of the disciplinary dynamics within ergonomics applied to robotics, without the presumption of strictly labeling a certain approach, for instance, as multidisciplinary or interdisciplinary-especially when the yearly innovations move the boundaries between diverse fields and perspectives. Both ergonomics and robotics (in parallel and in synergy) are rooted in different types of disciplinary integration to understand and innovate HRSs. Each methodology inherited from a disciplinary field should become an asset to address the complexities of systems involving humans and robots more effectively (and every educational path can be oriented to help in this task). This is a compulsory step to consider-by discussing with team mates with heterogeneous expertise, by involving diverse stakeholders in co-design processes, and by teaching students how to pay attention to problems while they learn the perspectives of diverse disciplines-the skills, limitations, needs, and goals of the user in a real-world context before designing or evaluating a robot.

References

1. Popper, K.: Conjectures and refutations: The growth of Scientific Knowledge. Routledge, London, United Kingdom (1963)
2. Re, A.: Interdisciplinary research and pratice in ergonomics. J. Bio. Res. Bollettino della Società Italiana di Biologia Sperimentale 85(1), (2012)
3. Michalec, O., O'Donovan, C., Sobhani, M.: What is robotics made of? The interdisciplinary politics of robotics research. Hum.Ities Soc. Sci. Commun. **8**(1) (2021)
4. Alvargonzález, D.: Multidisciplinarity, interdisciplinarity, transdisciplinarity, and the sciences. Int. Stud. Philos. Sci. **25**(4), 387–403 (2011)
5. Stember, M.: Advancing the social sciences through the interdisciplinary enterprise. Soc. Sci. J. **28**(1), 1–14 (1991)
6. Klein, J.T.: Boundary discourse of cross disciplinary and cross-sector research: refiguring the landscape of science. Minerva **61**(1), 31–52 (2023)
7. Newman, J.: Promoting interdisciplinary research collaboration: a systematic review, a critical literature review, and a pathway forward. Soc. Epistemol. **38**(2), 135–151 (2024)
8. Fiore, S.M.: Interdisciplinarity as teamwork: how the science of teams can inform team science. Small Group Res. **39**(3), 251–277 (2008)
9. Habib, M.K.: Interdisciplinary mechatronics engineering science and the evolution of human friendly and adaptive mechatronics. Interdiscip. Mechatron. pp **1**, 17 (2013)
10. Zaga, C.: Toward a future beyond disciplinary divides. In: Designing interactions with robots, pp. 170–181. Chapman and Hall/CRC, London, United Kingdom (2024)
11. Haselton, M.G., Bryant, G.A., Wilke, A., Frederick, D.A., Galperin, A., Frankenhuis, W.E., Moore, T.: Adaptive rationality: an evolutionary perspective on cognitive bias. Soc. Cogn. **27**(5), 733–763 (2009)
12. Caterino, M., Rinaldi, M., Di Pasquale, V., Greco, A., Miranda, S., Macchiaroli, R.: A human error analysis in human-robot interaction contexts: evidence from an empirical study. Machines **11**(7), 670 (2023)
13. O'Neill, J.L.: Accessibility for all abilities: how universal design, universal design for learning, and inclusive design combat inaccessibility and ableism. J. Open Access L. **9**, 1 (2021)
14. Bequette, B., Norton, A., Jones, E., Stirling, L.: Physical and cognitive load effects due to a powered lower-body exoskeleton. Hum. Factors **62**(3), 411–423 (2020)
15. Monari, E., Avallone, G., Valori, M., Agostini, L., Chen, Y., Palazzi, E., Vertechy, R.: Physical ergonomics monitoring in human-robot collaboration: a standard-based approach for hand-guiding applications. Machines **12**(4), 231 (2024)
16. Gualtieri, L., Fraboni, F., Brendel, H., Pietrantoni, L., Vidoni, R., Dallasega, P.: Updating design guidelines for cognitive ergonomics in human-centred collaborative robotics applications: an expert survey. Appl. Ergon. **117**, 104246 (2024)
17. Firescu, V., Gaşpar, M.-L., Crucianu, I., Rotariu, E.: Collaboration between humans and robots in organizations: a macroergonomic, emotional, and spiritual approach. Front. Psychol. **13**, 855768 (2022)
18. Lang, A., Hignett, S.: Ergonomics/human factors in healthcare: a vision for the future. In: Congress of the international ergonomics association, pp. 50–57. Springer (2018)
19. Kim, M., Oh, K., Choi, J., Jung, J., Kim, Y.: User-centered HRI: HRI research methodology for designers. Mix. Hum.-Robot. Interact.,pp 13–33 (2011)
20. Lindblom, J., Alenljung, B., Billing, E.: Evaluating the user experience of human-robot interaction. Hum.-Robot. Interact.: Eval. Methods Their Stand, 231–256 (2020)
21. Bertolaso, M., Capone, L., Rodríguez-Lluesma, C.: Digital humanism: a human-centric approach to digital technologies. Springer, Switzerland AG (2022)
22. Vemuri, N., Thaneeru, N.: Enhancing human-robot collaboration in industry 4.0 with AI-driven HRI. Power Syst. Technol **47**(4), 341–358 (2023)
23. Baber, C., Golightly, D., Waterson, P.: The cybernetic return in human factors and ergonomics, Elsevier (2019)

24. Guastello, S.J.: Human factors engineering and ergonomics: a systems approach. CRC Press, Boca Raton, Florida, United States (2023)
25. Proia, S., Carli, R., Cavone, G., Dotoli, M.: Control techniques for safe, ergonomic, and efficient human-robot collaboration in the digital industry: a survey. IEEE Trans. Autom. Sci. Eng. **19**(3), 1798–1819 (2021)
26. Czaja, S.J., Nair, S.N.: Human factors engineering and systems design. Handbook of human factors and ergonomics, pp. 38–56. (2012)
27. Hollnagel, E.: Human factors/ergonomics as a systems discipline? "the human use of human beingsâŁž revisited. Appl. Ergon. **45**(1), 40–44 (2014)
28. Gorichanaz, T.: Toward humanity-centered design without hubris. In: Extended abstracts of the CHI conference on human factors in computing systems, pp. 1–7. (2024)
29. Walker, G.H., Stanton, N.A., Salmon, P.M., Jenkins, D.P., Rafferty, L.: Translating concepts of complexity to the field of ergonomics. Ergonomics **53**(10), 1175–1186 (2010)
30. Coronado, E., Mastrogiovanni, F., Indurkhya, B., Venture, G.: Visual programming environments for end-user development of intelligent and social robots, a systematic review. J. Comput. Lang. **58**, 100970 (2020)
31. Morgan, S., Dibble, C., Susner, M., Wolcott, T., Wolcott, D., Largier, J.: Robotic biomimicry demonstrates behavioral control of planktonic dispersal in the sea. Mar. Ecol. Prog. Ser. **663**, 51–61 (2021)
32. Taşar, B., Tatar, A.B., Tanyıldızı, A.K., Yakut, O.: Fimec tremor stabilization spoon: design and active stabilization control of two DOF robotic eating devices for hand tremor patients. Med. Biol. Eng. Comput. **61**(10), 2757–2768 (2023)
33. Young, A.J., Ferris, D.P.: State of the art and future directions for lower limb robotic exoskeletons. IEEE Trans. Neural Syst. Rehabil. Eng. **25**(2), 171–182 (2016)
34. Lathouwers, E., Díaz, M.A., Maricot, A., Tassignon, B., Cherelle, C., Cherelle, P., Meeusen, R., De Pauw, K.: Therapeutic benefits of lower limb prostheses: a systematic review. J. Neuroeng. Rehabil. **20**(1), 4 (2023)
35. Zhang, G., Hansen, J.P.: Telepresence robots for people with special needs: a systematic review. Int. J. Hum.-Comput. Interact. **38**(17), 1651–1667 (2022)
36. Liu, L., Guo, F., Zou, Z., Duffy, V.G.: Application, development and future opportunities of collaborative robots (cobots) in manufacturing: A literature review. Int. J. Hum.-Comput. Interact. **40**(4), 915–932 (2024)
37. Figliano, G., Manzi, F., Tacci, A.L., Marchetti, A., Massaro, D.: Ageing society and the challenge for social robotics: a systematic review of socially assistive robotics for mci patients. PLoS ONE **18**(11), 0293324 (2023)
38. Cornelio, P., Haggard, P., Hornbaek, K., Georgiou, O., Bergström, J., Subramanian, S., Obrist, M.: The sense of agency in emerging technologies for human-computer integration: a review. Front. Neurosci. **16**, 949138 (2022)
39. Beckerle, P., Castellini, C., Lenggenhager, B.: Robotic interfaces for cognitive psychology and embodiment research: a research roadmap. Wiley Interdiscip. Rev.: Cogn. Sci **10**(2), 1486 (2019)
40. Zhang, C., Chen, J., Li, J., Peng, Y., Mao, Z.: Large language models for human-robot interaction: a review. Biomim. Intell. Robot., p 100131 (2023)
41. Boy, G.A.: Introduction: a human-centered design approach. The handbook of human-machine interaction: a human-centered design approach, pp. 1–20. (2011)
42. Laschi, C., Mazzolai, B.: Handbook on Soft Robotics, pp. 13–33. Springer, Cham (2024)
43. Jacobs, S.: Biomimetics: a simple foundation will lead to new insight about process. Int. J. Des. Nat. Ecodynamics **9**(2), 83–94 (2014)
44. Laschi, C., Mazzolai, B., Cianchetti, M.: Soft robotics: technologies and systems pushing the boundaries of robot abilities. Sci. Robot. **1**(1), 3690 (2016)
45. Hussein, A., Ghignone, L., Nguyen, T., Salimi, N., Nguyen, H., Wang, M., Abbass, H.A.: Characterization of indicators for adaptive human-swarm teaming. Front. Robot. AI **9**, 745958 (2022)

46. Brayshaw, G., Ward-Cherrier, B., Pearson, M.J.: A neuromorphic system for the real-time classification of natural textures. In: 2024 IEEE international conference on robotics and automation (ICRA), pp. 1070–1076. IEEE (2024)
47. Xiloyannis, M., Alicea, R., Georgarakis, A.-M., Haufe, F.L., Wolf, P., Masia, L., Riener, R.: Soft robotic suits: state of the art, core technologies, and open challenges. IEEE Trans. Rob. **38**(3), 1343–1362 (2021)
48. Ortiz, J., Di Natali, C., Caldwell, D.G.: Xosoft-iterative design of a modular soft lower limb exoskeleton. In: International symposium on wearable robotics, pp. 351–355. Springer (2018)
49. Jørgensen, J., Bojesen, K.B., Jochum, E.: Is a soft robot more "naturalâŁž? exploring the perception of soft robotics in human-robot interaction. Int. J. Soc. Robot. **14**(1), 95–113 (2022)
50. Arnold, T., Scheutz, M.: The tactile ethics of soft robotics: designing wisely for human-robot interaction. Soft Rob. **4**(2), 81–87 (2017)
51. Friedman, B., Kahn, P., Borning, A.: Value sensitive design: theory and methods. Univ. Wash. Tech. Rep. **2**(8), 1–8 (2002)
52. Longo, F., Padovano, A., Umbrello, S.: Value-oriented and ethical technology engineering in industry 5.0: a human-centric perspective for the design of the factory of the future. Appl. Sci **10**(12), 4182 (2020)
53. Friedman, B., Hendry, D.G.: Value sensitive design: shaping technology with moral imagination. Mit Press, Cambridge, MA (2019)
54. Petrushin, A., Freddolini, M., Barresi, G., Bustreo, M., Laffranchi, M., Del Bue, A., De Michieli, L.: Iot-powered monitoring systems for geriatric healthcare: overview. Internet Things Hum.-Cent.Ed Des.: Appl. Elder. Healthc, 99–122 (2022)
55. Grassi, L., Canepa, D., Bellitto, A., Casadio, M., Massone, A., Recchiuto, C.T., Sgorbissa, A.: Diversity-aware verbal interaction between a robot and people with spinal cord injury. In: 2023 32nd IEEE international conference on robot and human interactive communication (RO-MAN), pp. 2512–2519. IEEE (2023)
56. Borning, A., Muller, M.: Next steps for value sensitive design. In: Proceedings of the SIGCHI conference on human factors in computing systems, pp. 1125–1134. (2012)
57. Frank, D.-A., Chrysochou, P., Mitkidis, P., Ariely, D.: Human decision-making biases in the moral dilemmas of autonomous vehicles. Sci. Rep. **9**(1), 13080 (2019)
58. Madhav, A.S., Tyagi, A.K.: Explainable artificial intelligence (XAI): connecting artificial decision-making and human trust in autonomous vehicles. In: Proceedings of third international conference on computing, communications, and cyber-security: IC4S 2021, pp. 123–136. Springer (2022)
59. Thaler, R., Sunstein, C.: NUDGE: improving decisions about health, wealth, and Happiness, vol. 47, (2009)
60. Calboli, S.: Robot nudgers. what about transparency? In: International conference on software engineering and formal methods, pp. 291–301. Springer (2022)
61. Chong, I., Proctor, R.W.: On the evolution of a radical concept: affordances according to gibson and their subsequent use and development. Perspect. Psychol. Sci. **15**(1), 117–132 (2020)
62. Liao, Y..-C., Todi, K., Acharya, A., Keurulainen, A., Howes, A., Oulasvirta, A.: Rediscovering affordance: a reinforcement learning perspective. In: Proceedings of the 2022 CHI conference on human factors in computing systems, pp. 1–15. (2022)
63. Kolko, J.: Thoughts on interaction design. Morgan Kaufmann, Burlington, MA (2010)
64. Aprile, W., Avanzini, F., Barbero, J., Bardy, B., Boulic, R., Bullot, N., Bülthoff, H., Cadoz, C., Camurri, A., Carozzino, M.: Enaction and enactive interfaces: a handbook of terms, (2017)
65. Barresi, G., Caldwell, D.G., Mattos, L.S.: Natural surgeon interfaces: perspectives and examples of intuitive laser control systems in the μralp project. In: Congress of the Italian society of ergonomics and human factors, pp 111–123. Springer (2022)
66. Markovich, T., Honig, S., Oron-Gilad, T.: Closing the feedback loop: the relationship between input and output modalities in human-robot interactions. In: Human-friendly robotics 2019: 12th international workshop, pp. 29–42. Springer (2020)

67. Gallagher, S.: The 4es and the 4as (affect, agency, affordance, autonomy) in the meshed architecture of social cognition. In: Handbook of embodied psychology: thinking, feeling, and acting, pp. 357–379. (2021)
68. Bettoni, M.C., Castellini, C.: Interaction in assistive robotics: a radical constructivist design framework. Front. Neurorobot. **15**, 675657 (2021)
69. Nowak, M., Castellini, C., Massironi, C.: Applying radical constructivism to machine learning: a pilot study in assistive robotics. Constr. Found. **13**(2), 250–262 (2018)
70. Khalid, H.M.: Embracing diversity in user needs for affective design. Appl. Ergon. **37**(4), 409–418 (2006)
71. Nagamachi, M.: Kansei/affective Engineering. CRC Press, Boca Raton (2016)
72. Coronado, E., Venture, G., Yamanobe, N.: Applying kansei/affective engineering methodologies in the design of social and service robots: a systematic review. Int. J. Soc. Robot. **13**(5), 1161–1171 (2021)
73. Corti, L., Stefano, N.D., Bertolaso, M.: Artificial emotions: toward a human-centric ethics. Int. J. Soc. Robot. **15**(12), 2039–2053 (2023)
74. Liu, C., Rani, P., Sarkar, N.: Affective state recognition and adaptation in human-robot interaction: a design approach. In: IEEE/RSJ international conference on intelligent robots and systems, pp. 3099–3106. IEEE (2006)
75. Gervasi, R., Barravecchia, F., Mastrogiacomo, L., Franceschini, F.: Applications of affective computing in human-robot interaction: state-of-art and challenges for manufacturing. Proc. Inst. Mech. Eng., Part B: J. Eng. Manuf. **237**(6–7), 815–832 (2023)
76. Saraiva, M., Ayanoğlu, H., Özcan, B.: Emotional design and human-robot interaction. In: Emotional design in human-robot interaction: theory, methods and applications, pp. 119–141. (2019)
77. Ayanoğlu, H., Duarte, E.: Emotional design in human-robot interaction. Springer, Cham (2019)
78. Zhu, H., Yu, C., Cangelosi, A.: Affective human-robot interaction with multimodal explanations. In: International conference on social robotics, pp. 241–252. Springer (2022)
79. Parasuraman, R.: Neuroergonomics: research and practice. Theor. Issues Ergon. Sci. **4**(1–2), 5–20 (2003)
80. Ayaz, H., Dehais, F.: Neuroergonomics. Handbook of human factors and ergonomics, pp. 816–841. (2021)
81. Barresi, G.: Neuroergonomics for human-centered technological contexts. In: Digital environments and human relations, pp. 61–85. Springer, Cham (2024)
82. Falk, T.H., Guger, C., Volosyak, I.: Brain-computer interfaces: recent advances, challenges, and future directions. Handbook of human-machine systems, pp. 11–22. (2023)
83. Zhu, Y., Johnson, C., Chang, S..-H.J., Mehta, R.K.: Neuroergonomics metrics to evaluate exoskeleton based gait rehabilitation. In: 2020 IEEE international conference on systems, man, and cybernetics (SMC), pp. 3447–3453. IEEE (2020)
84. Barresi, G., Olivieri, E., Caldwell, D.G., Mattos, L.S.: Brain-controlled AR feedback design for user's training in surgical HRI. In: 2015 IEEE international conference on systems, man, and cybernetics, pp. 1116–1121. IEEE (2015)
85. Iarlori, S., Monteriú, A., Perpetuini, D., Filippini, C., Cardone, D.: Affective computing and mental workload assessment to enhance human-machine interaction. Frontiers Media SA, (2024)
86. Madduri, M.M., Burden, S.A., Orsborn, A.L.: Biosignal-based co-adaptive user-machine interfaces for motor control. Curr. Opin. Biomed. Eng. **27**, 100462 (2023)
87. Mehta, R.K., Parasuraman, R.: Neuroergonomics: a review of applications to physical and cognitive work. Front. Hum. Neurosci. **7**, 889 (2013)
88. Ciminaghi, F., Angioletti, L., Rovelli, K., Balconi, M.: Collaborative intelligence and hyperscanning: exploring ai application to human-robot collaboration through a neuroscientific approach. In: International conference on extended reality, pp. 334–341. Springer (2024)
89. Del Vecchio, A., Castellini, C., Beckerle, P.: Peripheral neuroergonomics-an elegant way to improve human-robot interaction? Front. Neurorobot. **15**, 691508 (2021)

90. Yanai, I., Lercher, M.: Renaissance minds in 21st century science, Springer (2020)
91. Salmon, P.M., Read, G.J., Walker, G.H., Stevens, N.J., Hulme, A., McLean, S., Stanton, N.A.: Methodological issues in systems human factors and ergonomics: perspectives on the research-practice gap, reliability and validity, and prediction. Hum. Factors Ergon. Manuf. Serv. Ind. **32**(1), 6–19 (2022)
92. Safin, S., Pintus, P., Elsen, C.: Ergonomics in design and design in ergonomics: issues and experience in education. Work **66**(4), 917–931 (2020)
93. Sun, X., Houssin, R., Renaud, J., Gardoni, M.: A review of methodologies for integrating human factors and ergonomics in engineering design. Int. J. Prod. Res. **57**(15–16), 4961–4976 (2019)
94. Naeini, H.S., Mosaddad, S.H.: The role of ergonomics issues in engineering education. Procedia Soc. Behav. Sci. **102**, 587–590 (2013)
95. Demirel, H.O., Ahmed, S., Duffy, V.G.: Digital human modeling: a review and reappraisal of origins, present, and expected future methods for representing humans computationally. International Journal of Human-Computer Interaction **38**(10), 897–937 (2022)
96. International organization for standardization: Robotics—Safety requirements — Part 2: Industrial robot applications and robot cells. https://www.iso.org/standard/73934.html
97. Association for advancing automation: updated ISO 10218: Answers to frequently asked questions (FAQs). https://www.automate.org/robotics/blogs/updated-iso-10218-faq Accessed 2025-12-15
98. Floridi, L.: Ai as agency without intelligence: on chatgpt, large language models, and other generative models. Philos. Technol. **36**(1), 15 (2023)
99. Jarrahi, M.H., Lutz, C., Newlands, G.: Artificial intelligence, human intelligence and hybrid intelligence based on mutual augmentation. Big Data Soc. **9**(2), 20539517221142824 (2022)
100. Barresi, G., Pacchierotti, C., Laffranchi, M., De Michieli, L.: Beyond digital twins: Phygital twins for neuroergonomics in human-robot interaction. Front. Neurorobot. **16**, 913605 (2022)
101. Zhou, H., Lv, H., Wang, R., Wu, H., Yang, G.: Revitalizing human-robot interaction: phygital twin driven robot avatar for China-Sweden teleoperation. Chin. J. Mech. Eng. **36**(1), 124 (2023)
102. Make UK, Sage: Making it smarter: Global lessons for accelerating automation and digital adoption in UK manufacturing. Technical report, Make UK (August 2025). https://www.makeuk.org/insights/reports/making-it-smarter-global-lessons-accelerating-automation-digital-adoption-uk
103. International Federation of Robotics: The impact of robots on productivity, employment and jobs. Positioning paper, International Federation of Robotics (2017). https://ifr.org/img/office/IFR_The_Impact_of_Robots_on_Employment.pdf
104. Aaltonen, I., Salmi, T., Marstio, I.: Refining levels of collaboration to support the design and evaluation of human-robot interaction in the manufacturing industry. Procedia CIRP **81**, 1393–1398 (2019). https://doi.org/10.1016/j.procir.2019.04.039

Application Areas

Human–Robot Interaction in Industrial Contexts

Patricia Helen Rosen, Rebecca Erlebach, Eva Heinold, Marta Lagomarsino, Sascha Wischniewski, Andrea Maria Zanchettin, Arash Ajoudani, and Marta Lorenzini

Abstract Technological advances in industrial robotics have significantly changed modern working environments, enabling greater autonomy, flexibility and productivity. Here the emergence of advanced robots designed for direct interaction with humans play a substantial role. By integrating emerging technologies such as AI, machine learning, and sophisticated interaction design, advanced robots are reshaping industries. All while increasing efficiency and customization and remaining cost-effective. Despite their benefits, the use of robotic systems also brings challenges, particularly in the area of occupational safety and health. After an overview of the evolution of different robotic systems, a description of industrial work tasks, related working conditions and highlighted risk and opportunities associated with automation, this chapter examines key safety factors that are essential for close human–robot interaction and assesses their impact on workers. Two case studies illustrate how the practical application of advanced robotic systems in industrial settings affects workplace dynamics. The chapter further highlights the role of worker expectations and experience in relation to advanced robotics. By combining empirical evidence with real-world applications, this chapter underscores the importance of adaptable, user-friendly robotic systems that prioritize worker well-being alongside operational efficiency. In doing so, it contributes to the broader dialogue on sustainable, ergonomic innovation in the workplace and highlights the critical role of human-centered design in the future of industrial robotics.

Keywords Industrial robotics · Advanced robotics · Cobots · Automation · Industrial tasks

P. H. Rosen (✉) · R. Erlebach · E. Heinold · S. Wischniewski
Human Factors, Ergonomics Group, Federal Institute for Occupational Safety and Health (BAuA), Dortmund, Germany
e-mail: rosen.patricia@baua.bund.de

M. Lagomarsino · A. Ajoudani · M. Lorenzini
Human–Robot Interfaces and Interaction (HRII), Istituto Italiano di Tecnologia, Genoa, Italy

A. M. Zanchettin
Politecnico di Milano, Dipartimento di Elettronica, Informazione e Bioingegneria (DEIB), Milan, Italy

G. Barresi and S. Wischniewski (eds.), *Ergonomics in Robotics: Advances and Innovations*, Springer Tracts in Advanced Robotics 162,
https://doi.org/10.1007/978-3-032-19374-2_6

1 Introduction

Technological developments are constantly changing the way we live and work. The introduction of innovative work equipment, such as advanced robotics, has the potential to fundamentally change the design and execution of human work.

Industrial robots have been used as automation solutions in industry for many years. One of the first industrial robots is considered to be a development by George Devol in the mid-1950s. While subsequent iterations of these systems have certainly helped to increase their technological sophistication, fundamental system characteristics, such as complete separation from any human operator, have remained the same.

In 2006, a significant technological advancement was presented for the first time—interactive lightweight robots, also known as collaborative robots or cobots. Compared with the so far existing robots, these new systems had a significantly different outward shape and structure and were designed specifically for industrial assembly applications in closer distance to humans [1]. In the following years, innovations in the field of sensor and actuator technology have constantly produced new generations of lightweight robots. Manufacturing companies hoped that the use of lightweight robots will enable them to respond to greater demands for flexibility, such as shorter product cycles and small volumes, and thus remain competitive [1]. The use of robotic systems is also often motivated by the desire to respond more efficiently and flexibly to greater product customization, without significantly increasing production costs [2]. In principle, these new forms of robots can operate without a safety fence, opening up innovative (hybrid) forms of semi-automated interactions between humans and robots. By integrating advanced programming, machine learning, and artificial intelligence (AI), along with sophisticated interaction design, these robots enable far more complex interactions than were ever possible with traditional fenced and less advanced systems. These advances have significant implications for human factors research and workplace design, as they give rise to new interaction paradigms and design requirements. This chapter will present an overview of the history of industrial robotics and a description of industrial tasks and their accompanying working conditions, illustrating the setting under which these technologies are deployed today and which opportunities and challenges arise from a human factor's perspective with a special focus on occupational safety and health (OSH). Furthermore, this chapter will present basics on close interaction and safety modes as well as relevant findings from human factors perspectives on industrial robotics.

2 Evolution of Robots in Industrial Settings

"Industrial robotics is a discipline concerning robot design, control and applications in industry" [3]. In the year 1954 the patent application was filed for a technological development by George Devol resembling the first industrial robot. A few years later, in 1961, the first robot was produced within a company set up by Devol and Joseph Engelberger and was immediately installed in an American automotive company where it extracted parts from a die-casting machine. More robots were installed in the US in the following years, mainly for spot-welding of cars. In the late 1960's the first industrial robots were deployed in Japan and the first Asian company was licensed to produce their own version of the robot developed by Devol. Meanwhile, the first robotic installation was also seen in Sweden. An industrial robot carrying out a simple pick-and-place task [4]. A few years later, a cheaper electro-hydraulic robot was developed in Norway, mainly for spray painting, which was originally developed for the company's internal use but was then further developed into a commercial product.

The first industrial robots were pneumatic or hydraulic and intended for larger loads. Later developments in the 1970s brought forward electric robots and also one of the first programming languages for robots were written at that time [3]. Although microprocessors and other components were being used in that time more frequently, this second generation of industrial robots were still very application-specific. It was very difficult to use the same robot for a different task, unless substantial modifications of the controller and software reprogramming were performed. In next iterations of robotic systems, user interfaces and motion planning became more advanced, robots could be paired with computer aided design (CAD) software or databases and use data from vision or perception systems leading to robotic capabilities for more complex tasks, such as tactile inspections or assembly operations. Further, progress was made in kinematic structures increasing robotic speed and accuracy. Hardware components started to be provided and industrial robots were equipped with cameras, force sensors or laser scanners and control software became more complex [4].

Despite the described evolution of robotic systems, efforts in the field of industrial robotics still focused on large, heavy machines primarily designed for applications, such as assembly lines in the automotive industry. These robots were powerful, with a focus on increasing productivity and reducing costs, but lacked the agility and versatility needed for more delicate tasks. With a push for flexibility of robotic systems and adaptability to a larger variety of tasks, a transition to new types of robots began [5].

In the year 2006, a German company presented the first "light weight robot", which was developed jointly in a research activity with the German aerospace center (DLR). Soon afterwards the first systems came onto the market, selling systems with a significant upgrade in sensor and actuator technology, less payload and a simplified programming environment. The main target group for this technology have been small- and medium sized companies, in theory enabling humans to work in closer proximity to the robot than before [6].

Today, industrial robots are defined based on the definition of the International Organization for Standardization as an "automatically controlled, reprogrammable multipurpose manipulator, programmable in three or more axes, which can be either fixed in place or fixed to a mobile platform for use in automation applications in an industrial environment" (ISO 8373:2021). Important determinants of an industrial robot in this definition are the specific terms used. Reprogrammable refers to the system's ability to have motions or auxiliary functions changed without physical alterations, multipurpose refers to its adaptability to different applications with physical alteration, manipulator meaning an arrangement of segments, jointed or sliding relative to one another, and axis referring to the direction used to specify motion in a linear or rotary mode [6].

Industrial robots may also be classified along their mechanical structure. According to the International Federation of Robotics (IFR) classification, there are six different mechanical structures:

- **Cartesian robots**, which are manipulators with three prismatic joints and whose axes form a Cartesian coordinate system,
- **SCARA** (Selective Compliance Assembly Robot) **robots**, which are systems that have two parallel rotary joints providing compliance in selected planes,
- **Articulated robots** with three or more rotary joints,
- **Parallel** or **delta robots** whose arms form a closed loop structure using links,
- **Cylindrical robots**, which must have at least one rotary joint and one prismatic joint to form a cylindrical coordinate system using its axes,
- and **polar robots** or **spherical robots**, which have two rotary joints and one prismatic joint to form a polar coordinate system.

Typically, an industrial robot has three basic components: manipulator, controller and tooling.

- **Manipulator**: The manipulator consists of the robot's base and arm, powered by electrical, hydraulic, or pneumatic systems, enabling motion across multiple degrees of freedom.
- **Controller**: The controller defines the robot's trajectory, converts it into joint movements, and includes reprogrammable user interfaces for tasks using command devices and sensors.
- **Tooling**: Tooling, or end-effectors, are devices like vacuum cups or welding guns that enable the robot to perform specific jobs.

The complexity of these components varies between different robots and applications [1].

To further enable the possibility of HRI, a new form of industrial robots has emerged in the last years: Collaborative robots. The term "collaborative robot" (cobot, in short) has been introduced by Prof. J. E. Colgate and coworkers in 1996 [7]. Working in strict collaboration with General Motors, Prof. Colgate and his research group at the Northwestern University were looking for a device to improve Human Factors and Ergonomics (HFE) in the workplace, without introducing additional safety risks.

These interactive and advanced robots are able to engage in complex joint tasks with humans in shared workspaces, leveraging advanced sensors and online communication modalities. This allows for safe, fluid interaction with dynamic task allocation [8]. If the robot application and the operator are within the same safeguarded space, one can referr to a collaborative task; if one or more collaborative tasks are given, the term collaborative application is applicable, according to the recent uptade of terminology in the ISO standard [9].

Although it is difficult to reliably present the automation potential of different occupations, sources estimate that, according to a relevant report, more than 40% of occupations are at least half automatable [10]. It follows that, at least in the manufacturing industry, there might be a lot of room for growth in interactive installations. As a matter of fact, in the last 5 years, the market of these advanced robots has been constantly growing. According to the IFR, robot installations have hit a new record level in 2022, surpassing the record figure in 2021. Industrial robotics is a thriving and growing market, which stretches into many different industries. In 2022, the largest customer of robots has been the electronics industry with 28% of all units installed, followed by the automotive industry with 25% of all installations and the metal and machinery industry with 12%. While industrial robots are already a booming technological trend, the IFR expects even more growth in the next three years. Following a steady incline in demand, annual installations of industrial robots are expected to increase by 7% each year in a three year forecast [11]. Given this trend, it becomes highly important to understand the impact robotic systems have on todays and future working conditions.

3 Industrial Tasks and Working Conditions

With the evolution of industrial robotics and the emergence of advanced robotic systems, it is crucial to consider how these technologies are integrated into the workplace. The structure of industrial tasks and the conditions under which they are performed play a key role in determining the effectiveness and impact of robotic systems. By examining the categorization of tasks and their cognitive and/or physical demands, it is possible to understand how automation affects productivity, safety, and worker well-being. This perspective highlights opportunities to improve efficiency while addressing potential risks, ensuring that the use of robots supports sustainable and human-centered workplaces.

3.1 Overview of Industrial Tasks

Industrial tasks can be found in different economic activities dealing with the commercial extraction, processing and further processing of raw materials or intermediate products into material goods. Examples of industrial tasks like welding, transportation or assembly can be found for example in logistics or manufacturing. Today, many industrial tasks are being automated using advanced robotic systems. According to the IFR, the majority of robotic applications are found in the manufacturing sector [11]. This is supported by findings of the third European Survey of Enterprises on New and Emerging Risks (ESENER-3, 2019). The survey further shows that direct interaction between humans and robots strongly differs between sectors. 28% of the enterprises where direct human–robot interaction occurred belong to the manufacturing sector [12] (see Fig. 1). The second largest group can be found in the automotive domain [12], where industrial tasks like assembly dominate robotic automation [13].

However, the actual prevalence of robots that directly interact with humans across European enterprises is currently still quite low in all industries. According to the ESENER-3 survey, on average, 3.5% of the interviewed enterprises reported using robots with direct interaction capabilities, with the highest prevalence in Slovakia (8.7%), Denmark (6.9%) and the Czech Republic (6.7%). Conversely, the lowest prevalence was reported in Greece (1.6%), Cyprus (1.0%) and Serbia (0.9%) [12].

In general terms, work tasks can be categorized according to their primary focus and nature. The task taxonomy for AI-based systems and advanced robotics for the automation of tasks (see Fig. 2) gives an overview on types and characteristics of work tasks. Parts of the taxonomy were developed by the German Federal Institute for Occupational Safety and Health's program 'Occupational Safety and Health in the

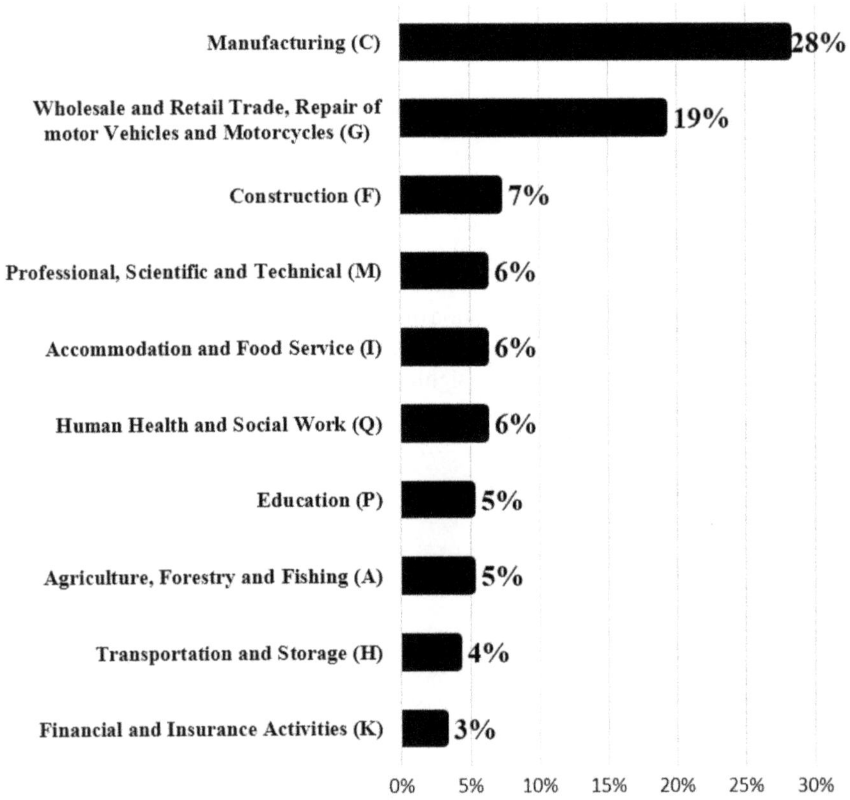

Fig. 1 NACE (v. 2) code of enterprises with HRI according to ESENER-3, published in [12]

Digital World of Work'. The main approach of the taxonomy is to distinguish between object-related tasks, information-related tasks, and person-related tasks and whether the task performed is a physical or cognitive action, regardless of the industry in which they occur. Further, tasks can be differentiated by their routine nature: traditionally automation technologies have been used to take over routine task. As more complex technologies emerge, also more and more non-routine tasks can be automated.

Applying the taxonomy to industrial tasks shows that those tasks are mainly physical, object-related tasks [15]. Historically, industrial robots have been caged and fixed, designed for the automation of object-related, physical tasks that require speed, precision or the lifting of heavy loads. However, the current trend is shifting towards interactive robots and lightweight robots that enable closer and more complex HRI [8]. The combination of more elaborate types of hardware and software, as well as advances in sensor technology allows for these changes, broadening the scope of tasks that can be performed by robotic systems. Modern information and communication technologies (ICT) are mainly used for the automation of cognitive tasks that do not require physical handling of objects or people. These include desktop computers, mobile devices, wearables, and smartwatches. To some extent, ICT enable certain robotic systems, when integrated into those, also to perform parts of cognitive

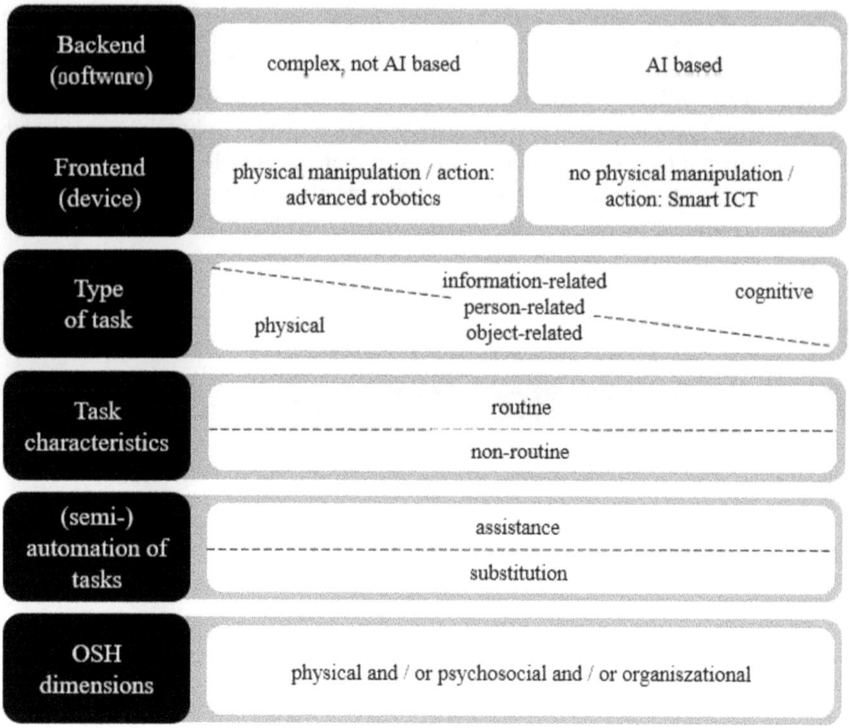

Fig. 2 Taxonomy for AI-based systems and advanced robotics for the automation of tasks, published in [14]

tasks. However they are less prevalent in industrial settings [16]. Within industrial settings, robotic tasks tend to be in one of three major categories: handling, welding or assembly [13]:

Handling comprises a wide range of processes, including grasping, transporting, packaging, palletizing, and polishing. Applications such as product testing and pick-and-place operations are examples of tasks that involve manipulating a part for subsequent steps in the manufacturing process. Robotic material handling offers significant advantages by reducing the physical effort required from workers in lifting and moving materials. Additionally, it is advantageous in situations where materials cannot be handled by humans due to hygiene requirements or safety hazards.

Welding, while a prominent area for human–robot interaction, it faces practical limitations in production environments due to the inherent complexity and uncertainty of the welding process. When used however it can support workers in a physically straining task.

Assembly is another key area for robotic applications, with interactive robots most commonly used in hybrid assembly cells within manufacturing lines. These setups combine human and robotic efforts to optimize efficiency and precision. Assembly tasks often require repeated motion while handling workpieces. Robotic systems are highly efficient in automating repetition and offer physical support to the worker [13].

Apart from the above mentioned frequently occurring tasks supported by robotic systems, Fig. 3 provides an overview on additional industrial tasks which are already today automated by robotic systems. However, there is a much wider range of tasks they are used for across the industry. Pedersen and colleagues provide a more detailed overview on industrial task categories, automated by robotic systems [17]:

This more nuanced overview allows for more accurate categorization of robotic tasks in the industrial sector, however, it too, is far from exhaustive, as the technology continues to evolve.

Logistic tasks	Assistive tasks	Service tasks
Transportation	Machine tending	Maintenance, repair, overhaul
Part feeding (units)	Assembly	Cleaning
Part Feeding (single)	Inspection	
	Process	

Fig. 3 General task categories in three domains, suitable for robotic systems

3.2 Risks and Challenges of Automation

The introduction of robotic systems, especially those with direct interaction capabilities, present several risks and challenges for workers and working conditions. Cognitive risks associated with automation can be broadly divided into two areas: increased monotony from monitoring tasks and potential overload from complex user control interfaces. For example, workers in the electronics industry often report unfavorable levels of cognitive strain when working with robotic systems, which require constant attention and quick reaction times. It has been shown that workers who spend more than 50% of their time monitoring robotic processes are at higher risk of experiencing burnout and stress symptoms [15]. Additionally, the introduction of such systems often necessitates significant training efforts, as employees must adjust both technically and psychologically to the new demands.

According to the ESENER-3 survey, there were several most commonly cited challenges, such as prolonged sitting, the need for training, and repetitive movements associated with the deployment of robotic systems. There were also concerns about blurring the boundaries between work and personal life and increased work intensity, which were overall lowest in prevalence among the discussions, but high in absolute numbers nonetheless. However, the possible risks associated with robotic systems in industrial setting are diverse, and highly dependent on the type of robotic system as well as the automated task. In the taxonomy for the automation of tasks (see Fig. 2), the risks and challenges are divided into three major categories: physical, psychosocial and organizational. For each category major risks can be identified based on the review of according human factors studies [14, 15]. The main risks of each category are presented in Fig. 4. Especially psychosocial risks are addressed and can be supported by the results of the ESENER-3 survey [12, 14, 15]. Organizations using HRI report a higher rate of fear of job loss than those not using robots (31% vs. 23%). In addition, the need for training is significantly higher in companies with HRI (87%). The introduction of robots which directly interact with humans increases the likelihood that the need for training will be discussed as a potential challenge by 1.45 times compared to non-HRI work environments [12].

Regarding all possible risks and challenges associated with automation of tasks through industrial robotic system one factor must not be overlooked. Complex interdependencies in hybrid systems may occur [18]. Combining human labor with automated processes requires careful balancing to maintain efficiency and worker safety and satisfaction. Risks and challenges arising from interdependencies of factors such as human centered design, task allocation, training etc. must be considered, too. As an example: continuous training might be needed to operate the system safely, but this could contribute to an increased mental workload. Hence, next to direct effects on safety and health, interdependencies between impact factors must be considered too [18].

Physical risks	Psychosocial risks	Organizational risks
Mechanical failures and errors	Automation bias	Loss of job control
Hazardous environments	Mental workload and stress	Work intensification
Ergonomic issues	Fear of job loss	Reduction in social support
	Skill degredation	Poor task and responsibility allocation
	Interaction challenges	

Fig. 4 Physical, psychosocial and organizational risk categories

3.3 Working Conditions in Robotic Workplaces

Working conditions for industrial tasks, especially those involving robotic systems, are characterised by several factors, including job characteristics and health indicators. In general, there are several psychological theories and models that can be applied to understand and investigate working conditions and their influence on worker well-being. Two such models are the Job-Demand-Control (JDC) model and the Job-Demands-Resources (JD-R) model, which have gained widespread popularity [19–21]. According to the JDC model, job strain is a result of high job demands, such as work intensity or overload, coupled with low job control, while the JD-R model categorizes job characteristics into job demands and resources. This suggests that the most strenuous jobs are those with high demands and few resources to compensate. Whether new and emerging technologies such as robotic systems are designed as demands or resources for employees and workers is a critical question discussed in literature.

The impact of technology depends not only on the technology itself, but also on the skills of the workforce and the characteristics of the tasks. Automation solutions primarily address routine tasks, potentially leading to job polarization [22]. Job polarization is a term used to describe the phenomenon that has first been seen in the US labor market. It refers to the increasing concentration of employment in higher and lower wage jobs, while middle-skill jobs dwindle in numbers [23]. This can also be seen in workplaces using robotic systems. Here, high-skill jobs tend to focus on routine cognitive tasks, while low-skilled jobs focus on simpler support activities, which are often not cost-effective for robotic automation [22]. In general, there is also a tendency for higher skill levels to be found in higher skill jobs and vice versa. This however, cannot be directly applied to workplaces using robotic systems. Here, workers in robotic jobs tend to have lower formal qualifications, even in high-skilled jobs.

Those employed in low-skilled roles who utilize robots report the highest degree of technical difficulty and consistently higher ratings for the factors of "constant repetition of work steps" and "work steps determined by machines". While aspects of monotony are generally more prevalent in low-skill jobs, it is an exception that robot users with high-skill jobs report the highest levels of machine-determined work [23]. The time pressure reported by robot users is significantly higher than by other machine users, yet remains at the average level across all industries. The level of decision latitude or job control exhibited by robot users is consistently lower than the overall average. This is particularly evident with regard to aspects such as work organization, work speed, and workload. When looking at health-related outcomes, robot users report a higher prevalence of burnout symptoms, diminished autonomy in controlling their workload, and greater autonomy in determining their task approach. Additionally, they demonstrate divergent patterns of support from colleagues and supervisors when compared to non-robot users. Persons who work with robots in high skill level jobs experience less physical strain but more periods of prolonged sitting at work than other machine operators. Furthermore, those persons report fewer headaches and backaches, but a higher incidence of heart pain and respiratory problems [22].

This shows that the current state of research on robotics in the workplace only partially support the polarization hypothesis of digitalization. Modern production technologies are associated with reduced decision latitude across all skill levels, suggesting that even highly skilled jobs are not immune to the effects of reduced job control [22]. As stated by the JDC or JDR model, the persisting imbalance between job stressors, such as low job control or high levels of workload, and job resources may lead to negative effects of wellbeing and health. A number of studies provide insights into the long-term association between the exposure to these psychosocial working conditions and health outcomes such as cardiovascular diseases, depression and other mental disorders. In particular, there is evidence that the unfavorable combinations of demands and resources as described in the JDC and JDR model can lead to cardiovascular diseases (coronary heart disease and stroke) and depression in the long term [24]. This highlights the necessary for prospective and human-centered work design to address the risks associated with reduced job control and increased time pressure in robotic workplace, especially in low-skilled jobs. The leading idea should be to turn robotic systems, like other automation technologies, into a job resource rather than a stressor [25], to avoid negative long-term effects on health outcomes. If this can be achieved, modern industrial robotic system might have the potential to enhance both productivity and human-centred production work, thus stimulating healthy workplaces.

4 Close Interaction and Safety Modes

When it comes to integrating robots into a shared workspace, safety of the worker is the first and essential priority. It needs to be ensured already during the design of the workspace. As of now safety standards and guidelines are available for both the robot manufacturer and the system integrator.

From the manufacturers' perspective the ISO 10218-1 (Robots and robotic devices—Safety requirements for industrial robots, Part 1: Robots) suggests a list of features an industrial robot together with its controller should provide to the system integrator to handle safety [9]. On the other hand, ISO 10218-2 (Robots and robotic devices—Safety requirements for industrial robots, Part 2: Robotic systems and integration) contains guidelines to design and integrate the safety functionalities for a robotic installation. At the time of writing, the current ISO 10218-2:2011 lacks sufficient details in providing the guidelines for safety in HRI and the specific aspects are demanded to the technical specification ISO/TS 15066 (Robots and robotic devices—Collaborative robots) [26].

The document ISO/TS 15066 indeed contains systematic guidelines for the safety of the operator working alongside an industrial robot and defines four possible collaborative operations as well as possible ways to switch from one to another. Among the four possible methods, speed and separation monitoring (5.5.4) as well as power and force limiting (5.5.5) are probably the most relevant [26]. The former essentially limits the velocity of the robot as a way to calibrate the distance to the human operator: the closer the human, the slower the robot. In practice, the robot has to stop its motion before the distance to the operator falls under a specified threshold. To achieve this, safety devices such as laser scanners or light barriers are installed and a simple logic is triggered to command the robot to stop or reduce its speed. In most of the practical applications speed and separation monitoring is commended and for the robot to halt completely when the worker enters the working area around the robot, whose radius is set proportional to its stopping time.

Power and force limitations are crucial for safe HRI. It accounts for biomechanical models of the human body as well as two possible contact situations between the robot and the worker: transient and quasi-static contacts. The former is intended to handle situations in which the person's body part can retract from the robot without being clamped or trapped. On the other hand, quasi-static contacts concern the most conservative situations in which the person's body part is trapped between a moving part of the robot and another part of the work cell. In the first situation the kinetic energy transferred from the robot to the human's body part has to be guaranteed within a given threshold (specific for each body part itself), in the latter it is the pressure applied to the body part by the robot to be maintained within given bounds. In both cases, based on a simplistic two-mass model of the contact, all quantities can be related to the velocity of the robot which is then controlled to stay within certain boundaries. As a consequence, industrial robots capable of satisfying power and force limiting are necessarily light-weight (a reduced mass implies a reduced kinetic energy for a given speed) and with smooth or padded surfaces (the absence of sharp edges prevents high pressure values) [26].

From an application perspective, the first interaction modality (speed and separation monitoring) always existed in industry. Modern standards are simply allowing the halting motion of the robot to be executed on a software basis (Safe Operating Stop, SOS), rather than using abrupt interruptions of the supplied power (Safe Torque Off, STO), see IEC 60204-1 [27]. In most applications, speed and separation monitoring severely limits the possibility of tight collaboration between the operator and the robot and typically translates in an increased footprint of the robotic application. As a matter of fact, this interaction modality is typically implemented in a very conservative way: the separation distance is usually overestimated by the distance of the worker to the robot maximum reach without taking care of the actual position of the robot. In addition, the corresponding robot velocity are only selected among few possible values (e.g. full-speed, limited speed or standstill).

At the same time, the limitation of power and force is bounding the maximum speed of the robot (Safety Limited Speed, SLS) to a value that would not harm the human in case of quasi-static or transient contacts. The actual maximum speeds for the robot to be accounted for, either for quasi-static or transient contract, depend on the actual application and are selected through the risk assessment procedure. Though the technical specification allows to account the two cases in different ways, it is difficult, in practice, to completely exclude the worst-case possibility (quasi-static contact) due to fixtures, working surfaces and other elements of the layout that could trap the person's body part. As a consequence, quasi-static thresholds are mostly adopted due to the difficulty to exclude trapping possibility either a priori, or at production time using safety-rated sensors.

From a more technical perspective, speed and separation monitoring allows the robots to constantly operate at their maximum, apart when human intervention is needed, while their possible presence in the workspace of the robot is detected by safety guards. In turn, power and force limiting operations do not necessarily require safety guards, as the velocity of the robot is always maintained lower than the value that would possibly cause mild and non-permanent injuries to the worker. Clearly, this option represents a suboptimal strategy when no humans are present in the workspace of the robot. For their complementarity, speed and separation monitoring is preferred when the application is fully autonomous with sporadic and short interventions of the worker, while power and force limiting is adopted in frequent or persistent workspace sharing between the robot and the worker. Borderline situations, as for example sporadic but durative interventions of the human, have to be analyzed accurately a hybrid solution could represent the best strategy: the robot operates in speed and separation monitoring when humans are sufficiently far or absent, and then possibly shift to power and force limiting when the person further approaches the workspace of the robot. As a drawback, safety guards have to be integrated with the application, causing an increased cost.

5 Human-Centered Perspectives on Advanced Robotics

Despite significant advancements, many fixed robot frameworks still underutilize the full potential of advanced robotics. Current systems impose rigid safety constraints on coexistence and workspace sharing and often struggle to adapt to unexpected situations, such as minor changes in object placement or humans entering a safety zone. These events typically halt the industrial process, requiring human intervention to resume operations.

In addition, current safety standards tend to adopt conservative measures that restrict the robot speed, force, and interaction range, as described in the previous section. These standards, while essential to protect workers in close proximity to robots and prevent accidents, inadvertently reduce operational efficiency and might diminish the return on investment, particularly for small and medium-sized enterprises (SMEs). For instance, limitations on robot movement or speed in shared workspaces can slow down operations, reducing the productivity gains that automation typically promises [28]. This creates a barrier to the widespread adoption of advanced robotic systems, especially in industries where profitability margins are tight.

Alongside fixed interactive robots, mobile robots are becoming increasingly popular. They combine the manual functionality and dexterity of traditional manipulators with the mobility and operational freedom of mobile bases. This blend gives mobile robots much greater flexibility and functionality, allowing for a wider range of applications and potentially overcoming some of the limitations of fixed robots. However, while there are established standards for fixed manipulators and mobile platforms, there is currently a lack of standards for mobile robots. These devices have summed ranges of motion and degrees of freedom, requiring specialized control systems to coordinate the movement of both the base and the arm effectively. In addition, they have a larger workspace and modified load capacity [29]. Therefore, dedicated guidelines and rules are necessary. In this context, the European project SOPHIA (Socio-Physical Interaction Skills for Cooperative Human–Robot Systems in Agile Production) has contributed to standardization efforts by identifying gaps and developing strategies to update existing standards. Importantly, these standards should be designed to enhance rather than restrict the capabilities of mobile manipulators. At the same time, control algorithms and user interfaces need to be developed to optimize their use in both close proximity and remote operations. It is crucial to enable not only the control of the entire robot but also the independent operation of the mobile platform and the arm.

In general, the dynamic capabilities of advanced robotic systems, either fixed or mobile, are rarely utilized, and they are seldom designed for direct socio-physical interaction with humans to perceive, understand, and respond to their needs and mental states thus enabling more productive work.

From a human factors perspective, the core challenge is to find the optimal trade-off between human health/safety risks and system productivity. To this aim, a two-pronged approach is required. The first step is to develop advanced planning and control algorithms that optimize safety parameters online based on the proximity and behavior of humans. These algorithms would online adapt the robot actions to ensure worker safety while maintaining higher performance levels when human presence is minimal or potential contact is non-hazardous. Furthermore, recent research suggests that intrinsically safe physical contact between the human and the robot could be beneficial to some industrial problems, opening new possibilities for interaction. Some attempts in this direction will be presented in the next subsection.

The second step is to update existing safety standards to reflect these technological advancements. Rather than relying on static, one-size-fits-all safety measures, future standards should include more flexible, context-aware and human-in-the-loop approaches to robot control.

While safety remains the paramount requirement in human–robot interaction, it is also crucial to ensure that the operator's physical and mental well-being is at least maintained or even enhanced, not compromised, by the interaction [30]. Robots have been designed to take on the physically demanding, hazardous, and repetitive tasks of industrial processes, reducing risks and improving the experience of operators in semi-automated production lines. However, improper use of this technology can lead to the opposite effect. Sudden, unpredictable robot movements and excessively long work cycles can lead to negative consequences for workers, such as stress. Similarly, poorly configured robots can inadvertently promote, rather than prevent, incorrect postures and compensatory behaviors, which over time may result in physical injuries. Consequently, there are two crucial requirements to focus on. First, it is essential to design the robot's behavior to be intuitive, smooth, and pleasant for the operator, ensuring optimal sharing of the workspace. Second, the continuous and accurate monitoring of the human's physical and mental state must be deployed and leveraged to adapt the robot's behavior in response (e.g., by modifying the pace of interaction, adapting human–robot distances, or reconfiguring movement trajectories).

These measures, integrated with the previously mentioned strategies to optimize and update safety standards ensure that, alongside productivity, health and well-being of the worker are safeguarded.

5.1 Case Studies of Human–Robot Interaction

Several studies have explored innovative strategies to optimize the trade-off between system productivity and workers health and safety by implementing adaptive and reconfigurable HRI frameworks. Concerning the safety prescriptions listed in ISO 10218-1 and detailed in ISO/TS 15066, the academic community has continuously worked towards the development of the most efficient way to implement these requirements by suggesting more advanced, yet compliant, strategies. For example, the modulation of the speed of the robot on a continuous basis has proven to further enhance the interaction capabilities with respect to the simple logic widely adopted to stop the robot when the operator appears in its close proximity [31]. Based on this idea, an optimal strategy for instantaneously selecting the robot's maximum velocity based on the perceived position of the worker and the robot's dynamic capabilities has shown dramatic reductions in cycle time compared to the industrial baseline [31].

Current limitations in achieving seamless and fluent operations in collaborative settings are likely due to the sharp distinction between speed and separation monitoring, and power and force limiting. While modern collaborative robotics applications typically implement both, either one or the other is active at any time. Merging the two, i.e. allowing the robot to operate at maximum speed while guaranteeing that its speed in future and possible contacts will comply with power and force limiting prescriptions, has been investigated [32]. The results show the possibility to operate at higher speeds and/or lower separating distances, while still adhering with standards. The following sections provide insights into two case studies deploying these innovative approaches.

5.1.1 A Case Study for a Mobile Cobot in Human–Robot Conjoined Actions

At a company in the Netherlands operating within the manufacture of gears, bearings and other industry elements, employees are often required to perform repetitive lifting tasks involving products of many different sizes and high weights. Several lifting aids, including overhead cranes and forklifts, are provided by the company but these are seldom used for smaller loads since they are often regarded as bulky and impractical. Moreover, the frequent lifting of items with different properties makes adjusting the equipment manually both time-consuming and inconvenient. As a result, many employees indicated that they only use lifting aids when a load is too heavy to lift manually, and thus execute heavy lifting very frequently, often done with improper posture. A small internal study showed that this causes a high incidence of elbow, shoulder and lower back injuries. In particular, bursitis (inflammation of one or more bursae) of the shoulder is the most common one. In addition to the high risk of physical injuries to workers, this situation has increased machine idle times, which in turn raises costs and reduces competitiveness in the international market.

To cope with this scenario, a mobile cobot equipped with a user interface was proposed, which can follow human loco-manipulation commands to perform industrial tasks (KAIROS-MAN). The methodological development and experimental validation in laboratory settings is detailed in [33, 34], and initial pilot studies in end-user companies have demonstrated promising results [35]. The system includes a high-payload 6-DoF velocity-controlled robotic arm and an omni-directional mobile platform. Thanks to an easy-to-use and intuitive admittance user interface located at the robot end-effector, the user's wrenches are translated into loco-manipulation commands, generated by a whole-body controller. This assistive device is thus able to reduce workers' physical workload by adapting gravity support and compensating the carried object, while allowing for quick, easy and precise object control. Moreover, it can be moved with little effort, to be relocated rapidly to different workstations and have adaptable end-effectors to accommodate different objects of various sizes. To enable the operator to seamlessly manage both navigation and manual materials handling tasks, the KAIROS-MAN user interface includes two operational modalities. The *locomotion* mode prioritizes the wheel movements and allows the user to guide the robotic system to the desired workstation through the admittance user interface taking advantage of the robot's omni-directional locomotion skills. Instead, the *manipulation* mode gives priority to the robotic arm movements and allows fine object manipulation. Figure 5 shows the use case setup [36].

By using rather than avoiding human–robot contact, the proposed mobile cobot equipped with the KAIROS-MAN user interface strikes an effective balance between

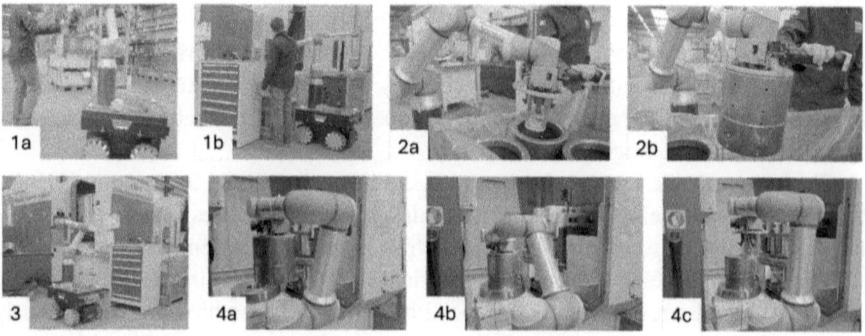

Fig. 5 Use case pictures of omindirectional mobile cobot platform [36]

productivity and physical ergonomics. Indeed, thanks to its agile maneuverability and fast reconfigurability, it allows them to avoid overloading their bodies with heavy loads and adopting incorrect postures without affecting work times and operational efficiency. Furthermore, in addition to the fact that this device is inherently safe as it is already certified for industrial use, the implemented controller provides an additional level of safety by accounting for the interaction forces with the worker and the environment and responding accordingly to prevent any damage to them.

5.1.2 A Case Study of Adaptive Human-in-the-Loop Automation for Worker Well-Being and Performance Enhancement

At a key supplier to the European automotive industry many tasks involve the repetitive manipulation and inspection of metal components. Here, a use case has been selected that involves the release of laminated electrical steel stator or rotor stacks. This task requires workers to grasp the heavy metal part typically with one hand, transfer it from a pallet to a hitting table, perform a forceful motion to de-glue the stack, inspect the quality of the produced part, and remove the annealing rod. While this task seems common for the shop floor, an internal study highlighted several significant biomechanical risks for workers. The bending and twisting motions required to lift and transfer the part from the pallet to the worktable as well as the repeated impacts on the worktable, can place significant strain on the worker's musculoskeletal system, especially if non-ergonomic postures are adopted and maintained for prolonged periods. Indeed, the repetitive and frenetic nature of the task increases physical and mental health risks, which can accumulate over time and cause long-term damage.

To mitigate these risks, a partial automation solution was implemented and tested with shop floor workers, which allocates the most physically demanding portions of the task to the robot and exploits the worker's cognitive abilities and flexibility. A fixed cobot handled the stack manipulation and hitting motion required to separate the laminated parts, thus avoiding manual lifting and repetitive physical exertion.

Instead, the worker was responsible for quality checking of the laminated product through visual inspection and the removal of the annealing rod, critical steps that are difficult to automate with the same level of accuracy and sensitivity. The removal process, in particular, became challenging due to the unpredictable rod placement after the robot's shaking motion.

The most notable aspect of the proposed approach was the focus on preserving the workers' well-being and comfort, avoiding the cause of stress due to the robot's integration into the workflow. To achieve this, an online decision-making algorithm was developed to select the most appropriate robot speed online according to the psycho-physical stress levels experienced by the operator, which are continuously monitored during the interaction by analyzing the heart rate variability. Moreover, the workers' position and attention towards the robot were tracked using a stereo camera. This not only ensured that the robot avoided any harmful interaction with the worker but also allowed for the creation of adaptive safety zones around the operator based on their attention level. If the worker was distracted or focused on another task without directly observing the robot's movement, the algorithm increased the human–robot separation distance to enhance the worker's perceived safety. Furthermore, the system could detect when the worker had completed the quality check before proceeding with the next task, optimizing safety and overall system performance. The methodology for adapting the proximity level and reactivity of robot trajectories has been presented and rigorously validated in laboratory experiments in and [35], respectively. Pilot experiments at the end-user company [34] demonstrated that this approach has the potential to optimise physical and cognitive workload and mitigate mental stress caused by frenetic and repetitive activities, fostering smoother and more effective human–robot interaction. The according use case set up is shown in Fig. 6.

5.2 Expectations and Experience in HRI

HRI in industrial environments is a multifaceted domain in which worker expectations play a critical role. When workers interact with robotic systems, they bring with them a set of beliefs, thoughts, and prior experiences. These preconceived notions significantly shape their expectations about the capabilities and operations of these systems. From a design perspective, incorporating these insights into the development of new robotic systems is essential. Understanding usability expectations can have a profound impact on how workers interact with and perceive robotic systems, ultimately affecting the systems' success and acceptance [34, 37].

Worker expectations in HRI are influenced by several factors, including prior experiences with technology and individual comfort levels with robotic systems. Workers' expectations of advanced robots vary considerably depending on their previous experience with the technology. Studies show that experienced users place a high value on adaptability and controllability. In a case study from the metal industry, 78% of respondents indicate that the ability to customize advanced robots was critical to their

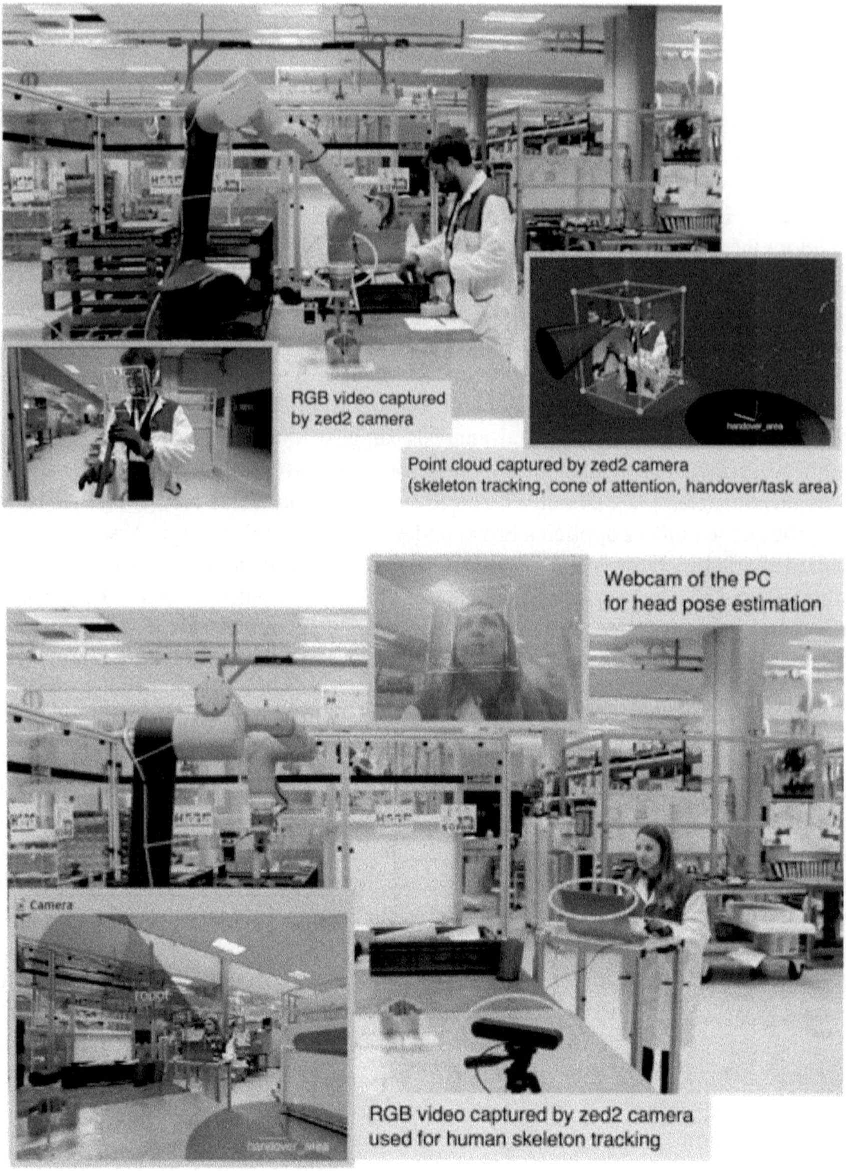

Fig. 6 Use case including signal monitoring technology [36]

acceptance. Conversely, less experienced users emphasized the importance of user-friendly introductions, such as intuitive user interfaces and step-by-step tutorials, to build confidence in interacting with robotic systems [37].

These expectations can affect how easily workers adapt to new systems, how efficiently they use those systems, and their overall satisfaction with the technology. For example, a worker who has had positive experiences with user-friendly, intuitive robotic systems may have higher expectations for new systems and be more likely to adopt them. Conversely, workers with negative experiences or those who are less familiar with the technology may approach new systems with skepticism or resistance. Culture and prior exposure to technology can also influence workers' expectations independently of direct experience [37]. There seems to be some evidence that demographic variables, such as gender or age, as well as cultural aspects and education influence people's general propensity to embrace and use new technologies. Results of a cross-sectional analysis suggest that younger people, men and higher-educated individuals tend to have more positive attitudes towards adopting new technologies [38]. Though it has to be stated, that these results are not robot specific as the authors applied a broad understanding of technology. Assessing individual characteristics specifically in relation to robotic systems seems to support to some extend the importance of education. However, diametral results were found for age and gender. Here women showed a more positive attitude towards the investigated robotic systems and older individuals were more willing to interact [39]. A conclusive influence of cultural and demographic variables on robot acceptance remains unclear and further research in this field is needed.

Experienced users and inexperienced users might also differ in their preferences regarding usability of a robotic systems, which has great relevance not only for system designers but also for practitioners deploying different robotic systems. Experienced users place higher value on factors like individualization, controllability, and conformity with user expectations. These aspects empathize a robot's ability to adapt to user specific needs, while staying controllable during an interaction. Experienced users also seem to have more nuanced expectations of robotic systems, which are informed by their previous exposure. Overall, they seem to priorities how systems integrate with their workflow and minimize disruptions. Inexperienced users seem to priorities learnability and self-descriptiveness in a robotic system, indicating a preference for systems that are intuitive and easy to understand without extensive training or experience [34, 37, 40].

When it comes to expectations towards the system, different degrees of experience still matter, however, there are unifying factors. Users tend to expect a robot to be controllable and generally suitable for the task for which they are going to use it. Furthermore, in industrial uses, robots are expected to be exceptionally error tolerant; meaning that they are both robust towards error, as well as offer adequate options to handle errors should they occur [37, 40].

Overall, these different needs should be considered in design actions as well as during implementation into real-world applications accordingly. Especially the initial implementation phase of robotic systems into real-world applications can be challenging. The presented description of differences in users and their expectations towards robotic usability might help practitioners to prioritize actions and implementation iterations. Recognizing and addressing these different expectations during the design phase is important. Designing with user expectations in mind can lead to more intuitive and user-friendly systems, reducing the learning curve and increasing overall satisfaction. This user-centered approach is a cornerstone of effective HRI design and emphasizes the need for thorough research into worker expectations and experiences especially in relation to more adaptive systems.

5.2.1 Measuring Worker Expectations

A valuable source for assessing worker expectations related to the general design of interaction principles of a robotic system can be found in the IsoMetrics questionnaire [41]. The questionnaire is based on the standard ISO 9241-110:2020 (en) Ergonomics of human-system interaction—Part 110: Interaction principles, which covers the ergonomics of human-system interaction and related human-centered design processes.

The ISO 9241-110:2020 outlines seven key facets of system design, known as interaction principles. These principles form the basis for the IsoMetrics-based questionnaire and include suitability for the task, self-descriptiveness, controllability, conformity with user expectations, error tolerance, suitability for learning, and user engagement. Each principle addresses a specific aspect of usability, ensuring a comprehensive evaluation of a system's performance from the user's perspective.

Studies using the IsoMetrics questionnaire and ISO 9241-110:2020 principles have validated this approach for evaluating expectations related to system interaction principles as well as the actual perceived usability of robotic systems from the worker's perspective [37, 40, 42]. These studies have shown that worker's experience significantly influences their expectations related to the specific interaction principles of a system. Experienced workers tend to have higher expectations and provide higher ratings for all seven interaction principles compared to inexperienced workers. This suggests that a system's usability rating can be heavily influenced by the user's prior experience with similar technologies.

Higher ratings from experienced workers may indicate that prior experience leads to more refined and higher expectations for new systems. This underscores the importance of considering experience levels in HRI design. It must be considered that experienced users might demand more advanced features and greater efficiency from new systems. Incorporating worker expectations into the design and evaluation process is important for developing successful robotic systems.

One possibility to ensure that the final product meets the actual needs and expectations of its users is to involve end users early on in the development process. Involving employees in the design and development phases can provide valuable insight into their needs and expectations. This can happen using surveys, focus groups, workshops, and usability testing. Another method would be to establish continuous feedback loops with end users.

The version of the robotic IsoMetrics questionnaire adapted to robotic systems, provides not only an extensive list of design recommendations for system designers for each interaction principle but also a set of evaluation criteria for an ongoing system evaluation, allowing to provide specific feedback on the robot's abilities and features to system designers and integrators from end-users. The interactions principles could even serve for structuring more qualitative methods for user feedback, such as workshops or interviews.

These regular updates based on user feedback can help designers and system integrators to make iterative improvements to the system and to ensure that the system remains relevant and user-friendly. Furthermore, when workers feel their input is valued and their needs are addressed, they are more likely to embrace new technologies and integrate them into their workflows. This can lead to long-term adoption, improved productivity, and potential economic benefits [37, 40].

Focusing on key interaction principles ensures the proper fit for the system's purpose. The interaction principles outlined in the ISO standard 9241-110:2020 can guide the design and iterative evaluation process. This may lead to systems that can be customized to meet individual preferences and requirements. This approach recognizes that different users can have different needs and expectations.

Empirical studies have demonstrated the importance of these strategies. For example, research shows that systems designed with a focus on user expectations and experience levels tend to receive higher usability ratings [37, 40]. These studies suggest that systems that meet user expectations and provide high levels of controllability and error tolerance are more likely to be accepted and effectively used by workers.

Understanding and addressing worker expectations and experience in HRI is essential for the successful implementation of robotic systems in industrial environments. By incorporating these insights into the design process, developers can create systems that are intuitive, user-friendly, and aligned with the needs their users. This user-centered approach not only improves system usability, but also fosters greater worker acceptance and satisfaction, thus ultimately leading to a more effective, efficient, and sustainable HRI.

6 Conclusion

Looking ahead, the future of industrial robotics lies in achieving an optimal balance between technological advances, industry standards and human-centered approaches. As highlighted in this chapter, interactive robots have already begun to redefine workplace dynamics by increasing flexibility, efficiency and customization across industries. However, their true potential can only be realized by addressing key challenges, including ergonomic design, safety, and worker acceptance as a whole.

By prioritizing worker safety, well-being, and engagement, organizations can unlock the full potential of interactive robots and transform workplaces into more inclusive, efficient, and adaptable environments. Emerging technologies, such as

AI-enhanced robotics, adaptive algorithms, and wearable user interfaces promise to further enhance the possibilities for interaction between humans and robots, enabling seamless integration across industries. These advances could particularly benefit sectors that require high precision and adaptability, such as healthcare.

At the same time, fostering a skilled workforce will be critical to the successful adoption of these technologies. Investing in comprehensive training programs and designing intuitive systems that appeal to different levels of technical expertise will ensure a smooth transition to automated work environments. In addition, addressing the broader societal implications of robot adoption—including job displacement, ethical considerations, and workforce equity—will be critical to shaping a sustainable and inclusive future of work.

Ultimately, the success of HRI depends on adopting a human-centered perspective that not only drives technological innovation, but also improves the quality of work and life for all. By aligning technological developments with the needs and expectations of workers, organizations can foster environments that are not only productive, but also sustainable and safe.

References

1. Singh, B., Sellappan, N., Kumaradhas, P.: Evolution of industrial robots and their applications. Int. J. Emerg. Technol. Adv. Eng. **3**(5), 763–768 (2013)
2. Colim, A., et al.: Lean manufacturing and ergonomics integration: defining productivity and wellbeing indicators in a human–robot workstation. Sustainability **13**(4), 1931 (2021)
3. Wallén, J.: The History of the Industrial Robot. Linköping University Electronic Press (2008)
4. Gasparetto, A., Scalera, L.: A brief history of industrial robotics in the 20th century. Adv. Hist. Stud. **8**, 24–35 (2019)
5. Hägele, M., Nilsson, K., Pires, J.N., Bischoff, R.: Industrial robotics. In: Springer Handbook of Robotics, pp. 1385–1422. (2016)
6. Robotics, I.F.: Industrial Robotics (2024). Available from: https://ifr.org/industrial-robots
7. Colgate, J.E., Wannasuphoprasit, W., Peshkin, M.A.: Cobots: robots for collaboration with human operators. ASME Int. Mech. Eng. Congr. Exposition **15281**, 433–439 (1996)
8. Onnasch, L., Roesler, E.: A taxonomy to structure and analyze human–robot interaction. Int. J. Soc. Robot. **13**(4), 833–849 (2021)
9. ISO 10218-1:2025: Robotics — Safety requirements — Part 1: Industrial robots. International Organization for Standardization, Geneva (2025)
10. McKinsey, A.: Future That Works: Automation, Employment, and Productivity. Executive Summary. McKinsey Global Institution (2017)
11. Robotics, I.F.: World Robotics—Industrial Robots (2024). Available from: https://ifr.org/wr-industrial-robots
12. Wischniewski, S., Heinold, E., Rosen, P.H.: Results from the Third European Survey of enterprises on new and emerging risks on human–robot interaction. In: Black, E.N.L., Neumann, W.P., Noy, I. (eds.) Proceedings of the 21st Congress of the International Ergonomics Association (IEA 2021). Volume V: Methods & Approaches, pp. 343–346. Springer International Publishing: Online-Kongress (2021)
13. Villani, V., Pini, F., Leali, F., Secchi, C.: Survey on human–robot collaboration in industrial settings: safety, intuitive interfaces and applications. Mechatronics **55**, 248–266 (2018)
14. Rosen, P.H., Heinold, E., Fries-Tersch, E., Wischniewski, S.: Advanced robotics and automation: implications for occupational safety and health. In: EU-OSHA (ed.) European Agency for Safety and Health at Work Bilbao.

15. Rosen, P.H., et al.: Artificial intelligence, advanced robotics and the automation of tasks at work: taxonomy, policies and strategies in Europe. In: EU-OSHA Policy Brief. European Agency for Safety and Health at Work, Bilbao (2022)
16. Rosen, P.H., et al.: Taxonomy for the classification of AI-based ICT and robotics and OSH impacts. In: 11th International Conference WOS.net 2022. Focus on Humans in a Technological World, p. 54. Universidade do Minho, Olhao, Portugal (2022)
17. Pedersen, M.R., Nalpantidis, L., Andersen, R.S., Schou, C., Bøgh, S., Krüger, V., Madsen, O.: Robot skills for manufacturing: From concept to industrial deployment. Robot. Comput.-Integr. Manuf. **37**, 282–291 (2016)
18. Kadir, B.A., Broberg, O., da Conceição, C.S.: Current research and future perspectives on human factors and ergonomics in Industry 4.0. Comput. Ind. Eng. **137** (2019)
19. Bakker, A.B., Demerouti, E.: The job demands-resources model: state of the art. J. Manag. Psychol. **20**(7), 743–757 (2007)
20. Bakker, A.B., Demerouti, E.: Multiple levels in job demands-resources theory: implications for employee well-being and performance. In: In: Diener, E., Tay, L. (eds.) Handbook of Well-Being. DEF Publishers, Salt Lake City, UT (2018)
21. Karasek, R.A.: Job demands, job decision latitude, and mental strain: implications for job design. Adm. Sci. Q. **24**(2), 285–308 (1979)
22. Hartwig, M., Rosen, P., Wischniewski, S.: Working conditions of industrial robot operators-an overview of technology dissemination, job characteristics, and health indicators in modern production workplaces. IEEE Trans. Hum.-Mach. Syst. **54**(2), 145–151 (2024)
23. Jaimovich, N., Siu, H.E.: Job polarization and jobless recoveries. Rev. Econ. Stat. **102**(1), 129–147 (2020)
24. Niedhammer, I., Bertrais, S., Witt, K.: Psychosocial work exposures and health outcomes: a meta-review of 72 literature reviews with meta-analysis. Scand. J. Work Environ. Health **47**(7), 489 (2021)
25. Demerouti, E.: Turn digitalization and automation to a job resource. Appl. Psychol. **71**(4), 1205–1209 (2022)
26. Standardization, I.O.: Technical Specification: ISO/TS 15066: Robots and Robotic Devices: Collaborative Robots. ISO (2016)
27. Commission, I.E.: Safety of Machinery—Electrical Equipment of Machines—Part 1: General Requirements (IEC 60204-1). IEC, Geneva, Switzerland (2005)
28. Heinold, E., Rosen, P.H., Wischniewski, S.: Advanced Robotic Automation: Comparative Case Study Report, p. 57. European Agency for Safety and Health at Work, Bilbao (2023)
29. Goel, R., Gupta, P.: Robotics and industry 4.0. A roadmap to industry 4.0. In: Nayyar, A., Kumar, A. (eds.) Smart Production, Sharp Business and Sustainable Development. Springer, New Delhi (2020)
30. Lorenzini, M., Lagomarsino, M., Fortini, L., Gholami, S., Ajoudani, A.: Ergonomic human–robot collaboration in industry: a review. Front. Robot. AI **9**(813907) (2023)
31. Zanchettin, A.M., Lacevic, B.: Safe and minimum-time path-following problem for collaborative industrial robots. J. Manuf. Syst. **65**, 686–693 (2022)
32. Lucci, N., Lacevic, B., Zanchettin, A.M., Rocco, P.: Combining speed and separation monitoring with power and force limiting for safe collaborative robotics applications. IEEE Robot. Autom. Lett. **5**(4), 6121–6128 (2020)
33. Giammarino, A., Gandarias, J.M., Balatti, P., Leonori, M., Lorenzini, M., Ajoudani, A.: SUPER-MAN: SUPERnumerary robotic bodies for physical assistance in hu-MAN–robot conjoined actions. Mechatronics **103**(103240) (2024)
34. Lagomarsino, M., Lorenzini, M., De Momi, E., Ajoudani, A.: PRO-MIND: Proximity and Reactivity Optimization of Robot Motion to Tune Safety Limits, Human Stress, and Productivity in Industrial Settings. IEEE Trans. Robot. 41, 2067–2085 (2025)
35. Lagomarsino, M., Lorenzini, M., De Momi, E., Ajoudani, A.: Robot trajectory adaptation to optimise the trade-off between human cognitive ergonomics and workplace productivity in collaborative tasks. In: IEEE/RSJ International Conference on Intelligent Robots and Systems (IROS), pp. 663–669. IEEE (2022)

36. Deliverable 9.6, Validation and demonstration report of Industrial Pilot Lines 1–4. 2022. In: Socio-Physical Interaction Skills for Cooperative Human–Robot Systems in Agile Production (SOPHIA) grant agreement No. 871237, European Union's Horizon 2020 Research and Innovation Programme (H2020-ICT-2019-2/ 2019-2023). Available from: https://cordis.europa.eu/project/id/871237/results
37. Funk, M., et al.: How does experience influence usability expectations in human–robot interaction? Results of a worker evaluation in an industrial setting. In: 21st International Conference on Advanced Robotics (ICAR), pp. 271–276 (2023)
38. Rojas-Mendez, J.I., Parasuraman, A., Papadopoulos, N.: Demographics, attitudes, and technology readiness: a cross-cultural analysis and model validation. Mark. Intell. Plan. **35**(1), 18–39 (2017)
39. He, Y., He, Q., Liu, Q.: Technology acceptance in socially assistive robots: scoping review of models, measurement, and influencing factors. J. Healthc. Eng. **2022**(1), 6334732 (2022)
40. Niehaus, S., et al.: Expectations of usability aspects for human–robot interactions in manufacturing settings are changed by experience. In: 56th International Symposium on Robotics (ISR), p. 7. VDE, Stuttgart (2023)
41. Gediga, G., Hamborg, K.-C., Düntsch, I.: The IsoMetrics usability inventory: an operationalization of ISO 9241–10 supporting summative and formative evaluation of software systems. Behav. Inf. Technol. **18**(3), 151–164 (1999)
42. Heinold, E., Rosen, P., Wischniewski, S.: Usability questionnaire for robotic systems based on the ISO 9241-110. IEEE Robot. Autom. Lett. (2025)

Humans and Robots in Agriculture

Juan Pablo Vásconez, Federica Masci, George Adamides, and Ingrid Nicole Vásconez

Abstract Research in agricultural robotics has garnered increasing attention over the past decade. Simultaneously, Human-Robot Interaction (HRI) has emerged as a possible solution to increase automation and reduce the lack of human labor in the field. This chapter presents a study of research trends in HRI focusing on Artificial Intelligence (AI), human factors, and ergonomics in collaborative agricultural tasks. For example, we talk about the importance of robotics and the application of the different AI-based methods for HRI in field applications. Moreover, we emphasize situation awareness, stress, sleepiness, sleep quality, and distraction. We also explore various crucial and innovative topics related to HRI in agriculture, including safety, human factors, ergonomics, awareness, productivity, and management. This analysis shows that HRI paradigms can be effectively applied in agricultural applications, yielding promising production, safety, and farm management benefits, such as increasing productivity and reducing human workload.

Keywords Human-robot interaction · Human-robot collaboration · Agriculture robotics · Ergonomics · Human factors · Situation awareness

J. P. Vásconez (✉)
Faculty of Engineering, Universidad Andres Bello, 7500971 Santiago, Chile
e-mail: juan.vasconez@unab.cl

F. Masci
Department of Biosystems, KU Leuven, Leuven, Belgium
e-mail: federica.masci@kuleuven.be

G. Adamides
Department of Rural Development, Agricultural Research Institute, Nicosia, Cyprus
e-mail: gadamides@ari.moa.gov.cy

I. N. Vásconez
Centro de Biotecnología Vegetal, Facultad de Ciencias de la Vida, Universidad Andres Bello, Santiago, Chile
e-mail: i.vasconezhurtado@uandresbello.edu

G. Barresi and S. Wischniewski (eds.), *Ergonomics in Robotics: Advances and Innovations*, Springer Tracts in Advanced Robotics 162,
https://doi.org/10.1007/978-3-032-19374-2_7

1 Introduction

Human-robot interaction (HRI) has emerged as novel areas of study in recent years [1–3]. A key attribute of HRI is its multidisciplinary nature, emphasizing the study, design, and evaluation of robotic systems capable of interacting seamlessly with humans. HRI aims to find and develop strategies for designing and implementing robots within human-centered environments. Within this field, various approaches have been developed, demonstrating substantial implications for agricultural applications [4–6]. Specifically, HRI methods are crucial in intricate agricultural processes that resist full automation, such as fruit harvesting and pruning. The adoption of HRI strategies has demonstrated utility by facilitating collaboration between a field worker and one or more robots. This collaboration aims to enhance productivity, alleviate human workload and fatigue, and elevate safety and farm management practices, among other benefits [6–8]. The increasing global demand for food due to population growth, coupled with the migration of labor to urban areas, positions HRI applications as a potential solution to address the shortage of labor in the agricultural sector [7].

The research field of HRI in agricultural environments has witnessed significant growth. This expansion owes much to the rapid advancement of Artificial Intelligence (AI) algorithms and the Internet of Things (IoT) in recent years and the concurrent development of increasingly potent and portable hardware and software for robotics applications. Furthermore, the increasing development of more precise sensors tailored to robotics, coupled with improved cost accessibility, has rendered the use of robots an appealing choice for field applications [9]. The evolution of intelligent sensors grounded in AI has enabled the detection of various human aspects in field scenarios, including position, social space, human posture, cognitive state, and even emotional state [5, 7]. This information becomes valuable for a robot, allowing it to infer the human's intentions, discern the need for assistance in field tasks, and even engage in verbal and non-verbal communication techniques to better cooperation in field activities. However, the journey towards achieving fluid, efficient, natural, and safe cooperative agricultural processes through HRI is still a substantial undertaking [10].

One of the main contributions of HRI strategies in agriculture lies in their potential to significantly enhance agricultural productivity. Robotic systems can be used to greatly boost agricultural productivity by increasing the efficiency of various operations due to their precision, reliability, and capacity to carry out repetitive and heavy tasks such as harvesting, pruning, monitoring, and fruit or vegetable transport [11]. This enables farmers to make better-informed decisions and carry out targeted interventions [12]. Agricultural robots have demonstrated to provide high-performance support during labor shortages during peak seasons (e.g., harvesting) and can operate in hazardous conditions (e.g., pesticide spraying), thereby minimizing risks to human workers [9]. However, agricultural robots must navigate and operate in complex, unstructured, and unpredictable environments with volatile physical and atmospheric conditions (e.g., lighting and shading). To address such challenges, HRI is being used to link the advantages of humans (e.g., eye-hand coordination) with the robotic system (e.g., repeatability).

In this chapter, we present a comprehensive study on humans and robots in agriculture, with a particular focus on key aspects such as HRI, AI, human factors, ergonomics, situation awareness, production, workload, and safety. Drawing from this information, we engage in a discussion that highlights current trends, the associated benefits and outlines potential future challenges within this dynamic area of research.

2 Human-Robot Interaction (HRI) in Agriculture

First, we have to clarify that, according to the international standard ISO 8373:2012, when we refer to agricultural robotic systems, we refer to non-industrial systems (i.e. service robots) [4]. The interaction between field workers and robots in agriculture scenarios can manifest in physical, cognitive, and social dimensions, occurring either in shared workspaces or through remote teleoperation. The primary objective of HRI in agricultural environments is to communicate using verbal or non-verbal cues with robots to optimize agricultural production processes, coordinate work and decision-making, and enhance or supervise the safety of field operations (see Fig. 1). Typically, an agricultural robot may function as an autonomous or semi-autonomous system, addressing challenging and repetitive tasks at various stages of the agricultural process such as harvesting, transport, pruning, monitoring, spraying of chemicals, water irrigation, and transplanting [4, 13–17]. The level of autonomy from the HRI point of view in agriculture will be high if humans and robots have a fluid, natural, and efficient communication and coordination. However, achieving fluid interaction in field applications is hard, since agricultural robots need to face the complex challenge of operating in intricate environments, such as those featuring uneven and dynamic terrain, obstacles, trees, and plants [4].

Fig. 1 Concept art of human-robot interaction in agriculture using verbal and non-verbal communication

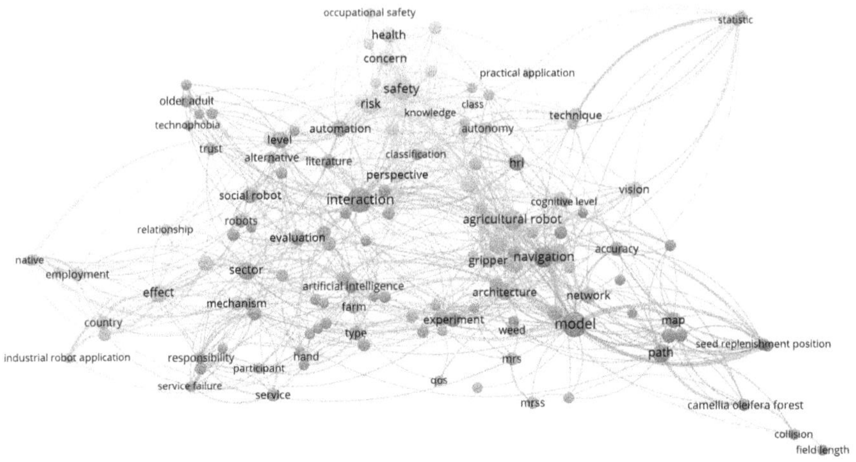

Fig. 2 Most frequent terms found in the literature related to human-robot interaction and collaboration in agriculture

In Fig. 2, we provide an illustration summarizing the most frequent terms found in the literature concerning Human-Robot Interaction and Collaboration in agriculture. Notably, the prominent terms include interaction, collaboration, safety, agricultural robot, social robot, navigation, model design, HRI architecture, and applications along with their effects. Additionally, several human factors were identified, encompassing trust, level of autonomy, risk, health, safety, and cognitive level. Ergonomic factors were also observed, such as architectural parameters, interaction mechanisms, and health. In the subsequent sections, we emphasize and provide further explanations for most of these key factors.

As part of HRI, collaborative robotics in agriculture has grown in recent years as a possible solution to the lack of agricultural labor [19, 20]. Collaborative robotics is a part of human-robot interaction, where humans can collaborate with service robots in three ways, coexistence, cooperation, and collaboration.

In the case of coexistence, the robot performs tasks independently of human workers while sharing the same environment. Coexistence strategies have proven to be very useful in perception tasks such as crop monitoring, weed detection, precise irrigation, and automated harvesting, which are activities that the robot can perform as low-level tasks when interacting with humans that are performing high-level tasks such as crop management. On the other hand, cooperative interactions involve humans and robots working towards a common goal while performing tasks that may be different or complementary.

Cooperative interactions involve humans and robots working towards a shared goal but performing different and complementary tasks. For example, in the context of manual harvesting of fruit from trees, such as apple, avocado, tangerine, or lemon trees, humans can be dedicated to the task that requires more manual dexterity, which is harvesting, while mobile service robots can be used specifically for fruit transport tasks, which is the most repetitive and physically demanding task. In this way, the productivity of the harvesting processes can be increased while the workload of human workers is reduced. There is evidence that the lack of manpower for harvesting often causes fruit to fall to the ground because it is not collected in time, so this cooperative task can be of great help for these processes [4, 5, 7].

Collaboration is the most advanced form of HRI, where humans and robots work together by sharing the workspace to perform joint tasks. Collaborative robots can safely interact with humans in close proximity and respect their socially acceptable zones, often utilizing advanced sensors to perceive their surroundings and adapt in a versatile manner to the actions or intentions of field workers [4, 5, 7]. In tasks related to precision agriculture, collaborative robots have proven to be very useful in reducing the use of pesticides through selective spraying which can be performed collaboratively with humans, thus minimizing environmental impact while increasing efficiency. These tasks can be carried out in a teleoperated manner or even while sharing the workspace with human field workers, representing several benefits and opportunities from the HRI perspective.

In [18], authors elaborate on another four types of human-robot collaboration in agriculture: separate, sequential, simultaneous, and supportive. In Separate, human(s) and robot(s) are kept apart; the human and the robot do not share workspaces, tools, or work pieces. In Sequential, human and robot tasks are complete consequently. In simultaneous, human and robot tasks are executed concurrently and are focused on achieving separate task goals. In supportive collaboration, the human and the robot work together to complete a common task. According to the type of collaboration and level of autonomy, human-robot interaction aspects such as situation awareness, ergonomics, and safety are taken into consideration for designing the functionality (sensing, planning/decision-making, actuation) [18, 21, 22].

3 Integrating AI in HRI in Agriculture

The rapid development of artificial intelligence (AI) is transforming HRI paradigms in agriculture. This is in part because robot sensory systems have greatly benefited from AI algorithms that have increased their perception capabilities to levels not thought possible a few years ago. Supervised machine learning algorithms (ML) such as support vector machine (SVM), K-nearest neighbors (KNN), Decision trees, random forests, as well as deep learning (DL) algorithms such as artificial neural networks (ANN) and Convolutional Neural Networks (CNNs) has proven to be fundamental for robots to distinguish obstacles, scenarios, field workers and their actions during field tasks [23–29]. AI and supervised vision-based perception systems have been able to be used to detect and classify plants, identify fruit ripeness, and fruit and leaf disease, and monitor plant health in real-time, allowing for more accurate and faster decision-making in the field. For example, several CNNs such as MobileNet, ResNet, Inception V2, GoogLeNet, Xception, AlexNet, MobileNet, and VGGNet have shown outstanding results for image classification problems, whilst

Faster R-CNN, Single Shot Multi-box detector (SSD), and You Only Look Once (YOLO) have demonstrated to be very useful in object detection problems in agricultural scenarios [23, 24, 30]. Unsupervised Learning has been also used in field applications for robots to characterize deviations in crop health, as well as to detect robot failures without the need for labeled data. Moreover, several terrain classification techniques have been used for robots based on clustering approaches such as K-means and Gaussian Mixture Models (GMM). Finally, reinforcement learning (RL) approaches have been used to develop robot control algorithms able to predict optimal trajectories based on past navigation data that was learned by experience in simulated scenarios [23–29]. The environment complexity of field terrain and environment constraints of agricultural fields, combined with the high variability of human behaviors and preferences makes reinforcement learning techniques such as Q-learning or Deep Q-learning feasible solutions to help robots to better interact and collaborate with field workers. For example, legged or humanoid robots, whose mathematical models are not always easy to define in a closed form, can greatly benefit from reinforcement learning algorithms to learn to solve tasks such as navigation and path following naturally, adapting to field conditions in a fluid and natural way [26–29, 31, 32]. Finally, the use of large language models (LLMs) in HRI applications has the potential to revolutionize communication and decision-making in agricultural field tasks. LLMs allow robots to comprehend and respond to intricate verbal communication, grasp subtle human directives, and modify their replies according to agricultural contextual factors [33–38]. This feature facilitates more intuitive and efficient interactions, improving teamwork between humans and robots across many field tasks such as harvesting and transport. This learning capability that the robot can achieve using AI is done based on experience, and although it is computationally expensive and time-consuming, it can open up new possibilities for robots to navigate highly complex terrains such as mountains to support humans in harvesting. As AI continues to advance, agricultural robots are expected to become increasingly autonomous and more capable of learning to interact with humans, which will further boost sustainability and efficiency in agriculture [33–38].

Generative artificial intelligence (generative AI) has enormous potential to transform HRI in agriculture due to its great capacity to generate content, images, text, movement patterns, and simulations [39, 40]. For example, generative AI allows the creation of representations of agricultural scenarios based on certain contexts, which can be useful for the development of visual human-robot interfaces. This could have a positive impact on the level of adaptability that people have when interacting with robots within agricultural environments, making robots capable of displaying dynamic information on a screen, which can contain valuable information about crops or even information about possible simulated emotions of the robot can be displayed for more natural interaction [33–35, 40, 41]. Additionally, verbal communication with the robot is key for cooperation and coordination between humans and robots, so the use of generative AI can provide natural language tools so that humans and robots can communicate bi-directionally to explain their actions, intentions, results, possible problems, personalized recommendations, among others [33–38]. In summary, generative AI opens up a world of possibilities to improve human-robot interaction and collaboration in agriculture by enabling the creation of customized visual communication interfaces, up to the use of verbal and non-verbal language to interact and solve problems naturally with field workers. These technologies can greatly help in facilitating adoption and improving the adaptation curve of field workers in agricultural tasks, which can allow for a smoother and more efficient transition to these technologies. It is to be noticed that the visual representations that can be provided uisng generative AI can be used to represent the field situation, process states and indicators, material requirements, and problem warning messages, among others [33–38].

4 Ergonomics in Agriculture

Ergonomics in agriculture encompasses physical, cognitive, and organizational dimensions, which are essential for HRI strategies. Physical ergonomics is based on the premise that physical compatibility between humans and robots, as well as tasks and environments, is important for safe and effective interaction. On the other hand, cognitive ergonomics minimizes mental workload and enhances decision-making, to foster good HRI in agriculture. Finally, organizational ergonomics analyzes the management of robots within existing operational structures and the structural consequences they may have on the people interacting and working with them.

4.1 Physical Ergonomics

Physical ergonomics emphasizes establishing physical compatibility among humans, robots, tasks, and settings to enhance the safety and efficiency of agricultural operations. Physical ergonomics entails the design of robots with suitable dimensions, configurations, and motions to mitigate human strain, discomfort, workload, and injury risk in jobs such as product loading, harvesting, and pruning. Vision-based systems are essential for obtaining data regarding an individual's location, posture, and particular activities. Semantic action recognition is significant, as it enables robots to deduce human actions from location, posture, and the items involved. A vision-based system created at [5] utilizes convolutional neural networks to ascertain the position and posture of field workers, along with their interactions with equipment and crops, including fruits. Using this technology, it has been shown that the actions that human beings carry out can be inferred based on their posture and the objects with which they interact. The results of semantic recognition of human actions based only on objects detected during avocado harvesting processes obtain an average accuracy of between 41% and 80% depending on the activity to be performed [6]. On the other hand, the semantic recognition of the action for the same application using objects and also the posture of the human improves its accuracy from 57% up to 99% [5]. When inferring field worker action, ergonomics are important because depending

Fig. 3 Sample of detection of human posture in tasks related to research and harvest of crops in a greenhouse

on the placement of the arms, legs, and the angle of the back, it can be determined what action the field worker is performing. It is also possible to detect if the action is performed correctly or even if the position of the field worker's body is affecting any part of the body such as the back or neck. It has been reported that field workers may have body aches depending on the posture they perform, and the time and number of times they perform that activity during the day. For example, in greenhouse crop harvesting and study applications such as the one shown in Fig. 3, it is observed that the posture of workers or researchers is often associated with several hours with high back and neck angles of inclination, which can cause pain or discomfort. In this context, developing an intelligent system capable of recognizing the actions and postures of the field worker can be useful so that a robot can interact with and support the field workers if they require it.

Ergonomics are essential in these situations, as posture analysis helps ascertain whether employees are executing tasks properly and identify positions that may result in discomfort or harm. Field studies, like those on greenhouse crop harvesting, indicate that workers frequently assume postures with significant back and neck flexion angles, potentially leading to pain or discomfort when maintained for prolonged durations. The creation of intelligent systems that can identify workers' postures and behaviors may allow robots to aid people and reduce ergonomic hazards.

4.2 Cognitive Ergonomics

Cognitive ergonomics seeks to reduce mental stress and improve decision making for both workers and robots in the field. This facilitates fluid and organic contact, which is achieved through intuitively designed interfaces that respond to the attributes of each individual in the domain. Cognitive factors that can help ergonomics include mental workload, emotional state, stress level, drowsiness, sleep quality, distraction, among others. These factors can be monitored in field workers using AI to then develop ergonomic interfaces that allow workers to maintain desirable cognitive factors in field tasks. For example, a well-designed and optimized interface can allow workers to pay more attention to high-importance field tasks while avoiding distractions [6, 45].

Human Factors in agriculture The high complexity of agricultural tasks, together with the randomness and variability of human factors, makes HRI strategies dependent on systems capable of monitoring and processing data and signals related to humans and their behavior. Identifying and analyzing human variables involves deducing essential elements of human behavior, cognition, and perception that can influence how effectively and safely a robot will collaborate with humans in agricultural settings. Detecting and even predicting human factors is crucial to the design of HRI systems to make them intuitive, efficient, and safe. The following section examines the application of human-robot interaction and collaboration and human factors in contexts of agricultural production.

Situation Awareness in agriculture In agriculture, Situational Awareness (SA) has proven to be crucial for various human roles involved in robotic systems, including supervisor, operator, mechanic, teammate, and bystander. SA consists of two primary components: the farmer's awareness of the robot's actions and the robot's awareness of the farmer's actions [42, 43]. In the first scenario, SA is about the understanding that the farmer has regarding the location, surroundings, activities, and status of the agricultural robot. For the latter case, SA is related to the knowledge that the agricultural robot has of the operator's status, his/her activities, surrounding limitations, and the possible actions to take in case the human operator requires assistance. The use of RGB cameras, 3D stereo vision, LiDAR sensors, radar, and thermography combined with data mining, computer vision, and AI has demonstrated the ability to enhance situational awareness in agricultural environments by improving object detection, obstacle avoidance, crop monitoring, and environmental mapping for field robotics [5, 7, 44, 45].

To measure SA in the field, the Situation Awareness Global Assessment Technique (SAGAT) can be used [42]. Such explicit methods are of high validity and objective because they allow direct assessment of the participant's knowledge of the situation [46]. It is key to identify in the field five awareness categories: location, activities, surroundings, status, and overall mission [47]. It has been found that the 3D maps were more effective in providing good location and status awareness, while video-centric interfaces are more effective in providing good surroundings and activity awareness. In [48], the authors proposed 3D reconstruction and different onboard sensor technologies, specifically stereo-vision, LIDAR, radar, and thermography (and combinations of the aforementioned technologies) to automatically detect obstacles and humans in agricultural scenarios to increase SA. A combination of LIDAR and stereo-vision was demonstrated to help detect traversable ground by producing dense maps at relatively high frequency and overcoming poor lighting conditions, shadows, and lack of texture. On the other hand, an integrated system of radar and stereo-vision was helpful in 3-D detection, while a combination of thermal imaging and 3D-stereo-vision have improved significantly the features for detecting humans and animals as well as other machinery, stones, and poles. The authors in [49] used a head-mounted display (HMD) and Augmented Reality (AR) enabled SA measurement in robot teleoperation. They adopted the Situation Present Assessment Method (SPAM) for real-time measurement of SA. In [50] the authors used the SAGAT and the Situation Awareness Rating Technique (SART) questionnaires to evaluate situation awareness in HRI during real-time tasks.

In [51], the authors presented the development and evaluation of suitable interfaces with enhanced HRI awareness to the farmer to effectively tele-operate a semi-autonomous vineyard robotic sprayer. A value added from their work was the placement of a camera on top of the end-effector sprayer to provide accurate target identification and spraying verification, thus improving SA. Similarly, placing a camera at the back-top of the robot provided peripheral vision and enabled the operator to locate obstacles around the robot wheels, thus improving location and surroundings awareness. In addition, [52] proposed a mathematical programming framework to develop a formal approach to human-robot collaboration. In a simulated grove environment, they showed a robotic harvesting system that seeks human collaboration only to service any self-diagnostic component-level inefficiencies. They observed that human interaction was higher during the early stages of harvesting. As a result, the human collaborator was not required to maintain situation awareness and could be engaged in supervising multiple robotic harvesters.

Production and Human's Workload in agriculture The examination of production efficiency and its impact on physical workload in agricultural processes offers valuable insights for enhancing resource management on farms. While numerous works in the literature address workload, only a few specifically consider human workload in the context of HRI applications in agriculture. For instance, in the application detailed in [53], a collaborative scenario involving several humans and a mobile robot for fruit collection tasks was developed. During the process, field workers focused on fruit harvesting, the most complex task, while the mobile robot handled the repetitive task of transporting fruits from the harvesting area to the transport vehicle, as illustrated in Fig. 4. The study demonstrated that is possible to efficiently harvest a quantifiable amount of fruit (production) within a specified time while calculating the energy consumption based on the metabolic rate information (physical workload). The authors compare the only-humans scenario with the humans-and-robots collaboration scheme. Results indicated that employing mobile robots in harvesting tasks increased production by 23–85% through the human-robot collaborative strategy, with the field worker's workload increasing only by 1–16% in this scenario. This implies a significant production benefit with minimal additional workload [53].

Other works explored the use of mobile robots capable of transporting fruits and field workers while they pick fruits from a platform on the robot during navigation in the field [54]. These studies demonstrated that the productivity of harvesting processes can increase by up to 58% by employing a cooperative strategy based

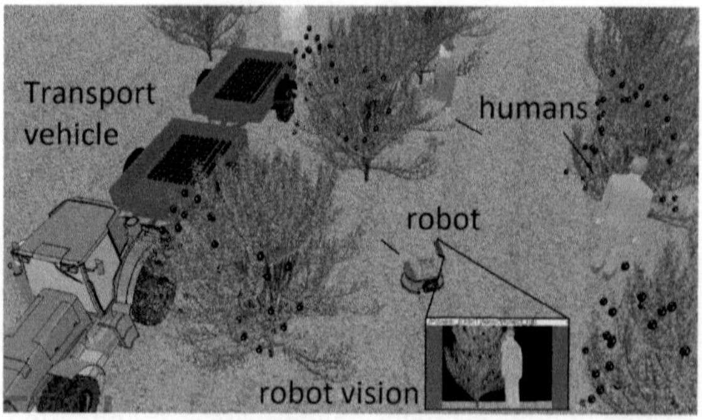

Fig. 4 Avocado fruit harvesting process concept using human-robot interaction strategy. The field workers perform harvesting, whilst the robots perform transport tasks

on shared workspace coordination with the robot. Furthermore, some works offer guidelines related to workload that are applicable in agriculture. For instance, in [55], the authors investigate various workload evaluation methods based on variables such as sensitivity, suitability, and generalization capacity. These methods use biological metrics like electroencephalogram (EEG), heart rate variability, respiratory rate, galvanic skin response, blink rate, and speech response time. The study emphasizes that these methods can be designed for specific tasks under properly monitored conditions and environments.

4.3 Organizational Ergonomics

HRI-based organizational ergonomics in agriculture focuses on the incorporation and evaluation of the impact of cooperative robots in workflows and management related to agricultural processes. The idea is to evaluate using process management in the operational structures and work dynamics of field tasks such as plowing, sowing, irrigation, chemical dosing, and harvesting. One of the first visible advantages is the optimization of the distribution of tasks between humans and robots, using, for example, robots for repetitive activities such as transport and monitoring, while workers concentrate on tasks that require high manual dexterity, as well as decision-making and supervision. Worker training programs in agriculture are essential to optimize employee competencies and organizational ergonomics of farms, facilitating effective collaboration with robots and promoting a culture that perceives the robot as a helper and not as a simple tool. Analysis of the management of the flows of different agricultural processes can allow for more efficient integration between humans and robots, identifying and resolving bottlenecks or inefficiencies generated by tasks that are difficult to fully automate. For example, robots adapted with vision and AI systems can assist in harvesting by pre-classifying the size and growth status of crops, as well as their health and hydration status. This reduces the physical load on workers, and can be key to decision-making and harvest prediction. However, it is essential to address ergonomic risks, such as mental fatigue or stress derived from the adaptation of cooperative robots to new dynamic workflows, which according to the latest trends, are driven by AI so that they can adapt to humans and to the field process flows in a fluid and natural way, favoring their organizational ergonomics.

5 Novel HRI Fields of Application in Agriculture

Field applications of HRI span diverse fields and applications, from harvesting, which we reviewed in previous sections, to the novel fields such as dairy and forestry sectors. Below, we focus on these areas, highlighting the importance of ergonomics and human factors related to HRI.

5.1 Dairy Sector

From farmer to fork, the food chain has developed into a very complex industry. Every link in the food chain is incorporating industrial robots to boost output, produce goods of higher quality, meet or exceed consumer expectations, and provide better solutions to prevent work-related diseases. The dairy industry, where technologically advanced solutions have been implemented for a variety of purposes, is another very productive area of application for HRI in agriculture. It is common knowledge that early detection of health issues in the herd can provide significant animal benefits [56], including decreased discomfort, pain, and an extended lifespan, as well as significant benefits for the herdsman in the form of increased milk production, decreased feed costs, and decreased veterinary costs. Lately, a group of researchers [57] have developed a system that has the potential to improve dairy cow welfare by enabling continuous and automated monitoring of key health parameters, allowing for early detection of health issues and timely intervention. This system utilizes computer vision algorithms to analyze the 3D video data and extract relevant information about the cows' body condition, mobility, and weight. Additionally, the system estimates the cows' weight, providing valuable information for monitoring their growth and overall health. The use of a single 3D video capture device makes the monitoring process more efficient and cost-effective compared to traditional methods that require multiple devices or manual measurements. HRI has also been important for the development of new solutions to improve labor efficiency and prevent work-related disorders. It is well known that the physical demands of milking in tethered systems include lifting heavy objects, moving and carrying equipment, and awkward working positions, all of which increase the risk of developing Musculoskeletal Disorders (MSDs) [58]. In loose-housing systems, the monotonous and repetitive milking work is thought to increase the risk of upper extremity MSDs, particularly in the shoulders and wrists/hands [59]. This repetitive workload is present not only in large, industrialized dairy farms, such as those commonly found in the United States where herd sizes are substantial but also in smaller operations, like many in Italy, which still require sustained physical effort and high muscular load during milking tasks. Studies have shown that both environments pose significant ergonomic challenges, with large U.S. dairies often featuring fast-paced, high-intensity work and smaller Italian dairies presenting consistent, though less intense, physical demands [60, 61]. Since the first units were installed in the Netherlands in the early1990 s Automatic Milking Systems (AMS) have become widespread throughout Europe. In these facilities, groups of 50–70 cows are milked by a robotic arm that is installed in a milking box that is situated in the cow pen. The robotic arm records the placement of each cow's teat, attaches the milking unit, and collects the milk without the assistance of a human. The popularity of automated milking systems has surged in the US and Canada, primarily driven by the challenges associated with sourcing skilled labor for traditional cow-milking parlors [62]. In response to this labor shortage, the optimization of barn design and the integration of automated feeding systems have proven to be effective strategies for enhancing labor efficiency within automated milking dairy facilities [63]. Highcapacity milking parlors with a high throughput of cows per person per hour and automatic milking systems, in which manual labor is replaced by a milking robot, are the two main directions in which high-tech milking equipment is being developed. In a milking parlor with high capacity, the combination of automation and efficient labor utilization allows a single operator to milk multiple cows. In these parlors, a lone employee can milk as many as 125 cows within an hour, leaving only a few seconds to prepare the udder for milking. AMS replaces all manual labor involved in milking by employing robots. These systems typically feature either a one-stall setup, where a single robot arm operates a single milking stall, or a multiple-stall configuration, where a single robot arm handles several milking stalls simultaneously. However, it is important to note that the cows must voluntarily use the automatic milking machine, resulting in significant variation in the interval between milkings [64]. However, there are still several difficulties and AMS tendencies, such as: a) the technological performance, composition, and hygienic quality of raw milk obtained from robotically milking cows; b) the requirement for proper AMS management and upkeep to guarantee peak performance; c) the significance of cow welfare and comfort in the development and use of AMS; d) the capacity of AMS to increase labor productivity and lower labor requirements; e) the demand for additional study into the economic viability of AMS and their effects on milk production and quality [65].

5.2 *Forestry Sector*

Logging tasks are some of the most physically demanding in forestry, placing workers at significant risk for musculoskeletal disorders (MSDs). Activities like tree felling, delimbing, and bucking require repetitive movements, lifting heavy loads, and maintaining awkward postures, which collectively increase the physiological load and potential for MSDs among workers. Studies employing wearable sensors, such as the Zephyr Bioharness, have shown that felling tasks, in particular, require loggers to maintain a trunk inclination greater than 60 °C for an extended time, which correlates with increased load on the lumbar spine and elevated heart rates due to strenuous physical effort [66].

The integration of robotics and automation in forestry operations is increasingly viewed as essential for alleviating the physical demands and health risks that forestry workers face, particularly concerning musculoskeletal disorders (MSDs). Traditional logging tasks, such as tree felling, delimbing, and bucking, expose workers to repetitive and high-intensity physical exertion, which can contribute to significant biomechanical strain, particularly on the lumbar spine, thereby increasing the risk of MSDs [67, 68].

Recent advancements in mobile robotic systems have enabled precision forestry applications, such as accurate positioning of harvesters and single-tree navigation, which are already helping to alleviate some of these physical burdens by automating high-risk tasks [69]. In China, forest industrial robots introduced for tasks like cone picking, stump excavation, and logging have shown the potential for further reducing manual labor intensity [70]. Furthermore, studies such as La Hera et al.[71] demonstrate the viability of fully autonomous machines for essential tasks in forestry, including the use of unmanned forwarders equipped with computer vision and autonomous navigation for tasks like log collection and transportation. This system not only achieves efficient and safe forwarding but also holds promise for scaling autonomous forestry operations in challenging environments.

Advancements in automation, particularly in crane trajectory planning and control systems, also show promise for reducing the high physiological load associated with the manual handling of timber and logging equipment. Experimental platforms testing control systems and motion-planning algorithms aim to eventually manage these tasks with minimal human intervention, allowing workers to shift from high-risk physical roles to monitoring and managing automated systems [72]. Additionally, robotics research addressing logistics challenges in loading, unloading, and palletizing has further informed potential applications in forestry, suggesting that such automation could significantly reduce reliance on human muscle power, thus enhancing operational efficiency and reducing training time. These developments contribute to a more sustainable and health-conscious approach to forestry operations [73].

These technological advancements collectively suggest a shift toward an operational model where automation not only boosts productivity but also serves as a preventive strategy against occupational health risks such as MSDs. Autonomous systems, such as those developed by La Hera et al.[71], demonstrate the feasibility of fully automating strenuous tasks, providing a dual benefit of improving productivity and safety in forestry, ultimately paving the way for a sustainable, technology-driven future in forestry.

6 Safety in Agriculture

Safety is of critical importance when discussing aspects of HRI strategies in agricultural tasks, since humans are part of the loop but also need to protect the environment. Agriculture can be a hazardous industry, with potential risks to human workers due to various factors such as dangerous machinery, sharp tools, chemical exposure, and working in challenging environmental conditions. Besides, as in most industries, in agriculture MSDs are the most prevalent of all non-fatal illnesses [74, 75]. HRI in agriculture plays a crucial role in improving safety in agricultural operations and focuses on enhancing working conditions, agility, efficiency, productivity, and safety in the agricultural sector. The integration of robots in agricultural tasks can help reduce the risk of accidents and injuries to human workers. By automating repetitive and physically demanding tasks, robots can minimize human exposure to hazardous conditions, such as working with heavy machinery or handling toxic chemicals. Additionally, robots can be equipped with sensors and algorithms to detect and respond to potential safety hazards in real-time, promoting protective measures. Previous recent reviews of HRI in agriculture dealt with HRI concepts related to safety-ergonomics and productivity from management perspectives. HRI aspects on safety include among others robot navigation planning, interaction potential, and movement prediction. [76] approached safety from the social-aware robot navigation which takes into consideration naturalness, comfort, and sociability- meaning the farm workers perform activities unaffected by the simultaneous presence of robots that may be doing other activities. Agriculture-related HRI research is ongoing, with a focus on ensuring efficient teamwork and cooperation between humans and robots while giving human workers' health and safety a top priority. Robots must be equipped with advanced sensor systems to detect and respond to the presence of humans in their vicinity. Safety mechanisms, such as collision avoidance systems and emergency stop functions, are essential to prevent collisions and accidents between robots and humans [4]. To maintain safety, it is essential to distribute tasks between humans and robots properly. Tasks that are hazardous, physically demanding, or involve exposure to harmful substances should be assigned to robots. As a result, human workers can concentrate on jobs that call for greater cognition, judgment, and flexibility, where robots might still fall short [4, 8]. In addition, Humans and robots must have a good understanding of each other's actions, locations, and surroundings to prevent accidents and coordinate effectively. Technologies like semantic action recognition and advanced sensing help enhance situation awareness [8]. It's essential to provide human employees with the right training, and robot operators must follow specific safety procedures. Workers need to be taught how to interact safely with robots and comprehend the capabilities and limitations of the robotic systems they are using [4]. Lastly, for effective HRI in agriculture, intuitive and user-friendly interfaces are crucial. It should be simple for human workers to communicate with robots and get precise information about their intentions and actions [4].

7 Discussion and Future Trends

Based on the study presented up to this point and foresight studies review in the agricultural field [12, 77], we present below a discussion related to HRI strategies in agriculture, regarding possible applications, benefits, and future trends.

- The agricultural sector is undergoing a significant transformation toward Agriculture 4.0, where digital tools such as IoT, AI, and robotics are expected to optimize productivity, resource management, and crop quality. This shift aims to reduce reliance on traditional, labor-intensive methods, enhancing efficiency and sustainability. However, achieving widespread implementation will require addressing current barriers, including technical constraints, infrastructure limitations, and socio-economic challenges [12, 78].
- The reliance on big data, AI, and machine learning in modern agriculture will make data literacy a critical skill for future agricultural workers. Workers will need to interpret and apply insights derived from data to optimize crop yields, reduce waste, and improve overall farm efficiency, indicating a shift towards roles that blend agricultural knowledge with data science and IT skills [12].
- The use of AI algorithms, supervised, unsupervised, and reinforcement learning, opens many doors for HRI applications in agriculture. For example, computer vision algorithms have greatly benefited from convolutional neural networks for classification and detection applications, which are indispensable for robot perception for fluid and natural interaction and cooperation with humans. Likewise, generative algorithms and neural networks, such as large-scale large language models (LLMs), have revolutionized the development of verbal and non-verbal communication methods. These technologies allow human language to be interpreted, generated, and contextualized more accurately, facilitating more natural interactions between humans and intelligent systems.
- Autonomous systems, including robots and unmanned vehicles, are expected to handle physically demanding and hazardous tasks, such as harvesting on steep slopes or working in extreme weather. This shift not only reduces the physical strain on workers but also minimizes workplace injuries, making agricultural work safer and allowing workers to focus on supervisory and analytical tasks instead [77].
- With the growing complexity of smart farming, new specialized roles will emerge within interdisciplinary teams, including precision agriculture specialists, drone operators, data analysts, and robotic technicians. This will lead to a restructured workforce where collaboration between agricultural experts, engineers, and IT professionals becomes essential [77].
- Both, robots and human workers will need to develop the ability to collaborate effectively with autonomous tractors operating in the field. This entails establishing streamlined processes for assigning specific tasks to each agent within a given production process. The success of these interactions and collaborations hinges on the efficient allocation of responsibilities among the various entities involved, ensuring a harmonious and productive workflow in agricultural operations.

- The incorporation of HRI technologies into agriculture requires a profound knowledge of situational awareness for each task. The convergence of AI and advanced sensor technologies, particularly in vision-based systems, underscores the critical role of heightened situational awareness in optimizing agricultural processes. The ability of robots to comprehend and respond to dynamic environments is a testament to the strides made in enhancing situational awareness. Vision-based technologies, such as semantic recognition of actions, enable robots to discern and interpret human activities, locations, and interactions with their surroundings. This, in turn, empowers them to respond proactively to the ever-changing conditions in the agricultural landscape.
- Ergonomics, encompassing the design of systems for optimal human use and well-being, plays a crucial role in enhancing the efficiency, safety, and overall comfort of HRI in agricultural operations. The application of vision-based technologies, such as semantic recognition of actions, showcases a deliberate effort to align HRI with ergonomic principles. By accurately assessing human posture, location, and interactions with the environment, robots equipped with semantic recognition capabilities not only streamline tasks but also contribute to minimizing physical strain on agricultural workers. This symbiotic relationship between humans and robots, when crafted with a keen focus on ergonomic principles, holds the promise of revolutionizing work practices in agriculture. The pursuit of ergonomic excellence in HRI demands ongoing research, collaboration between stakeholders, and a commitment to designing technologies that complement the human workforce. As we navigate this intersection of technology and agriculture, a conscientious approach to ergonomics not only optimizes productivity but also fosters a sustainable and human-centered evolution in the agricultural landscape.
- One of the foremost challenges in the development of HRI systems for agriculture lies in creating robotic systems capable of seamless cooperation with humans. The pivotal goal is to ensure that the interaction is not only intuitive but also easily learnable for field workers. Effectively addressing the dynamics of verbal and non-verbal communication between humans and robots is crucial. A proficient and comfortable interaction hinges on clear communication channels, making the cooperation between individuals and robots both efficient and user-friendly. This emphasis on communication becomes paramount to fostering a harmonious and productive working relationship in agricultural settings.
- In the context of the autonomy level in agricultural processes, the forthcoming emphasis will not solely revolve around the challenging goal of achieving complete autonomy. Instead, the focus is anticipated to shift towards fostering full cooperation within agricultural processes. The objective will be to enhance the synergy between humans and robots, aiming for a level of cooperation that is both seamless and natural. This shift acknowledges the complexities often associated with achieving full autonomy in agricultural tasks while recognizing the potential for optimizing efficiency through a harmonious collaboration between human workers and robots.
- In the same way that mobile robots contribute to heightened productivity and alleviate human workload, the integration of aerial robots, akin to drones, holds immense

potential. These airborne robots can assist both human workers and mobile robots in real-time environmental monitoring, keeping track of various agents and assessing factors such as fruit density and health in the field. Beyond observation, aerial robots can be instrumental in tasks like fruit analysis and the precise application of pesticides and fertilizers. The deployment of such aerial robots stands to enhance the overall efficiency and precision of agricultural operations.

- Over the long term, individuals working in the agricultural sector, especially those involved in HRI, will likely find that their skill set needs to evolve. One crucial aspect of this evolution will be the ability to collaborate with a fleet of robots effectively. As technology continues to advance and autonomous systems become integral to agricultural processes, cultivating the skill of seamless cooperation with these robots will be essential for maximizing efficiency and productivity on the field. This includes understanding how to interface with, instruct, and coordinate tasks with multiple robots to harness their collective capabilities, ultimately contributing to the optimization of agricultural operations.
- Finally, another of the biggest challenges is focused on making HRI systems economically feasible for agricultural production companies. That is, in the medium or long term, the use of robots for agricultural applications should compensate for your investment with an increase in production and a decrease in workload. Another key aspect is that several crops need to be harvested in a short time to avoid spoilage, pests, or even adverse effects of the weather. In this case, the use of HRI strategies could help reduce possible losses in the harvest, which would make its use attractive to consumers.

8 Conclusions

In conclusion, the study of HRI in agriculture has evolved into a dynamic and promising field over the past decades. This multidisciplinary domain focuses on developing and evaluating robotic systems that can collaborate seamlessly with humans, optimizing various agricultural tasks, enhancing productivity, and improving overall farm management. The integration of HRI techniques in agricultural processes, such as spraying, harvesting, and irrigation, has demonstrated tangible improvements, leading to increased production efficiency. Advancements in AI algorithms, robust hardware, and accessible sensors have played a pivotal role in enhancing HRI research trends. Vision-based technologies, like semantic action recognition, enable robots to understand human activities, fostering improved collaboration between humans and robots. Looking ahead, the potential contributions of HRI in agriculture are vast. Anticipated developments include higher levels of automation, improved safety measures, enhanced productivity, greater situational awareness, and improved ergonomics. These advancements are expected to address current challenges and reshape agricultural practices by making cooperation between humans and robots more fluid, natural, and economically viable. Challenges remain, such as creating user-friendly interfaces, efficient communication channels, and encouraging cooperative robot interactions. The integration of drones and aerial robots into agricultural processes is also envisioned to further increase efficiency and effectiveness in tasks like monitoring and analysis. While HRI techniques present promising benefits in terms of production, safety, and farm management, it is essential to consider the broader societal impact of designs. The application of HRI in agriculture has addressed various challenges, but emerging issues such as sustainability, an aging population, and privacy and security require a more-than-human-centered design approach. This approach considers the interdependence of humans, non-human actors, and the environment, emphasizing a holistic perspective on design for sustainability and social responsibility, recently termed "humanity-centered design".

Funding This work was supported by ANID (National Research and Development Agency of Chile) under Fondecyt Iniciación 2024 Grant 11240105, Fondecyt de Postdoctorado N°3250059, ANID AC3E CIA250006, and ANID–Millennium Science Initiative Program–NCN2024_047.

References

1. Vasconez, J.P., Cheein, F.A.: Sleepiness detection for cooperative vehicle navigation strategies. In: 5th International Conference on Control, Decision and Information Technologies (CoDIT), pp. 940–945. IEEE (2018)
2. Vasconez, J.P., Cheein, F.A.: Finding a proper approach to obtain cognitive parameters from human faces under illumination variations. In: 5th International Conference on Control, Decision and Information Technologies (CoDIT), pp. 946–951. IEEE (2018)
3. Vasconez, J.P., Viscaino, M., Guevara, L., Cheein, F.A.: A fuzzy-based driver assistance system using human cognitive parameters and driving style information. Cogn. Syst. Res. **64**, 174–190 (2020)
4. Vasconez, J.P., Kantor, G.A., Cheein, F.A.A.: Human–robot interaction in agriculture: a survey and current challenges. Biosyst. Eng. **179**, 35–48 (2019b). URL: http://www.sciencedirect.com/science/article/pii/S1537511017309625, https://doi.org/10.1016/j.biosystemseng.2018.12.005
5. Vasconez, J., Admoni, H., Cheein, F.A.: A methodology for semantic action recognition based on pose and human-object interaction in avocado harvesting processes. Comput. Electron. Agric. **184**, 106057 (2021)
6. Vasconez, J.P., Salvo, J., Auat, F.: Toward semantic action recognition for Avocado harvesting process based on single shot multibox detector. In: 2018 IEEE International Conference on Automation/XXIII Congress of the Chilean Association of Automatic Control (ICA-ACCA), pp. 1–6. IEEE (2018)
7. Vasconez, J.P., Guevara, L., Cheein, F.A.: Social robot navigation based on HRI non-verbal communication: a case study on avocado harvesting. In: Proceedings of the 34th ACM/SIGAPP Symposium on Applied Computing, pp. 957–960 (2019a)
8. Adamides, G., Edan, Y.: Human-robot collaboration systems in agricultural tasks: a review and roadmap. Comput. Electron. Agric. **204**, 107541 (2023)
9. Edan, Y., Adamides, G., Oberti, R.: Agriculture automation. In: Springer Handbook of Automation, pp. 1055–1078. Springer (2023)
10. Vásconez, J.P., Basoalto, F., Briceño, I.C., Pantoja, J.M., Larenas, R.A., Rios, J.H., Castro, F.A.: Comparison of path planning methods for robot navigation in simulated agricultural environments. Procedia Comput. Sci. **220**, 898–903 (2023)

11. Moysiadis, V., Benos, L., Karras, G., Kateris, D., Peruzzi, A., Berruto, R., Papageorgiou, E., Bochtis, D.: Human-robot interaction through dynamic movement recognition for agricultural environments. AgriEngineering **6**(3), 2494–2512 (2024)
12. Abbasi, R., Martinez, P., Ahmad, R.: The digitization of agricultural industry—a systematic literature review on agriculture 4.0. Smart Agric. Technol. **2**, 100042 (2022)
13. Guevara, L., Wariyapperuma, P., Arunachalam, H., Vásconez, J., Hanheide, M., Sklar, E.: Robot-assisted fruit harvesting: a real-world usability study. In: 2024 IEEE 20th International Conference on Automation Science and Engineering, pp. 2517–2523. IEEE (2024)
14. Urvina, R.P., Guevara, C.L., Vásconez, J.P., Prado, A.J.: An Integrated route and path planning strategy for skid-steer mobile robots in assisted harvesting tasks with terrain traversability constraints. Agriculture **14**(8), 1206 (2024)
15. Moya, V., Quito, A., Pilco, A., Vásconez, J.P., Vargas, C.: Crop detection and maturity classification using a YOLOv5-based image analysis. Emerg. Sci. J. **8**(2), 496–512 (2024)
16. Estrada, J.S., Vásconez, J.P., Fu, L., Cheein, F.A.: Deep learning based flower detection and counting in highly populated images: a peach grove case study. J. Agric. Food Res. **15**, 100930 (2024)
17. Adamides, G., Katsanos, C., Parmet, Y., Christou, G., Xenos, M., Hadzilacos, T., Edan, Y.: HRI usability evaluation of interaction modes for a teleoperated agricultural robotic sprayer. Appl. Ergon. **62**, 237–246 (2017)
18. Adamides, G., Edan, Y.: Advances in human-robot collaboration in agricultural robotics. In: Advances in Agri-food Robotic, Burleigh Dodds Science Publishing (2024)
19. Vásconez, J.P., Basoalto, F., Briceño, I.C., Pantoja, J.M., Larenas, R.A., Rios, J.H., Castro, F.A.: Comparison of path planning methods for robot navigation in simulated agricultural environments. Procedia Comput. Sci. **220**, 898–903 (2023)
20. Yerebakan, M.O., Hu, B.: Human-robot collaboration in modern agriculture: a review of the current research landscape. Adv. Intell. Syst, 2300823 (2024)
21. Guevara, L., Wariyapperuma, P., Arunachalam, H., Vasconez, J., Hanheide, M., Sklar, E.: Robot-assisted fruit harvesting: a real-world usability study. In: 2024 IEEE 20th International Conference on Automation Science and Engineering (CASE), pp. 2517–2523. IEEE (2024)
22. Rysz, M.W., Mehta, S.S.: A risk-averse optimization approach to human-robot collaboration in robotic fruit harvesting. Comput. Electron. Agric. **182**, 106018 (2021)
23. Estrada, J.S., Vasconez, J.P., Fu, L., Cheein, F.A.: Deep Learning based flower detection and counting in highly populated images: a peach grove case study. J. Agric. Food Res. **15**, 100930 (2024)
24. Pilco, A., Moya, V., Quito, A., Vásconez, J.P., Limaico, M.: Image processing-based system for apple sorting. J. Image Graph 12(4), (2024)
25. Beloev, I., Kinaneva, D., Georgiev, G., Hristov, G., Zahariev, P.: Artificial intelligence-driven autonomous robot for precision agriculture. Acta Technologica Agri. **24**(1), 48–54 (2021)
26. Zha, J.: Artificial intelligence in agriculture. J. Phys. Conf. Ser. **1693**(1), 012058 (2020)
27. Sharma, S., Verma, K., Hardaha, P.: Implementation of artificial intelligence in agriculture. J. Comput. Cogn. Eng. **2**(2), 155–162 (2023)
28. Jha, K., Doshi, A., Patel, P., Shah, M.: A comprehensive review on automation in agriculture using artificial intelligence. Artif. Intell. Agric. **2**, 1–12 (2019)
29. Benos, L., Moysiadis, V., Kateris, D., Tagarakis, A.C., Busato, P., Pearson, S., Bochtis, D.: Human-robot interaction in agriculture: a systematic review. Sensors **23**(15), 6776 (2023)
30. Vásconez, J.P., Vásconez, I.N., Moya, V., Calderón-Díaz, M.J., Valenzuela, M., Besoain, X., Seeger, M., Auat, C.F.: Deep learning-based classification of visual symptoms of bacterial wilt disease caused by Ralstonia solanacearum in tomato plants. Comput. Electron. Agric. **227**, 109617 (2024)
31. Vásconez, J.P., Barona López, L.I., Valdivieso Caraguay, Á.L., Benalcázar, M.E.: Hand gesture recognition using EMG-IMU signals and deep q-networks. Sensors **22**(24), 9613 (2022)
32. Vásconez, J.P., López, L.I.B., Caraguay, Á.L.V., Benalcázar, M.E.: A comparison of EMG-based hand gesture recognition systems based on supervised and reinforcement learning. Eng. Appl. Artif. Intell. **123**, 106327 (2023)

33. Zhu, H., Qin, S., Su, M., Lin, C., Li, A., Gao, J.: Harnessing large vision and language models in agriculture: a review, (2024). arXiv:2407.19679 arXiv preprint
34. Didde, A.R.: Enhancing Agricultural Efficiency: The Role of Robotics, Spatial Mapping, and Multi-modal Language Models in Crop Management, (2024)
35. Zeng, F., Gan, W., Wang, Y., Liu, N., Yu, P.S.: Large language models for robotics: a survey, (2023). arXiv:2311.07226 arXiv preprint
36. Li, J., Xu, M., Xiang, L., Chen, D., Zhuang, W., Yin, X., Li, Z.: Foundation models in smart agriculture: Basics, opportunities, and challenges. Comput. Electron. Agric. **222**, 109032 (2024)
37. De Clercq, D., Nehring, E., Mayne, H., Mahdi, A.: Large language models can help boost food production, but be mindful of their risks. Front. Artif. Intell. **7**, 1326153 (2024)
38. Awais, M., Alharthi, A.H. S. A., Kumar, A., Cholakkal, H., Anwer, R.M.: AgroGPT: Efficient agricultural vision-language model with expert tuning, (2024). arXiv:2410.08405 arXiv preprint
39. Vásconez, J.P., Schotborgh, E., Vásconez, I.N., Moya, V., Pilco, A., Menéndez, O., Guamán, R., Guevara, L.: Smart delivery assignment through machine learning and the Hungarian algorithm. Smart Cities **7**(3), 1109–1125 (2024)
40. Zhang, Z., Pan, A., Li, X., Luo, Y.: Large-model and generative-intelligence agricultural robot systems. In: 2023 International Annual Conference on Complex Systems and Intelligent Science (CSIS-IAC), pp. 752–759. IEEE (2023)
41. Krupitzer, C.: Generative artificial intelligence in the agri-food value chain-overview, potential, and research challenges. Front. Food Sci. Technol. **4**, 1473357 (2024)
42. Endsley, M.R.: Design and evaluation for situation awareness enhancement. In: Proceedings of the Human Factors Society annual meeting, pp. 97–101. Sage Publications Sage CA, Los Angeles, CA (1988)
43. Endsley, M.R.: Situation awareness. In: Handbook of Human Factors and Ergonomics, pp. 434–455. (2021)
44. Urvina, R.P., Guevara, C.L., Vásconez, J.P., Prado, A.J.: An integrated route and path planning strategy for skid-steer mobile robots in assisted harvesting tasks with terrain traversability constraints. Agriculture **14**(8), 1206 (2024)
45. Vasconez, J.P., Carvajal, D., Auat Cheein, F.: On the design of a human-robot interaction strategy for commercial vehicle driving based on human cognitive parameters. Adv. Mech. Eng 11(7), (2019)
46. Gatsoulis, Y., Virk, G.S., Dehghani-Sanij, A.A.: On the measurement of situation awareness for effective human-robot interaction in teleoperated systems. J. Cogn. Eng. Dec. Making **4**(1), 69–98 (2010)
47. Drury, J.L., Keyes, B., Yanco, H.A.: LASSOing HRI: analyzing situation awareness in map-centric and video-centric interfaces. In: 2nd ACM/IEEE International Conference on Human-Robot Interaction (HRI), pp. 279–286. IEEE (2007)
48. Reina, G., Milella, A., Rouveure, R., Nielsen, M., Worst, R., Blas, M.R.: Ambient awareness for agricultural robotic vehicles. Biosys. Eng. **146**, 114–132 (2016)
49. Zhang, G., Minakata, K., Hansen, J.P.: Enabling real-time measurement of situation awareness in robot teleoperation with a head-mounted display. In: 50th Nordic Human Factors and Ergonomics Society (NES) Conference: Future Work: Digitalisation and Innovation, (2019)
50. Dini, A., Murko, C., Yahyanejad, S., Augsdörfer, U., Hofbaur, M., Paletta, L.: Measurement and prediction of situation awareness in human-robot interaction based on a framework of probabilistic attention. In: IEEE/RSJ International Conference on Intelligent Robots and Systems (IROS), pp. 4354–4361. IEEE (2017)
51. Adamides, G.: User Interfaces for Human-Robot Interaction: Application on a Semi-Autonomous Agricultural Robot Sprayer. Doctoral dissertation, Open University of Cyprus (2016)
52. Rysz, M.W., Mehta, S.S.: A risk-averse optimization approach to human-robot collaboration in robotic fruit harvesting. Comput. Electron. Agric. **182**, 106018 (2021)

53. Vasconez, J.P., Auat Cheein, F.A.: Workload and production assessment in avocado harvesting processes using HRI strategies based on action recognition. Biosyst, Eng (2022)
54. Bergerman, M., Maeta, S.M., Zhang, J., Freitas, G.M., Hamner, B., Singh, S., Kantor, G.: Robot farmers: autonomous orchard vehicles help tree fruit production. IEEE Robot. Autom. Mag. **22**, 54–63 (2015). https://doi.org/10.1109/MRA.2014.2369292
55. Heard, J., Harriott, C.E., Adams, J.A.: A survey of workload assessment algorithms. IEEE Trans. Human-Mach. Syst. **48**, 434–451 (2018)
56. Green, L., Huxley, J., Banks, C., Green, M.: Temporal associations between low body condition, lameness and milk yield in a UK dairy herd. Prev. Veter. Med. **113**, 63–71 (2014). https://www.sciencedirect.com/science/article/pii/S0167587713003036, https://doi.org/10.1016/j.prevetmed.2013.10.009
57. Hansen, M.F., Smith, M.L., Smith, L.N., Abdul Jabbar, K., Forbes, D.: Automated monitoring of dairy cow body condition, mobility and weight using a single 3D video capture device. Comput. Ind. **98**, 14–22 (2018). https://doi.org/10.1016/j.compind.2018.02.011
58. Masci, F., Rosecrance, J., Mixco, A., Cortinovis, I., Calcante, A., Mandic-Rajcevic, S., Colosio, C.: Personal and occupational factors contributing to biomechanical risk of the distal upper limb among dairy workers in the Lombardy region of Italy. Appl. Ergon. **83**, 102796 (2020). https://doi.org/10.1016/j.apergo.2018.12.013
59. Masci, F., Crespi, E., Pernigotti, E., Tassoni, M., Rosecrance, J., Colosio, C.: Carpal tunnel syndrome among milking parlor workers in Northern Italy: a comparison of screening approaches. La Medicina del Lavoro **110,** 271–277 (2019). https://doi.org/10.23749/mdl.v110i4.8359
60. Mixco, A., Masci, F., Brents, C.A., Rosecrance, J.: Upper limb muscle activity among workers in large-herd industrialized dairy operations. Front. Public Health **4**, 134 (2016)
61. Masci, F., Mixco, A., Brents, C.A., Murgia, L., Colosio, C., Rosecrance, J.: Comparison of upper limb muscle activity among workers in large-herd US and small-herd Italian dairies. Front. Public Health **4**, 141 (2016)
62. Koning, De., Rodenburg, K., J.: Automatic milking: state of the art in Europe and North America. In: Automatic Milking, a Better Understanding, pp. 25–37. (2004)
63. Pitkäranta, J., Kurkela, V., Huotari, V., Posio, M., Halbach, C.E.: Designing automated milking dairy facilities to maximize labor efficiency. Veter. Clin. Food Animal Pract. **35**, 175–193 (2019)
64. Hogeveen, H., Ouweltjes, W.: Sensors and management support in high-technology milking. J. Anim. Sci. **81**, 1–10 (2003)
65. Cogato, A., Brščić, M., Guo, H., Marinello, F., Pezzuolo, A.: Challenges and tendencies of automatic milking systems (AMS): a 20-years systematic review of literature and patents. Animals Open Access J. **11**, 356 (2021). https://doi.org/10.3390/ani11020356
66. Masci, F., Spatari, G., Giorgianni, C.M., Bortolotti, S., Rosecrance, J., Colosio, C.: A wearable device to assess the spine biomechanical overload in a sample of loggers. In: Proceedings of the 21st Congress of the International Ergonomics Association (IEA 2021) Volume V: Methods & Approaches, vol. 21, pp. 162–170. Springer (2022)
67. Masci, F., Spatari, G., Giorgianni, C.M., Pernigotti, E., Antonangeli, L.M., Bordoni, V., Magenta Biasina, A., Pietrogrande, L., Colosio, C.: Hand-wrist disorders in chainsaw operators: a follow-up study in a group of Italian loggers. Int. J. Environ. Res. Public Health **18**(14), 7217 (2021)
68. Masci, F., Spatari, G., Bortolotti, S., Giorgianni, C.M., Antonangeli, L.M., Rosecrance, J., Colosio, C.: Assessing the impact of work activities on the physiological load in a sample of loggers in sicily (Italy). Int. J. Environ. Res. Public Health **19**(13), 7695 (2022)
69. Rossmann, J., Schluse, M., Schlette, C., Buecken, A., Krahwinkler, P., Emde, M.: Realization of a highly accurate mobile robot system for multi purpose precision forestry applications. In: 2009 International Conference on Advanced Robotics, pp. 1–6. IEEE (2009)
70. Guo, X., Lu, H.: Recent development of forest industrial robot in China. In: 2010 International Conference on Intelligent Computation Technology and Automation, vol. 2, pp. 984–987. IEEE (2010)
71. La Hera, P., Mendoza-Trejo, O., Lindroos, O., Lideskog, H., Lindbäck, T., Latif, S., Songyu, L., Karlberg, M.: Exploring the feasibility of autonomous forestry operations: results from the first experimental unmanned machine. J. Field Robot. **41**(4), 942–965 (2024)

72. Ortiz Morales, D., Westerberg, S., La Hera, P.X., Mettin, U., Freidovich, L., Shiriaev, A.S.: Increasing the level of automation in the forestry logging process with crane trajectory planning and control. J. Field Robot. **31**(3), 343–363 (2014)
73. Echelmeyer, W., Kirchheim, A., Wellbrock, E.: Robotics-logistics: challenges for automation of logistic processes. In: IEEE International Conference on Automation and Logistics, pp. 2099–2103. IEEE (2008)
74. Benos, L., Bechar, A., Bochtis, D.: Safety and ergonomics in human–robot interactive agricultural operations. Biosyst. Eng. **200**, 55–72 (2020). URL: https://www.sciencedirect.com/science/article/pii/S1537511020302506, https://doi.org/10.1016/j.biosystemseng.2020.09.009
75. Vásconez, J.P., Calderón-Díaz, M., Briceño, I.C., Pantoja, J.M., Cruz, P.J.: A behavior-based fuzzy control system for mobile robot navigation: design and assessment. In: International Conference on Advanced Research in Technologies. Information, Innovation and Sustainability, pp. 412–426. Springer, Cham (2023)
76. Anagnostis, A., Benos, L., Tsaopoulos, D., Tagarakis, A., Tsolakis, N., Bochtis, D.: Human activity recognition through recurrent neural networks for human-robot interaction in agriculture. Appl. Sci. **21**, 20 (2021)
77. Prager, S., Wiebe, K.: Foresight for agriculture: past ghosts, present challenges, and future opportunities, (2020)
78. Rojas, J.I.S., Pantoja, J.M., Matamala, M., Briceño, I.C., Vásconez, J.P., Romero-Conrado, A.R.: An IoT-based prototype for optimizing agricultural irrigation: a case study in the Biobío region of Chile. Procedia Comput. Sci. **238**, 1009–1014 (2024)

Social Robotics and Human–Robot Communication

Guy Laban and Emily S. Cross

Abstract This chapter explores the ergonomic, cognitive, and psychological mechanisms underpinning human–robot communication, focusing on the role of social robots in shaping human perception, behaviour, and socio-emotional engagement. By analysing factors such as robot design, embodiment, and communication modalities, the chapter highlights how these elements influence user interaction and the perception of robots as social entities. It examines the integration of verbal and non-verbal communication modalities, addressing challenges in reciprocity and robots' behavioural appropriateness. The discussion extends to the implications of adaptive technologies like large language models, emphasising the balance between responsiveness and confidentiality in human–robot interactions. Furthermore, the chapter underscores the potential of social robots in workplace and therapeutic settings to enhance efficiency, safety, and emotional well-being of human interactants. By addressing user expectations and ethical considerations, the chapter seeks to provide a comprehensive framework for advancing social robotics applications in fostering meaningful human–robot relationships.

G. Laban
Department of Industrial Engineering and Management, Ben-Gurion University of the Negev, Beer Sheva, Israel

School of Brain Sciences and Cognition, Ben-Gurion University of the Negev, Beer Sheva, Israel

The Azrieli National Center for Autism and Neurodevelopment Research, Ben-Gurion University of the Negev, Beer Sheva, Israel

G. Laban (✉) · E. S. Cross
School of Psychology and Neuroscience, University of Glasgow, Glasgow, UK
e-mail: laban@bgu.ac.il

E. S. Cross
e-mail: emily.cross@gess.ethz.ch

E. S. Cross
MARCS Institute for Brain, Behaviour and Development, Western Sydney University, Sydney, New South Wales, Australia

Professorship for Social Brain Sciences, ETH Zurich, Zurich, Switzerland

G. Barresi and S. Wischniewski (eds.), *Ergonomics in Robotics: Advances and Innovations*, Springer Tracts in Advanced Robotics 162,
https://doi.org/10.1007/978-3-032-19374-2_8

Keywords Social robots · Human–robot interaction · Perception · Social cognition · Behaviour · Emotional wellbeing · Cognitive ergonomics

1 Introduction

As most readers of this volume will be well acquainted with, a robot can be defined as a programmable, multifunctional manipulator designed to move material, parts, tools or devices through variable programmed motions to achieve a variety of tasks [1]. The concept of robots that have a social (*anthropomorphic* or *zoomorphic*) appearance or behave in a social (human-like or animal-like) way has become increasingly prominent through post-modern science fiction literature, art, and cinema [2]. Yet, the widely-used definition for *social robots* in applied (non-fictional) settings was proposed only in the last century, and specifies that such robots are autonomous machines that interact and communicate with humans or other agents by following social behaviours and rules relevant to the specific role for which they have been designed [3]. Social robots are often expected to perform a variety of complex tasks in a wide range of human-centred contexts and social settings, or are expected to be able to engage humans in some form of social interaction. These tasks can range from socially simple actions, such as delivering drinks to guests at a cocktail party, to socially complex activities that involve expressive communication, such as engaging in conversations and discussing people's lives and emotions. As the potential integration of social robots is increasingly discussed across various social settings, including commerce, services, healthcare, education, and even in households [2], researchers are exploring how these machines might assume diverse roles and responsibilities while identifying the factors that distinguish social robots from non-social industrial machines. Accordingly, in this chapter we discuss the ergonomic, cognitive, and psychological mechanisms underlying human–robot communication, focusing on how social robots shape human perception, behaviour, and socio-emotional engagement through their design, embodiment, behaviour and communication modalities.

2 Social Robot Perception: Between Person Perception and Object Perception

When we first meet someone new, we often look for physical or other visual cues in that person's appearance to establish a *social perception*, as these are highly accessible cues [4]. On these occasions people tend to rely on their senses of sight and hearing to gather information. This means that people are more likely to notice a person's physical appearance and other visual cues before noticing other, less visible aspects of that person's identity, such as their personality or interests [5]. Additionally, physical and visual cues tend to be more salient than other types of cues, meaning that they stand out more in our perceptual field [4]. For example, a person's clothing or body language might be more noticeable than their accent or the way they speak, and these might be more noticeable than their personality or actions. This salience makes physical and visual cues easier to process and remember. Due to previous knowledge and our experience in the world, we tend to make assumptions regarding one's role in the world and form a social perception based on these visual features [6].

When perceiving non-human objects in the world, we also look for visual features that enable us to further understand them [7]. *Object perception* involves both bottom-up and top-down processes. *Bottom-up* processes involve the processing of sensory information from the environment, such as color, shape, texture, and motion, to create a basic representation of an object. *Top-down processes* involve the use of prior knowledge, expectations, and context to help interpret the sensory information and fill in missing details [8]. Moreover, object perception also involves the use of perceptual grouping principles, such as proximity, similarity, and continuity (i.e., *Gestalt laws of grouping*), to organize the sensory information into meaningful objects [9]. The ability to perceive objects is essential for everyday tasks, for recognizing objects' purpose, usability, and navigating our environment [7]. As such, objects' appearances tend to follow the concept of *"Form Follows Function"* [10], which states that an object's design (originally addressed to buildings) should be determined by its intended purpose and function [11].

Considering these principles in relation to robots, users rely on the physical features of robots to classify these machines, interpret their intentions and uses, and determine whether they serve a social purpose. Users look at design features such as shape and *embodiment* (i.e., the location and the character of the body in the world, and how this body structures and enables experience [12]). Thus, robots are often designed and engineered to convey their purpose and capabilities via their physical manifestation. In contrast, industrial (or, simply non-social) robots are ultimately engineered and designed according to the *ergonomic* (i.e., the processes of designing and arranging products and systems to fit their users and/or their hosting organization [13]) requirements of the industrial task (e.g., moving materials or assembling parts) such as safety, efficiency, human operation, as well as considering related mechanical considerations [14]. While these factors would still be considered important for developing and designing a social robot, ergonomic factors and requirements related to the social nature of the task or the social settings where the robot is designed to operate would be crucial considerations. Accordingly, social robots are often designed with principles of *cognitive ergonomics* in mind, which are concerned with mental processes during people's interactions with these machines, such as perception, attention, memory, reasoning, decision-making, learning, and motor responses, among others [15]. Different features of the design (e.g., layout and elements), system (e.g., modality and robotic behaviour), performance (e.g., success and failure), and implementation (e.g., organizational measures) are designed to engage cognitive processes (e.g., perception and reasoning) that affect the perception of social robots in a variety of social situations (e.g., communicating, learning, playing) and social settings (e.g., education, health care, the household) [2].

These features are being used as *affordances*, clues to the ways that people understand and negotiate technology in everyday lives [16]. These clues provide individuals with visual means to establish a relationship between an object and its environment. *Perceived affordances* [17] in the robot's design support users to clearly understand the robot's capabilities without needing any further explanation (e.g., a robot with legs should be able to walk, or a robot with arms should be able to grab and manipulate, or carry stuff). More specifically related to social robots (and differentiating these from non-social robots) are *social affordances* [18]. These are clues for perceived opportunities for social interaction that are present in the environment, and pertain to the ways in which robots are designed and programmed to be perceived as social entities that invite social interactions. Social affordances can include properties like the robot's appearance, movement, and behaviour. Accordingly, social affordances also enable users to attribute social qualities to robots (e.g., friendliness, warmth, empathy), experience social elements related to the robot (e.g., trust, acceptance, enjoyment), and respond socially towards robots (e.g., talk, learn from, seek companionship) [19]. For example, a robot with a *humanoid* (i.e., a robot resembling a human body in shape [2]) appearance and expressive movements may be perceived as more social, approachable, and interactive than a robot with a more mechanical appearance (i.e., *product-oriented* [20]) and limited movements (c.f., [21]).

3 Attributing Social Meaning to Robots via Social Cognition

Beyond usability features of social robots (i.e., affordances), it is important to consider that the quality of being social is predominantly associated with living organisms [22]. Therefore, when we attribute social qualities to robots, we could be looking for social traits that are common across social living agents–humans or animals [23], depending on the robot's role. The application of this idea in interactions with artificial agents (like social robots) is conveyed in the "*Computers as Social Actors*" (CASA) paradigm [24]. This theoretical framework explains that people tend to attribute social characteristics, such as intentions and emotions, to computers, artificial agents, and other forms of technology and media. This is based on the idea that people interact with technology in ways similar to how they interact with other people, leading them to perceive the technology as having similar traits. When following the social affordances of social robots (as well as other artificial agents) people tend to mindlessly attribute social characteristics to these agents and perceive and react to these accordingly. This theory has informed the design of social artificial agents to enhance trust, social interaction quality, and rapport building with robots and other artificial agents (e.g., [25]). Moreover, several authors have posited that the cognitive process of observing a robot in a social context should more or less resemble the process of observing another social being (human or animal), due to the way the human brain is wired to process social information. Our brains have evolved to be highly attuned to *social cues* that signal the intentions and emotions of others [26, 27]. These are nonverbal or verbal behaviours, like facial expressions, body language, making eye contact, nodding, vocal cues, and using gestures, that people use to communicate information and convey meaning in social interactions [26]. When we observe a social entity, such as a human, an animal, or, for the scope of this chapter, a robot, several brain regions are activated in the process of *social cognition* [27].

This process begins with the *fusiform gyrus*, which is a brain region associated with facial recognition that is reliably activated when we look at faces [28, 29], whether human or robotic [30, 31]. This is followed by activation of the *superior temporal sulcus*, which is associated with interpreting social cues provided by movement and posture for understanding social actions [32, 33], as well as language perception [32, 34]. As we observe a social being, the *amygdalae*, which process emotional information, also become active, allowing us to interpret the emotions of others [35–37]. The *inferior parietal lobule*, which is implicated in understanding others' intentions, is activated next, helping us to infer what the social entity is thinking or feeling [38–40], also when this entity is a robot [31, 41–43]. The *inferior frontal gyrus*, which is associated with speech processing (among other things), also becomes active, allowing us to understand verbal and nonverbal social cues of both humans [44, 45] and robots [31, 46–48]. Finally, the *temporo-parietal junction* and the *medial prefrontal cortex*, which assist with *theory of mind* [49] processes (i.e., the capacity to understand others by ascribing mental states to them [50, 51]), are activated when we try to infer the mental states of others [52–54], including robots [31, 48, 55, 56]. Through this process, complementary networks and systems within the human brain enable us to understand social cues provided by social agents (whether human or robotic), and make inferences about their intentions, emotions and mental states.

Accordingly, we tend to interpret the appearance and behaviours of robots in terms of human-like (or animal-like, depending on the robot) characteristics such as emotions, intentions, and mental states [57]. However, it is worth noting that the level of activation of these regions is not the same when observing a robot as when observing a human, as the level of complexity and realism of the robot can affect the activation of these regions, among several other important factors (see [58–60] for further discussion). Moreover, the likelihood of attributing social qualities and mental states to robots is influenced by a variety of other factors, including the age, motivation, and prior experiences/beliefs of the human observer, as well as the behaviour, appearance, and framing of the robot (see review [59]). Therefore, we should further consider the different factors that shape our social perception and social behaviour when interacting with robots.

4 Facilitating Robot Communication via Robotic Communication Modalities

As social robots are increasingly deployed across diverse settings, they must engage users through a variety of communication modalities to foster effective and socially meaningful interactions. The ability to employ both verbal and non-verbal channels is critical, enabling robots to better adapt to human communicative expectations and social cues. However, while this expanding range of input and output modalities offers new avenues for engaging users, it also presents specific ergonomic and technical challenges.

Verbal communication is a cornerstone of social interaction [61], and recent advances in natural language processing enable robots to interact with human users through increasingly nuanced dialogue. Robots equipped with sophisticated speech recognition and synthesis can extend beyond simple commands to facilitate ongoing conversational exchanges [62]. This capability is especially relevant for social robots, where natural conversation can enhance user comfort and engagement, and should match the robot's embodied design [63]. However, the ability to navigate conversational nuances—such as shifts in topic, emotional tone, and user intent—remains challenging [64]. Misinterpretations or delayed responses can undermine the interaction, emphasizing the need for continuous development in language processing that is adaptive and context-sensitive. Non-verbal cues, including gaze, facial expressions, body language, and gestures, play a fundamental role in human social cognition and significantly influence how social robots are perceived [65]. Robots that can perform contextually appropriate gestures and maintain eye contact can enhance user engagement, establishing a greater sense of presence and responsiveness [2]. Yet, aligning these non-verbal cues with verbal responses poses a notable challenge. A mismatch in timing or lack of synchronization can disrupt the flow of interaction, leading to decreased user satisfaction [66]. Additionally, designing robots to perform culturally appropriate non-verbal behaviours is essential, as non-verbal expectations vary widely across user demographics and interaction contexts [67]. A multimodal approach, where verbal and non-verbal communication channels work in tandem, is crucial for achieving more lifelike and socially intuitive HRIs. Effective multimodal communication requires precise synchronization between verbal and non-verbal outputs, allowing robots to respond in ways that appear natural and contextually relevant [68, 69]. However, achieving seamless coordination between these modalities remains a complex task, as technical limitations often prevent robots from providing fully integrated responses. Misalignments between speech, gaze, and gesture can diminish the perceived social competence of the robot, underscoring the need for robust systems that manage these modalities in a cohesive way (e.g. [70]).

The variety of settings in which social robots operate necessitates the ability to adapt communication modalities based on context. A robot interacting with patients in a healthcare setting, for instance, must utilize a different communication approach than one used in educational or personal assistant roles [2]. This adaptability not only enhances interaction quality but also aligns with user expectations for appropriate and supportive social responses [71]. However, ensuring that robots can interpret and respond to contextual data without compromising user privacy is a significant challenge [72].

The expansion of communication modalities presents an opportunity to enhance the interaction experience, especially in settings where emotional engagement and social support are essential [73]. Advanced sensory inputs such as tactile feedback and proximity sensing can support nuanced interactions, allowing robots to respond attentively to user presence. This capability is particularly valuable in therapeutic and collaborative settings, where heightened sensitivity to user needs is critical [74]. By leveraging these expanded modalities, social robots can more effectively meet ergonomic and emotional requirements, promoting more positive and meaningful human–robot interactions (HRIs).

5 The Role of Embodiment and Social Cues in Human–Robot Communication

Physical face-to-face interactions convey information through unrestricted verbal expression, gestures and facial displays, and through continuous collaboration between both parties of the interaction. The presence and interaction of these features together are unique to physical face-to-face interactions, whereas other modes of interaction are limited in extracting and demonstrating a wide range of social cues [75]. In the context of information technologies, the *media richness theory* (MRT) [76] explains that a communication medium's ability to reproduce the information sent through it is driven by its ability to communicate a complex message adequately. Hence, social interactions typically perform better through media with the capacity to convey richer social cues, like gestures and body language [76]. This idea emphasizes the importance of *social embodiment* (i.e., the states of the body, such as postures, arm movements, and facial expressions that arise and are manipulated during social interactions and play a central role in social information processing [77]) and *physical embodiment* (i.e., having a physical body or physical instantiation [78]) in interactions with artificial agents (and particularly in HRIs). The theory of *embodied cognition* explains (in the context of social cognition) that the body has an active and significant role in understanding an agent's mind and cognitive capacities. When an object (for the scope of this chapter, an artificial agent) is embodied, it takes on a physical representation in the world that describes its capabilities, roles, and features. On the other hand, when the object is *disembodied*, it takes on an abstract representation that might limit our understanding of it [79]. As humans, we find it easier to understand similar bodily experiences to our own embodied experiences [80].

When interacting with artificial agents, people perceive and respond to their reciprocated behaviour in different ways depending on the agents' embodiment [63, 81]. People tend to perceive robotic agents more socially when using embodied socially relevant information through body language and when the agent can express itself using a wide range of social cues [63, 81], such as voice (e.g. [82]), gaze (e.g. [82, 83]), facial expression (e.g. [84]) or even touch (e.g. [85–87]). Moreover, the effect of embodiment is even more distinct when it is physical, considering the wider variations and freedom of the agent's behaviour and reactions [88]. Therefore, embodied robots that utilise different social cues and are physically present may be perceived as more social and responsive than a robot that does not have these capabilities.

Nevertheless, while embodiment and the presence of a wide variety of social cues would encourage users to interact with social robots more socially, it could also hamper the interaction. The "*uncanny valley*" [89] is a phenomenon where a robot or other non-human entity that is almost, but not quite, human-like in appearance can cause a sense of unease or revulsion in some people. This theory suggests that as the appearance and behaviour of a robot become more human-like, people's social (as well as emotional) response to the robot becomes increasingly positive until it reaches a point where the robot is almost, but still not quite, human-like. At this point, people's social and emotional responses become increasingly negative, creating a "valley" in their social perception and response to the robot. In other words, increased (human) realism of a robotic agent does not necessarily imply acceptance. This theory has been tested and studied in various social settings using different types of measurements [90] further explaining that people may be more likely to perceive a robot with a highly human-like appearance as uncanny or creepy.

6 Communication Reciprocity and Robots' Social Behaviours

As the physical features of robots hold an important role in the psychology of social HRI, so does the robot's behaviour. Since social robots are designed to perform specific tasks and functions, their behaviour can influence how they are perceived. In order to take on social roles, robots need to be able to partake in *two-way interaction*, expecting a robot to be able to respond to a human in a social manner [2]. Hence, at their most desirable form, social robots should be engineered to initiate social interactions (c.f., [91]), understand and respond to humans' speech or social cues (c.f., [74, 92, 93]), provide feedback (see review [94]), engage in turn-taking (see review [64]), collaborate (see [95]), and adapt (see [96]).

Users often expect a certain degree of *behavioural reciprocity* when interacting with other humans (see [97]), and accordingly, when interacting with artificial agents whose objective is to fulfil a social role, such as social robots [98]. Behavioural reciprocity is a response or reaction from an interactant that can encourage or attenuate the interaction and the relationship in general [97]. It could be an appropriate verbal response, but could also include a variety of social cues that are being manipulated during an interaction, like eye contact, interpersonal distance, gestures of affiliation (such as nods, sounds, and mimicry), touch, posture, and hand gestures [99]. Throughout an interaction, both interactants interpret and react to each other's behaviours and disclosed content. They attribute meaning to these behaviours and use them to balance and regulate their behaviour to achieve an *equilibrium* by reciprocating behaviours. When the equilibrium is not balanced, individuals tend to socially withdraw from the interaction [97]. Accordingly, humans would expect a robot to show reciprocity in their behaviour, and if an equilibrium is not achieved it might affect the extent to which a robot is perceived as social, and the extent to which individuals socially interact with it. However, users may attribute reciprocal meaning to robot behaviours that would not typically be considered social or reciprocal, provided these behaviours meet their needs [100]. For instance, previous studies have shown that people tend to engage more with robots when experiencing negative emotions [101], by providing a channel for disclosure [102] or when they lack opportunities to discuss their emotional challenges with others [103, 104].

7 Users' Expectations as a Facilitator of Human–Robot Communication

The role of reciprocity sets the stage for one more behavioural factor that is crucial for understanding users' communication with social robots. The novelty effect is a common problem in social robotics, and long-term studies have often found a reduced engagement with various robotic platforms over time [105]. As social robots are a new emerging technology that is exciting for most, users often have higher expectations of social robots and experience dissonance when a social robot's performance is not in line with their expectations. Accordingly, when users interact with robots over time, they tend to perceive it as being less social as the interaction (or interactions) go on as their expectations of the robot are not being fulfilled. Previous studies show that even household robotic devices that are not particularly social (like the Roomba vacuum cleaner) suffer from the novelty effect [106], with users being excited about the robotic device at first and using it less as they get familiar with it. Accordingly, it should be noted that social robotic behaviour is not solely driven by social factors, but also has to do with general *competency* (i.e., the efficiency of performance) and *user experience* (i.e., user's perceptions of utility, ease of use, and efficiency of intelligent interfaces [107], including social robots [108]). Like other robotic devices, a social robot should be able to complete the task it was designed and developed to complete. A previous study demonstrated that when robots take on anthropomorphic embodiment, their failures and lack of competency are even more pronounced [109].

This represents the intersection between a robot's visual appearance and its functionality, as the robot's appearance often conveys affordances about its capabilities. If the robot fails to meet users' expectations based on these affordances, it is likely to be perceived as incompetent [66]. There is, in fact, a certain dissonance that often occurs in HRIs when a person is not sure how to behave or interact with the robot, or, more importantly, when the robot's behaviour or functionality is not in line with the person's expectations or preconceived notions about robots. This dissonance has previously been termed the *social robot paradox* [110]. The idea of robots helping with all aspects of daily life, as depicted in science fiction, has fuelled our imagination about the possibilities of the future [111]. These ideas are even amplified when considering the physical and social affordances of robots, setting high expectations for robotic behaviour. Although these fictional robots continue to be a distant vision, they have influenced our understanding and expectations of what autonomous technology could potentially achieve [2, 110].

This dissonance extends beyond basic aspects of competency and user experience, arising also when a social robot's appearance is inconsistent with the social context or norms associated with its behaviour. Similar to the uncanny valley effect [89], when a social robot is designed in a way that suggests it is suited for social interaction (e.g., a humanoid robot intended for verbal communication with humans) but then exhibits behaviours that are not social (e.g., an inability to sustain a conversation), it is often perceived as less social, leading people to withdraw from the interaction (c.f., [112]). This idea is augmented by the *mind perception* theory [113]. While the uncanny valley is focused mostly on the aesthetic and visual features of the robot, mind perception provides an additional approach focusing on the social role and functioning of the robot. The theory explains that people often ascribe two key and distinct dimensions of mind to other social agents around them – "*agency*" (referring to an agent's ability to act independently) and "*experience*" (referring to an agent's ability to sense and feel) [114]. Therefore, people often perceive and react to other social entities, human or non-human, based on the moral judgment of their behaviour – whether these can be responsible for their actions (i.e., demonstrating agency), and whether they can understand others and show patience (i.e., demonstrating experience). When one's actions do not meet these definitions of social behaviour, we would be less likely to react to this agent in a social way [113]. Thus, a social robot would be perceived as more social when behaving in a socially moral way, following the rules and expectations of human behaviour. Previous studies show further support for mind perception via people's behaviour towards robots as well as affecting their social attitudes in different settings. For example, people are more likely to follow a robot's instructions when it asks them to participate in a task that is revocable rather than irrevocable [115]. Another study showed that people were more likely to change their decisions in a decision-making task following advice from a robot that was presented via logical reasoning rather than via a judgmental fallacy [116]. Hence, when the robot's behaviour is appropriate for its role and consistent with users' social expectations of it, people may perceive and behave more socially towards it.

8 Ergonomics of Social Robots' Communication When Working with People

Building on the discussion of social cognition and the cues that shape human perception of robots, we turn to examine how social robots contribute specifically to ergonomics within workplace settings. Unlike industrial robots strictly confined to repetitive tasks, social robots in these environments are designed with both cognitive and physical ergonomic principles in mind, aiming to support human well-being, safety, and efficiency through effective communication and adaptive behaviour.

Cognitive ergonomics in the workplace benefits significantly from the interactive communication abilities of social robots. Social robots engage with workers through verbal and nonverbal signals that provide adaptive and context-sensitive support, thus reducing mental load [117]. For example, robots deployed in customer-facing roles in hospitality or healthcare not only handle routine inquiries, but communicate in ways that facilitate workflow. By effectively interpreting human speech and body language, social robots can respond with personalised answers, directions, or reminders, allowing employees to focus on more complex, judgement-based responsibilities. This adaptive, socially attuned communication style enables the robot to become an active partner in managing tasks, mitigating cognitive strain by freeing employees from repetitive tasks and sustaining engagement, which is crucial in high-stress environments [118]. In physically demanding environments, such as assembly lines or logistics, social robots play a critical role in supporting safe HRI. Designed to enhance both safety and efficiency, these robots use embodied social cues—like gestures, adaptive movements, and spatial awareness—to communicate their actions and intentions to human co-workers, fostering a safer, more coordinated workflow [14]. Unlike traditional robots, which operate mechanically and are often perceived as rigid or detached, socially assistive robots are engineered to engage with humans intuitively, using cues such as eye contact and vocal interaction [2]. For example, collaborative robots (cobots) in assembly lines can monitor their surroundings, issuing vocal alerts or adjusting their movements to avoid human contact when workers are nearby [119]. By aligning their behaviours with human expectations, these robots foster ergonomic safety by reducing the risk of musculoskeletal strain and minimising incidents, which are common in fast-paced, physically intensive environments [120].

In addition to customer service and safety roles, social robots in collaborative environments offer support by interpreting and responding to real-time human input [100, 121]. In logistics, cobots equipped with speech recognition and contextual understanding provide verbal feedback or clarification when workers issue commands, enhancing task coordination. For example, robots can confirm instructions, ask follow-up questions to clarify ambiguous commands, or notify workers if they detect obstacles, ensuring that tasks are completed efficiently and safely [119]. The interactive communication between workers and social robots ensures that workers' cognitive and physical needs are met while enabling a smoother workflow.

The role of communication in HRI within workplaces extends beyond functional exchanges. Social robots create a more inclusive and engaging work environment by offering friendly, supportive interactions that can uplift morale and reduce feelings of isolation in repetitive or solitary roles [2]. This sense of social presence, achieved through communication features like eye contact, personalised responses, and even humour, can transform the robot into a collaborative partner rather than merely a tool [57], thereby enhancing the ergonomic quality of the work environment by addressing both physical and psychosocial needs. Social robots' communication-centred approach aligns with the principles of ergonomic safety, task efficiency, and well-being, fostering a positive, balanced work environment for human workers.

9 Social and Affective Robots' Communication for Providing Socio-Emotional and Health Support

While social robots are proving to be increasingly valuable in structured workplace environments, their potential extends well beyond task-specific applications. In diverse social settings, these robots play a crucial role in providing socio-emotional support, fostering companionship, and enhancing overall well-being. Social and affective robots are now used to support individuals' psychological health, offering companionship, empathy, and interactive support tailored to various emotional needs.

Following from the previous section, it remains uncontroversial that social robots do not (yet) offer the same opportunities as humans for social interactions [58, 60], they can nonetheless afford valuable opportunities for social engagement with human users when introduced in specific contexts (e.g., [63]), and in careful, ethically responsible ways [122]. A growing evidence base documents how social robots might function as autonomous tools to support psychological health interventions [123] and mental health [124, 125], physical therapy, rehabilitation and physical health [126], and other means to amplify or support human therapeutic efforts [127]. Moreover, social robots are being equipped with technologies such as sensors, cameras, and processors, which promote the collection of human data (such as where a person is standing, where they are looking, what they are saying, etc.) with high fidelity, as well as support on-line, on-going analysis of a human interactant's behaviour [74].

Research into the application of social robots in psychosocial health interventions highlights how social robots that take on different forms of embodiment and design can benefit different interventions. For example, robots like Paro, which take on a zoomorphic pet- or cuddly toy-like embodiment, hold value for interventions when used with appropriate target populations (e.g., [128]), including older adults in care homes and people with cognitive impairments (e.g., dementia; see review [129]). These can also support older adults that are suffering from mood disorders or even just feeling low due to the loneliness experienced in old age. For example, a study with 20 older adults by Chen and colleagues [130] found that after 8 weeks of intervention using the robot Paro participants showed significant improvements in mental well-being (including decreasing rates of depression and loneliness and improving quality of life over time). These findings are further supported in the literature. A scoping review of 29 studies by [131] found that previous studies using Paro provided evidence that this robot reduces negative emotions in patients, improves their social engagement, and generally promotes a positive mood, atmosphere, and quality of care experience. Moreover, these effects are grounded in people's objective experience of emotional well-being as recent studies show how interactions with these companion robots positively affect people's physiology. A study by [85] documents the psychophysiological benefits of interacting with a companion robot like Paro, demonstrating that stroking Paro reduces pain perception and salivary oxytocin levels in a sample of 83 healthy adults. Accordingly, *companionship* is one of the key forms of emotional support that social robots are capable of offering.

However, social robots with a more human-like embodiment may be effective in delivering health interventions through verbal communication or demonstration, as well as providing emotional support in scenarios that require active user participation. Some studies show how social robots could provide socio-emotional support to individuals who suffer from different mental health issues and psychopathologies and/or related symptoms (see [125]). A scoping review with 30 papers that were published since 2015 showed that social robots have been studied with populations suffering from varying levels of cognitive impairments and dementia, people suffering from neurodevelopmental disorders such as autism spectrum disorder (ASD), attention deficit disorders (ADHD) and intellectual disability, as well as mental illness such as schizophrenia and depression [132].

For example, a mixed-method study with 180 low-income and socially isolated older adults in South Korea showed that after interacting with the humanoid social robot Hyodol for 3 months (i.e., providing reminders and guiding different exercises and recreational activities), participants showed reduced depressive symptoms and improved health-related quality of life [133]. Another example is a study where the social robot Nao (SoftBank Robotics) was employed to mediate a single-session loving-kindness meditation and walking meditation, oriented to counter symptoms of depression among young people ($N = 142$). The study reports that the social robotic interventions were successful in evoking state openness, with the former also exerting a positive effect on valence [134]. In terms of anxiety disorders, one prominent example is a study by [135] with two participant cohorts (one cohort including 21 participants that were not screened for anxiety, and another cohort including 22 participants who reported receiving treatment for anxiety) that demonstrated a significant reduction in 6 anxiety measures after participating in a deep breathing intervention guided by the social robot Ommie. A previous study by [136] showed the benefits of employing social robots for minimising social tensions and anxieties, describing that participants with higher social anxiety tended to feel less anxious and demonstrated lower tensions when knowing that they would interact with robots as opposed to humans. In a previous paper [124], we discussed similar benefits of using social robots for diagnosing and treating individuals suffering from severe and chronic forms of anxiety, such as post-traumatic stress disorder (PTSD). We highlighted how social robots can help overcome various logistical and social barriers that trauma survivors encounter when monitoring symptoms and seeking mental health interventions. Similar interventions have also been successfully tested with children, supporting their mental health assessment [137, 138].

Beyond supporting people who are suffering from clinically diagnosed psychopathologies, social robots could also provide emotional support via self-managed interventions to healthy individuals that might experience difficult emotional situations and stressors in their daily lives. Previous studies have examined the use of social robots in emotionally supportive settings showing meaningful outcomes in terms of cognitive change and affect. In a long-term experiment performed by our research team, 39 participants conversed with the social robot Pepper twice a week for 5 weeks (10 sessions in total), disclosing to the robot about general everyday experiences. We found that participants self-disclosed more to the robot as the sessions progressed, perceiving the robot to be more socially competent and comforting over time. The repeated interactions also led to improved mood (after each session, and over time) and decreased feelings of loneliness [102]. Moreover, a secondary analysis of the study reveals that people who report higher scores of introversion are more likely to disclose more to a social robot (i.e., in terms of the duration and length of their disclosures). Moreover, the results of the study highlight that when experiencing negative emotions, people are likely to disclose more to a social robot [101]. We replicated this study with a sample of informal caregivers, who often experience high levels of emotional distress. Our findings replicated the previous results [103, 104] and showed that caregiver participants felt less lonely and stressed, were more accepting of their caregiving situation, positively reappraised their caregiving situation and experienced reduced feelings of blame towards others [104]. These results demonstrate that people can establish meaningful relationships with social robots and highlight the value of social robot-led interventions with individuals living with considerably difficult life situations. Social robots could potentially elicit rich interactions with stressed individuals over time, acquire relevant information from their disclosures, and support their emotional well-being. Robotic interventions for people's well-being rarely take place in people's homes and are often conducted in laboratories. One successful example is a study employing the social robot Jibo (NTT) as a positive psychology coach to improve students' psychological well-being in students' on-campus housing. The study results describe a positive effect on students' psychological well-being with positive mood change and students expressing their motivation to change their psychological well-being [139]. Another interesting example deployed QTrobot (LuxAI) in university halls for a long-term emotion regulation intervention, showing how students' interactions with the robot in everyday settings improved their interpretation of emotional challenges [140] .

10 The Future of Human–Robot Communication Using Generative AI

Adaptive human–robot communication is advancing as large language models (LLMs) increasingly enhance robots' capacities for responsive and context-aware interactions, supporting both linguistic and non-linguistic exchanges [62]. These AI-driven improvements empower robots to respond to users with greater sensitivity, allowing for emotionally attuned and contextually relevant replies that support nuanced interactions. This adaptability enhances the robot's role in settings where verbal communication is central, as it allows for responses that feel personalized and empathetic [100]. By tuning into social cues, both verbal and non-verbal, LLM-powered robots are able to foster rapport, encouraging users to share more openly and engage more deeply. However, such advancements introduce new ergonomic challenges. As robots become more adept at simulating human-like interaction, they may risk being perceived as overly aware, which could compromise the user's sense of confidentiality. This balance between adaptivity and confidentiality is essential, as users must feel both understood and secure in their interactions. Kim et al. [141] noted that users increasingly expect social robots to exhibit non-verbal cues that match their enhanced conversational abilities. Yet, mismatches in synchronization or culturally inappropriate gestures may hinder engagement, as any inconsistency can disrupt the flow of interaction and detract from the robot's perceived emotional depth and social competence. This delicate balance is critical for fostering user satisfaction and engagement, as these nuanced interactions directly impact the robot's role in social interactions and its ability to create meaningful connections. Spitale et al. [70] found that while users appreciated the robot's adaptive responses, they also expressed discomfort if interactions felt too intrusive, highlighting a potential trade-off between empathy and privacy. This issue is especially pronounced in settings where trust is fundamental to open communication. In response, designers are exploring ways to ensure that while robots maintain adaptivity, they also convey a consistent, trustworthy interaction style that respects users' expectations for privacy and discretion. In another study, robots communicating using LLMs were found to be too verbose or overly pragmatic, thus disrupting the fluidity of interaction. Moreover, participants in the same study expressed concerns about possible biases and stereotypes in the robot's adaptive responses [142]. Researchers and roboticists are encouraged to test and refine their interaction prompts by conducting trials through online interactions (e.g., see [143]). Such trials would also provide essential data for comparing disembodied agents (e.g., ChatGPT) with embodied agents, such as social robots. These valuable comparisons of agents that interact using the same models but differ in embodiment will offer crucial insights into the deployment of LLMs and generative AI in social robots and other embodied agents, highlighting differences in human behaviour and perceptions of these agents compared to their more readily accessible counterparts. While LLM integration marks a transformative step in adaptive human–robot communication, ensuring that these capabilities are ergonomically aligned with user expectations will be vital.

11 Conclusions

In this chapter, we have thoroughly examined the intricate ergonomic, cognitive, and psychological aspects that influence and shape human perception and behaviour in the context of HRIs. Furthermore, we have outlined the ways in which social robots and HRIs utilise these factors to offer socio-emotional support. Human–robot interaction is complex and multifaceted, involving both the physical features and behaviour of the robot, as well as the context of the interaction and users' expectations and experiences. The ability of a robot to demonstrate rich social cues while understanding and responding to humans' speech and social cues can enhance its social perception, and further affect the quality of the interaction. However, users also expect social robots to share the same environment, display thoughts and feelings, be socially aware, provide social support, and demonstrate autonomy. Nevertheless, it is important to remember that the robot's appearance and behaviour must align with users' expectations of it, otherwise users will perceive it as less social and would not interact with it socially. Overall, understanding the social perception of robots is important for the development and design of social robots that can meet users' needs and expectations and foster positive interactions between humans and robots.

The ability of a robot to engage users through multiple communication modalities, both verbal and non-verbal, is crucial for fostering effective and meaningful interactions. These modalities must be carefully synchronized to avoid disruptions that could diminish user engagement and satisfaction. This multimodal approach highlights the importance of cognitive ergonomics, as robots can adaptively respond to the user's emotional and social cues, promoting a more immersive and supportive interaction. In workplace settings, for example, social robots enhance ergonomic safety and efficiency by using embodied social cues to communicate intentions and promote seamless human-robot collaboration, which addresses both cognitive and physical ergonomic needs.

As social robots advance with the integration of generative AI and large language models, they are increasingly capable of adaptive and context-aware responses. These advancements enhance robots' sensitivity to social cues, supporting deeper engagement and rapport-building. However, challenges remain in balancing adaptivity with confidentiality, as users must feel understood yet secure in interactions. This evolution underscores the need for ongoing research to address ethical considerations, ensuring that social robots align with user expectations for trustworthy, empathetic, and respectful interaction. The state of the art on social robots is continuing to make progress toward developing machines to contribute to the greater good of society. To that end, increasing research effort is being invested to develop our social assistants of tomorrow, and emerging results speaking to how robots might be able to support human psychosocial, socio-emotional, and physical function are promising. While these technologies could support individuals by being their companions, they should also help them to connect with others and overcome their social and emotional barriers. Social robotics can undoubtedly contribute to improving people's quality of life, but the need remains for more methodologically rigorous and ethically sound research into how social robots might interact with humans in a sensitive, timely and nuanced manner.

Acknowledgements The authors gratefully acknowledge funding from the European Union's Horizon 2020 Research and Innovation Programme under the Marie Skłodowska-Curie to ENTWINE, the European Training Network on Informal Care (Grant agreement no. 814072), the European Research Council (ERC) under the European Union's Horizon 2020 Research and Innovation Programme (Grant agreement no. 677270 to EC), and the Leverhulme Trust (PLP-2018-152 to EC).

References

1. International Organization for Standardization: ISO 8373: Robots and robotic devices—vocabulary. Tech. rep. International Organization for Standardization (2012). https://www.iso.org/obp/ui/#iso:std:iso:8373:ed-2:v1:en Standardization: ISO 8373: Robots and robotic devices—vocabulary. Tech. rep. International Organization for Standardization (2012). https://www.iso.org/obp/ui/#iso:std:iso:8373:ed-2:v1:en
2. Henschel, A., Laban, G., Cross, E.S.: What makes a robot social? A review of social robots from science fiction to a home or hospital near you. Curr. Robot. Rep. **2**, 9–19 (2021). https://doi.org/10.1007/s43154-020-00035-0
3. Breazeal, C.: Toward sociable robots. Robot. Auton. Syst. **42**(3), 167–175 (2003)
4. Zebrowitz, L.A., Montepare, J.M.: Social psychological face perception: why appearance matters. Soc. Pers. Psychol. Compass **2**(3), 1497 (2008). https://www.ncbi.nlm.nih.gov/pmc/articles/PMC2811283/
5. Moskowitz, G.B., Gill, M.J.: Person perception. In: Reisberg, D. (ed.) The Oxford Handbook of Cognitive Psychology, chap. 58. Oxford University Press, Oxford. United Kingdom (2013). https://academic.oup.com/edited-volume/34404/chapter/291773946
6. Evans, J.S.B.: Dual-processing accounts of reasoning, judgment, and social cognition. Annu. Rev. Psychol. **59**, 255–278 (2008). https://doi.org/10.1146/annurev.psych.59.103006.093629
7. Johnson, S.P.: Object perception. In: Oxford Research Encyclopedia of Psychology, Oxford University Press, Oxford, United Kingdom (2018)
8. Palmer, S.E., Rosch, E., Chase, P.: Canonical perspective and the perception of objects. In: Long, J., Baddeley, A. (eds.) Attention and Performance IX, pp. 135–151. Lawrence Erlbaum Associates, Hillsdale, NJ (1981)
9. Spelke, E.S.: Principles of object perception. Cogn. Sci. **14**(1), 29–56 (1990)
10. Sullivan, L.H.: The Tall Office Building Artistically Considered. J.B. Lippincott Co., Philadelphia, PA (1896)
11. Papanek, V.J.: Design for the Real World: Human Ecology and Social Change [Paperback]. Academy Chicago, Chicago, IL (1985)
12. Cromby, J.: Embodiment. In: Teo, T. (ed.) Encyclopedia of Critical Psychology, pp. 550–555. Springer, New York, NY, New York, NY (2014). https://doi.org/10.1007/978-1-4614-5583-7_89
13. Wickens, C.D., Gordon, S.E., Liu, Y., Lee, J.: An Introduction to Human Factors Engineering, Pearson Higher Education Limited (2013). 2nd edn
14. Gualtieri, L., Rauch, E., Vidoni, R., Matt, D.T.: Safety, ergonomics and efficiency in human-robot collaborative assembly: design guidelines and requirements. Procedia CIRP **91**, 367–372 (2020)
15. McLeod, R.W.: Introduction. In: Designing for Human Reliability, pp. 1–11 (2015)
16. Gibson, J.J.: The theory of affordances. In: Shaw, R.E., Bransford, J. (eds.) Perceiving, Acting, and Knowing: Toward an Ecological Psychology, pp. 67–82. Lawrence Erlbaum Associates, Hillsdale, N.J. (1977)
17. Norman, D.A.: Affordance, conventions, and design. Interactions **6**(3), 469–473 (1999). https://dl.acm.org/doi/10.1145/301153.301168
18. Gaver, W.W.: Situating action II: affordances for interaction: the social is material for design. Ecol. Psychol. **8**(2), 111–129 (1996)
19. Awaad, I., Kraetzschmar, G.K., Hertzberg, J.: The role of functional affordances in socializing robots. Int. J. Soc. Robot. **7**(4), 421–438 (2015). https://doi.org/10.1007/s12369-015-0281-3

20. Kwak, S.S., Kim, J.S., Choi, J.J.: Can robots be sold? The effects of robot designs on the consumers' acceptance of robots. In: ACM/IEEE Int. Conf. Hum.-Robot. Interact, pp. 220–221. (2014). https://doi.org/10.1145/2559636.2563684
21. Prakash, A., Rogers, W.A.: Why some humanoid faces are perceived more positively than others: effects of human-likeness and task. Int. J. Soc. Robot. **7**(2), 309–331 (2015). https://doi.org/10.1007/s12369-014-0269-4
22. Gosling, S.D., John, O.P.: Personality dimensions in nonhuman animals: a cross-species review. Curr. Dir. Psychol. Sci. **8**(3), 69–75 (1999). https://doi.org/10.1111/1467-8721.00017
23. Argyle, M., Little, B.R.: Do personality traits apply to social behaviour? J. Theory Soc. Behav. **2**(1), 1–33 (1972)
24. Nass, C., Steuer, J., Tauber, E.R.: Computers are social actors. In: Proceedings of the SIGCHI Conference on Human Factors in Computing Systems, p. 72–78. CHI '94, Association for Computing Machinery, New York, NY, USA (1994)
25. Meyer, J., Miller, C., Hancock, P., de Visser, E., Dorneich, M.: Politeness in machine-human and human–human interaction. In: Proceedings of the Human Factors and Ergonomics Society Annual Meeting, vol. 60, pp. 279–283 (2016)
26. Greene, D.J., Zaidel, E.: Hemispheric differences in attentional orienting by social cues. Neuropsychologia **49**(1), 61–68 (2011)
27. Zaki, J.: Cue Integration: a common framework for social cognition and physical perception. Perspect. Psychol. Sci. **8**(3), 296–312 (2013). https://doi.org/10.1177/1745691613475454
28. Iidaka, T.: Role of the fusiform gyrus and superior temporal sulcus in face perception and recognition: an empirical review. Jpn. Psychol. Res. **56**(1), 33–45 (2014)
29. Kanwisher, N., McDermott, J., Chun, M.M.: The fusiform face area: a module in human extrastriate cortex specialized for face perception. J. Neurosci. **17**(11), 4302–4311 (1997). https://www.jneurosci.org/content/17/11/4302
30. Gobbini, M.I., Gentili, C., Ricciardi, E., Bellucci, C., Salvini, P., Laschi, C., Guazzelli, M., Pietrini, P.: Distinct neural systems involved in agency and animacy detection. J. Cogn. Neurosci. **23**(8), 1911–1920 (2011). https://direct.mit.edu/jocn/article/23/8/1911/5161/Distinct-Neural-Systems-Involved-in-Agency-and
31. Rauchbauer, B., Nazarian, B., Bourhis, M., Ochs, M., Prévot, L., Chaminade, T.: Brain activity during reciprocal social interaction investigated using conversational robots as control condition. Philos. Trans. R. Soc. B: Biol. Sci. **374**(1771) (2019)
32. Deen, B., Koldewyn, K., Kanwisher, N., Saxe, R.: Functional organization of social perception and cognition in the superior temporal sulcus. Cerebral Cortex **25**(11), 4596–4609 (2015). https://academic.oup.com/cercor/article/25/11/4596/2367585
33. Shultz, S., Lee, S.M., Pelphrey, K., Mccarthy, G.: The posterior superior temporal sulcus is sensitive to the outcome of human and non-human goal-directed actions. Soc. Cogn. Affect. Neurosci. **6**(5), 602–611 (2011). https://academic.oup.com/scan/article/6/5/602/1658766
34. Beauchamp, M.S.: The social mysteries of the superior temporal sulcus. Trends Cogn. Sci. **19**(9), 489–490 (2015)
35. Amunts, K., Kedo, O., Kindler, M., Pieperhoff, P., Mohlberg, H., Shah, N.J., Habel, U., Schneider, F., Zilles, K.: Cytoarchitectonic mapping of the human amygdala, hippocampal region and entorhinal cortex: intersubject variability and probability maps. Anat. Embryol. **210**(5–6), 343–352 (2005). https://doi.org/10.1007/s00429-005-0025-5
36. Baxter, M.G., Croxson, P.L.: Facing the role of the amygdala in emotional information processing. Proc. Natl. Acad. Sci. **109**(52), 21180–21181 (2012). https://doi.org/10.1073/pnas.1219167110
37. Maren, S.: Long-term potentiation in the amygdala: a mechanism for emotional learning and memory. Trends Neurosci. **22**(12), 561–567 (1999)
38. Fogassi, L., Ferrari, P.F., Gesierich, B., Rozzi, S., Chersi, F., Rizzolotti, G.: Parietal lobe: from action organization to intention understanding. Science **308**(5722), 662–667 (2005). https://pubmed.ncbi.nlm.nih.gov/15860620/
39. Patri, J.F., Cavallo, A., Pullar, K., Soriano, M., Valente, M., Koul, A., Avenanti, A., Panzeri, S., Becchio, C.: Transient disruption of the inferior parietal lobule impairs the ability to attribute

intention to action. Curr. Biol.: CB **30**(23), 4594–4605 (2020). https://pubmed.ncbi.nlm.nih.gov/32976808/
40. Rizzolatti, G., Ferrari, P.F., Rozzi, S., Fogassi, L.: The inferior parietal lobule: where action becomes perception. In: Novartis Foundation, Chadwick, D.J., Diamond, M., Goode, J. (eds.) Percept, Decision, Action: Bridging the Gaps. Novartis Foundation Symposium, vol. 270, pp. 129–145. John Wiley & Sons, Ltd, Chichester (2006). https://doi.org/10.1002/9780470034989.ch11
41. Chaminade, T., Zecca, M., Blakemore, S.J., Takanishi, A., Frith, C.D., Micera, S., Dario, P., Rizzolatti, G., Gallese, V., Umiltà, M.A.: Brain response to a humanoid robot in areas implicated in the perception of human emotional gestures. PLoS ONE 5(7), (2010)
42. Cross, E.S., Liepelt, R., Hamilton, A.F. D. C., Parkinson, J., Ramsey, R., Stadler, W., Prinz, W.: Robotic movement preferentially engages the action observation network. Hum. Brain Mapp **33**(9), 2238–2254 (2012)
43. Cross, E.S., Riddoch, K.A., Pratts, J., Titone, S., Chaudhury, B., Hortensius, R.: A neurocognitive investigation of the impact of socializing with a robot on empathy for pain. Philos. Trans. R. Soc. B: Biol. Sci. **374**(1771), 20180034 (2019). https://doi.org/10.1098/rstb.2018.0034
44. Greenlee, J.D., Oya, H., Kawasaki, H., Volkov, I.O., Severson, M.A., Howard, M.A., Brugge, J.F.: Functional connections within the human inferior frontal gyrus. J. Comp. Neurol. **503**(4), 550–559 (2007). https://doi.org/10.1002/cne.21405
45. Rogers, J.C., Davis, M.H.: Inferior frontal cortex contributions to the recognition of spoken words and their constituent speech sounds. J. Cogn. Neurosci. **29**(5), 919 (2017)
46. Di Cesare, G., Vannucci, F., Rea, F., Sciutti, A., Sandini, G.: How attitudes generated by humanoid robots shape human brain activity. Sci. Rep 10(1), (2020)
47. Hogenhuis, A., Hortensius, R.: Domain-specific and domain-general neural network engagement during human–robot interactions. Eur. J. Neurosci. **56**(10), 5902–5916 (2022)
48. Jastrzab, L.E., Chaudhury, B., Ashley, S.A., Koldewyn, K., Cross, E.S.: Beyond human-likeness: socialness is more influential when attributing mental states to robots. iScience **27**(6), 110070 (2024). https://doi.org/10.1016/j.isci.2024.110070
49. Premack, D., Woodruff, G.: Does the chimpanzee have a theory of mind? Behav. Brain Sci. **1**(4), 515–526 (1978)
50. Apperly, I.A., Butterfill, S.A.: Do humans have two systems to track beliefs and belief-like states? Psychol. Rev. **116**(4), 953–970 (2009)
51. Baron-Cohen, S.: Natural Theories of Mind: Evolution, Development and Simulation of Everyday Mindreading, pp. 233–251. Basil Blackwell, Cambridge, MA, US (1991). Precursors to a theory of mind: Understanding attention in others
52. Krause, L., Enticott, P.G., Zangen, A., Fitzgerald, P.B.: The role of medial prefrontal cortex in theory of mind: a deep rTMS study. Behav. Brain Res. **228**(1), 87–90 (2012)
53. Saxe, R., Wexler, A.: Making sense of another mind: the role of the right temporo-parietal junction. Neuropsychologia **43**(10), 1391–1399 (2005)
54. Saxe, R., Kanwisher, N.: People thinking about thinking peopleâŁ¯: the role of the temporo-parietal junction in "theory of mind.âŁž In: Berntson, G.G., Cacioppo, J.T. (eds.) Social Neuroscience, 1st edn., pp. 171–182. Psychology Press, New York (2004)
55. Özdem, C., Wiese, E., Wykowska, A., Müller, H., Brass, M., Overwalle, F.V.: Believing androids - fMRI activation in the right temporo-parietal junction is modulated by ascribing intentions to non-human agents. Soc. Neurosci. **12**(5), 582–593 (2017). https://doi.org/10.1080/17470919.2016.1207702
56. Klapper, A., Ramsey, R., Wigboldus, D., Cross, E.S.: The control of automatic imitation based on bottom-up and top-down cues to animacy: insights from brain and behavior. J. Cogn. Neurosci. **26**(11), 2503–2513 (2014). https://doi.org/10.1162/jocn_a_00651
57. Hortensius, R., Hekele, F., Cross, E.S.: The perception of emotion in artificial agents. IEEE Trans. Cogn. Dev. Syst. **10**(4), 852–864 (2018)
58. Cross, E.S., Hortensius, R., Wykowska, A.: From social brains to social robots: applying neurocognitive insights to human-robot interaction. Philos. Trans. R. Soc. B: Biol. Sci. **374**(1771), 20180024 (2019)

59. Cross, E.S., Ramsey, R.: Mind meets machine: towards a cognitive science of human-machine interactions. Trends Cogn. Sci. **25**(3), 200–212 (2021)
60. Hortensius, R., Cross, E.S.: From automata to animate beings: the scope and limits of attributing socialness to artificial agents. Ann. N. Y. Acad. Sci. **1426**(1), 93–110 (2018)
61. Scholl, W.: The socio-emotional basis of human interaction and communication: How we construct our social world. Soc. Sci. Inf. **52**(1), 3–33 (2013)
62. Zhang, C., Chen, J., Li, J., Peng, Y., Mao, Z.: Large language models for human–robot interaction: a review. Biomim. Intell. Robot. **3**(4), 100131 (2023)
63. Laban, G., George, J.N., Morrison, V., Cross, E.S.: Tell me more! Assessing interactions with social robots from speech. Paladyn J. Behav. Robot. **12**(1), 136–159 (2021). https://doi.org/10.1515/pjbr-2021-0011
64. Skantze, G.: Turn-taking in conversational systems and human-robot interaction: a review. Comput. Speech Lang. **67**, 101178 (2021)
65. Wallkötter, S., Tulli, S., Castellano, G., Paiva, A., Chetouani, M.: Explainable embodied agents through social cues. ACM Trans. Hum.-Robot. Interact. (THRI) **10**(3), 27 (2021). https://doi.org/10.1145/3457188
66. Tian, L., Oviatt, S.: A taxonomy of social errors in human-robot interaction. ACM Trans. Hum.-Robot. Interact. (THRI) 10(2), (2021). https://doi.org/10.1145/3439720
67. Lim, V., Rooksby, M., Cross, E.S.: Social robots on a global stage: establishing a role for culture during human-robot interaction. Int. J. Soc. Robot. **13**(6), 1307–1333 (2021). https://doi.org/10.1007/s12369-020-00710-4
68. Mavridis, N.: A review of verbal and non-verbal human-robot interactive communication. Robot. Auton. Syst. **63**(P1), 22–35 (2015)
69. Maurtua, I., Fernández, I., Tellaeche, A., Kildal, J., Susperregi, L., Ibarguren, A., Sierra, B.: Natural multimodal communication for human-robot collaboration. Int. J. Adv. Rob. Syst. **14**(4), 1–12 (2017). https://doi.org/10.1177/1729881417716043
70. Spitale, M., Axelsson, M., Gunes, H.: VITA: A Multi-modal LLM-based System for Longitudinal, Autonomous, and Adaptive Robotic Mental Well-being Coaching (2023). http://arxiv.org/abs/2312.09740
71. Cantucci, F., Falcone, R., Marini, M.: Redefining user expectations: the impact of adjustable social autonomy in human-robot interaction. Electronics 2024, **13**(1), 127 (2023). https://www.mdpi.com/2079-9292/13/1/127/
72. Lutz, C., Schöttler, M., Hoffmann, C.P.: The privacy implications of social robots: Scoping review and expert interviews. Mob. Media Commun. **7**(3), 412–434 (2019). https://doi.org/10.1177/2050157919843961
73. Calvo, R., D'Mello, S., Gratch, J., Kappas, A., Castellano, G., Gunes, H., Peters, C., Schuller, B.W.: Multimodal affect recognition for naturalistic human-computer and human-robot interactions. In: The Oxford Handbook of Affective Computing (2014). https://academic.oup.com/edited-volume/28057/chapter/212016009
74. Spitale, M., Axelsson, M., Jeong, S., Tuttosi, P., Stamatis, C. A., Laban, G., Lim, A., Gunes, H.: Past, Present, and Future: A Survey of The Evolution of Affective Robotics For Well-being. IEEE Trans. Affect. Comput. (2025). https://doi.org/10.1109/TAFFC.2025.3567740
75. Bavelas, J.B., Hutchinson, S., Kenwood, C., Matheson, D.H.: Using face-to-face dialogue as a standard for other communication systems. Can. J. Commun. **22**(1) (1997). https://doi.org/10.22230/cjc.1997v22n1a973
76. Daft, R.L., Lengel, R.H.: Organizational information requirements, media richness and structural design. Manage. Sci. **32**(5), 554–571 (1986)
77. Barsalou, L.W., Niedenthal, P.M., Barbey, A.K., Ruppert, J.A.: Social embodiment. Psychol. Learn. Motiv. Adv. Res. Theory **43**, 43–92 (2003)
78. Ziemke, T.: What's that thing called embodiment? Proc. Annu. Meet. Cogn. Sci. Soc. **25**(25), 1305–1310 (2003). https://escholarship.org/uc/item/60w6v9jz
79. Anderson, M.L.: Embodied cognition: a field guide. Artif. Intell. **149**(1), 91–130 (2003)
80. Meltzoff, A.N., Prinz, W.: The Imitative Mind: Development, Evolution and Brain Bases. Cambridge University Press (2002). https://www.cambridge.org/core/books/imitative-mind/2B0FC5BC23D1D3E4E6322B02DB45C1A9

81. Deng, E., Mutlu, B., Mataric, M.J.: Embodiment in socially interactive robots. Found. Trends Robot. **7**(4), 251–356 (2019)
82. Skantze, G., Oertel, C., Hjalmarsson, A.: User feedback in human-robot interaction: Prosody, gaze and timing. In: Proceedings of the Annual Conference of the International Speech Communication Association, INTERSPEECH, pp. 1901–1905. (2013)
83. Babel, F., Kraus, J., Miller, L., Kraus, M., Wagner, N., Minker, W., Baumann, M.: Small talk with a robot? The impact of dialog content, talk initiative, and gaze behavior of a social robot on trust, acceptance, and proximity. Int. J. Soc. Robot. **13**(6), 1485–1498 (2021). https://doi.org/10.1007/s12369-020-00730-0
84. Rawal, N., Stock-Homburg, R.M.: Facial emotion expressions in human-robot interaction: a survey. Int. J. Soc. Robot. **14**(7), 1583–1604 (2022). https://doi.org/10.1007/s12369-022-00867-0
85. Geva, N., Uzefovsky, F., Levy-Tzedek, S.: Touching the social robot PARO reduces pain perception and salivary oxytocin levels. Sci. Rep. **10**(1), 1–15 (2020). https://www.nature.com/articles/s41598-020-66982-y
86. Sefidgar, Y.S., MacLean, K.E., Yohanan, S., Van Der Loos, H.F.H., Croft, E.A., Garland, E.J.: Design and evaluation of a touch-centered calming interaction with a social robot. IEEE Trans. Affect. Comput. **7**(2), 108–121 (2016)
87. Shiomi, M., Nakata, A., Kanbara, M., Hagita, N.: Robot reciprocation of hugs increases both interacting times and self-disclosures. Int. J. Soc. Robot. (2020). https://doi.org/10.1007/s12369-020-00644-x
88. Li, J.: The benefit of being physically present: a survey of experimental works comparing copresent robots, telepresent robots and virtual agents. Int. J. Hum Comput Stud. **77**, 23–37 (2015)
89. Mori, M., MacDorman, K.F., Kageki, N.: The uncanny valley. IEEE Robot. Autom. Mag. **19**(2), 98–100 (2012)
90. Laakasuo, M., Palomäki, J., Köbis, N.: Moral uncanny valley: a robot's appearance moderates how its decisions are judged. Int. J. Soc. Robot. **13**(7), 1679–1688 (2021). https://doi.org/10.1007/s12369-020-00738-6
91. Shi, C., Shiomi, M., Kanda, T., Ishiguro, H., Hagita, N.: Measuring communication participation to initiate conversation in human-robot interaction. Int. J. Soc. Robot. **7**(5), 889–910 (2015). https://doi.org/10.1007/s12369-015-0285-z
92. Powell, H., Laban, G., George, J.N., Cross, E.S.: Is deep learning a valid approach for inferring subjective self-disclosure in human-robot interactions? In: Proceedings of the 2022 ACM/IEEE International Conference on Human-Robot Interaction, pp. 991–996. HRI '22, IEEE Press, Sapporo, Japan (2022). https://doi.org/10.5555/3523760.3523921
93. Powell, H., Laban, G., Cross, E. S.: A multimodal neural network for recognizing subjective self-disclosure towards social robots. IEEE/RSJ International Conference on Intelligent Robots and Systems (IROS), 850–857 (2025). https://doi.org/10.1109/IROS60139.2025.11246044
94. Axelsson, A., Buschmeier, H., Skantze, G.: Modeling feedback in interaction with conversational agents-a review. Front. Comput. Sci. **4**, 22 (2022)
95. Kontogiorgos, D., Pelikan, H.R.: Towards adaptive and least-collaborative-effort social robots. In: ACM/IEEE International Conference on Human-Robot Interaction, pp. 311–313. IEEE Computer Society (2020)
96. Churamani, N., Kalkan, S., Gunes, H.: Continual learning for affective robotics: why, what and how? In: 29th IEEE International Conference on Robot and Human Interactive Communication. RO-MAN 2020, pp. 425–431. Institute of Electrical and Electronics Engineers Inc., Naples, Italy (2020)
97. Derlega, V.J., Harris, M.S., Chaikin, A.L.: Self-disclosure reciprocity, liking and the deviant. J. Exp. Soc. Psychol. **9**(4), 277–284 (1973)
98. Zonca, J., Folsø, A., Sciutti, A.: The role of reciprocity in human-robot social influence. iScience **24**(12), 103424 (2021)

99. Patterson, M.L.: Compensation in nonverbal immediacy behaviors: a review. Sociometry **36**(2), 237–252 (1973)
100. Laban, G., Cross, E.S.: Sharing our emotions with robots: why do we do it and how does it make us feel? IEEE Trans. Affect. Comput. (2024). https://doi.org/10.1109/TAFFC.2024.3470984
101. Laban, G., Kappas, A., Morrison, V., Cross, E.S.: Opening up to social robots: how emotions drive self-disclosure behavior. In: 2023 32nd IEEE International Conference on Robot and Human Interactive Communication (RO-MAN), pp. 1697–1704. IEEE, Busan, Republic of Korea (2023). https://ieeexplore.ieee.org/document/10309551/
102. Laban, G., Kappas, A., Morrison, V., Cross, E.S.: Building long-term human-robot relationships: examining disclosure, perception and well-being across time. Int. J. Soc. Robot. **16**(5), 1–27 (2024). https://doi.org/10.1007/s12369-023-01076-z
103. Laban, G., Morrison, V., Kappas, A., Cross, E.S.: Informal caregivers disclose increasingly more to a social robot over time. In: CHI Conference on Human Factors in Computing Systems Extended Abstracts, pp. 1–7. Association for Computing Machinery, New York, NY, USA (2022). https://doi.org/10.1145/3491101.3519666. CHI EA '22
104. Laban, G., Morrison, V., Kappas, A., Cross, E.S.: Coping with emotional distress via self-disclosure to robots: an intervention with caregivers. Int. J. Soc. Robot. (2025). https://doi.org/10.1007/s12369-024-01207-0
105. Smedegaard, C.V.: Reframing the role of novelty within social HRI: from noise to information. In: Proceedings of the 14th ACM/IEEE International Conference on Human-Robot Interaction, pp. 411–420. IEEE Press (2019) HRI '19
106. Sung, J., Christensen, H.I., Grinter, R.E.: Robots in the wild: understanding long-term use. In: 4th ACM/IEEE International Conference on Human-Robot Interaction (HRI), pp. 45–52. ACM, New York, NY, USA (2009)
107. Norman, D.A., Miller, J., Henderson, A.: What you see, some of what's in the future, and how we go about doing it. In: Conference Companion on Human Factors in Computing Systems—CHI '95, p. 155 (1995). https://doi.org/10.1145/223355.223477
108. Shourmasti, E.S., Colomo-Palacios, R., Holone, H., Demi, S.: User experience in social robots. Sensors 21(15), (2021)
109. Kontogiorgos, D., Pereira, A., Gustafson, J.: Grounding behaviours with conversational interfaces: effects of embodiment and failures. J. Multimodal User Interfaces **15**(2), 239–254 (2021). https://doi.org/10.1007/s12193-021-00366-y
110. Duffy, B.R., Joue, G.: The paradox of social robotics: a discussion. In: AAAI Fall 2005 Symposium on Machine Ethics. Hyatt Regency, (2005)
111. Broadbent, E.: Interactions with robots: the truths we reveal about ourselves. Annu. Rev. Psychol. **68**(1), 627–652 (2017)
112. Gompei, T., Umemuro, H.: A robot's slip of the tongue: effect of speech error on the familiarity of a humanoid robot. In: Proceedings - IEEE International Workshop on Robot and Human Interactive Communication, pp. 331–336. (2015)
113. Gray, K., Young, L., Waytz, A.: Mind perception is the essence of morality. Psychol. Inq. **23**(2), 101–124 (2012)
114. Gray, H.M., Gray, K., Wegner, D.M.: Dimensions of Mind Perception. Science **315**(5812), 619 LP –619 (2007). http://science.sciencemag.org/content/315/5812/619.abstract
115. Salem, M., Lakatos, G., Amirabdollahian, F., Dautenhahn, K.: Would you trust a (faulty) robot?: effects of error, task type and personality on human-robot cooperation and trust. In: ACM/IEEE International Conference on Human-Robot Interaction, pp. 141–148. (2015). https://doi.org/10.1145/2696454.2696497
116. Polakow, T., Laban, G., Teodorescu, A., Busemeyer, J.R., Gordon, G.: Social robot advisors: effects of robot judgmental fallacies and context. Intel. Serv. Robot. **15**(5), 593–609 (2022). https://doi.org/10.1007/s11370-022-00438-2
117. Aoki, M., Rea, F., Jirak, D., Sandini, G., Yanagi, T., Takamatsu, A., Bouet, S., Yamamura, T.: On the influence of social robots in cognitive multitasking. Int. J. Hum.Oid Robot. **19**(6) (2022). https://www.worldscientific.com/worldscinet/ijhr

118. Belanche, D., Casaló, L.V., Flavián, C., Schepers, J.: Service robot implementation: a theoretical framework and research agenda. Serv. Ind. J. **40**(3–4), 203–225 (2020). https://doi.org/10.1080/02642069.2019.1672666
119. García, O.R., Chaminade, T., Thurow, K., Taesi, C., Aggogeri, F., Pellegrini, N.: COBOT applications-recent advances and challenges. Robotics **12**(3), 79 (2023)
120. Lorenzini, M., Lagomarsino, M., Fortini, L., Gholami, S., Ajoudani, A.: Ergonomic human-robot collaboration in industry: a review. Front. Robot. AI **9**, 813907 (2023). https://pmc.ncbi.nlm.nih.gov/articles/PMC9893795/
121. Semeraro, F., Griffiths, A., Cangelosi, A.: Human-robot collaboration and machine learning: A systematic review of recent research. Robot. Comput.-Integr. Manuf. **79**, 102432 (2023)
122. Lee, M., Sin, J., Laban, G., Kraus, M., Clark, L., Porcheron, M., Cowan, B.R., Følstad, A., Munteanu, C., Candello, H.: Ethics of conversational user interfaces. In: Extended Abstracts of the 2022 CHI Conference on Human Factors in Computing Systems, pp. 1–7. CHI EA '22, Association for Computing Machinery, New York, NY, USA (2022). https://doi.org/10.1145/3491101.3503699
123. Robinson, N.L., Cottier, T.V., Kavanagh, D.J.: Psychosocial health interventions by social robots: systematic review of randomized controlled trials. J. Med. Internet Res. **21**(5), 1–20 (2019)
124. Laban, G., Ben-Zion, Z., Cross, E.S.: Social robots for supporting post-traumatic stress disorder diagnosis and treatment. Front. Psychiatry **12** (2022). https://doi.org/10.3389/fpsyt.2021.752874
125. Scoglio, A.A.J., Reilly, E.D., Gorman, J.A., Drebing, C.E.: Use of social robots in mental health and well-being research: systematic review. J. Med. Internet Res. **21**(7), e13322 (2019)
126. Langer, A., Levy-Tzedek, S.: Emerging roles for social robots in rehabilitation. ACM Trans. Hum.-Robot. Interact. (THRI) 10(4), (2021). https://doi.org/10.1145/3462256
127. Laban, G., Morrison, V., Cross, E.: Social robots for health psychology: a new frontier for improving human health and well-being. Eur. Health Psychol. **23**(1), 1095–1102 (2024). https://www.ehps.net/ehp/index.php/contents/article/view/3442
128. Robinson, H., MacDonald, B., Kerse, N., Broadbent, E.: The psychosocial effects of a companion robot: a randomized controlled trial. J. Am. Med. Dir. Assoc. **14**(9), 661–667 (2013)
129. Koh, W.Q., Felding, S.A., Budak, K.B., Toomey, E., Casey, D.: Barriers and facilitators to the implementation of social robots for older adults and people with dementia: a scoping review. BMC Geriatr. **21**(1), 1–17 (2021). https://doi.org/10.1186/s12877-021-02277-9
130. Chen, S.C., Moyle, W., Jones, C., Petsky, H.: A social robot intervention on depression, loneliness, and quality of life for Taiwanese older adults in long-term care. Int. Psychogeriatr. **32**(8), 981–991 (2020)
131. Hung, L., Liu, C., Woldum, E., Au-Yeung, A., Berndt, A., Wallsworth, C., Horne, N., Gregorio, M., Mann, J., Chaudhury, H.: The benefits of and barriers to using a social robot PARO in care settings: a scoping review. BMC Geriatr. **19**(1), 232 (2019). https://doi.org/10.1186/s12877-019-1244-6
132. Guemghar, I., de Oliveira Padilha, P.P., Abdel-Baki, A., Jutras-Aswad, D., Paquette, J., Pomey, M.P.: Social robot interventions in mental health care and their outcomes, barriers, and facilitators: scoping review. JMIR Ment. Health **9**(4) (2022). https://mental.jmir.org/2022/4/e36094
133. Lee, O.E.K., Nam, I., Chon, Y., Park, A., Choi, N.: Socially assistive humanoid robots: effects on depression and health-related quality of life among low-income, socially isolated older adults in South Korea. Journal of Applied Gerontology (2022). https://doi.org/10.1177/07334648221138283
134. Huang, I.S., Cheung, Y.W., Hoorn, J.F.: Loving-kindness and walking meditation with a robot: countering negative mood by stimulating creativity. Int. J. Hum Comput Stud. **179**, 103107 (2023)
135. Matheus, K., Vázquez, M., Scassellati, B.: A social robot for anxiety reduction via deep breathing. In: 2022 31st IEEE International Conference on Robot and Human Interactive Communication, pp. 89–94. RO-MAN (2022)

136. Nomura, T., Kanda, T., Suzuki, T., Yamada, S.: Do people with social anxiety feel anxious about interacting with a robot? AI Society **35**(2), 381–390 (2020). https://doi.org/10.1007/s00146-019-00889-9
137. Abbasi, N.I., Laban, G., Ford, T., Jones, P.B., Gunes, H.: Robotising psychometrics: validating wellbeing assessment tools in child-robot interactions. In: 2024 33rd IEEE International Conference on Robot and Human Interactive Communication (ROMAN), pp. 1651–1658. IEEE, Pasadena, CA (2024). https://ieeexplore.ieee.org/document/10731253/
138. Abbasi, N.I., Laban, G., Ford, T., Jones, P.B., Gunes, H.: A longitudinal study of child wellbeing assessment via online interactions with a social robot. ACM Trans. Hum.-Robot. Interact **14**(3), 51 (2025). https://doi.org/10.1145/372212
139. Jeong, S., Aymerich-Franch, L., Arias, K., Alghowinem, S., Lapedriza, A., Picard, R., Park, H.W., Breazeal, C.: Deploying a robotic positive psychology coach to improve college students' psychological well-being. User Modeling and User-Adapted Interaction, (2022). https://doi.org/10.1007/s11257-022-09337-8
140. Laban, G., Wang, J., Gunes, H.: A robot-led intervention for emotion regulation: from expression to reappraisal. IEEE Trans. Affect. Comput. 1–15 (2026). https://doi.org/10.1109/TAFFC.2026.3657604
141. Kim, C.Y., Lee, C.P., Mutlu, B.: Understanding large-language model (LLM)-powered human-robot interaction. In: ACM/IEEE International Conference on Human-Robot Interaction, pp. 371–380. (2024). https://doi.org/10.1145/3610977.3634966
142. Spitale, M., Axelsson, M., Gunes, H.: Appropriateness of LLM-Equipped Robotic Well-being Coach Language in the Workplace: A Qualitative Evaluation. IEEE Int. Workshop Robot Human Commun. RO-MAN, 1571–1578 (2024). https://doi.org/10.1109/RO-MAN60168.2024.10731269
143. Laban, G., Laban, T., Gunes, H.: LEXI: large language models experimentation interface. In: Proceedings of the 12th International Conference on Human-Agent Interaction (HAI '24). ACM, Swansea, United Kingdom (2024). https://doi.org/10.1145/3687272.3688296

Human-Centered Robotics for Surgery

Rocco Moccia, Raffaella Mancino, Emanuele Ruffaldi, Massimiliano Simi, Rossella Onofrio, and Fanny Ficuciello

Abstract Surgical ergonomics is influenced by many aspects associated with posture, vision, hand motion, and haptic feedback. Specific ergonomics aspects can induce long-term effects in the surgeon directly affecting the performance. One of the major disadvantages of Minimally Invasive Surgery (MIS) remains the non-ergonomic design of the entire surgical environment. This condition results in a significant reduction in the quality of interventions and discomfort for the medical operator. Modern teleoperated surgical robots offer an optimized console, in which the surgeon can operate in a seated position with specific arm support and a 3D view of the operating site. The da Vinci surgical console offers ergonomic improvements over standard endoscopic surgery. Modern robotic platforms stand beyond the da Vinci System, offering more advanced ergonomic solutions in terms of haptic feedback and visual perception. Due to the introduction of more advanced technology, surgery has changed, resulting in increased system complexity at the technical, human and organizational levels, which may lead to higher variability of patient outcomes due to new error pathways. Adopting a proactive approach enables the prior identification of critical factors and the design of safer socio-technical systems, thanks to a multi-level perspective. This chapter addresses the fundamental elements of surgical ergonomics, considering the factors that affect performance and healthiness during traditional procedures, with specific reference to MIS and microsurgery. This baseline is then compared to the use of surgical robots by evaluating the effect of design

R. Moccia
Center for Advanced Biomaterials for Healthcare, Istituto Italiano di Tecnologia, Napoli, Italy

R. Mancino
Università di Genova, Genova, Italy
e-mail: raffaella.mancino@edu.unige.it

E. Ruffaldi (✉) · M. Simi
Medical Microinstruments, Inc., Pisa PI, Italy
e-mail: Emanuele.Ruffaldi@mmimicro.com

R. Onofrio
School of Management, Politecnico di Milano, Milano, Italy

F. Ficuciello
Università degli Studi di Napoli Federico II, Napoli, Italy

G. Barresi and S. Wischniewski (eds.), *Ergonomics in Robotics: Advances and Innovations*, Springer Tracts in Advanced Robotics 162,
https://doi.org/10.1007/978-3-032-19374-2_9

decisions on operator ergonomics. In addition, studies propose proactive methodologies that assess the personal and organizational factors of surgical performance in this complex socio-technical system.

Keywords Laparoscopy · Microsurgery · Teleoperation · da Vinci · Robotic surgery

1 Introduction

Ergonomics in Surgical Robotics encompasses physical, cognitive, and organizational aspects. It strives to optimize the human-robot interaction, ensuring surgeons' physical comfort, cognitive support, and efficient organizational processes. This multidisciplinary approach enhances surgical outcomes and surgeon well-being in the context of robotic surgery's evolving complexity.

Surgical ergonomics plays a crucial role in the performance and well-being of surgeons, as it encompasses various factors such as posture, vision, hand motion, and haptic feedback. These aspects can have long-term effects on surgeons, directly influencing their performance. Over the years, surgery has undergone a significant transformation, with technological developments opening the way for more accurate and less invasive operations. The progress of surgery has brought about major changes, notably in terms of ergonomics for the surgeon, from traditional open procedures to the development of robot-assisted methods. A specific feature of ergonomics in this scenario is the physical connection between surgeons and surgical systems during procedures.

Physical Ergonomics involves the design and adaptation of the physical characteristics of the surgical environment, such as console ergonomics, instrument design, and operating room setup, to improve surgeon comfort, fatigue reduction, and overall performance. Physical ergonomics seeks to establish a safer, more efficient, and user-friendly interface between surgeons and robotic systems by taking into account the biomechanics and ergonomics of tool manipulation, body posture, and body motions. Surgical robots might increase surgeon wellness, improve surgical accuracy, and ultimately lead to better patient outcomes by incorporating ergonomic concepts. Open surgical treatment was the typical procedure in the early days of surgery, necessitating extensive incisions to access the operating site. Due to restricted illumination and exposure, surgeons had difficulty attaining adequate visualization. The ergonomic demands of maintaining proper posture and hand movements for an extended amount of time resulted in physical strain and fatigue. Furthermore, microsurgical interventions requiring a microscope necessitate precise manipulation of fragile structures using delicate equipment. Surgeons must keep their hands steady and their hand-eye coordination stable. Ergonomics becomes crucial in microsurgery since any short tremor or discomfort might have serious consequences for the procedure's outcome [1]. Robot-assisted surgery (RAS) has been created as an innovative solution in response to the drawbacks of conventional surgical techniques, starting from laparoscopic surgery. Robotic devices, such as the widely used da Vinci® Surgical System, have made a significant difference in ergonomics. The consoles provided a comfortable sitting position, ergonomic arm supports, and simple controls, all of which reduced physical strain during procedures. Furthermore, the use of three-dimensional visualization technologies increased depth perception, allowing for better spatial awareness and more precise surgical maneuvers. Beyond the da Vinci® system, the area of surgical robots has expanded, with newer platforms offering even more sophisticated ergonomic solutions. These improvements have not only revolutionized surgical practice but have also addressed the challenge of visibility in traditional laparoscopic and microsurgical methods. The emergence of high-definition, three-dimensional imaging robotic systems has greatly enhanced visualization, allowing surgeons to have a more thorough and detailed view of the operative site. The transition from manual surgery to RAS impacts ergonomics through improved visualization. Surgeons work from a console, which relieves tension on the back, neck, and shoulders. Intuitive controls and ergonomic arm supports reduce the need for repetitive and forceful motions, reducing the risk of musculoskeletal problems. Furthermore, stable camera orientation and filtered hand tremors reduce mental and physical strain. Advanced imaging enhances spatial awareness by improving the navigation of complex structures and instrument use. Overall, this transition from manual to RAS presents a new paradigm shift in visualization, necessitating clear ergonomic aspects. The introduction of novel surgical visualization devices might have significant drawbacks. Visual information presented on displays or monitors at the surgeon's station is used in RAS. The display system's quality and clarity, including resolution, color accuracy, and depth perception, might impact the surgeon's ability to view the operating site properly. There may be limits in the capacity to modify the magnification or zoom level of the camera view and connect to the limited field of vision, depending on the specific robot-assisted surgical system. Additionally, the manipulation of surgical tools undergoes significant changes in ergonomics when transitioning from open surgery to RAS. In open surgery, doctors use their hands to control equipment, which might be limited by the extent of incisions and the physical dexterity necessary. RAS uses specific instruments that have greater accuracy and a wider range of motion. These devices are controlled by surgeons through a console, allowing for more complex motions and access to difficult-to-reach places. These tools are especially intended to improve ergonomics. They frequently include ergonomic handles containing features like ergonomic grips, adjustable angles, and simple controls. Articulating devices in robot-assisted surgical systems can emulate the natural movement of the human wrist. With this capability, the surgeon can perform complicated maneuvers with more dexterity and precision. These instruments increase ergonomics by enabling more natural and pleasant tool use by emulating the range of motion of the human wrist. In general, the design of the console and the positioning of controls can influence the ergonomics of tool manipulation. Poorly designed consoles or awkward hand positions may lead to discomfort, fatigue, and potential musculoskeletal issues for the surgeon over extended periods of use. One of the limitations of RAS is the absence of direct tactile feedback. Surgeons rely on visual and haptic feedback provided by the robotic system, but the loss of direct touch sensation can make it challenging to assess tissue characteristics and subtle variations. Surgeons must rely on visual cues and indirect feedback, which may require adaptation and potentially impact the overall tactile experience during surgery. Characterization of human errors in this context can significantly improve performance by concentrating on ergonomic aspects that may contribute to errors or discomfort during the surgical process. By analyzing the errors that occur in robot-assisted surgery and their relationship with ergonomic variables, potential sources of fatigue, strain, or cognitive overload faced by surgeons can be identified. These findings may be used to influence the development of robotic systems with enhanced user interfaces, control mechanisms, and haptic feedback, allowing surgeons to operate comfortably and effectively throughout the process. These instructions may include suggestions on body posture, hand placement, and instrument manipulation to lessen the risk of repetitive strain injuries or musculoskeletal illnesses.

It is an ongoing, decades-long effort to investigate healthcare as one of the most complex socio-technical systems because of its intrinsic characteristics. Indeed, in healthcare organizations, even the simplest of procedures turns out to be the result of interdependent processes, subject to several requirements and needing multiple human interactions; all put in a context of rapid technological evolution and organizational change. Through the lenses of complex socio-technical systems theory, it became clear the opportunity to learn from high-risk industries; mainly to understand and adapt methodologies and techniques aimed at reducing the influence of human and organizational factors on patient safety in an environment characterized by increasing technological complexity, the multiplicity of interacting professionals, and a new conceptualization of human errors, conceived as a trigger for the manifestation of system-level “latent failures”.

The operating room and, in particular, robotic surgery is paradigmatic under this perspective: increasingly challenging procedures, multi-professional staff with a very high level of interdependence, sophisticated technological equipment and devices, time constraints, and occurrence of unexpected situations since the patient is under induced unstable and critical conditions. Minimally Invasive Surgery (MIS) and, in particular, robotic surgery presents many characteristics found in high-risk industries, such as advanced technology, complex socio-technical systems, the presence of highly standardized protocols and procedures, and quality and safety systems. Robotic surgery is intended to overcome the limitations of laparoscopic surgery, such as 2D vision, inconsistent instrument movements, unnatural surgeon positions, dissociation between vision and instrument control, and inability to perform microsutures. Using a computer and remote handling system, a surgeon can reproduce the movement of the human hand during surgery. The surgeon’s experience remains fundamental for the assessment of the patient, the selection of information, and the execution of the operation. Proper diagnosis and assessment of the status, condition, and risk class of the patient are crucial to ensure safe robotic surgical procedures. For some patients, robotic surgery is not suitable, unnecessarily expensive, or riskier than traditional methods, while for others, in safe conditions, it may be more precise and effective than laparoscopic surgery. Over the last decade, new robot-assisted surgical procedures have been developed in the fields of oncology, gynecology, orthopaedics, and maxillofacial, thoracic, pediatric, ophthalmologic, and also cardiac surgery [2]. However, this advanced technology may be associated with quality and safety issues due to its high degree of human-machine interaction and procedural complexity. Assessment of its clinical efficacy and patient safety implications is still early and has produced contrasting results and recommendations, with reports of several different limitations and problems associated with specific robotic surgical procedures. For example, some authors report problems due to poor communication between the operating surgeon and the rest of the surgical team, particularly the surgical assistant. Indeed, robotic surgery disrupts the existing workflow and changes the roles of every team member [3]. Technical failures are reported as well, mainly due to system malfunction (e.g. instrumentation) or collision of the robotic arms with the patient, surgeon or each other [4]. Therefore, in this complex context of robotic surgery, the application of principles of cognitive and organizational ergonomics is an urgent and emerging field. In this Chapter, we introduce the discipline of "Human Reliability Analysis" (HRA) and briefly describe some contributions to the ergonomics discipline in healthcare, particularly in robotic surgery.

Cognitive Ergonomics is concerned with mental processes, such as perception, memory, reasoning, and motor response, as they affect interactions among humans and other elements of a system. (Relevant topics include mental workload, decision making, skilled performance, human-computer interaction, human reliability, work stress, and training as these may relate to human-system design) [5]. Organizational Ergonomics is concerned with the optimization of sociotechnical systems, including their organizational structures, policies, and processes. (Relevant topics include communication, crew resource management, work design, design of working times, teamwork, participatory design, community ergonomics, cooperative work, new work paradigms, virtual organizations, telework, and quality management) [5]. The primary objective of this chapter is to present a comprehensive overview of the various facets of ergonomics in robotic surgery. Section 2 examines the evolution from conventional open surgery to advanced robot-assisted procedures, focusing on physical ergonomics aspects. It delves into the comprehensive spectrum of surgical interventions, encompassing traditional open surgery, minimally invasive techniques, and microsurgery, while highlighting the emergence of state-of-the-art robotic platforms. Special emphasis is placed on exploring the physical ergonomic dimensions intrinsic to all surgical procedures, with particular attention given to aspects such as visualization, posture, and tool manipulation. It explores the influence of physical ergonomics on console design, tool manipulation, and operating room setup. Section 3 delves into the detailed analysis of the impact of cognitive, and organizational aspects in robotic surgery. HRA plays a crucial role in intricate healthcare systems, such as robotic surgery. It proactively identifies and addresses human and organizational factors that impact patient safety. Through the examination of Influencing Factors (IFs), HRA aligns with cognitive and organizational ergonomics, ultimately improving patient safety in minimally invasive and robotic surgeries. Ultimately, Sect. 4 offers an overview of the entire chapter.

2 Physical Ergonomics

Physical ergonomics is vital in surgery, optimizing the surgeon's interaction with the surgical environment through design improvements in operating rooms, instruments, and workstations. It aims to enhance comfort, reduce fatigue, and improve overall performance during surgical procedures. In traditional open surgery, in which a surgeon performs with handed equipment through a wide cut, surgeons encountered several challenging circumstances during surgical procedures, including poor lighting, difficult-to-grasp instruments, excessive table height or width, repetitive motions, postural discomfort, and working close to other operating room team members. Particularly, open surgery-related discomfort for the surgeon is fairly prevalent, ranging from 66 to 94% as exposed in [6]. These types of illnesses could be related to the surgeon's physical condition and are frequently observed in the hands, arms, neck, low back, and shoulders, according to several works in the literature [7].

MIS, including laparoscopy, has revolutionized surgical procedures. It involves using small incisions to insert surgical instruments into the abdomen, resulting in less post-operative pain, shorter hospital stays, fewer complications, reduced blood loss, and minimal scarring for patients. In MIS, the image of the surgical site is created by a camera attached to the viewing end of a wide-angle optical endoscope inserted into a body cavity. In MIS, a screen often displays a view of the inside organs and tissue, allowing the surgeon to establish a diagnosis, visually recognize internal organs, and conduct surgery on them. This is a significant improvement over the direct view of the operative field in open surgery, and visualization is critical to comprehend the surgeon's ergonomics. However, the indirect viewing of the operative area alters eye-hand coordination and offers a 2D image of the inside organs, resulting in a low depth perception. The surgeon's posture is affected by the 2D vision as they need to stand and hold long instruments while looking at the display. Moreover, an assistant holding the endoscope controls the direction and orientation of the surgeon's vision. As a result, the surgeon must adjust to a low-resolution vision of the field with fluctuating illumination that is continually subject to movement by the assistant. Under these circumstances, manipulating and positioning activities take an excessively long time to be completed. Papers in literature utilize systematic reviews and statistical analyses to assess surgeons' physiological conditions across different procedures, examining parameters such as muscular fatigue, heart rate, and blood pressure. These studies contribute to a deeper understanding of the physical demands imposed on surgeons during surgeries. For instance, research in [8] considers electromyography data and medical indices to gain insights into surgeons' muscular activity and health status during open surgery and MIS procedures. It reveals a significant incidence of work-related musculoskeletal conditions among surgeons, with MIS surgeons experiencing neck discomfort, upper and lower extremity pain, numbness, and exhaustion compared to open surgery. For surgeons performing MIS, specific ergonomics guidelines have been established to address these factors. The guidelines recommend positioning laparoscopic monitors between 10 and 30° below eye level, with the image displayed 15 to 45° below eye level. A screen height of 160 cm is also suggested to support ergonomic positioning.

As previously exposed, surgical vision is critical in this circumstance, and certain surgical specialties require imaging capabilities that an endoscope cannot provide, as well as optical magnification provided by an operating microscope. Microsurgery is a surgical procedure that adopts magnification and is now employed in many surgical procedures, including general surgery, ophthalmology, orthopedic surgery, gynecological surgery, otolaryngology, neurosurgery, oral and maxillofacial surgery, plastic surgery, and pediatric surgery. Reconstructive procedures, impossible to execute with regular surgery, represent the typical application of microsurgery. The surgeon performs microsurgery on very small structures using miniaturized instruments that are small enough to accurately grip and manipulate tissue. Visual perception in this situation requires fundamentally different paradigms than MIS. The operating microscope is now a regular feature in modern operating rooms, and it plays a key role in the success of many of the most complex and challenging surgical treatments. The settings for the microscope are often included on an adjustable handle and a foot pedal that the surgeon adjusts throughout the surgery. By separating the main optical channel with a mirror, a second binocular system is offered to an assistant. This image is frequently shown on a video monitor for the operating room team. Modern surgical microscopes with ergonomic designs increase surgeons' physical comfort, reducing fatigue and the risk of musculoskeletal injuries. This is especially relevant for reconstructive surgeons, who are at significant risk of work-related musculoskeletal problems due to maintaining unergonomic and unpleasant postures during protracted operations. However, despite enormous technological breakthroughs, modern microscopes indeed have drawbacks. The majority of these concerns are connected to the microscope's movement and angulation. The influence of surgical discomfort on long-term working abilities has grown in relevance in recent decades, owing to the increase in surgical complexity and duration [9]. Furthermore, this type of microscope has a very bulky structure that takes up a lot of space in the operating room and inhibits the surgical team's motions. Neurosurgical operations, in particular, necessitate the surgeon to see the anatomy from multiple viewpoints. Direct binocular vision typically requires the surgeon to bend the neck for extended periods, and the assistant frequently confronts identical constraints. This position causes the surgeon discomfort and fatigue, which can lead to poor surgical outcomes. For these reasons, a novel optical tool called the exoscope has been introduced into the operating room. The high-definition exoscope system became an alternative to the operating microscope. The exoscope offers a much more compact and easier solution for camera manipulation than the operating microscope. Additionally, higher-quality completely digitalized image processing is provided, coupled with a posture that is more comfortable for the surgeon. Exoscopes, unlike microscopes, transmit images onto external mobile screens rather than fixed lenses. Nonetheless, due to the distance between the surgeon and the display, exoscopes have limits in terms of visual field depth and visual quality at greater magnification. These constraints are greatly addressed by exoscopes and microscopic cameras attached to robotic arms. Superior spatial resolution, geometric correctness, dexterity, quicker maneuverability, and a steady motion are all features of robot-guided exoscopes.

In general, the vision of the surgical field had a significant impact on and characterized the introduction of robots into the surgical scenario. The robots increase the precision of tissue manipulation activities and the quality of surgical methods, resulting in better results and fewer revision surgeries. Medical robots give higher dexterity, motion scaling, heightened precision, high-definition, and expanded 3D images compared to conventional laparoscopy, allowing for increased accuracy and visibility within the patient's body. The robot introduces new ergonomics principles, the majority of which are based on teleoperated technologies. The surgical robot includes a console, a patient-side cart, and a vision cart. The surgeon operates the surgical console, which is located outside of the sterile area, with two master controls and pedals. A 3D endoscope gives a view of the surgical site, which is processed by the vision cart and presented to the surgeon through a stereo visor, with the tooltips aligned with the surgeon's hands, which hold the master controllers. Finally, the patient side is the component of the system that is in direct contact with the patient. In the da Vinci® robotic platform [10], the surgeon is sitting, their forearms rest on pads, the manipulation is greatly simplified, and the 3D display restores the important depth perception aspect seen in open surgical approaches. The da Vinci robots solve the issue of surgical visualization in part by allowing the surgeon to observe the 3D-operating field utilizing binocular endoscopic vision at 10–15× magnification. Some modern robotic solutions include a head-mounted display (HMD) that can display real-time 3D images of the surgical field. This approach allows the surgeon to have a natural posture, with no constraints on visualization.

Another crucial element in the definition of ergonomic conditions in surgery is manipulation. In open surgery the surgeon performs and develops their skills using common, stable equipment like forceps, scissors, clamps, etc., which were specifically designed for universal usage. The introduction of the operating microscope brings with it a wide variety of micro-instruments. Microforceps, needle holders, vessel dilators, dissecting scissors, and vessel clamps are among the tools used for micro-suturing and anastomosis. The later surgical paradigms, which utilize surgical video from an endoscope or an exoscope, introduced new ergonomic concerns to human tissue manipulation. The traditional laparoscopic tools incorporate an axial or pistol gripper with a circular shaft that hosts the actuation mechanism and a terminal part composed of a two-jawed forceps-like tool for tissue manipulation. The most common diameter for laparoscopic tools is 5 mm, however they can range from 2 to 12 mm. The majority of laparoscopic devices only have 4 Degrees of Freedom (DOFs) of movement: in/out, up/down, left/right, and rotation. The tool movements are limited to 4 DOFs by small instruments called trocars, implanted at the fulcrum point coincident with the insertion point in the human body. Nonetheless, MIS and microsurgical procedures have significant limitations, primarily because of restricted tactile perceptions, which may degrade intervention quality and create significant obstacles in this type of surgery. The surgeon's unintentional use of excessive force is the primary cause of tissue harm in this case. Furthermore, all humans have physiological tremor, which is damped by holding the instrument near the tip as feasible.

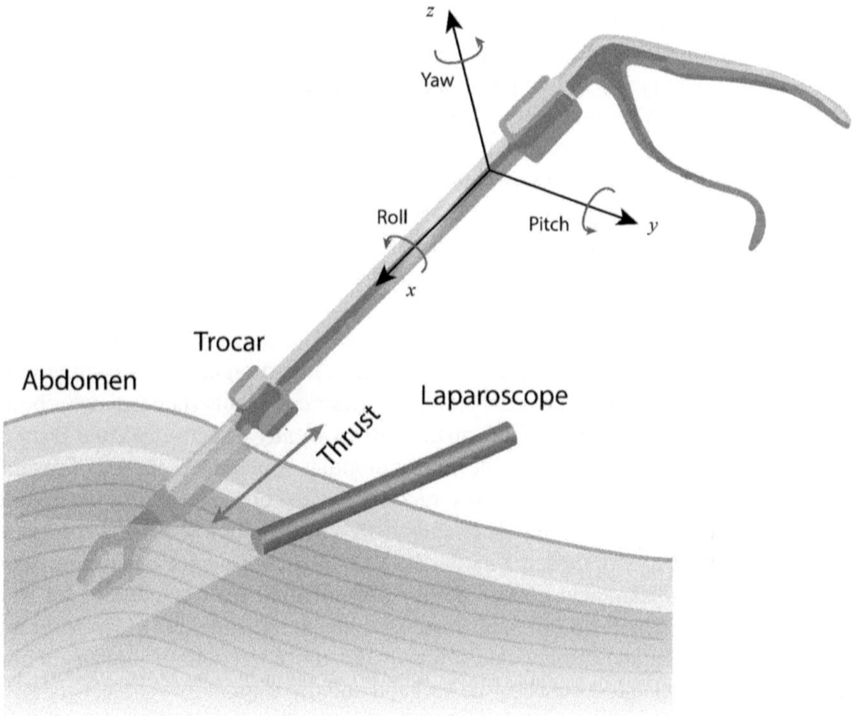

Fig. 1 Laparoscopic tool movements and the "fulcrum effect"

However, any condition requiring an increased undamped instrument will result in a larger tremor amplitude. Excessive force applied by the surgeon, as in MIS, can distort the object in the instrument's jaw, causing tissue injury. Furthermore, traditional open surgery and microsurgery allow surgical tools to move freely with no restrictions on the surgeon's hand mobility. On the contrary, MIS limits the range of motion, causing a "fulcrum effect" on the laparoscopic tools and compelling the surgeon to move the hands in the opposite direction of the instrument's movement (Fig. 1).

In this regard, robotic technology also attempts to address MIS challenges. As previously stated, the surgical tools on the robotic patient-side are controlled by the surgeon on the master side. The aforementioned da Vinci system introduces a typical example of a teleoperated surgical console, based on an ergonomic working position and on different instruments' definition and control. The patient system contains surgical instruments with EndoWrist®technology that have an 8 mm diameter and are around 60 cm long. The Endowrist allows for movement in seven axes and rotation of $\sim 360°$, as well as the use of a wide range of robotic devices such as needle holders, electrocautery forceps, cold forceps, gripping forceps, and bipolar dissectors. The master device has also been enhanced in this design, allowing the surgeon to effortlessly control the surgical instrument. The Master Tool Manipulator (MTM) is an 8 DoFs manipulator that only considers the first 7 DoFs, with the addition of a fourth non-actuated joint allowing for more natural movement. The system's control transfers the surgeon's finger motions directly to the robot's end-effector, simplifying control and relieving the surgeon of the "fulcrum effect" imposed by laparoscopic surgery. Despite all of the benefits outlined above, the surgeon's position during the interaction with the robots brings additional ergonomic challenges. Furthermore, surgical instruments still have restricted DoFs for accessing various organs. Finally, the lack of haptic feedback limits the surgeon's sensation of touch, as well as the safety and telepresence during surgery.

New teleoperated robotic devices have been used in a range of surgical operations to address these issues. Other systems follow the same design, with the surgeon in physical touch with the master console. CMR Surgical debuts the Versius®robotic surgical system [11]. A robotic platform that differs greatly from the da Vinci system. The patient side, in particular, is made up of a modular structure composed of several independent robotic arms. The Versius surgeon console is open and may be used while standing or sitting, introducing significantly different ergonomics aspects compared to the previously exposed device. Finally, the Versius robotic device control has been moved to onboard handheld control devices, eliminating the requirement for foot pedal controllers. The head-up display on the console transmits 3D video from the endoscopic camera, and the open design provides more freedom of movement and smoother communication between the surgeon and the team.

The Medtronic company introduced a robotic system named HUGO™ RAS System [12]. The system consists of a surgical console that the surgeon uses while seated. Adjusting the position of the viewer, the armrest, and the pedal unit will result in the most ergonomic position. To control the robot, the surgeon uses "pistol grip" manipulators, similar to da Vinci's MTMs, which include infrared contact sensors that allow the robot to move while showing the surgeon's hand. The visualization component includes two monitors. A passive 3D-HD screen permits the surgeon to see the surgical screen while wearing 3D glasses, and a touchscreen surgeon interactive display allows the surgeon to assign the hooked instruments to the right/left hand, configure the motion scaling, and choose visual filters and camera rotation. The platform also contains a head tracking system, where a vision device reveals the position of the 3D glasses to the console and enables robotic control. The system allows better positioning of the surgeons into the console.

Finally, TransEnterix, Inc. (Morrisville, NC, USA) proposes the Senhance Surgical System, a novel robotic system aimed at becoming an advanced tool for the majority of laparoscopic surgeries (Fig. 2) [13]. The surgeon's open console is one of the device's unique features. The surgeon can maintain a sitting posture, using an eye-tracking system that moves the laparoscope ensuring continuous camera control. The master system seeks to improve and mimic laparoscopic procedures. The console incorporates haptic robotic manipulators because of these factors, and they have a design that is quite similar to that of laparoscopic equipment. Verbal communication between the team is also facilitated by the surgeon's console being close to the patient. The complete system may be deployed in a regular operating theatre, elim-

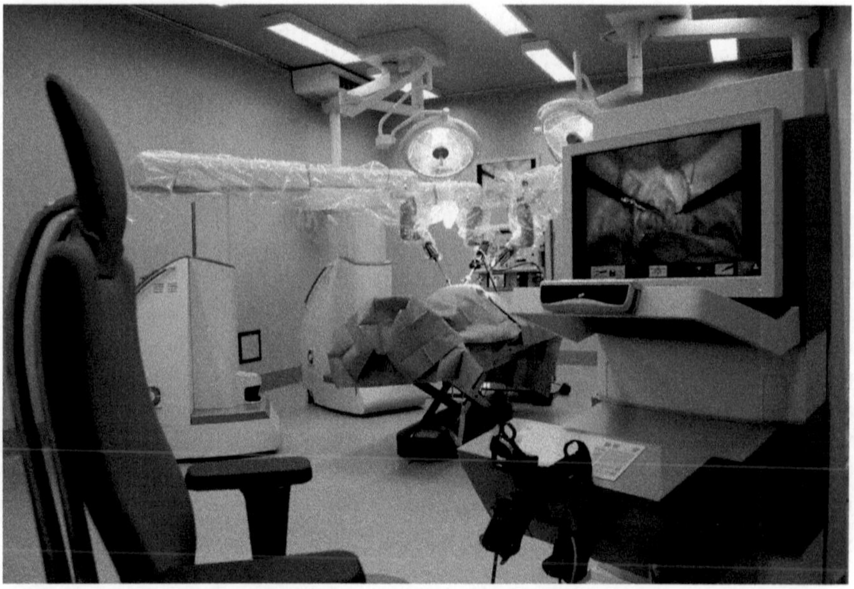

Fig. 2 The operating room setup with the Senhance surgical system

inating the need to create a specialized facility. The surgical console in each of the aforementioned devices is composed of two manipulators that the surgeon can utilize by gripping them from their end effectors. One of the advanced surgical robot for complex microsurgical procedures is Symani® Surgical System (Medical Microinstruments (MMI) S.p.A, Italy) [14]. The main advantage of this device in terms of ergonomics concerns the surgeon's position. The master console is composed of a surgical chair on which surgeons maintain a more comfortable and natural posture during surgical procedures [15, 16] in comparison to traditional microsurgery. The ergonomic design of the system helps reduce strain and fatigue on the surgeon's body, minimizing the risk of musculoskeletal injuries. The Symani Surgical System employs a free-hand master device that the surgeon holds and manipulates. There are no restrictions on master tools that have no connection to a manipulator's end effector. These ergonomic portable devices are meant to mirror the surgeon's hand movements, turning them into accurate movements of the microinstruments [17, 18]. As much as the microscope increases human perception, Symani provides two digits motion scaling and motion filtering, that reduce the effect of human hand tremor and reduce the need for stabilized control of the hands as in traditional microsurgery. The Symani system employs cutting-edge 3D visualization technology to provide high-definition, enlarged pictures of the surgical site when integrated with surgical microscopes. This increased visualization eliminates the need for surgeons to physically lean in or strain their eyes during procedures, increasing their comfort and decreasing eye fatigue.

Robot-assisted image-based catheter systems might be briefly mentioned. A robot-assisted image-guided catheter is a medical device that combines robotic technology with innovative imaging methods to help medical professionals carry out catheter-based treatments with more control and precision. These systems combine imaging techniques, robotic arms or other devices, and navigational aids to direct the insertion and movement of catheters inside the body. High-resolution images are used to create a 3D model of the patient's anatomy, guiding catheter navigation. The surgeon operates the robotic system from a control console. The console typically includes hand controllers that translate the provider's movements into robotic motions. The robotic arms or mechanisms precisely manipulate the catheter based on the provider's commands. Throughout the procedure, real-time imaging feedback is displayed to the provider, enabling continuous visualization of the catheter's position and alignment relative to the target area. Several advanced robot-assisted image-guided catheter systems have been developed for various medical procedures. CorPath GRX System by Corindus Vascular Robotics is designed for robotic-assisted percutaneous coronary intervention (PCI). The system includes a console with joysticks and touchscreen controls for the guidewire, guide catheter, and balloon/stent delivery system [19].

3 Cognitive and Organizational Ergonomics

3.1 Why Human Reliability Analysis (HRA) in Surgery and in Particular in Robotic Surgery?

Human factors and Ergonomics studies paved the way for the investigation of the influence of human aspects in complex socio-technical systems. In this direction, the patient's condition, and the technical skills and performance of the individual surgeon, for specific operations, are extensively studied in literature as the primary risk factors of surgical outcome [20]. Furthermore, with the recent advances in surgical technology, several studies focused on the association between surgical outcomes and surgical technology [21]. Moreover, several studies tried to identify personal and organizational factors that influence the surgical outcome [22]. On the same line, a relevant contribution comes from retrospective analyses, such as incident reporting analyses [23]. Moreover, the recent spread of taxonomies of non-technical skills (NTS), has promoted the creation and the application of observational tools, which allow the assessment of the NTS during the surgical procedures and assessment of the surgical team according to a real-time analysis approach [24–26]. Accordingly, this phenomenon has been already investigated from different perspectives that seem they do not interact with each other.

Although nowadays the benefits of transferring and applying to healthcare services risk analysis methods, traditionally implemented in high-risk industries, are fully recognized in the patient safety literature [27, 28], healthcare organizations are still characterized by reactive culture in their approach to safety [29]. Current practices are predominantly based on the analysis of adverse events (ex-post analysis), and approaches looking to the anticipation and elimination of vulnerabilities in the system are not widely adopted by practitioners. It can be argued that a proactive culture of patient safety is still in its infancy. This is one reason why many scholars, who investigated risk management issues in healthcare, proposed, the application of HRA recognizing the positive effect that the HRA applications brought to the high-risk industry in terms of reduction of human errors giving a strong contribution to cognitive and organizational ergonomics [27, 28, 30–32]. The systemic view of safety could well meet the healthcare needs in terms of management of complexity and human error control, as key requirements towards higher safety performance. According to this school of thought, the HRA discipline aims at investigating and quantifying the human contribution to safety within the technical and organizational context. The main assumption is that human behavior is influenced by the working environment, as the comprehensive combination of technological and social components and their respective interactions. Below two main definitions of HRA are reported:

"The Human Reliability Analysis is a method by which human reliability is estimated. Human reliability is the probability that a person correctly performs some system-required activity in a required time period and performs no extraneous activity that can degrade the system. In carrying out an HRA, it is necessary to identify those human actions that can have an effect on system reliability or availability" [33]. *"HRA is concerned with a particular type of prediction, namely the prediction of the likelihood that a human action may fail, usually expressed in terms of the probability of human erroneous action"* [34].

The purpose of HRA is then the identification of the weaknesses in the system that create vulnerability to errors. HRA contributed to a radical improvement in several safety-critical industries, such as nuclear, aviation, and process industries, giving hope about the potential that the adoption of HRA techniques in healthcare might have on the achievement of higher patient safety performance and higher safety of healthcare workers [28, 32]. Two important features of the HRA methodologies well fit with the need to improve ergonomics and safety in complex surgical contexts. The first one regards the HRA anticipatory pattern rather than the retrospective one. In detail, estimating the probability of error allows us to provide ex-ante analysis before incidents or adverse events analysis. Hence, HRA techniques are anticipatory approaches, and they represent an innovation in healthcare, which is historically focused on retrospective approaches such as incident reporting, Root Cause Analysis (RCA), or systematic patient record review. The second one is the application of taxonomies of Influencing Factors (IFs) (1) - i.e. aspects of behavior and context that impact human performance [35]. These factors are exactly the set of aspects of cognitive and organizational ergonomics.

By means of the selection and quantification of a set of IFs, several HRA techniques, in the context of a Probabilistic Risk Assessment (PRA) procedure, support the specification of the principal (or most likely) accident scenarios that can lead to human errors and unintended consequences; on this basis, they estimate the probability of occurrence of such errors and thus the contribution to the accident event.

Several different taxonomies of IFs have been proposed (cfr. Definition 1) in HRA industrial literature, in particular, different taxonomies have been developed ad-hoc for specific complex socio-technical contexts such as aviation and nuclear sectors. The reason for that is that the IFs tend to be standardized at the higher level of the classification hierarchy, where factors are generally related to the basic factors of a generic working system (software, hardware, environment, liveware - SHEL) [36]. In the meanwhile, at a lower level of classification, the same taxonomies of IFs differentiate a lot, since they are connected to more contingent and detailed characteristics of the specific task or working system. This is also confirmed by the recent open debate in the industrial literature about which IFs should be used in HRA and what is the appropriate number of IFs to include in a method or analysis. In fact, *"There is a considerable range in the number of IFs provided by individual HRA methods, ranging from single factor models such as time reliability curves, up to 50 or more IFs in some current HRA models"* [35]. Nevertheless, the introduction of HRA in healthcare is still an emerging field, and only in recent years have some authors tried to import HRA techniques and industrial applications to study human performance in some clinical settings [31, 32]. As Lyons claimed, *"Though the scarcity of HRA techniques in healthcare is likely to be due in some part to the safety culture, much is likely to be due to a lack of awareness of the usefulness of the techniques and their applicability to the problem of human error in the clinical context"* [32].

Definition 1 Nomenclature used for "Influencing Factors" (IFs). Throughout their qualitative and quantitative phases, many HRA methods make use of Influencing Factors (IFs), i.e. all those factors characterizing the work context (personal, environmental, and organizational) that can influence human performance. These factors can be external (linked to the external working conditions, i.e. aspects of the context) and internal (intrinsic characteristics of humans, i.e. aspects of behavior). Several HRA techniques in high-risk industries provide a modeling approach and parameters to evaluate the influence of contextual factors on the reliability of human actions. The contextual factors may vary in nature, number, and modelization. In most cases, these factors come with some coefficients that modulate the probability of error occurrence. The terminology about context factors differs from one HRA technique to another. Some HRA techniques label these factors as "Performance Shaping Factors" (PSFs) - THERP technique—[33], "Influencing Factors (IFs)"-CREAM technique —[34], or "Error Promoting Conditions (EPCs)" - HEART technique [37]. There is not a standard set of these factors; they present a considerable overlapping belonging to different HRA techniques, despite each method bringing with it a slightly different emphasis and slightly different sets or taxonomies of PSFs/IFs/EPCs. This note is intended to specify that in this manuscript we will always call them "Influencing Factors" (IFs), apart from when they strictly refer to their specific HRA technique.

The growing interest in HRA in surgery showed the emergent need to implement proactive analyses for human error reduction in different surgical contexts. Indeed, modern surgery is largely heterogenous, in terms of technology implemented, the composition of the surgical team, standardized protocols and procedures, such as the difference between traditional open surgery and MIS. Consequently, a key question facing surgical units relates to the identification of factors that can influence surgical outcomes. Research efforts across the world have sought to identify factors that may influence surgical performance, and with recent advances in surgical technology, several studies have focused on the association between surgical outcomes and surgical technology and the technical skills and types of training of surgeons. Although the growing interest in ergonomics in surgery, and in particular in laparoscopic surgery, to the best of authors' knowledge scientific literature on proactive analyses of human errors in robotic surgery is lacking but emerging field.

3.2 Recent Studies of HRA in Robotic Surgery

In order to generate advanced knowledge in a form that can be directly used to enhance the measurement and improvement approaches to surgical performance and patient safety, in this Chapter, we quote one of our studies [38] which aimed at designing a taxonomy of IFs for HRA application in surgery starting from Human Factors literature to identify and select personal and organizational factors categories that shape surgical performance and validated in Italy and Denmark. It is worth mentioning that the IFs outlined in this section for surgical applications are categorized according to a specialized taxonomy developed for assessing human error in robotic surgical procedures. The starting point was the literature review, aimed at identifying the highest number of applicable personal and organizational factors, with particular attention to (i) factors investigated in ergonomics studies; (ii) factors reported in frameworks derived from incident reporting analyses; (iii) the semantic categorization of skills into technical and non-technical. The main result was a preliminary list of factors, grounded in literature evidence, to be used as a knowledge base. Ten macro-categories of factors resulted from the literature review:

- Noise and background talk are not related to the task.
- Safety culture and safety climate.
- Standardization.
- Communication and teamwork.
- Experience and team training.
- Leadership.
- Staffing and team member familiarity.
- Workload.
- Fatigue.
- Equipment HMI and space design.

The final validated taxonomy of the personal and organizational factors that influence human performance in healthcare and in particular in surgery according to the HRA approach is reported in the Table 1 [38].

More in general, the 21 IFs on surgeon's performance can be used as a reference for setting safety priorities in different surgical procedures. Particular attention

Table 1 Validated list of IFs for HRA applications in surgery [38]

SHEL Component	IF category [39]	IF items
Environment	Noise & back-ground talk not related to the task	**Noise and ambient talk:** continuous or sudden noise; team members talking in the background or coming and going and moving around in a noisy way
		Music: Presence of background music in the operating room
	Distractions	**Noisy use of social media:** team members talking about and obtrusively sharing social media content
		Verbal interruptions: verbal Interruptions that are either untimely or not patient relevant
	Safety culture and climate	**Poor management of errors and threats to patient safety:** failure to share information promptly and openly about errors and threats to patient safety
Software	Standardization	**Poor guidelines, procedures or checklists:** guidelines, procedures or checklists are inadequate: lacking, too complex, or not at right level
		Improper use of procedures and checklists: the improper use, or non-use, of the WHO checklist (or similar), protocols and procedures
Hardware	Equipment, HMI and space design	**Inadequate ergonomics of equipment and work place:** equipment and workplace not designed to optimize usability and reduce operator fatigue and discomfort
		Poor use of the technology: lack of ability to use relevant technologies
Liveware	Communication and team-work	**Rude talk and disrespectful behaviours:** derogatory remarks, behaviours showing a lack of respect of OR team members, shouting and harsh tones of voice
		Unclear or failed communication: communication that should have been given was not or was inadequate or was misunderstood and not corrected
		Poor coordination: failure in coordinating team activities; failure to anticipate the needs of the lead surgeon or lead anaesthetist (surgeon at the console in robotic surgery)
		Poor decision making: failure to consider, select and communicate options; inadequacy or delay in implementing and reviewing decisions
		Poor situation awareness: failure to gather and/or to integrate information or failure to use information to anticipate future tasks, problems, and states of the operation
	Experience and Team Training	**Lack of experience of surgical team colleagues:** lack of experience of within surgical team, with the surgical procedure or technology
		Lack of experience of anaesthetics team colleagues: lack of experience within anaesthetic team, with the anaesthetic procedure or technology
	Fatigue	**Fatigue:** mental fatigue or physical fatigue
	Leadership	**Poor leadership:** failure to set and maintain standards or to support others in coping with pressure

(continued)

Table 1 (continued)

SHEL Component	IF category [39]	IF items
	Staffing and team member familiarity	**Team member unfamiliarity:** team members unfamiliar with each other and each other's competencies
	Workload	**Time pressure:** psychological stress resulting from experiencing a need to get things done in less time than is required or desired
		Preoperative emotional stress: stress induced by factors not directly related to the team, or to the characteristics and evolution of the intervention, e.g., responsibility for the budget and for other hospital objectives, organizational problems of the department, other critical patients, lawsuits

should be paid to minimally invasive and robotic surgeries, where a higher level of vulnerability of surgical teams to different sources of disturbance has been highlighted in the present study. Indeed, the HEP in mini-invasive surgical procedures was assessed as relatively more vulnerable to "noise and ambient talk" and "verbal interruptions". Moreover, the study showed that the surgeon's technical performance is more vulnerable to personal and organizational risk factors in an MIS than in an open surgery context. More specifically, the study highlights the primary importance of team factors such as verbal interruptions and noise and ambient talk. The adoption of a proactive and design-based approach to patient safety enables prior identification of the most critical factors and a more effective and ergonomic design of the system, thus allowing one to avoid the costs associated with adverse outcomes.

However, to the best of the authors' knowledge, there is a lack of scientific literature addressing proactive analysis of human errors in robotic surgery. We believe that improving awareness of IFs on surgeons' performance and their variability may prevent compromises in safety tasks and instead allow priority to plan, diagnosis, and checklists. This might represent a key factor in improving patient safety in surgery. Moreover, in this complex socio-technical context, as argued by Karsh and colleagues [40–42], cross-level interactions in complex socio-technical systems should be taken into consideration to fully understand the causal mechanisms that link personal and organizational factors to patient safety. However, proper implementation of the human factors engineering paradigm in healthcare [40] requires the development of a robust methodological approach supported by suitable techniques. In this direction, in another study [43] we proposed a new methodology to support multi-level meso-ergonomic investigations in surgery. A modified version of the specific technique "human error assessment and reduction technique" - HEART; [37]) was developed and tested by applying it to a robot-assisted radical prostatectomy (RARP) procedure [44]. More specifically, a structured methodology for incorporating uncertainties related to the influence of personal, team, and organizational factors (so-called influencing factors, or IFs) on the surgeon's human error probability (HEP) was proposed; the cumulative effect of multiple IFs on the expected patient outcome was finally assessed by combining a dynamic event tree (DET) model of the surgical procedure and Monte Carlo simulation. This study proposed a methodology for performing meso-ergonomics analysis of surgical procedures, based on a modified version of the HEART technique. The pilot application involved the analysis of the impact of personal, group, and organizational factors on patient outcomes of the Robot-Assisted Radical Prostatectomy procedure [44]. The study offers significant contributions to practice as it sheds light on the most relevant IFs that impair surgeons' performance and patient outcomes. The results reveal that team-related factors have the greatest impact on patient outcome variability. Going into details of these studies is not the purpose of this chapter but mentioning these contributions is only meant to emphasize the emerging need for fostering theoretical knowledge and practical expertise on HRA among healthcare practitioners as well. These studies have relevant practical implications. Indeed, while ongoing training programs clearly focus on granting the technical proficiency of the surgeon, they are not able to reduce the surgeon's vulnerability to the most critical IFs. In this regard, the effectiveness of the most common training practices seems to be limited to the increase in surgeons' awareness. Finally, when it comes to the comparison of different surgical techniques, results show that surgeons implementing mini-invasive surgery are more likely to experience detrimental effects of IFs on performance than surgeons involved in open surgery. Overall, these studies offered a demonstration of the importance of developing specific knowledge for fostering a proper application of HRA in healthcare and contributing to improving cognitive and organizational ergonomics in increasingly complex socio-technical systems.

4 Conclusions

This chapter provides a comprehensive overview of ergonomics in surgical robotics, encompassing physical, cognitive, and organizational dimensions. It emphasizes the crucial role of ergonomics in optimizing human-robot interaction, ensuring surgeon comfort, cognitive support, and streamlined organizational processes. The transition from traditional open surgery to robot-assisted methods is explored, highlighting the significant improvements in physical ergonomics, including posture, visualization, and tool manipulation. The influence of cognitive and organizational factors is examined, with a focus on HRA as a proactive approach to identifying and mitigating factors affecting patient safety. This chapter underscores the multidisciplinary approach of surgical ergonomics and its pivotal role in enhancing surgical outcomes and surgeon well-being amidst the evolving complexity of robotic surgery.

Physical ergonomics is vital in surgery, optimizing the surgeon's interaction with the surgical environment. Traditional open surgery can lead to discomfort for surgeons, affecting their physical well-being. MIS has benefits but presents ergonomic challenges due to indirect visualization. Microsurgery adds further complexities. Robotic surgical systems like the da Vinci Surgical System have addressed some of these challenges by providing better visualization and ergonomic advantages, enabling surgeons to work comfortably. Newer robotic platforms like Versius and HUGO offer improved ergonomics by allowing surgeons to sit and providing more freedom of movement. Symani Surgical System prioritizes surgeon comfort with its ergonomic design. In MIS and microsurgery, surgical tools can limit tactile perception and lead to excessive force usage. Robotic systems like da Vinci address this with advanced instruments and precise control, reducing strain on surgeons. In summary, physical ergonomics in surgery is crucial for surgeon well-being and patient outcomes. Robotic systems have introduced significant ergonomic improvements, allowing surgeons to work more comfortably and effectively.

Human factors and ergonomics research is crucial in understanding how human elements affect complex systems, notably in surgery. Factors like patient health, surgeon skills, and technology impact surgical outcomes. Personal and organizational factors also play a role. While risk analysis methods from high-risk industries are valuable in healthcare, they remain predominantly reactive. HRA offers a proactive approach to enhance patient safety by identifying system weaknesses. HRA has been successful in fields like nuclear and aviation, emphasizing IFs related to cognitive and organizational ergonomics. Applying HRA to surgery is an emerging field, with studies creating taxonomies of IFs. Proactive analysis of human errors, especially in robotic surgery, is a growing area. Understanding IFs in surgeon performance can prevent safety issues, and considering interactions in complex systems is crucial. In summary, human factors research in surgery identifies factors impacting outcomes. Applying proactive HRA techniques, focusing on IFs, is gaining traction and contributes to better ergonomics in complex surgical settings.

Acknowledgements We acknowledge the support of the BRIEF "Biorobotics Research and Innovation Engineering Facilities" project (Project identification code IR0000036) funded under the National Recovery and Resilience Plan (NRRP), Mission 4 Component 2 Investment 3.1 of Italian Ministry of University and Research funded by the European Union - NextGenerationEU.

References

1. Wee, I.J.Y., Kuo, L.-J., Ngu, J.C.-Y.: A systematic review of the true benefit of robotic surgery: ergonomics. Int. J. Med. Robot. Comput. Assist. Surg. **16**(4), e2113 (2020)
2. Al-Naami, M., Anjum, M., Al-Khayal, K., Alkharji, H.: Robotic general surgery experience: a gradual progress from simple to more complex procedures. Int. J. Med. Robot. Comput. Assist. Surg. **9**(4), 486–491 (2013)
3. Lai, F., Entin, E.: Robotic surgery and the operating room team. Proc. Hum. Factors Ergon. Soc. Annu. Meet **49**(11), 1070–1073 (2005)
4. Binder, J., Bräutigam, R., Jonas, D., Bentas, W.: Robotic surgery in urology: fact or fantasy. BJUI Int. **94**(8), 1183–1187 (2004)
5. Bridger, R.: Introduction to Human Factors and Ergonomics, CRC Press (2017). 4th ed
6. Catanzarite, T., Tan-Kim, J., Whitcomb, E., Menefee, S.: Ergonomics in surgery: a review. Female Pelvic Med. Reconstr. Surg. **24**(1), 1–12 (2018)
7. Epstein, S., Sparer, E.H., Tran, B.N., Ruan, Q.Z., Dennerlein, J.T., Singhal, D., Lee, B.T.: Prevalence of work-related musculoskeletal disorders among surgeons and interventionalists: a systematic review and meta-analysis. JAMA Surgery **153**(2), e174947–e174947 (2018)

8. Stucky, C., Cromwell, K., Voss, R., et al.: Surgeon symptoms, strain, and selections: systematic review and meta-analysis of surgical ergonomics. Annals of Medicine and Surgery **27**, 1–8 (2018)
9. Weinstock, R., Ainslie-Garcia, M., Ferko, N., Qadeer, R., Morris, L., Cheng, H., Ehlers, J.: Comparative assessment of ergonomic experience with heads-up display and conventional surgical microscope in the operating room. Clin. Ophthalmol. **15**, 347–356 (2021)
10. Da Vinci surgical System (Online). Available: https://www.intuitive.com/en-us/products-and-services/da-vinci
11. Versius Surgical System (Online). Available: https://cmrsurgical.com/versius
12. Symani Surgical System (Online). Available: https://www.medtronic.com/covidien/en-us/robotic-assisted-surgery/hugo-ras-system.html
13. Senhance Surgical System (Online). Available: https://www.senhance.com/us/home
14. Symani Surgical System (Online). Available: https://www.mmimicro.com/
15. Savastano, A., Rizzo, S.: A novel microsurgical robot: preliminary feasibility test in ophthalmic field. Transl. Vis. Sci. Technol. **11**(8), 13–13 (2022)
16. Barbon, C., Grünherz, L., Uyulmaz, S., Giovanoli, P., Lindenblatt, N.: Exploring the learning curve of a new robotic microsurgical system for microsurgery. JPRAS Open, (2022)
17. Lindenblatt, N., Grünherz, L., Wang, A., Gousopoulos, E., Barbon, C., Uyulmaz, S., Giovanoli, P.: Early experience using a new robotic microsurgical system for lymphatic surgery. In: Plastic and Reconstructive Surgery Global Open, vol. 10, no. 1 (2022)
18. Ballestín, A., Malzone, G., Menichini, G., Lucattelli, E., Innocenti, M.: New robotic system with wristed microinstruments allows precise reconstructive microsurgery: Preclinical study. Annals of Surgical Oncology, 1–9 (2022)
19. Kagiyama, K., Ueno, T., Mitsutake, Y., Yamaji, K., Ishimatsu, T., Sasaki, K.-I., Fukumoto, Y.: First experience of robotic-assisted percutaneous coronary intervention in Japan. Intern. Med. **58**, 01 (2020)
20. Crandall, M., Agarwal, S., Muskat, P., Ross, S., Savage, S., Schuster, K., Tominaga, G., et al.: Application of a uniform anatomic grading system to measure disease severity in eight emergency general surgical illnesses. J. Trauma Acute Care Surg. **77**(5), 705–708 (2014)
21. Parsons, J., Messer, K., Palazzi, K., Stroup, S., Chang, D.: Diffusion of surgical innovations, patient safety, and minimally invasive radical prostatectomy. JAMA Surg. **149**(8), 845–851 (2014)
22. ElBardissi, A.W., Sundt, T.M.: Human factors and operating room safety. Surg. Clin. N. Am. **92**(1), 21–35 (2012)
23. Vincent, C., Moorthy, K., Sarker, S., Chang, A., Darzi, A.: Systems approaches to surgical quality and safety: from concept to measurement. Ann. Surg. **239**(4), 475–482 (2004)
24. Lyk-Jensen, H.T., Jepsen, R.M., Spanager, L., Dieckmann, P., Østergaard, D.: Assessing nurse Anaesthetists' non-technical skills in the operating room. Acta Anaesthesiol. Scand. **58**(7), 794–801 (2014)
25. Mishra, A., Catchpole, K., McCulloch, P.: The oxford NOTECHS system: reliability and validity of a tool for measuring teamwork behaviour in the operating theatre. Qual. Saf. Healthc. **18**(2), 104–108 (2009)
26. Undre, S., Healey, A.N., Darzi, A., Vincent, C.A.: Observational assessment of surgical teamwork: a feasibility study. World J. Surg. **30**(10), 1774–1783 (2006)
27. Cagliano, A., Grimaldi, S., Rafele, C.: A systemic methodology for risk management in healthcare sector. Saf. Sci. **49**(5), 695–708 (2011)
28. Verbano, C., Turra, F.: A human factors and reliability approach to clinical risk management: Evidence from Italian cases. Saf. Sci. **48**(5), 625–639 (2010)
29. Hudson, P.: Applying the lessons of high risk industries to health care. Qual. Saf. Health Care Suppl **12**(Suppl 1), i7–i12 (2003)
30. Deeter, J., Rantanen, E.: Human reliability analysis in healthcare. In: Proceedings of the 2012 Symposium on Human Factors and Ergonomics in Health Care, pp. 45–51. Human Factors and Ergonomics Society (2012)

31. Lyons, M.: Towards a framework to select techniques for error prediction: supporting novice users in the healthcare sector. Appl. Ergon. **40**(3), 379–395 (2009)
32. Lyons, M., Adams, S., Woloshynowych, M., Vincent, C.: Human reliability analysis in healthcare: a review of techniques. Int. J. Risk Saf. Med. **16**, 223–237 (2004)
33. Swain, A.D., Guttmann, H.E.: Handbook of human-reliability analysis with emphasis on nuclear power plant applications. Final Report 8, (1983)
34. Hollnagel, E.: Cognitive Reliability and Error Analysis Method (CREAM), Elsevier (1998)
35. Boring, R.: How many performance shaping factors are necessary for human reliability analysis. In: 10th International Conference on Probabilistic Safety Assessment and Management 2010, PSAM 2010, vol. 2 (2010)
36. E. Edwards, "Man and machine: Systems for safety. In: Proceedings of British Airline Pilots Association Technical Symposium, British Airline Pilots Association, pp. 21–36 (1972)
37. Williams, J.: HEART - a proposed method for achieving high reliability in process operation by means of human factors engineering technology. In: Proceedings of a Symposium on the Achievement of Reliability in Operating Plant, Safety and Reliability Society (SaRS), NEC, pp. 21–36. (1972)
38. Onofrio, R., Trucco, P.: Human reliability analysis (HRA) in surgery: identification and assessment of influencing factors. Saf. Sci. **110**, 110–123 (2018)
39. Onofrio, R., Trucco, P., Torchio, R., Onofrio, A., Trucco, P.: Towards a taxonomy of influencing factors for human reliability analysis (HRA) applications in surgery. Procedia Manuf. **3**, 144–151 (2015)
40. Karsh, B.T., Holden, R.J., Alper, S.J., Or, C.K.L.: A human factors engineering paradigm for patient safety: designing to support the performance of the healthcare professional. BMJ Qual. Saf. **15**(suppl 1), i59–i65 (2006)
41. Karsh, B.T., Brown, R.: Macroergonomics and patient safety: the impact of levels on theory, measurement, analysis and intervention in patient safety research. Appl. Ergon. **41**(5), 674–681 (2010)
42. Karsh, B.T., Waterson, P., Holden, R.J.: Crossing levels in systems ergonomics: a framework to support "mesoergonomicâŁž inquiry. Procedia Manuf. **45**(1), 45–54 (2014)
43. Onofrio, R., Trucco, P.: A methodology for dynamic human reliability analysis in robotic surgery. Appl. Ergon. **88**, 103–150 (2020)
44. Galfano, A., Ascione, A., Grimaldi, S., Petralia, G., Strada, E., Bocciardi, A.: A new anatomic approach for robot-assisted laparoscopic prostatectomy: a feasibility study for completely intrafascial surgery. Eur. Urol. **58**(3), 457–461 (2010)

Human Factors in Robotics for Motor Rehabilitation, Assistance, and Prosthetics

Federico Tessari, Marek Sierotowicz, Sabine Thuerauf, Heloisa Seratiuk Flores, Jonas Grau Thomsen, Matteo Laffranchi, Lorenzo De Michieli, Emanuele Gruppioni, Satoshi Muraki, Letizia Leocani, Giampaolo Brichetto, Loredana Zollo, Claudio Castellini, and Giacinto Barresi

Abstract Among the challenges in biomedical robotics, assisting and restoring motor skills is a priority to enable individuals with impairments to perform activities of daily living (ADLs). Robots can enhance human mobility and interaction capabilities through solutions tailored to individual skills, limitations, and goals in specific contexts. For rehabilitation, end-effector robots and exoskeletons are often programmed to guide users in exercises designed to restore motor functions after severe incidents, such as strokes, or in cases of chronic and degenerative conditions. These devices can also be integrated into telerehabilitation systems. Assistive robots, on the other hand, improve the quality of life for individuals whose impairments

F. Tessari (✉)
Newman Laboratory for Biomechanics and Human Rehabilitation, Massachusetts Institute of Technology, Cambridge, MA, USA
e-mail: ftessari@mit.edu

M. Sierotowicz · C. Castellini
Department of Artificial Intelligence in Biomedical Engineering, Friedrich-Alexander-Universität Erlangen-Nürnberg, Erlangen, Germany

S. Thuerauf
Chair of Robotic Systems in the Sign of Demographic Change, Technichal University Dortmund, Dortmund, Germany

H. Seratiuk Flores · S. Muraki
Laboratory of Ergonomics for People of All Ages and Abilities, Faculty of Design, Kyushu University, Fukuoka, Japan

J. Grau Thomsen
Nordic Bionics,Copenhagen, Denmark

M. Laffranchi · L. De Michieli
Rehab Technologies Lab, Istituto Italiano di Tecnologia, Genoa, Italy

E. Gruppioni
Centro Protesi INAIL, Istituto Nazionale per l'Assicurazione contro gli Infortuni sul Lavoro, Bologna, Italy

L. Leocani
Institute of Experimental Neurology, Division of Neuroscience, Scientific Institute and University Ospedale San Raffaele, Milan, Italy

G. Barresi and S. Wischniewski (eds.), *Ergonomics in Robotics: Advances and Innovations*, Springer Tracts in Advanced Robotics 162,
https://doi.org/10.1007/978-3-032-19374-2_10

cannot be fully addressed through therapy, such as those due to aging. Devices like robotic arms, wheelchairs, walkers, and exoskeletons support ADLs and even facilitate social interactions through telepresence systems. Bionic prostheses, designed for limb replacement in cases of amputation or agenesis, could achieve a high level of embodiment–allowing users to perceive the prosthesis as part of their body. This need for a better integration between humans and robots exemplifies challenges common to all three domains. This chapter introduces fundamental aspects of cognitive ergonomics and human factors that designers, clinicians, and users must consider when using rehabilitative, assistive, or prosthetic systems. The ultimate goal is to provide a 'truly' human-centered motor assistance, and recovery, through robotic systems.

Keywords Assistive robotics · Rehabilitation robotics · Bionics · Prosthetics · Human factors

1 Introduction

Human-robot interaction (HRI) offers especially relevant solutions when it is designed to enhance the capabilities of people with motor impairments. This is possible thanks to the opportunity of having an intelligent machine able to precisely (and carefully) extend the physical range of actions of the human body. Robotic systems can be used to restore motor functions through exercises, implementing rehabilitation procedures established by the clinician according to the patient's conditions, and moving the impaired limb in repetitive training to rebuild patients' motor control (when possible). When therapy cannot fully restore the individuals' skills, people with motor impairments can also employ assistive robots to perform activities of daily living (ADLs)—such as eating, walking, and personal care—and, consequently, to improve the quality of their lives. Finally, prosthetic devices aim at replicating and replacing limbs to perform ADLs seamlessly and comfortably, and providing the user with higher independence.

G. Brichetto
Area Scientifica Genova, Fondazione Italiana Sclerosi Multipla, Genoa, Italy

L. Zollo
Unit of Advanced Robotics and Human-Centered Technologies, Università Campus Bio-Medico di Roma, Rome, Italy

C. Castellini
Institute of Robotics and Mechatronics, German Aerospace Center (DLR), Oberpfaffenhofen, Germany

G. Barresi
Bristol Robotics Laboratory, School of Engineering, CATE, University of the West of England, Bristol, UK

This chapter will introduce these topics from the human factors perspective, with special attention to cognitive ergonomics issues related to user experience and mental workload as examples to highlight the challenges in the human-centered design of these robotic systems. The chapter will also introduce examples of perspectives and issues in this heterogeneous domain, including: neurotechnologies in HRI for motor recovery and assistance; robotic-digital strategies for engaging the users in training activities; distributed and remote systems; the collection of user needs in co-design paradigms; special applications as prosthetics for sports; and the ethical issues due to attitudes like technoableism. These topics integrate the discussion on the ergonomics of robotic devices for motor recovery and assistance and highlight the multi-faceted aspects of the adoption of smart mechatronic systems to help people in their daily life.

2 Robots for Motor Rehabilitation and Assistance

Several robotic systems emerged as potential 'intelligent' and automated tools to support human motor functions [1]. In the context of this chapter, they can be classified into two primary purposes: rehabilitation and assistance.

Motor rehabilitation robots are specifically designed to restore physical functions in people with (at least partially) treatable impairments in moving their own body. These devices are commonly used with patients recovering from neurological or musculoskeletal injuries due to traumatic accidents (stroke, traumatic brain injury, joint dislocations or tendon tears, for instance) or chronic and degenerative conditions, e.g., Multiple Sclerosis (MS) or Parkinson's Disease (PD). Aging can lead to issues requiring rehabilitation as well. Rehabilitation robots aim to facilitate motor recovery by guiding repetitive, task-specific movements, which are known to drive processes like neural plasticity and improve motor outcomes over time [2, 3].

When rehabilitation cannot fully restore individual skills, **motor assistive robots** can be adopted to enhance the physical capabilities of people facing temporary or permanent motor limitations such as the ones caused by chronic and degenerative disorders. These intelligent machines are designed to support and enhance intentional user movements in a variety of daily life contexts when certain tasks would be difficult or impossible to accomplish. Such assistive robots consequently promote greater autonomy and functionality in ADLs, with potential impacts on the overall quality of life of individuals with amputations or congenital birth defects, as well as elderly people who cannot move properly [4].

Even if assistive and rehabilitative robots target different aspects of motor function, they both share several features. A first categorization divides them into wearable mechatronic devices (like exoskeletons and exosuits) and machines able to support or guide (or even fully replace) individual movements (like end-effectors, robotic arms, robotic walkers, and automated wheelchairs). The proposed categorization considers as criterion the user level of physical interaction with the device. However, alternatives are also possible, such classifying the devices based on their functional use, e.g., rehabilitation, assisted manipulation or assisted mobility.

Among the wearable robotic systems for both motor recovery and assistance, we find exoskeletons and exosuits.

- **Exoskeletons** act as rigid external frames for sustaining the posture and supporting or guiding the movements of the human body, aligning with the joints of human anatomy thanks to a design mimicking their morphology and range of motion. Exoskeletons exist for both upper and lower limbs and can serve both rehabilitative and assistive purposes according to their scope [5, 6]. Their rigid structure provides mechanical support and a wide range of functionalities. For instance, they can move the human body through intention detection solutions by inferring—from the general posture or some biosignals—the commands of a person with spinal cord injury, whose movements are incapacitated in the body districts affected by this condition. However, exoskeletons are often limited by the time-consuming process of donning and doffing and are primarily used in clinical settings. These devices can adopt multiple types of design, e.g., the one proposed by Heo et al. [7] in which they presented a lower-limb training exoskeleton that enhances muscle power during cycling, offering potential applications to walking rehabilitation.
- **Exosuits**, in contrast to exoskeletons, are soft or semi-soft wearable systems designed to provide partial support to the user [8, 9]. The exosuits conform better to the body soft tissues, such as skin and muscles, offering a more comfortable fit and easier alignment with human kinematics. While they typically have less mechanical power and offer fewer functionalities than exoskeletons (especially for sustaining the user's body or for autonomously initiating the movements of the limbs without any residual activity of the user), exosuits are generally safer, quicker to wear and have broader applications. These applications include use both inside and outside the clinical environment, offering higher chances for real-life contexts adoption.

End-effector robots and robots for assisted manipulation and mobility constitute examples of non-wearable devices providing functions of motor recovery and assistance.

- **End-Effector Robots for Rehabilitation** resemble industrial robotic systems, featuring rigid structures and (typically) serially connected links. End-effector robots are equipped with specialized tools such as grippers, suction devices, or magnetic attachments to assist with both rehabilitation and daily tasks [10, 11]. Unlike exoskeletons, end-effector robots interact with users only at a specific point, minimizing issues related to wearability and joint alignment. While these robots are highly functional and relatively easy to set up, their high costs and lack of portability restrict them to clinical use.
- **Robots for Assisted Manipulation** are designed to assist users with impairments in performing everyday tasks such as grasping, holding, or manipulating objects [12]. Among them, we can find robotic arms typically mounted on

wheelchairs or workstations. They can be controlled through a range of user interfaces, including joysticks, switches, eye-tracking cameras, or even brain-computer interfaces (BCIs), enabling users to execute tasks independently. We can also consider—as part of this category—assistive robotic tools, like spoons, which are designed to actively mitigate the essential tremor for helping a person to eat. In this case, the device is fully held by the user, which employs it for performing ADLs with a really specific contribution of the machine [13].
- **Robots for Assisted Mobility** include devices designed to assist gait and locomotion, including robotic wheelchairs and powered walkers [14, 15]. These systems build on their passive counterparts by integrating motors and sensors to provide active support for movement. Also in this case, their control can depend on a variety of user interfaces, from joysticks to switches, from eye-tracking cameras to BCIs (in general the latter case is still an experimental solution). Robotic walkers are particularly useful for individuals with permanent mobility impairments or older adults, offering enhanced safety and mobility. In this context, robotic devices shaped as a cane-like object and designed to guide—through vibratory feedback—people with visual impairments [16], may be also considered as robotic tools for assisted mobility instead of manipulation.

In both mentioned macro-categories of rehabilitative and assistive robotic devices, physical ergonomics plays a critical role in determining user comfort, safety, and effectiveness of HRIs. Examples in this domain are adjustability and safety both in exoskeletons (rigid, mechanical frames) and exosuits (soft, flexible designs) [17]—of the contact points and range of motion of the end-effector robots for rehabilitation [18]. Also the selection of the most appropriate feedback modality enriching the exercises performed through an end-effector should be considered for avoiding any sensory discomfort [19]. About robotic arms for motor assistance, key physical ergonomic concerns are related, for instance, to reachability and stability, especially when the movement should be performed towards the users (as in feeding tasks) [20]. Similar issues can occur for assisted mobility devices [21], especially because of their direct contact with the user, as in the case of robotic wheelchairs and powered walkers. Furthermore, they can show difficulties due to abnormal weight distribution.

Psychological factors can surely impact HRIs in these domains too. We can consider, as an example, how rehabilitative robots possess capabilities for both assessment and training but they definitely need a certain level of clinical compliance in any exercise for providing reliable, quantitative evaluations of human conditions, for example, enhancing the likelihood of patient recovery. Consequently, these systems [22] facilitate patient recovery by employing various control strategies tailored to the patient's residual abilities and informed by the latest research on neuroplasticity. Furthermore, these approaches can be robot-driven or patient-driven, with active assistive training being particularly valuable. This method supports participants in moving their impaired limbs into desired postures during tasks like grasping, reaching, or walking, mirroring the principles of physical and occupational therapy. Specifically, the "assist-as-needed" approach is commonly used to prevent patients from becoming overly dependent on the robot to complete rehabilitation tasks. This is a demonstration of the need of considering the user from cognitive-behavioural perspectives of HRI in order to correctly implement physical therapies through robotic systems.

The next paragraphs will focus on cognitive-behavioural aspects of the interaction, pondering examples about user experience, usability, mental workload, and technology acceptance.

2.1 User Experience

User experience in robotic rehabilitation systems extends beyond usability and encompasses the overall interaction between the user and the technology. This interaction is multidimensional, involving physical, psychological, and social factors that influence how users perceive and engage with robotic systems.

Gombolay et al. [23] illustrated how embodiment (feeling an external object as part of one's body) in robotic systems can increase the effectiveness of robotic assistance, particularly in patient care. This concept of embodiment, where users perceive the robot as an extension of themselves, plays a pivotal role in enhancing the user's overall experience. The physical integration of robotic devices, such as exoskeletons and robotic suits, creates a sense of control and ownership, which can lead to improved therapeutic outcomes. Devittori et al. [24] further supported this notion by demonstrating that unsupervised robot-assisted therapy in clinical settings led to a meaningful increase in therapy dose and that participants expressed satisfaction with the device's usability.

However, providing effective feedback during the rehabilitation process is crucial for optimizing user experience, as evidenced by Fruchter et al. [25]. Feedback systems should be designed to enhance performance without overwhelming the user. Personalization of feedback, avoiding excessive feedback, and employing a hierarchy in feedback delivery are key factors in improving user engagement. Visual and auditory feedback, for example, can assist in error correction and skill acquisition, thus accelerating motor learning. Moreover, Laparidou et al. [26] conducted a thematic review of user experiences with robotic rehabilitation devices and found that despite technological and logistical challenges, users often found these devices physically and psychologically beneficial. Users also reported that positive therapeutic relationships with healthcare staff were essential for a successful rehabilitation experience.

A user experience in robotic rehabilitation is built on the integration of personalized, real-time feedback mechanisms and strong interpersonal relationships with caregivers, creating a holistic approach to therapy.

2.2 Usability

Usability plays a crucial role in the adoption and efficacy of robotic systems in both assistance and rehabilitation. Effective usability ensures that these technologies are not only accessible to a broad spectrum of users but also seamlessly integrated into therapeutic practices, enhancing patient outcomes. Several studies have provided insights into the various factors that influence usability in robotic systems for rehabilitation.

Robotic devices, such as those designed for upper limb rehabilitation, have garnered positive feedback for their ease of use, functionality, and perceived safety. For instance, Guillén-Climent et al. [27] assessed the usability of the MERLIN robotic system for upper limb rehabilitation in home settings and found it to be both safe and motivating, offering a viable option for extending clinical therapy. Such findings support the growing trend of integrating telerehabilitation and home-based therapy, thereby expanding the accessibility and duration of treatment.

Similarly, Zanatta et al. [28] noted that robotics has contributed to increased treatment intensity and improved patients' physical and psychosocial well-being. However, despite the generally positive reception, there are still significant technical and design challenges that hinder the usability of these systems. Issues such as physical strain during use and cognitive or communication deficits present barriers, limiting the broader application of robotic rehabilitation technologies. This sentiment was echoed in the work of Park et al. [29], which contrasted passive and active exoskeletons. Passive exoskeletons were deemed more suitable for stroke survivors capable of voluntary movement, offering smoother and more effective participation in therapy sessions compared to their active counterparts. This, however, might also be related to the still ongoing refinement and integration of actuation technologies which hinders the usability compared to the further developed passive solutions.

Furthermore, usability evaluations have highlighted the importance of user-centered design in ensuring the effectiveness of these devices. Meyer et al. [30] emphasized the need for dedicated usability evaluation studies, particularly in the context of augmentative and assistive technologies. Meyer's study underlined the importance of balancing qualitative and performance metrics when assessing usability. By prioritizing functionality, ease of use, and user involvement, robotics developers can address common technology acceptance challenges and promote wider adoption.

These findings suggest that while robotic rehabilitation technologies have made significant strides, there is still a need to address the usability challenges that may limit their long-term adoption and integration into clinical settings.

2.3 Mental Workload

Mental workload is a critical factor that influences the effectiveness of robotic rehabilitation technologies. The cognitive demands placed on users when operating assistive devices can affect both short-term performance and long-term adoption. Understanding the mental workload associated with these systems is therefore essential for designing technologies that are not only functional but also user-friendly.

Marchand et al. [31] highlighted the significant mental workload required to operate wearable assistive devices such as exoskeletons. The complexity of controlling these devices, combined with the physical effort of rehabilitation, can lead to high cognitive demands, making it challenging for users to fully benefit from the technology. The study identified methods such as dual-task paradigms and subjective assessments like the NASA-TLX, which have proven effective in measuring mental workload during the use of exoskeletons. However, these methods come with limitations, including subjectivity and complexity in interpretation.

Upasani et al. [32] focused on eye-tracking metrics as a tool to assess cognitive load in physical human-robot interactions. Their research demonstrated that gaze entropy and pupil diameter were reliable indicators of mental workload, providing valuable insights into how task difficulty affects cognitive demands. The study also found that increased task difficulty led to a shift in users' strategies, emphasizing robot monitoring over task engagement. This underscores the importance of designing rehabilitation systems that minimize unnecessary cognitive load and encourage active user participation.

Novak et al. [33] proposed that psychophysiological responses, such as skin conductance and heart rate, could be integrated into closed-loop systems to monitor users' psychological states during therapy. Such systems could adapt the difficulty of tasks based on real-time feedback, reducing mental strain and optimizing therapy sessions. Similarly, La Bara et al. [34] emphasized the need for psychometrically validated tools like System Usability Score (SUS) and NASA Task Load Index (NASA-TLX) to measure cognitive workload effectively, ensuring that rehabilitation technologies can be not only usable but also cognitively accessible.

In conclusion, mitigating the mental workload associated with robotic rehabilitation systems is essential for improving user performance and ensuring long-term adoption. By integrating real-time feedback mechanisms and reducing the cognitive demands placed on users, these technologies can be made more accessible, fostering better engagement and more effective rehabilitation outcomes.

2.4 Technology Acceptance

Technology acceptance models (TAMs) have long been used to explore the factors that influence the adoption of new technologies. When it comes to robotic rehabilitation systems, the challenge of acceptance is heightened by the complexity of these devices and their integration into therapeutic routines. Assistive devices acceptance is instead challenged by the need of integrating the technology in the users' daily life without making them perceive such technology as a source of social stigma. A deeper understanding of the factors influencing acceptance is critical to ensuring that these devices are used effectively in clinical practice.

Shore et al. [35] introduced three new constructs to better capture the experience of older adults with assistive technologies: Experiential Perception (EP), Self-Liberty (SL), and Quality of Life Enhancement (QOLE). These constructs (within the EXOSCORE framework for design and assessment) focus on the user's interaction with the robot, their perception of autonomy while using it, and the potential improvements in their daily lives. This focus on the lived experience is essential for ensuring that robotic systems not only meet functional requirements but also address the emotional and social needs of users, particularly older adults who may be resistant to adopting new technologies.

Luciani et al. [36] examined the acceptability of exoskeletons from the perspective of therapists, highlighting the importance of education and training in increasing adoption rates. A key barrier to acceptance, according to this study, is the fear that robotic systems create a disconnect between the therapist and the patient's body, leading to reduced trust in the technology. Luciani et al. suggested that integrating exoskeletons with multisensory feedback systems could help bridge this gap and increase therapists' perceived utility of the devices.

Klaic et al. [37] expanded on the idea of clinician adoption by applying the TAM framework to investigate the factors that influence rehabilitation clinicians' use of robotic devices. The study found that the complexity of these devices and the lack of adequate training limited their acceptance. Furthermore, Shore et al. [38] emphasized that older adults may experience stigma and frustration when using assistive robots, which underscores the need for user-centered design approaches to mitigate these feelings and encourage greater adoption.

The path to widespread acceptance of robotic systems in rehabilitation lies in addressing these concerns by ensuring that technologies are intuitive, that they enhance rather than replace human care, and that users feel empowered and supported in their use. The adoption of assistive robotic devices obviously faces further barriers for the adoption unless the devices propose a design with a seamless integration and a higher opportunities for personalization and customization that just a co-design approach can anticipate.

3 Replacing Limbs: Bionic Prosthetics

Bionic prostheses, specifically artificial limbs, are devices engineered to replace or augment lost or impaired extremities, restoring lost functionality and enhancing the quality of life for the wearer. As such, prostheses are generally classified as assistive rather than rehabilitative devices. The application of limb prostheses can be first differentiated into upper limb and lower limb replacements, each with distinct ergonomic and functional requirements. It is worth mentioning that we are excluding cosmetic prostheses, as we are focusing on devices that provide assistive functions.

Upper limb prostheses (ULPs), including artificial hands and arms, prioritize dexterity and fine motor control. Conversely, lower limb prostheses (LLPs), encompassing artificial feet and legs, focus on reliability and safety, often favoring simpler designs. It is worth mentioning that LLPs are more commonly used (49–95%) than ULPs (27–56%) by amputees [39].

ULP taxonomy is functionally defined by power source as either body-powered or self-powered. Body-powered prostheses utilize mechanical linkages and harnesses operated by the user's residual movements. Through these linkages, the operator transfers mechanical energy to the prosthesis. This approach is robust and simple, though intrinsically requiring unnatural compensatory movements. Through the direct mechanical connection, body-powered prostheses inherently provide force feedback. However, because of the involved lever, body-powered prostheses must navigate the trade-off of achievable force against the desired range of motion. Furthermore, they can realistically provide the user with typically one controllable degree of freedom. In contrast, self-powered ULPs are equipped with self-contained actuators. These devices offer greater functionality and versatility, as their interaction with the user's motion intent is shaped by the control system. Since self-powered ULPs should compensate for their greater complexity compared to their body-powered counterparts, they tend to embed more actuators to guarantee greater functionality. The more complex hardware and the underlying power sources can lead to increased device weight but they allow for more natural control strategies, thanks to peripheral nervous system interfaces, e.g., arrays of surface electromyographic sensors EMG combined with pattern recognition algorithms. Due to power efficiency considerations, and to prevent unwanted movements, most commercial manufacturers tend to choose non-backdrivable transmission mechanisms [40]. A downside of non-backdrivable actuations is that to finely control grasp forces you might need the inclusion of direct force sensors at the contact interface. Moreover, to prevent damage from bumping into objects, many commercial multi-grip hands feature passive elastic elements in series with the actuator.

LLPs are instead focused on gross motor skills like walking or balancing. Lower-limb amputations account for the majority of limb amputations worldwide, especially those due to devascularization, e.g., caused by complications from diabetes. It is worth mentioning that, in many cases, LLPs are not a viable option for such patients due to co-morbidities that might lead to issues such as skin breakage, bruising, and bleeding upon wearing a prosthetic device. In these cases, wheelchairs are still the preferable option. LLPs prioritize reliability because the potential risk associated with a device failure is more severe than for ULPs, potentially resulting in falls and injuries. Consequently, many commercial LLPs rely on purely mechanical components. Passive elements are commonly used in the knee and hip joints (e.g., hydraulic dampers and/or mechanical springs), and in the foot (e.g., composite leaf-spring heel and keel components) for energy storage and release [41, 42]. When it comes to the ankle joint, it has been demonstrated how built-in actuators in the ankle can lead to decreased metabolic costs. This is due to the ankle joint demanding more positive work during level-ground walking compared to the knee. However, the employment of active components in prosthetic ankles is still limited. In contrast, many commercial LLPs adopt a semi-active approach to the knee joint which applies variable resistant forces during the swing and stance phase of level-ground walking through—for example—magnetorheological or hydraulic systems. In activities like stair climbing, which demands more positive work from the knee, fully powered prostheses can improve life quality for those with proximal and/or bilateral amputations [43]. In support of this, Au et al. showed that the energy efficiency afforded by powered LLPs is superior compared to passive LLPs [44]. The effect is more significant for users with above-knee amputations, whose gait is otherwise generally less efficient. However, due to the requirements on reliability, LLPs feature mostly passive joints even for hemipelvectomy or hip disarticulation patients [42].

In this section, we will mainly focus on active prosthetic devices featuring electronic components and actuation, and on the technologies depicted in Fig. 1. Sensor modalities in human-machine interactions (HMIs) for self-powered prostheses include EMG sensors, which detect electrical activity from muscle contractions, and inertial measurement units (IMUs), which track limb orientation and movement. Additionally, force and pressure sensors measure the interaction of the prosthesis and external objects to provide feedback via invasive (implanted electrodes for residual nerve stimulation) or non-invasive interfaces (tactile, thermal or electrical surface stimulation) with the peripheral nervous system (PNS) for improving—for example—grip control and stability.

About non-invasive prosthetic systems, i.e., wearable prostheses that do not require to permanently "modify" the person through procedures like osseointegration, physical ergonomics issues are primarily related to fit, comfort, and alignment. As above, several physical issues in this area must be considered. For instance, it is critical to pay attention to the residual limb in terms of shapes and sizes, contact points and load distribution, especially in conditions of prolonged use that can force the person to alter their balance and to overly stress the intact limb over the prosthetic one [45]. In long-term use, an adequate (ergonomic) prosthetic fit increases the chance of distributing forces equally between the intact and the prosthetic limbs during the gait, reducing the risk of pathological consequences on the musculoskeletal system. The socket design is a priority in this context, working on materials and to the general compliance of the system [46]. Ergonomics should reduce the risk of skin irritation and musculoskeletal strain. This is particularly relevant for the spine—in case of lower limbs prostheses—and should consider also the fatigue effects of the socket to make the device more reliable during prolonged usage. Moreover, it is necessary to avoid, for instance, thermal effects [47] and conditions like dermatitis [48] on a confined and pressed stump within the closed space of the socket along a day, which affects the prosthesis usability and user experience. Anthropometric approaches [49], alongside the design of adjustable structures [50], offer key insights to improve the socket design in terms of ergonomics. This is definitely crucial for a prolonged use of prosthetic devices alongside factor requiring an appropriate balance—as in the case of the duration and the weight of the battery of a bionic leg [51].

As in the previous section, we will move from these physical aspects of HRI to examples of user experience, usability, and mental workload in the domain of prosthetics.

3.1 User Experience

Evaluating a prosthesis user's experience involves both subjective and objective metrics. Tools like the (Socket Comfort Score) SCS and the Trinity Amputation and Prosthesis Experiences Scales (TAPES) address both lower and upper limb prostheses. Furthermore, the Prosthesis Evaluation Questionnaire (PEQ) is a key tool with ten scales assessing prosthesis function, mobility, psycho-social aspects, and well-being for LLP users. Beyond questionnaires, quality-of-life improvements can be gauged

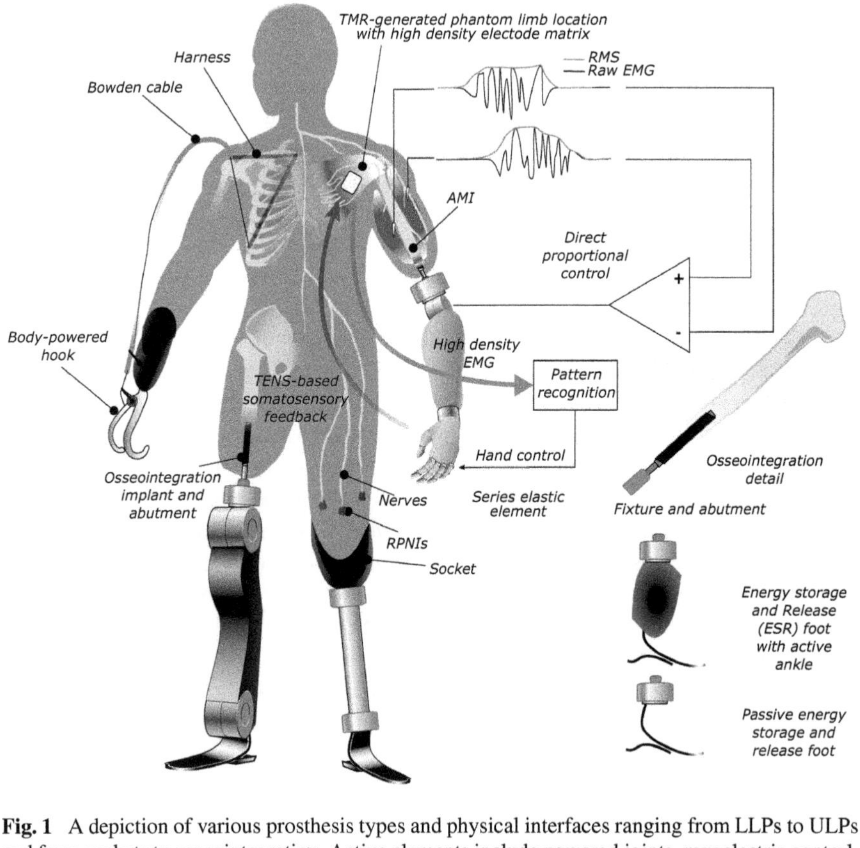

Fig. 1 A depiction of various prosthesis types and physical interfaces ranging from LLPs to ULPs and from sockets to osseointegration. Active elements include powered joints, myoelectric control, target muscle reinnervation (TMR), and somatosensory feedback devices

through measurable metrics such as gait velocity, metabolic costs, and changes in step activity for LLPs [52], and in the achievable performance in ADLs for ULPs.

3.1.1 Upper-Limb Prostheses

Despite constant improvements in the underlying technology, upper-limb prostheses still present non-negligible abandonment rates. Looking at the rejection rate over time, no significant effect of time could be determined, suggesting that technological progress has not yet met user needs. However, various studies underline how many of the principal factors weighing on the rejection ratio could be addressed by technology [53]. Such factors include comfort, weight, functionality, and level of amputation. Higher level of amputation, i.e., more proximal, could be addressed by high-density surface EMG, which can improve myocontrol even with reduced musculature [54] and by surgical techniques such as TMR and Regenerative Peripheral Nerve Interface (RPNI). Other factors are not as directly influenced by the underlying technology, such as those related to aesthetics and body representation, which is related to the concept of embodiment. Gender has a significant effect on the rejection rate of prostheses, with women showing a higher overall probability of rejection and a higher propensity for cosmetic, passive prostheses rather than for active ones [55]. Furthermore, a higher average rejection rate for ULPs is reported among pediatric patients compared to adults [53]. Anecdotal evidence and surveys seem to underline how multi-grip myoelectric devices do not seem to significantly improve user quality of life, indicating that functionality does not correlate with device complexity. The same seems to be true for device cost [56]. However, most literature does not consider devices with feedback—though successful examples of feedback restoration exist [57]—which can be expected to play a larger role in the upcoming decade.

3.1.2 Lower-Limb Prostheses

Besides reliability, the major issue reported by LLP users is the necessity for comfort. In LLPs more than in ULPs, a sub-optimal socket can lead to pain, bruising, and circulatory problems. In contrast to sockets, osseointegration, which consists of surgically implanting the load-bearing structure directly into the bone, transfers all interaction forces onto the skeleton. Osseointegration is more frequently adopted in LLPs than in ULPs, particularly due to the larger cross-section of the affected bones. This has many benefits since it relieves pressure on the soft tissues, but it relies on an invasive procedure and is not suitable for pediatrics. Osseointegration influences the perception of interaction forces with the environment since a stiffer interface leads to a higher mechanical transmission bandwidth. Even if powered prostheses are rare among LLPs, multiple studies showed positive effects of their use regarding long-term disorders, such as a reduction in the risk of knee osteoarthritis [58] and an increase in the self-selected walking speed [41]. However, self-powered prostheses tend to have a higher weight, increasing torque on the intact joints and metabolic expenditure. A further, rarely considered factor in the acceptance of active prostheses is the noise they may produce. The necessity for lower noise emissions may limit the available options for actuation, making certain forms of actuation, such as pneumatics, commercially unfeasible.

3.2 *Usability*

Prostheses users often don and doff their device daily, which can be a time-consuming part of one's daily routine. For LLP it is comparably easier to fit the device if the user has control of both hands. Furthermore, precise donning of the prosthesis is required for self-powered prostheses utilizing sensors.

Prostheses controlled through muscular activity show inconsistent behavior upon electrode slippage or if the residual limb's characteristics vary over time, which diminishes usability. In the case of pattern recognition control, usually employed in ULPs, periodic updates of the motion intent prediction model must also be considered. This process may require aid from experts, although, in modern commercial devices, it can be more easily executed through advanced Graphic User Interfaces (GUIs) and applications running on smart devices. Many commercial LLPs require explicit setting of the gait type on the part of the user, which influences usability as well as mental workload. Overall, because of a design that prioritizes reliability and repeatability, LLPs can be expected to have higher usability than ULPs.

3.3 *Mental Workload*

Because of the differing control types usually employed for active ULPs and LLPs, the expected mental workload changes greatly between these types of devices.

3.3.1 Upper-Limb Prostheses

Most forms of self-powered ULPs rely on muscular control, which allows the user to associate a prosthesis motion with the contraction of specific muscle groups. However, the association is fundamentally arbitrary, which can lead to some increase in the mental workload necessary to operate a ULP. Furthermore, many control systems on the market rely on sequential controls that allow switching between joints or types of grasps by quickly co-contracting agonist and antagonist muscles, and then controlling the excursion on the selected joint or grasping pattern by contracting the muscle groups separately. This form of control entails a higher mental workload. In this sense, it is possible to distinguish between proportional control vs. classification control, sequential vs. simultaneous, and direct vs. pattern recognition. Direct, proportional pattern recognition leads to the best outcomes in many standard functionality tests, with lower cognitive workload [59]. Such controllers, on the one hand, might be impractical for a higher number of degrees of freedom, which on the other hand have the potential to decrease the necessity for compensatory movements.

3.3.2 Lower-Limb Prostheses

Compared to ULPs, LLPs can rely on the assumption of quasi-periodic motions for most gait patterns. Therefore, active LLP control loops often consist of some form of gait type classification, in conjunction with a gait phase estimator. To this end, IMUs are usually employed, as well as joint encoders. Sometimes the type of gait or the current activity (such as cycling, or cross-country skiing) must be manually selected. However, the prosthesis usually acts automatically within a certain gait type, and once the user adapts their gait, the cognitive workload should be reduced, and likely minimal or optimal—though this aspect still needs further investigation. Volitional controls based on, e.g., surface EMG are less common in LLPs, but a positive trend is present in recent research [43]. Muscular activity could be affected by the user trying to stabilize and more DoFs could increase the mental workload. Nevertheless, muscle-based control of the prosthesis has the potential to enable natural, intuitive movements like those of able-bodied users [60].

3.4 Technology Acceptance and Trust

The level of trust in the prosthesis is mainly shaped by factors such as user background, familiarity with the device, and past prosthetic experiences, especially failures. Environmental interaction is affected by trust in LLPs more than in ULPs. Most users feel confident using LLP on standard surfaces, but this changes on slippery or uneven ground. Anecdotal evidence suggests that a user's perception of prosthesis performance is highly dependent on the user's trust [61]. In their study, Yamanaka et al. report that more than 70% of LLP users had a fall event in the past year and that more than 80% report a "near fall" event [62]. Daily use and adoption of LLPs are also deeply correlated with trust [63]. As will be explored in Sect. 3.5, the concept of embodiment is tightly related to the concept of trust, which is one of the principal reasons why embodiment is deemed as potentially advantageous.

3.5 Prosthetic Embodiment

Improving the prosthetic technology acceptance derives from the need to reduce the risk of abandonment of artificial limbs. As mentioned above, eliciting the phenomenon of embodiment of prosthetic limbs—the incorporation of an external object in the subjective experience and the underlying neurocognitive representation of one's body—seems an advantageous strategy to achieve a higher technology acceptance. This makes embodiment the object of a multidisciplinary study including neuroscience, engineering, and human factors analysis [64]. The concept of embodiment is constituted by three sub-aspects: the sense of self- and co-location (feeling to be where the incorporated item is), the sense of ownership (feeling that the incorporated item belongs to the user), the sense of agency (feeling in control of the incorporated item) [65]. Various bio-signals can be used to quantify embodiment ranging from skin temperature over histamine reactivity to cortical imaging [66]. However, the methodologies currently used for investigating it are typically based on subjective questionnaires and tests like the proprioceptive drift: estimating where the limb is after sessions of protocols based on the Rubber Hand Illusion (RHI). These tests are explored to understand how simultaneous multisensory stimulations (visually, on the artificial limb, and haptically, on the hidden biological limb) can lead to higher sense of body ownership [67, 68].

Ownership is often understood as the incorporation of co-occurring sensory feedback from various sources. In contrast, tool incorporation is the incorporation of non-co-occurring sensory feedback. The difference between these two can be quantified by measuring the difference in reaction time when co-occurring vs. non-co-occurring feedback is provided. This is called the Cross-modal Congruency Effect (CCE) [69]. However, co-occurrence can arguably be shaped surgically through, for example, TMR or Targeted Sensory Reinnervation (TSR), or by means of invasive nerve interfaces. Such interventions can have a complex impact on the individual. On one hand, as invasive solutions, they could be considered as lacking a design based on ergonomics. On the other hand, they are often performed with the purpose of minimizing the risk of phantom limb pains, improving the user experience of the individual. Regardless, current literature cannot adequately assess a potential transition from tool incorporation to ownership over time through specific training. However, many studies assessed effects in the short term. For instance, a study addressing specifically the effects of TMR and TSR confirmed how the integration of somatosensory feedback increases several outcome metrics that would be expected to improve with the emergence of embodiment, both in the sense of body representation and in the phenomenological sense [70]. Despite a relatively small sample of two participants, the study could detect significant improvements in the Fitts' law trials, and a significantly decreased gaze fixation on the prosthesis could be detected with the inclusion of feedback.

4 Perspectives on Key Advances and Issues in HRI for Motor Recovery and Assistance

The variety of solutions that can be explored and, if successful, adopted to integrate and empower medical robots for motor recovery and assistance can grow according to the insights emerging from the potential offered by futuristic innovations and unmet needs. The next sub-sections will describe a set of perspectives on key advances and issues in HRI that should be considered for enhancing rehabilitative, assistive, and prosthetic robotic systems in terms of ergonomics.

4.1 Neurotechnologies in HRI for Motor Recovery and Assistance

To optimize the interactions between people with motor impairments and rehabilitative, assistive, and prosthetic robotic systems, it is essential to explore key advances in human-robot interaction aimed at aligning the machine to its user. For instance, understanding how the nervous system reacts and adapts to direct stimuli plays a critical role in improving control strategies [71, 72]. Research into neural mechanisms guiding adaptation and assistance, such as the brain's plasticity and the role of peripheral nerves, offers valuable insights for enhancing the responsiveness of robotic systems [22]. Such aspects are generally related to the field of neuroergonomics—investigating human factors through the methods of neuroscience—and they consider neurotechnologies or influence (according to established principles like neuroplasticity without any real-time data into the loop of control) the central and peripheral nervous systems of the user. Neurotechnologies can be instrumental in enhancing various therapeutic approaches, improving quality of life, and restoring autonomy, for instance in the case of older adults according to their frailties through continuous assessment (based on wearable devices) and personalized interventions (based on robotic devices, for instance), minimising the progression of such disorders.

The integration of neurotechnologies into assistive and rehabilitative robotic tools also offers a profound impact on improving motor functions. These robots utilize principles of neuroplasticity to guide recovery, offering interactive feedback and physical guidance that aid motor relearning. By combining robotics with virtual and augmented reality environments (as will be discussed below), training becomes more engaging, motivating, and adaptive to individual needs thanks to the adaptation to neurophysiological processes (according to actually collected data or to neurobiological principles). Several examples of neurotechnologies for motor recovery can also be offered by BCIs for controlling assistive devices.

In this sub-section, we consider the example offered by one of the most promising neurotechnologies, working according to control principles that are strictly connected to both robotics and neuroscience. This is the case of the Functional Electrical Stimulation (FES). Solutions like FES are increasingly important [73, 74], as they help restore motor function by stimulating motor neurons and by extension muscles, thus improving the user's motion capability by adopting a robot-like assistance that directly affects the body of the individual in a control loop [75]. From a rehabilitation science perspective, engaging the user's own muscles—where possible—has definite advantages over providing assistive forces through external means. However, a better assessment of the spatio-temporal requirements for the assistance of useful movements or the rendering of complex sensory feedback is still required to obtain precise design guidelines for exonerves, i.e., FES-based neural interfaces. The implementation of force controls, especially, has to contend with highly time-varying behavior due to muscular fatigue. Furthermore, the trade-off between comfort and achievable force output is also of significant importance. For example, compensating for gravity in proximal joints, such as the shoulder, can require painful currents, so much so that in many cases external support might still be required. Similarly, injecting currents through a smaller transcutaneous electrode leads to higher discomfort, limiting the achievable electrode density and therefore the spatial resolution.

Since the peripheral nervous system, after initial anastomoses, becomes increasingly divergent in the distal direction, it is generally simpler to selectively trigger action potentials in any given targeted muscle by injecting currents at a distal site. Furthermore, selectivity is greater for implanted electrodes, compared to transcutaneous ones. Muscle contractions can therefore be achieved at lower stimulation currents. Implanted electrodes have the additional advantage of being less sensitive to changes in the kinematics since they are usually directly coupled to the nerves. In spite of this, various studies seem to demonstrate that, at least for larger muscle groups, electrical stimulation sums up linearly in the force output space [75], even for transcutaneous electrodes [76]. These considerations could make a surgical intervention preferable for certain patients. Achieving similar performance through non-invasive, transcutaneous electrodes requires further research in high-density stimulation, and potentially its application at the spinal level, thus increasing the potential for rehabilitation by engaging both efferent and afferent nerve connections and enforcing neural connectivity through trans-synaptic stimulation. Furthermore, the implementation of general-purpose models of the force output corresponding to a given stimulation capable of generalizing over various limb poses would also reduce the gap between implanted and transcutaneous electrodes. By aligning with the body's natural mechanics, these systems can offer more intuitive control, enhancing the user experience. However, leveraging the advantages offered by these solutions can often need, as highlighted above, a high level of compliance of the users. This is why the next sub-section will discuss how digital technologies can sustain the usage of robotic devices (and any system connected to them, like the neurotechnologies described in this paragraph).

4.2 Robotic-Digital Synergies for User Training

Another example of the fruitful synergy of robotic systems with other types of technological solutions comes from the digital systems enabling virtual and augmented settings, especially when they are shaped as serious games (games devised not only for leisure, as the exergames designed for stimulating the individual adherence to activities performed to improve skills) or they implement gamification strategies (adding engaging game-like features in activities that are not made for entertainment). These approaches merge robotics with digital tools to enhance the patient's engagement and, consequently, the clinical outcome of the rehabilitative process or the training in using an assistive or prosthetic device. Indeed, robots are not typically designed to promote intrinsic motivation in their use, leading to mental fatigue or boredom due to repetitive tasks in their control. Blending physical and digital items in the same setting seems quite advantageous, as observed in domains like Multiple Sclerosis [77], for increasing the patient's engagement by adding virtual and augmented scenarios while a rehabilitation robot is used, helping the patient to sustain the effort over time. Interestingly, robotic devices can also enhance these digital solutions by improving their haptic feedback with a wider range of stimulations, obtaining a true synergy during the use of both solutions (even in home-based settings, when robots will be less expensive).

In prosthetic training, similar types of virtual and augmented systems [78] for exergaming (and exergamification) can be adopted for enhancing prosthetic embodiment. Traditional approaches often use visuo-tactile techniques to obtain such a result during device use, but it is also possible to envision the stimulation of Virtual Hand Illusions (VHIs) to obtain a "preparatory" embodiment of the digital replica of a real prosthetic limb. Users can first interact with a virtual prosthetic limb before they physically receive one, helping them familiarise with the functionality and control of the device. This pre-training can significantly improve the ease with which they adopt their new prosthesis. This is also explored through self-regulation techniques in prosthetic embodiment training, such as slow-paced breathing exercises, which are showing promising results [68]. These studies are still preliminary explorations that can be considered as new inputs in the debate about the potential (not generally agreed) of game-like approaches to improve the prosthetic embodiment and control by transferring the results of the training into the actual experience during the daily use of the device. Finally, we must also consider the option of intertwining gamification approaches to the calibration itself of neurotechnologies for controlling assistive devices, as proposed by [79] about BCIs based on Motor Imagery detection through electroencephalographic (EEG) signals. Considering how the procedures for training a system to recognize EEG signals can be lengthy and monotonous, the authors proposed to overcome this issue through player-centered solutions based on BCI-controlled games. Through gamified feedback, avatars, and social interactions (with adaptive mechanics about game difficulty), it is possible to enhance the chance to align human and machine in effective and gratifying ways.

Such approaches highlight the opportunities offered by digital technologies in synergy with robotic devices for motor recovery and assistance. However, the range of digital solutions is not exclusively based on virtual and augmented environments, as the next sub-section will discuss.

4.3 Internet of Robotic Things and Telerobotics

In general, we should never fail to pay attention to the specific needs of frail people, as highlighted by paradigms of Digital Health (based on digital solutions for health) and Connected Health (emphasizing the interconnection of the mentioned solutions). This includes the medical applications of the Internet of Things (IoT, the network of physical objects collecting and exchanging data across the Internet and Internet-like infrastructures) [80] and the Internet of Robotic Things (IoRT, the IoT-based interconnection of distributed robotic devices for obtaining their coordinated activities). IoT and IoRT can offer advanced motor assistance and recovery solutions for vulnerable populations like the elderly [81]. Through the connection to solutions for home automation, remote assistance, and telerehabilitation, robotic technologies provide crucial, personalized support in the recovery and support of physical skills and daily living, as envisioned during the COVID-19 pandemic, when in-person care was restricted and examples of decentralized healthcare solutions like smart homes quickly emerged. If telerobotics offers substantial benefits in motor recovery for elderly patients who require rehabilitation after injury or surgery, the addition of IoRT capabilities can guide patients through personalized exercises based on the monitoring of their movements via embedded sensors while respecting privacy-by-design paradigms. These devices enable patients to perform physical therapy from the safety of their homes while clinicians monitor their progress remotely. For example, robots designed for diagnostics can track muscle activity and range of motion, providing therapists with real-time data to adjust the rehabilitation program [82]. This reduces the need for frequent clinic visits and enables continuous care even after the patient leaves the hospital.

Another interesting application of these technologies can be identified in prosthetics [83]. In this case (which can be extended to almost any kind of interaction technology located outside a research institute or a clinical center), IoRT can be used as a bridge to fill the gap between experimental innovations and real-world user needs. Indeed, laboratory-based testing typically lacks critical contextual and user-specific data (reflecting a general lack of ecological validity) that are necessary to refine and improve the design of assistive and prosthetic devices. IoRT (as part of technological ecosystems for Digital Health and Connected Health) can enable remote data collection, real-time monitoring, and technological reprogramming through cloud infrastructures, alongside information for the iterative re-design of the devices according to real-world data that would be difficult to obtain in a laboratory.

On the other hand, technological advances can also be used for extending the patient experience outside their home: this is the case of telepresence, when the robot moves or is moved in a remote location to make the user feel as if they were there [84]. For instance, we can consider the potential of such solutions for people with Amyotrophic Lateral Sclerosis (ALS), the motor neuron disease leading to progressive severe impairments across the body. Telepresence systems can be supported by assistive control solutions like user interfaces based on eye-tracking cameras (if the person is still able to control their gaze), enabling the exploration of remote environments, making the user meet people located where a robot is moving. This discloses opportunities of social connection for several categories with far less severe impairments too, experiencing anyway isolation and lack of autonomy.

This also highlights the importance of inclusion and analysis of user needs in human-centered design, as discussed in the next subsection.

4.4 User Needs and Co-Design

Co-design is a fundamental process to obtain truly human-centered robotic devices, including applications for motor rehabilitation and assistance. Co-design is essential to meet the real needs and preferences of users (patients and clinicians), technicians, and all stakeholders. These individuals are involved in iterative and participatory design procedures at every stage of product development. This leads to a final design that is effective and aligned with the user needs. This is a compulsory step towards technology transfer and market access as well. A co-design process typically follows several phases, as discussed in [85] with a "double diamond" design process where a first phase is dedicated to design the 'right thing' (choosing what solution is the best one to solve a certain problem) and a second phase must lead to design the 'thing right', i.e., in the right way for the user. Each phase involves several stakeholder to collect multiple perspectives on the problems and on the value of the proposed solutions. A practical example has been used in [86] while the general framework is depicted in Fig. 2. In co-design studies, an exploratory phase aims to gather detailed information on user needs through interviews, focus groups, and surveys.

For instance, clinicians can provide their expertise on rehabilitation goals and methods, while patients offer insights into their daily struggles, preferences, and limitations. By including these stakeholders, developers can identify critical user requirements and, possibly, explore the existing solutions according to the experience of clinicians and patients. During the development of rehabilitation exoskeletons, e.g., [86], crucial design criteria—such as allowing a wide range of movements while maintaining a contained size and weight—may emerge as key priorities from the perspective of the clinicians. Meanwhile, for the patients, other (potentially overlooked) priorities might arise, such as preferences about the device aesthetics (a more typical concern about assistive and prosthetic systems).

A second step can be labeled as a generative phase—oriented to design the 'right thing'—for creating mockups (later even basic prototypes) based on the collected user needs. Subsequently, such mockups (models of the product) are tested iteratively with feedback from users and selected stakeholders in order to identify the minimum viable product (MVP). For example, developers may create several iterations of a mockup representing a rehabilitation exoskeleton that balances wearability and function, allowing users to maintain natural movement while offering therapeutic benefits. Clinicians assess these for factors such as safety and ease of integration into clinical practices, while patients provide feedback on physical comfort.

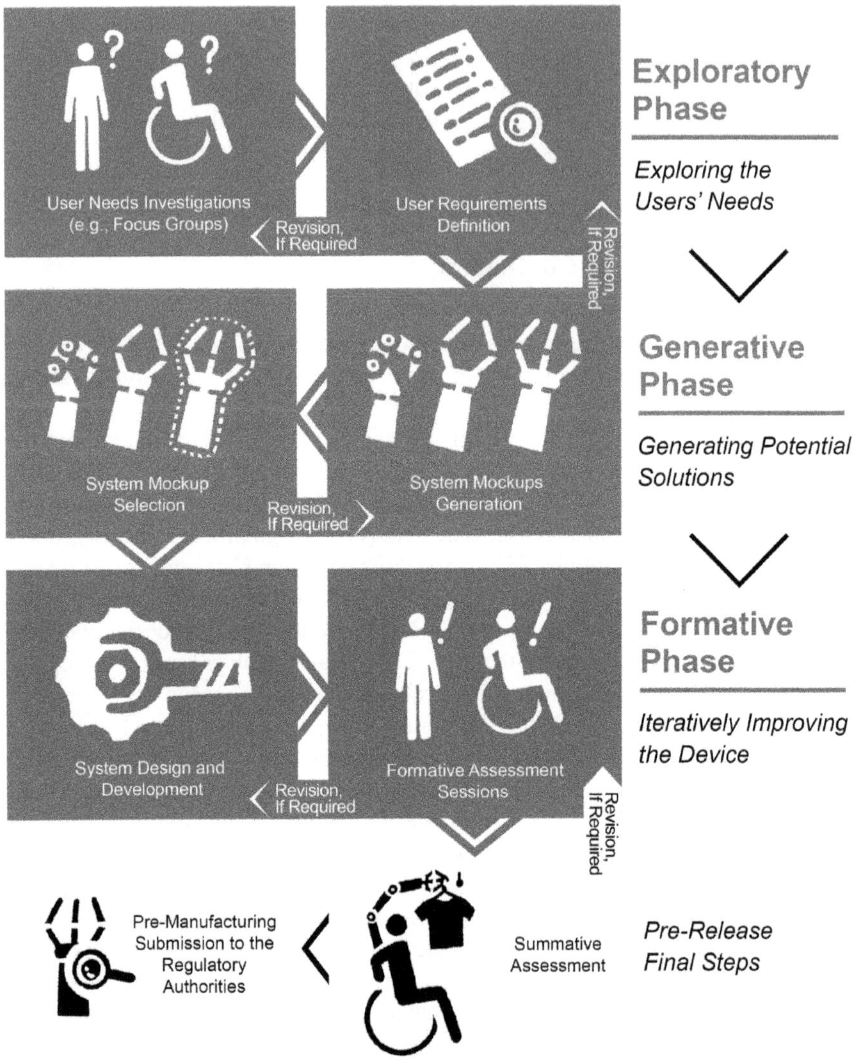

Fig. 2 Example of a co-design process based on three phases (from top to bottom, before the final steps towards the product launch) inspired by [86]

When the MVP is identified among the mockups or basic prototypes, the formative phase—devised to design the 'thing right'—can start. Accordingly, early prototypes are rigorously evaluated in real-world clinical environments. Patients use the devices under medical supervision, and their experiences are documented through usability tests and interviews. This phase is crucial for identifying safety concerns and checking how the device performs effectively under clinical conditions. This iterative testing allows for refinements in the hardware and software components in order to meet the standards for medical devices, such as IEC EN 62366-1:2015, which focuses on usability engineering. Developers also conduct risk assessments, following standards such as CEI EN 14971:2019, which proposes a formative assessment to be sure that the device minimizes potential hazards (alongside the recent Medical Device Regulation, MDR, in the European Union) [87]. This phase plays a fundamental role to make a device that is not only functionally sound but also safe, comfortable, and user-friendly (in this context especially in terms of accessibility, according to the impairments of the user). After the formative phase, a summative evaluation must be performed to ultimately verify the device in terms of usability and safety. If the summative evaluation is positive, the device can move to the final steps before the release.

Figure 2 depicts all these steps alongside the possibility (in certain cases) to revise the current iteration by repeating the previous phase.

To fully match the individual expectations, achieving a high level of human compatibility requires the incorporation of principles that highlight the peculiarities of each user and their context. For instance, the user age should always be considered during the co-design and the evaluation of assistive, prosthetic, and rehabilitative robots to improve the user experience and safety. This should be considered alongside the performance of a device that should adjust to the changes in the individual conditions and needs [88–90]. We should always and continuously consider the evolving needs of the user, even if this can increase the costs in research and development. This is certainly true not only for the older population, already mentioned in the section on the IoRT (and discussed by a chapter in the last section of this book), but also for the younger individuals. For example, about children and adolescents, the design of a robotic device should consider the potential for physical and cognitive growth, which can lead to an initially immature motor control and possible changes in wearability requirements. Engaging features for user training can be quite advantageous—as discussed is the robotic-digital synergies with serious games and gamification—to help the individual growth of skills to properly interact with a device for motor recovery and assistance. Wearability, comfort, and safety can be maintained by implementing adjustable features for avoiding to frequently change devices like exoskeletons, although this can be necessary in other cases, especially in the context of prosthetics. Special attention should be paid to the aesthetics of the devices in order to reduce the perception of social stigma among peers. Modularity and personalization can also help ensuring a certain degree of flexibility in the adjustments due to the development and to special applications like sports, a topic in the next section.

4.5 Special Applications in Prosthetics

The co-creative processes are essential/indispensable for achieving the specialization of the device's design itself, creating prostheses crafted to tend to specific needs beyond standard daily use. Historically, bespoke prosthetics have been created to cater to/address various purposes, such as upper limb prosthetics that enable a specific task involving the user's occupation. Today, given the many features of bionic limbs, the need to use a solution beyond the scope of what these advanced prostheses can deliver usually pertains to high-performance para-athletes [91]. Before considering robotic devices, it is advantageous to examine what kind of solutions are devised in this field in order to design bionic limbs accordingly. Among the prominent examples of specially made prostheses are running blades, made out of composite materials (often including carbon fiber), which are widely used in competitions such as the Paralympic Games. These running blades are specifically designed toward the practice of athletic activities and prioritize rapid energy storage and release. However, their dimensions and overall usability make them unfit for daily use. Beyond running blades, adapted prostheses are utilized across a variety of sports, including cycling, swimming, mountain climbing, fencing, skiing, and snowboarding. These sport-specific prostheses often seek to emulate the skill-related functionality of a biological limb, and in certain cases, they may even enhance performance beyond the capabilities of able-bodied athletes.

The confluence of knowledge and know-how in bespoke prosthetics and sports prosthetics, with the current improvements in robotic technology, can likely allow for advances in the development of novel prosthetic systems, increasing the range of opportunities for their users. This has been demonstrated by Zhou et al. [92], who presented the tests of a myoelectric artificial hand in activities like taijiquan, sword-play, and instrument fitness. Another example [93] describes a transfemoral prosthetic device for alpine skiing. This device was developed and tested on a ski slope according to the kinematics, dynamics, and strength analysis of skiers to obtain a model that—according to the biomechanics—should enable natural movements through a multi-axis artificial knee .

Although sport-specific prostheses are rarely employed outside elite environments, certain features from these devices can be integrated into everyday prostheses to support rehabilitation following an amputation, and current investigations focus on the opportunity to create robotic limbs accordingly. For instance, lower limb transtibial prostheses designed for cycling are typically simple, with no moving parts, allowing for lower production costs. Such devices could be especially beneficial in the early stages of rehabilitation, where patients are re-learning motor skills and their residual limb undergoes adjustments in shape and size, necessitating multiple prosthetic iterations. Cycling, particularly on a stationary bicycle, is less technically demanding than walking, as it requires no balance, making it a promising activity to promote confidence and physical readiness for walking. Research into the cycling prosthetic needs of patients at recreational and possibly rehabilitation levels is still ongoing [94]. While transtibial cycling prostheses are simple and can be used during rehabilitation, in the case of transfemoral amputations, powered and robotic prosthetic limbs can further benefit cycling practice. An example is the prosthesis developed by Lawson et al. [95] which is based on the need for coordinating the power delivery at the knee during cycling. This solution employed an algorithm to gauge the bicycle crankshaft angle and, thus, coordinating knee torque with user motion. It was implemented on a prosthesis prototype and tested with one trans-femoral amputee: the powered assistance reduced the asymmetry in motion and the user's metabolic cost.

Overall, the design of special prostheses can also lead to a better understanding of individual and general usability needs. These explorations can lead, anyway, to novel risks, even at an ethical level, as the next sub-section will discuss.

4.6 Ethical Issues

While co-design methodologies contribute significantly to the practical and functional aspects of wearable robotic devices, there are broader ethical concerns that must be addressed. In the examples in this subsection, approaches like the value-sensitive design (VSD)—based on the synergy of conceptual, empirical, and technical analyses—[96] can ensure that the devices are not only technically efficient but also promote dignity, autonomy, and inclusion. In general, ergonomics should prioritize the analysis of critical situations from the ethical point of view before proposing any potential design option. For instance, the idea of providing painful feedback [97] through a prosthetic device raises complex ethical and practical questions. Indeed, pain in a biological limb may serve as an important protective mechanism, alerting the person of potential injury or dangerous conditions. In this sense, depriving a prosthesis user of such feedback could put them at risk of harm to the prosthesis and to themselves. From this perspective, providing some form of discomfort or pain-like feedback might be beneficial in replicating the biological experience and enabling the user to react appropriately to potential hazards. However, intentionally inducing pain in a prosthetic user also presents ethical concerns. Pain is inherently unpleasant, and creating unnecessary discomfort in an individual could be seen as harmful or even cruel. Furthermore, the sensation of pain is highly subjective and complex: it would be difficult to replicate it accurately through artificial means. Potential options could be achieved by exploring how to provide informative feedback through alternative sensations. Accordingly, vibrations, pressure, or temperature warnings could serve as reliable and non-painful substitutes. These types of feedback could alert the user to dangerous conditions without causing physical distress, allowing them to respond to threats effectively while avoiding unnecessary discomfort.

Another critical issue, more related to relational and social aspects, is the technoableism, often related to the topic in the previous subsection. Technoableism [98] refers to the implicit or explicit framing of assistive technologies as tools designed to "fix" disabilities and push people towards a "normal" or non-disabled ideal condition. This view can reinforce societal biases about disability and can lead to the design of devices that prioritize able-bodied norms over individual user needs. In particular, the design of specialized prostheses for the Paralympics follows the need of obtaining the best possible performance, which can lead to the perception of para-athletes as 'super-powered cyborgs'. In the context of robotics for motor recovery and assistance, technoableism may occur when devices, especially wearable robotic systems, are designed with the goal of erasing or minimizing disability, rather than supporting diverse abilities and enabling users to live according to their own values and preferences. For example, prosthetic limbs that mimic the appearance and function of biological limbs may be designed to enhance the embodiment but they can also sustain the idea that the user's body is incomplete without the prosthetic device. A more inclusive and human-centered design approach should detect the potential impact of technoableism on bionic design, highlighting the potential for assistive technologies to empower individuals without necessarily conforming to able-bodied conditions. Instead of framing technology as a way to overcome disability, ergonomics must focus on enhancing the abilities of individuals in ways that are meaningful to them for improving the quality of their life. Such aspects must be considered while both investigating the user needs and proposing potential solutions like robotic devices enhancing individual skills.

5 Conclusion

This chapter introduced how robotics offers opportunities to enhance the lives of people with motor impairments. From helping them in rehabilitation to enabling independent performance in ADLs through assistive solutions, robots can significantly improve the autonomy of individuals by extending their capabilities in meaningful ways, even when they can constitute challenges in terms of user experience, usability, mental workload, and technology acceptance. Furthermore, advances in bionic prostheses highlight the potential for seamless integration of human and machine, enhancing independence through smart machines imitating limbs. These future prostheses will—however—require high levels of embodiment to overcome the risk of device abandonment. The presented developments require a greater synergy between engineers, clinicians, and ergonomists. Therefore, a multidisciplinary and interdisciplinary approach represents the key for improved function, comfort, and quality of life. Synergies are certainly necessary when multiple technological solutions converge, as in the cases of neurotechnologies, digital systems, or IoT with robotic devices for rehabilitation and assistance. All types of technological integration must be considered in the co-design of medical devices. By engaging individuals with motor impairments, clinicians, and related stakeholders in the co-design process, the resulting solutions are more likely to meet the actual needs of the users, easing the adoption of the technology in real life—even for supporting special activities like extreme sports. At the same time, ethical considerations must be carefully addressed. Human factors specialists must work to keep the focus, especially in this context, on supporting human dignity, agency, and diverse experiences rather than reinforcing narrow notions of "normal" ability. A truly human-centered approach that emphasizes user experience and ethical responsibility can definitely help expressing the full potential of robotics in this field, easing the deployment and the use of robots for motor recovery and assistance in real contexts of daily life.

References

1. Mohebbi, A.: Human-robot interaction in rehabilitation and assistance: a review. Curr. Robot. Rep. **1**(3), 131–144 (2020). https://doi.org/10.1007/s43154-020-00015-4
2. Morone, G., Riener, R., Mazzoleni, S.: Integrating robot-assisted therapy into neurorehabilitation clinical practice: where are we now? Where are we heading? NeuroRehabilitation **51**(4), 537–539 (2022). https://doi.org/10.3233/NRE-228025
3. Kwakkel, G., Kollen, B.J., Krebs, H.I.: Effects of robot-assisted therapy on upper limb recovery after stroke: a systematic review. Neurorehabil. Neural Repair **22**(2), 111–121 (2008). https://doi.org/10.1177/1545968307305457
4. Brose, S.W., Weber, D.J., Salatin, B.A., Grindle, G.G., Wang, H., Vazquez, J.J., Cooper, R.A.: The role of assistive robotics in the lives of persons with disability. Am. J. Phys. Med. Rehabil. **89**(6), 509–521 (2010). https://doi.org/10.1097/PHM.0b013e3181cf569b
5. Sacchini, A., Tessari, F., Vassallo, C., Maludrottu, S., De Momi, E., Laffranchi, M., De Michieli, L.: Sensorless impedance control for the TWIN lower limb exoskeleton: a preliminary study. In: 2022 9th IEEE RAS/EMBS International Conference for Biomedical Robotics and Biomechatronics (BioRob), pp. 1–7 (2022). https://doi.org/10.1109/BioRob52689.2022.9925376. ISSN: 2155-1782
6. Sankai, Y.: HAL: hybrid assistive limb based on cybernics. In: Siciliano, B., Khatib, O., Groen, F., Kaneko, M., Nakamura, Y. (eds.) Robotics Research, vol. 66, pp. 25–34. Springer (2010). https://doi.org/10.1007/978-3-642-14743-2_3. Series Title: Springer Tracts in Advanced Robotics
7. Heo, G.S., Lee, S.-R., Kwak, M.K., Park, C.W., Kim, G., Lee, C.-Y.: Motion control of bicycle-riding exoskeleton robot with interactive force analysis. Int. J. Precis. Eng. Manuf. **16**, 1631–1637 (2015). https://doi.org/10.1007/s12541-015-0214-y
8. Tricomi, E., Lotti, N., Missiroli, F., Zhang, X., Xiloyannis, M., Müller, T., Crea, S., Papp, E., Krzywinski, J., Vitiello, N., Masia, L.: Underactuated soft hip exosuit based on adaptive oscillators to assist human locomotion. IEEE Robot. Autom. Lett. **7**(2), 936–943 (2022) https://doi.org/10.1109/LRA.2021.3136240. Conference Name: IEEE Robotics and Automation Letters
9. Simpson, C., Huerta, B., Sketch, S., Lansberg, M., Hawkes, E., Okamura, A.: Upper extremity exomuscle for shoulder abduction support. IEEE Trans. Med. Robot. Bionics **2**(3), 474–484 (2020) https://doi.org/10.1109/TMRB.2020.3012471. Conference Name: IEEE Transactions on Medical Robotics and Bionics
10. Hogan, N., Krebs, H.I., Charnnarong, J., Srikrishna, P., Sharon, A.: MIT-MANUS: a workstation for manual therapy and training. I. In: Proceedings IEEE International Workshop on Robot and Human Communication, pp. 161–165. (1992). https://doi.org/10.1109/ROMAN.1992.253895
11. Rodgers, H., Bosomworth, H., Krebs, H.I., Wijck, F., Howel, D., Wilson, N., Aird, L., Alvarado, N., Andole, S., Cohen, D.L., Dawson, J., Fernandez-Garcia, C., Finch, T., Ford, G.A., Francis, R., Hogg, S., Hughes, N., Price, C.I., Ternent, L., Turner, D.L., Vale, L., Wilkes, S., Shaw, L.: Robot assisted training for the upper limb after stroke (RATULS): a multicentre randomised controlled trial. Lancet (London, Engl.) **394**(10192), 51–62 (2019). https://doi.org/10.1016/S0140-6736(19)31055-4
12. Dragomir, A., Pană, C.F., Cojocaru, D., Manta, L.F.: Human-machine interface for controlling a light robotic arm by persons with special needs. In: 2021 22nd International Carpathian Control Conference (ICCC), pp. 1–6. IEEE (2021). https://doi.org/10.1109/ICCC51557.2021.9454664
13. Taşar, B., Tatar, A.B., Tanyıldızı, A.K., Yakut, O.: Fimec tremor stabilization spoon: design and active stabilization control of two DoF robotic eating devices for hand tremor patients. Med. Biol. Eng. Comput. **10**, 2757–2768 (2023). https://doi.org/10.1007/s11517-023-02886-z
14. Sierra M., S.D., Jiménez, M.F., Frizera-Neto, A., Múnera, M., Cifuentes, C.A.: Control strategies for human-robot-environment interaction in assisted gait with smart walkers. Interfacing

Humans and Robots for Gait Assistance and Rehabilitation, pp. 259–286 (2022) https://doi.org/10.1007/978-3-030-79630-3_10
15. Juneja, A., Bhandari, L., Mohammadbagherpoor, H., Singh, A., Grant, E.: A comparative study of slam algorithms for indoor navigation of autonomous wheelchairs. In: 2019 IEEE International Conference on Cyborg and Bionic Systems (CBS), pp. 261–266. IEEE (2019). https://doi.org/10.1109/CBS46900.2019.9114512
16. Ye, C., Hong, S., Qian, X., Wu, W.: Co-robotic cane: A new robotic navigation aid for the visually impaired. IEEE Syst., Man, Cybern. Mag. **2**(2), 33–42 (2016) https://doi.org/10.1109/MSMC.2015.2501167
17. Chiaradia, D., Xiloyannis, M., Solazzi, M., Masia, L., Frisoli, A.: Chapter 4—rigid versus soft exoskeletons: interaction strategies for upper limb assistive technology. In: Rosen, J., Ferguson, P.W. (eds.) Wearable Robotics, pp. 67–90. Academic Press, (2020). https://doi.org/10.1016/B978-0-12-814659-0.00004-7
18. Pippo, I., Massone, V., Guazzotti, M., Gnetti, B., Albanese, G.A., Casadio, M., Berselli, G., Zenzeri, J.: Existing devices in wrist rehabilitation robotics: Mechanical design, control systems, and user requirements for commercialization. In: 2024 20th IEEE/ASME International Conference on Mechatronic and Embedded Systems and Applications (MESA), pp. 1–8. IEEE (2024). https://doi.org/10.1109/MESA61532.2024.10704903
19. Luzio, F.S., Lauretti, C., Cordella, F., Draicchio, F., Zollo, L.: Visual vs vibrotactile feedback for posture assessment during upper-limb robot-aided rehabilitation. Appl. Ergon. **82**, 102950 (2020). https://doi.org/10.1016/j.apergo.2019.102950
20. Leone, S., Giunta, L., Rino, V., Mellace, S., Sozzi, A., Lago, F., Curcio, E.M., Pisla, D., Carbone, G.: Design of a wheelchair-mounted robotic arm for feeding assistance of upper-limb impaired patients. Robotics **13**(3), 38 (2024). https://doi.org/10.3390/robotics13030038
21. Seguí-Verdú, I., Dunai, L., Antonio-Daviu, J.A.: Exploring ergonomic advancements and smart technologies in future walkers. In: 2024 IEEE 33rd International Symposium on Industrial Electronics (ISIE), pp. 1–8. IEEE (2024). https://doi.org/10.1109/ISIE54533.2024.10595750
22. Barresi, G., Zenzeri, J., Tessadori, J., Laffranchi, M., Semprini, M., De Michieli, L.: Neurogerontechnologies: applications and opportunities. In: Internet of Things for Human-Centered Design: Application to Elderly Healthcare, pp. 123–153. Springer (2022). https://doi.org/10.1007/978-981-16-8488-3_7
23. Gombolay, M., Yang, X.J., Hayes, B., Seo, N., Liu, Z., Wadhwania, S., Yu, T., Shah, N., Golen, T., Shah, J.: Robotic assistance in the coordination of patient care. Int. J. Robot. Res. **37**(10), 1300–1316 (2018). https://doi.org/10.1177/0278364918778344. Publisher: SAGE Publications Ltd STM
24. Devittori, G., Dinacci, D., Romiti, D., Califfi, A., Petrillo, C., Rossi, P., Ranzani, R., Gassert, R., Lambercy, O.: Unsupervised robot-assisted rehabilitation after stroke: feasibility, effect on therapy dose, and user experience. J. Neuroeng. Rehabil. **21**(1), 52 (2024). https://doi.org/10.1186/s12984-024-01347-4
25. Fruchter, D., Feingold Polak, R., Berman, S., Levy-Tzedek, S.: Hierarchy in algorithm-based feedback to patients working with a robotic rehabilitation system: toward user-experience optimization. IEEE Trans. Human-Mach. Syst. **52**(5), 907–917 (2022) https://doi.org/10.1109/THMS.2022.3170831. Conference Name: IEEE Transactions on Human-Machine Systems
26. Laparidou, D., Curtis, F., Akanuwe, J., Goher, K., Niroshan Siriwardena, A., Kucukyilmaz, A.: Patient, carer, and staff perceptions of robotics in motor rehabilitation: a systematic review and qualitative meta-synthesis. J. Neuroeng. Rehabil. **18**(1), 181 (2021). https://doi.org/10.1186/s12984-021-00976-3
27. Guillén-Climent, S., Garzo, A., Muñoz-Alcaraz, M.N., Casado-Adam, P., Arcas-Ruiz-Ruano, J., Mejías-Ruiz, M., Mayordomo-Riera, F.J.: A usability study in patients with stroke using MERLIN, a robotic system based on serious games for upper limb rehabilitation in the home setting. J. Neuroeng. Rehabil. **18**(1), 41 (2021). https://doi.org/10.1186/s1x2984-021-00837-z
28. Zanatta, F., Giardini, A., Pierobon, A., D'Addario, M., Steca, P.: A systematic review on the usability of robotic and virtual reality devices in neuromotor rehabilitation: patients' and

healthcare professionals' perspective. BMC Health Serv. Res. **22**(1), 523 (2022). https://doi.org/10.1186/s12913-022-07821-w
29. Park, J.H., Park, G., Kim, H.Y., Lee, J.-Y., Ham, Y., Hwang, D., Kwon, S., Shin, J.-H.: A comparison of the effects and usability of two exoskeletal robots with and without robotic actuation for upper extremity rehabilitation among patients with stroke: a single-blinded randomised controlled pilot study. J. Neuroeng. Rehabil. **17**(1), 137 (2020). https://doi.org/10.1186/s12984-020-00763-6
30. Meyer, J.T., Gassert, R., Lambercy, O.: An analysis of usability evaluation practices and contexts of use in wearable robotics. J. Neuroeng. Rehabil. **18**(1), 170 (2021). https://doi.org/10.1186/s12984-021-00963-8
31. Marchand, C., De Graaf, J.B., Jarrassé, N.: Measuring mental workload in assistive wearable devices: a review. J. Neuroeng. Rehabil. **18**(1), 160 (2021). https://doi.org/10.1186/s12984-021-00953-w
32. Upasani, S., Srinivasan, D., Zhu, Q., Du, J., Leonessa, A.: Eye-tracking in physical human-robot interaction: mental workload and performance prediction. Hum. Factors **66**(8), 2104–2119 (2024). https://doi.org/10.1177/00187208231204704. Publisher: SAGE Publications Inc
33. Novak, D., Ziherl, J., Olensek, A., Milavec, M., Podobnik, J., Mihelj, M., Munih, M.: Psychophysiological responses to robotic rehabilitation tasks in stroke. IEEE Trans. Neural Syst. Rehabil. Eng.: Publ. IEEE Eng. Med. Biol. Soc. **18**(4), 351–361 (2010). https://doi.org/10.1109/TNSRE.2010.2047656
34. La Bara, L.M.A., Meloni, L., Giusino, D., Pietrantoni, L.: Assessment methods of usability and cognitive workload of rehabilitative exoskeletons: a systematic review. Appl. Sci. **11**(15), 7146 (2021) https://doi.org/10.3390/app11157146. Number: 15 Publisher: Multidisciplinary Digital Publishing Institute
35. Shore, L., Eyto, A., O'Sullivan, L.: Technology acceptance and perceptions of robotic assistive devices by older adults—implications for exoskeleton design. Disabil. Rehabil. Assist. Technol. **17**(7), 782–790 (2022). https://doi.org/10.1080/17483107.2020.1817988
36. Luciani, B., Braghin, F., Pedrocchi, A.L.G., Gandolla, M.: Technology acceptance model for exoskeletons for rehabilitation of the upper limbs from therapists' perspectives. Sens. (Basel, Switz.) **23**(3), 1721 (2023). https://doi.org/10.3390/s23031721
37. Klaic, M., Fong, J., Crocher, V., Davies, K., Brock, K., Sutton, E., Oetomo, D., Tan, Y., Galea, M.P.: Application of the extended technology acceptance model to explore clinician likelihood to use robotics in rehabilitation. Disabil. Rehabil. Assist. Technol. **19**(1), 52–59 (2024). https://doi.org/10.1080/17483107.2022.2060356
38. Shore, L., Power, V., Eyto, A., O'Sullivan, L.: Technology acceptance and user-centred design of assistive exoskeletons for older adults: a commentary. Robotics **7** (2018) https://doi.org/10.3390/robotics7010003
39. Raichle, K.A., Hanley, M.A., Molton, I., Kadel, N.J., Campbell, K., Phelps, E., Ehde, D., Smith, D.G.: Prosthesis use in persons with lower- and upper-limb amputation. J. Rehabil. Res. Dev. **45**(7), 961–972 (2008). https://doi.org/10.1682/jrrd.2007.09.0151
40. Belter, J.T., Segil, J.L., Dollar, A.M., Weir, R.F.: Mechanical design and performance specifications of anthropomorphic prosthetic hands: a review. J. Rehabil. Res. Dev. **50**(5), 599–618 (2013). https://doi.org/10.1682/jrrd.2011.10.0188
41. Prost, V., Johnson, W.B., Kent, J.A., Major, M.J., Winter, A.G.: Biomechanical evaluation over level ground walking of user-specific prosthetic feet designed using the lower leg trajectory error framework. Sci. Rep. **12**(1), 5306 (2022). https://doi.org/10.1038/s41598-022-09114-y. Publisher: Nature Publishing Group
42. Houdek, M.T., Kralovec, M.E., Andrews, K.L.: Hemipelvectomy: high-level amputation surgery and prosthetic rehabilitation. American Journal of Physical Medicine & Rehabilitation **93**(7), 600–608 (2014). https://doi.org/10.1097/PHM.0000000000000068. Publisher: LWW
43. Windrich, M., Grimmer, M., Christ, O., Rinderknecht, S., Beckerle, P.: Active lower limb prosthetics: a systematic review of design issues and solutions. Biomed. Eng. Online **15**, 5–19 (2016). https://doi.org/10.1186/s12938-016-0284-9. Publisher: Springer

44. Au, S.K., Weber, J., Herr, H.: Powered ankle-foot prosthesis improves walking metabolic economy. Trans. Rob. **25**(1), 51–66 (2009). https://doi.org/10.1109/TRO.2008.2008747
45. Gailey, R., Allen, K., Castles, J., Kucharik, J., Roeder, M.: Review of secondary physical conditions associated with lower-limb amputation and long-term prosthesis use. J. Rehabil. Res. Dev. **45**(1) (2008) https://doi.org/10.1682/jrrd.2006.11.0147
46. Paternò, L., Ibrahimi, M., Gruppioni, E., Menciassi, A., Ricotti, L.: Sockets for limb prostheses: a review of existing technologies and open challenges. IEEE Trans. Biomed. Eng. **65**(9), 1996–2010 (2018). https://doi.org/10.1109/TBME.2017.2775100
47. Tyagi, B., Raj, A., Chandrakar, A.S., Sharma, G.S., Raj, T., Jain, A., Bhardwaj, L., Sahai, A., Sharma, R.S.: Fabrication of transfemoral prosthesis utilizing additive manufacturing and reverse engineering: a scoping review. Int. J. Interact. Des. Manuf. (IJIDeM) **18**(6), 3613–3631 (2024). https://doi.org/10.1007/s12008-024-01974-0
48. Frasier, K., Clark, B., Vingolu-Baur, J., Keane, J., Goldstein, L.B.: Prosthesis contact dermatitis with long-term prosthetic lining usage. J. Clin. (2023). https://doi.org/10.1016/j.jaad.2024.07.1147
49. Annisa, R., Hidayat, R., Winarso, K., Agustina, F., Salimah, S.F.: Ergonomics leg prosthetic redesign using an anthropometric approach. In: MATEC Web of Conferences, vol. 372, p. 08005 (2022). https://doi.org/10.1051/matecconf/202237208005. EDP Sciences
50. Ali, S.M., Nagmash, R.L., Alhadi, A.H.A., Zahraw, R.K., Abbas, A.N.A.: Design and modeling of a lower artificial limb adjustable socket. AIP Conf. Proc. 2977, (2023). https://doi.org/10.1063/5.0182138 AIP Publishing
51. Fagioli, I., Mazzarini, A., Livolsi, C., Gruppioni, E., Vitiello, N., Crea, S., Trigili, E.: Advancements and challenges in the development of robotic lower limb prostheses: a systematic review. IEEE Trans. Med. Robot. Bionics (2024). https://doi.org/10.1109/TMRB.2024.3464126
52. Richardson, A., Dillon, M.P.: User experience of transtibial prosthetic liners: a systematic review. Prosthet. Orthot. Int. **41**(1), 6–18 (2017) https://doi.org/10.1177/0309364616631343. Publisher: SAGE Publications Ltd STM
53. Biddiss, E.A., Chau, T.T.: Upper limb prosthesis use and abandonment: a survey of the last 25 years. Prosthet. Orthot. Int. **31**(3), 236–257 (2007). https://doi.org/10.1080/03093640600994581. Publisher: SAGE Publications Sage UK: London, England
54. Oliveira, D.S., Ponfick, M., Braun, D.I., Osswald, M., Sierotowicz, M., Chatterjee, S., Weber, D., Eskofier, B., Castellini, C., Farina, D., Kinfe, T.M., Del Vecchio, A.: A direct spinal cord–computer interface enables the control of the paralysed hand in spinal cord injury. Brain **147**(10), 3583–3595 (2024). https://doi.org/10.1093/brain/awae088
55. Resnik, L., Klinger, S., Gill, A., Ekerholm Biester, S.: Feminine identity and functional benefits are key factors in women's decision making about upper limb prostheses: a case series. Disabil. Rehabil. Assist. Technol. **14**(2), 194–208 (2019). https://doi.org/10.1080/17483107.2018.1467973
56. Kerver, N., Karssies, E., Krabbe, P.F., Sluis, C.K., Groen, H.: Economic evaluation of upper limb prostheses in the Netherlands including the cost-effectiveness of multi-grip versus standard myoelectric hand prostheses. Disabil. Rehabil. **45**(25), 4311–4321 (2023) https://doi.org/10.1080/09638288.2022.2151653. Publisher: Taylor & Francis
57. Zollo, L., Pino, G.D., Ciancio, A.L., Ranieri, F., Cordella, F., Gentile, C., Noce, E., Romeo, R.A., Bellingegni, A.D., Vadalà, G., Miccinilli, S., Mioli, A., Diaz-Balzani, L., Bravi, M., Hoffmann, K.-P., Schneider, A., Denaro, L., Davalli, A., Gruppioni, E., Sacchetti, R., Castellano, S., Lazzaro, V.D., Sterzi, S., Denaro, V., Guglielmelli, E.: Restoring tactile sensations via neural interfaces for real-time force-and-slippage closed-loop control of bionic hands. Sci. Robot. **4**(27), 9924 (2019). https://doi.org/10.1126/scirobotics.aau9924. https://www.science.org/doi/pdf/10.1126/scirobotics.aau9924
58. Morgenroth, D.C., Segal, A.D., Zelik, K.E., Czerniecki, J.M., Klute, G.K., Adamczyk, P.G., Orendurff, M.S., Hahn, M.E., Collins, S.H., Kuo, A.D.: The effect of prosthetic foot push-off on mechanical loading associated with knee osteoarthritis in lower extremity amputees. Gait Posture **34**(4), 502–507 (2011). https://doi.org/10.1016/j.gaitpost.2011.07.001

59. Cordella, F., Ciancio, A.L., Sacchetti, R., Davalli, A., Cutti, A.G., Guglielmelli, E., Zollo, L.: Literature review on needs of upper limb prosthesis users. Front. Neurosci. 10, (2016). https://doi.org/10.3389/fnins.2016.00209
60. Fleming, A., Liu, W., Huang, H.: Neural prosthesis control restores near-normative neuromechanics in standing postural control. Sci. Robot. **8**(83), 5758 (2023). https://doi.org/10.1126/scirobotics.adf5758. Publisher: American Association for the Advancement of Science
61. Valgeirsdóttir, V.V.: Lower-limb Prosthetics in the Age of Advanced Solutions: Understanding People's Needs and Future Benefits. University of Iceland, School of Health Sciences, Faculty of Psychology, Reykjavík (2021). https://doi.org/hdl.handle.net/20.500.11815/2886. Accepted 2022-02-08T12:21:56Z
62. Yamanaka, S., Tsuji, Y., Higaki, M., Suzuki, H.: Designing the sports prosthetic leg. In: Proceedings of the 2nd Augmented Human International Conference, pp. 1–4. ACM, Tokyo Japan (2011). https://doi.org/10.1145/1959826.1959841
63. Widehammar, C., Pettersson, I., Janeslätt, G., Hermansson, L.: The influence of environment: experiences of users of myoelectric arm prosthesis-a qualitative study. Prosthet. Orthot. Int. **42**(1), 28–36 (2018). https://doi.org/10.1177/0309364617704801. Publisher: SAGE Publications Ltd STM
64. Beckerle, P., Castellini, C., Lenggenhager, B.: Robotic interfaces for cognitive psychology and embodiment research: a research roadmap. Wiley Interdisc. Rev.: Cogn. Sci. **10**(2), 1486 (2019). https://doi.org/10.1002/wcs.1486
65. Toet, A., Kuling, I.A., Krom, B.N., Van Erp, J.B.: Toward enhanced teleoperation through embodiment. Front. Rob. AI **7**, 14 (2020). https://doi.org/10.3389/frobt.2020.00014
66. Segil, J.L., Roldan, L.M., Graczyk, E.L.: Measuring embodiment: a review of methods for prosthetic devices. Front. Neurorobot. **16**, 902162 (2022). https://doi.org/10.3389/fnbot.2022.902162
67. Riemer, M., Trojan, J., Beauchamp, M., Fuchs, X.: The rubber hand universe: on the impact of methodological differences in the rubber hand illusion. Neurosci. Biobehav. Revi. **104**, 268–280 (2019). https://doi.org/10.1016/j.neubiorev.2019.07.008
68. Barresi, G., Marinelli, A., Caserta, G., De Zambotti, M., Tessadori, J., Angioletti, L., Boccardo, N., Freddolini, M., Mazzanti, D., Deshpande, N., Frigo, C.A., Balconi, M., Gruppioni, E., Laffranchi, M., and De Michieli, L.: Exploring the embodiment of a virtual hand in a spatially augmented respiratory biofeedback setting. Front. Neurorobot. **15**, 683653 (2021). https://doi.org/10.3389/fnbot.2021.683653
69. Spence, C., Pavani, F., Driver, J.: Spatial constraints on visual-tactile cross-modal distractor congruency effects. Cogn., Affect., Behav. Neurosci. **4**(2), 148–169 (2004). https://doi.org/10.3758/cabn.4.2.148. Publisher: Springer
70. Marasco, P.D., Hebert, J.S., Sensinger, J.W., Beckler, D.T., Thumser, Z.C., Shehata, A.W., Williams, H.E., Wilson, K.R.: Neurorobotic fusion of prosthetic touch, kinesthesia, and movement in bionic upper limbs promotes intrinsic brain behaviors. Sci. Robot. **6**(58), 3368 (2021). https://doi.org/10.1126/scirobotics.abf3368. Publisher: American Association for the Advancement of Science
71. Yeoh, W.L., Choi, J., Loh, P.Y., Fukuda, O., Muraki, S.: Motor characteristics of human adaptations to external assistive forces. J. Roboti. Mechatron. **35**(3), 547–555 (2023) https://doi.org/10.20965/jrm.2023.p0547
72. Tessari, F., Hermus, J., Sugimoto-Dimitrova, R., Hogan, N.: Brownian processes in human motor control support descending neural velocity commands. Sci. Rep. **14**(1), 8341 (2024). https://doi.org/10.1038/s41598-024-58380-5
73. Cardoso, L.R., Bochkezanian, V., Forner-Cordero, A., Melendez-Calderon, A., Bo, A.P.: Soft robotics and functional electrical stimulation advances for restoring hand function in people with sci: a narrative review, clinical guidelines and future directions. J. Neuroeng. Rehabil. **19**(1), 66 (2022). https://doi.org/10.1186/s12984-022-01043-1
74. Habibollahi, Z., Zhou, Y., Jenkins, M.E., Garland, S.J., Friedman, E., Naish, M.D., Trejos, A.L.: Tremor suppression using functional electrical stimulation. IEEE Trans. Neural Syst. Rehabil. Eng. (2024). https://doi.org/10.1109/TNSRE.2024.3453222

75. Liao, Y.-W., Schearer, E.M., Perreault, E.J., Tresch, M.C., Lynch, K.M.: Multi-muscle FES control of the human arm for interaction tasks-stabilizing with muscle co-contraction and postural adjustment: a simulation study. In: IEEE/RSJ International Conference on Intelligent Robots and Systems, pp. 2134–2139. (2014). https://doi.org/10.1109/IROS.2014.6942849
76. Sierotowicz, M., Castellini, C.: Omnidirectional endpoint force control through functional electrical stimulation. Biomed. Phys. Eng. Express **9**(6), 065008 (2023). https://doi.org/10.1088/2057-1976/acf04b
77. Albanese, G.A., Bucchieri, A., Podda, J., Tacchino, A., Buccelli, S., De Momi, E., Laffranchi, M., Mannella, K., Holmes, M.W., Zenzeri, J., De Michieli, L., Brichetto, G., Barresi, G.: Robotic systems for upper-limb rehabilitation in multiple sclerosis: a SWOT analysis and the synergies with virtual and augmented environments. Front. Robot. AI **11**, 1335147 (2024). https://doi.org/10.3389/frobt.2024.1335147
78. Nissler, C., Nowak, M., Connan, M., Büttner, S., Vogel, J., Kossyk, I., Márton, Z.-C., Castellini, C.: Vita-an everyday virtual reality setup for prosthetics and upper-limb rehabilitation. J. Neural Eng. **16**(2), 026039 (2019). https://doi.org/10.1088/1741-2552/aaf35f
79. Atilla, F., Postma, M., Alimardani, M.: Gamification of motor imagery brain-computer interface training protocols: a systematic review. Comput. Human Behav. Rep. 100508, (2024). https://doi.org/10.1016/j.chbr.2024.100508
80. Petrushin, A., Freddolini, M., Barresi, G., Bustreo, M., Laffranchi, M., Del Bue, A., De Michieli, L.: Iot-powered monitoring systems for geriatric healthcare: overview. In: Internet of Things for Human-Centered Design: Application to Elderly Healthcare, pp. 99–122. Springer (2022). https://doi.org/10.1007/978-981-16-8488-3_6
81. De Michieli, L., Petrushin, A., Bustreo, M., Del Bue, A., Barresi, G.: Robots in geriatric care: a future with no return? In: Gerontechnology. A Clinical Perspective, pp. 143–161. Springer, (2023). https://doi.org/10.1007/978-3-031-32246-4_10
82. Tröbinger, M., Costinescu, A., Xing, H., Elsner, J., Hu, T., Naceri, A., Figueredo, L., Jensen, E., Burschka, D., Haddadin, S.: A dual doctor-patient twin paradigm for transparent remote examination, diagnosis, and rehabilitation. In: 2021 IEEE/RSJ International Conference on Intelligent Robots and Systems (IROS), pp. 2933–2940. IEEE (2021). https://doi.org/10.1109/IROS51168.2021.9636626
83. Wu, H., Dyson, M., Nazarpour, K.: Internet of things for beyond-the-laboratory prosthetics research. Phil. Trans. R. Soc. A **380**(2228), 20210005 (2022). https://doi.org/10.1098/rsta.2021.0005
84. Zhang, G., Hansen, J.P.: Telepresence robots for people with special need s: a systematic review. Int. J. Human-Comput. Inter. **38**(17), 1651–1667 (2022). https://doi.org/10.1080/10447318.2021.2009673
85. Bardaro, G., Antonini, A., Motta, E.: Robots for elderly care in the home: a landscape analysis and co-design toolkit. Int. J. Soc. Robot. **14**(3), 657–681 (2022). https://doi.org/10.1007/s12369-021-00816-3
86. Buccelli, S., Tessari, F., Fanin, F., De Guglielmo, L., Capitta, G., Piezzo, C., Bruschi, A., Van Son, F., Scarpetta, S., Succi, A., Rossi, P., Maludrottu, S., Barresi, G., Creatini, I., Taglione, E., Laffranchi, M., De Michieli, L.: A gravity-compensated upper-limb exoskeleton for functional rehabilitation of the shoulder complex. Appl. Sci. **12**(7), 3364 (2022). https://doi.org/10.3390/app12073364. Number: 7 Publisher: Multidisciplinary Digital Publishing Institute
87. Singh, J., Patel, P.: Methods for medical device design, regulatory compliance and risk management. J. Eng. Re. Rep. **26**(7), 373–389 (2024). https://doi.org/10.9734/jerr/2024/v26i71216
88. Gonzalez, A., Garcia, L., Kilby, J., McNair, P.: Robotic devices for paediatric rehabilitation: a review of design features. Biomed. Eng. Online **20**, 1–33 (2021). https://doi.org/10.1186/s12938-021-00920-5
89. Eguren, D., Cestari, M., Luu, T.P., Kilicarslan, A., Steele, A., Contreras-Vidal, J.L.: Design of a customizable, modular pediatric exoskeleton for rehabilitation and mobility. In: 2019 IEEE International Conference on Systems, Man and Cybernetics (SMC), pp. 2411–2416. IEEE (2019). https://doi.org/10.1109/SMC.2019.8914629

90. Sims, T., Cranny, A., Metcalf, C., Chappell, P., Donovan-Hall, M.: Participatory design of pediatric upper limb prostheses: qualitative methods and prototyping. Int. J. Technol. Assess. Health Care **33**(6), 629–637 (2017). https://doi.org/10.1017/S0266462317000836
91. Bragaru, M., Dekker, R., Geertzen, J.H.: Sport prostheses and prosthetic adaptations for the upper and lower limb amputees: an overview of peer reviewed literature. Prosthet. Orthot. Int. **36**(3), 290–296 (2012). https://doi.org/10.1177/0309364612447093
92. Zhou, X., Du, Z., Li, C., Zhang, H., Zhang, X., Hu, J.: Empirical research of sports assisted EMG mechanical bionic prosthesis. In: 2021 International Conference on Health Big Data and Smart Sports (HBDSS), pp. 130–136 (2021). https://doi.org/10.1109/HBDSS54392.2021.00034. IEEE
93. Demšar, I., Supej, M., Vidrih, Z., Duhovnik, J., Demšar, I., Duhovnik, M.S.-Z.V.-J.: Development of prosthetic knee for alpine skiing. Strojniški vestnik-J. Mech. Eng. **57**(10), 768–777 (2011) https://doi.org/10.5545/sv-jme.2011.017
94. Seratiuk Flores, H., Yeoh, W.L., Loh, P.Y., Morinaga, K., Muraki, S.: Biomechanical analysis of recreational cycling with unilateral transtibial prostheses. Prosthesis **5**(3), 733–751 (2023). https://doi.org/10.3390/prosthesis5030052
95. Lawson, B.E., Ledoux, E.D., Goldfarb, M.: A robotic lower limb prosthesis for efficient bicycling. IEEE Trans. Rob. **33**(2), 432–445 (2017). https://doi.org/10.1109/TRO.2016.2636844
96. Williams, K.: Prioritizing the patient: the ethical considerations of using value sensitive design for upper limb prosthetics. PBJ 9, (2019). https://doi.org/10.3389/fresc.2023.1335966
97. Osborn, L.E., Dragomir, A., Betthauser, J.L., Hunt, C.L., Nguyen, H.H., Kaliki, R.R., Thakor, N.V.: Prosthesis with neuromorphic multilayered e-dermis perceives touch and pain. Sci. Robot. **3**(19), 3818 (2018). https://doi.org/10.1126/scirobotics.aat3818
98. Shew, A.: Ableism, technoableism, and future AI. IEEE Technol. Soc. Mag. **39**(1), 40–85 (2020). https://doi.org/10.1109/MTS.2020.2967492

Technologies and Methods

Digital Human Modeling in Human–Robot Interaction

Martin Manns and Tadele Belay Tuli

Abstract In human–robot collaboration, there are different levels of interaction. At lower levels, collisions must be avoided. At a higher level, collaborative robots directly interact with humans, and human health and safety need to be assured when designing work systems with robots included. In the process, adequate models of human motions are required. The diversity of human motions has led to ongoing research activities that have provided a considerable range of models. This chapter provides an overview of full-body human models such as kinematics, dynamics, and skin models and their applications in robotics. Kinematic models provide an abstraction of the human skeleton structure and its joints, whereas skin models provide mappings from the skeleton models to skin surfaces. Both are required for motion simulation with most human motion models. A survey of analytical, physics and data-driven human motion models is given. The state of the art as well as recent research advances are summarized for the models, and models of singular human motions and motion corridor/density models are differentiated. Their relevance for different levels of human–robot collaboration and ergonomics is discussed. Applications of motion models in the collaborative robotics domain and their respective advantages and shortcomings are reviewed with a special focus on their applicability for safety, efficiency and user experience in digital twins. Finally, the chapter presents three application scenarios in assembly lines and material handling in automotive industry.

Keywords Human motion · Digital human model · Collaborative robot · Ergonomics · Human factor · Human–robot collaboration

M. Manns (✉) · T. B. Tuli
PROTECH - Institut für Produktionstechnik, Universität Siegen, Siegen, Germany
e-mail: martin.manns@uni-siegen.de

T. B. Tuli
e-mail: tadele-belay.tuli@uni-siegen.de

G. Barresi and S. Wischniewski (eds.), *Ergonomics in Robotics: Advances and Innovations*, Springer Tracts in Advanced Robotics 162,
https://doi.org/10.1007/978-3-032-19374-2_11

1 Digital Human Motion Models for Production Systems with Collaborative Robots

In the domain of production engineering, human models have been employed for a long time. Initially, specific aspects of humans have been modeled separately for different engineering process steps in order to keep data requirements and model complexity low. One example is workplace layout design, where a simple model for reachability analysis is a circle segment around the human that represents objects that are located in a way that matches a grasp difficulty level (c. [1]).

With the broad adoption of 3D CAx tools in the industry around the year 2000 c. [2], models of human behavior that have been early adopted in the industry mainly consider human pose modeling and analysis. A reason for 3D human models' dissemination to be slower than that of other CAx techniques may be that 3D modeling of humans in many cases, builds upon a skeleton with a large number of joints. Therefore, inverse kinematics and trajectory planning are harder than for a common robot arm. Resorting to manual human modeling of poses, trajectory planning became prohibitive for most engineering applications due to the needed time, cost and expertise, while it remained feasible, e.g., in the film industry (c. [3]). Recently, 3D game engines have entered the production engineering market. With them, models from computer games, film industry, and computer vision are finding their way into production engineering processes. Since such models are not directly applicable to industrial use cases, ongoing research and development projects are intending to close this gap.

Recently, the digital human model (DHM) has been employed for human–robot interaction (HRI), mainly to design collaborative robot and human workplace systems (see Fig. 1). Integrating digital human models for interacting with collaborative robots potentially contributes to individualizing human–robot collaboration by considering anthropometric and kinematic parameters, leading to human-centered workplace optimization and improved ergonomic assessments. DHM is crucial to investigate ergonomically feasible human–robot interaction [4]. For instance, to investigate if repetitive tasks in a production environment exceed workers' physical capacity [5]. In these cases, collaborative robots have been proposed to share workplaces and tasks with humans.

In this respect, it is important to guarantee safety and ensure acceptance of the human worker into the collaborative human–robot environment. In this respect, several safety models exist for risk assessment and management (cf. [21, 22]). These include hazard analysis, fault tree analysis, failure mode and effects analysis, hazard and operability study, fail-safe mechanisms, safety zones to minimize risks, collision avoidance, speed and separation monitoring, force limitation, human behavior prediction, context-aware systems, intention recognition, trust and transparency, and ethical decision-making (cf. [23]). Moreover, safety standards such as DIN ISO 10218 emphasize the general requirements for robot safety, as well as the safety of robot systems and integration. Collaborative robots that work alongside humans are standardized using DIN ISO/TS 15066, which applies safety models that limit the force and pressure that robots can exert on humans to prevent injuries. Designing an

a) b)
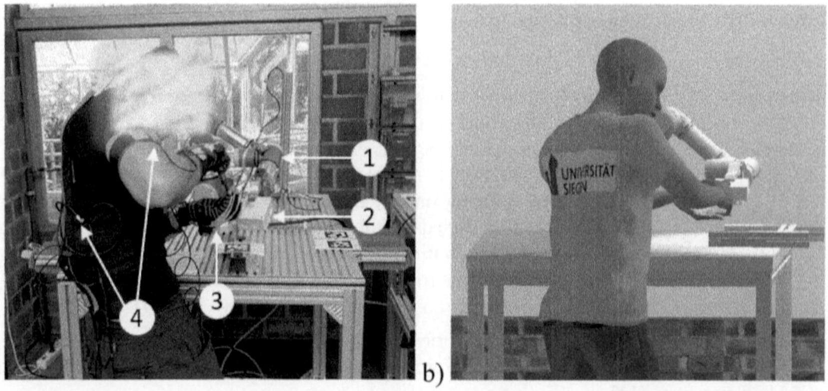

Fig. 1 Collaborative robot workplace design for enabling collaborative robots to adapt to human motion behavior using digital twin models. **a** Actual experimental setup (1—universal robot (UR3e™), 2—dummy object, 3—finger tracker (Manus VR Prime II™), 4—full body motion tracker (Xsens™ Link)); **b** Digital twin model using 3D game environment, Unity®

ergonomically optimal collaborative robot workplace system also requires ergonomic assessments (cf. [24, 25]). Table 1 summarizes digital human models and their implications in HRI.

2 Kinematic Models

Accurate human body motion modeling is beneficial for several robotic applications for enhancing human–machine interface, interaction, and collaboration [26]. Musculoskeletal models of human motion comprise rigid bodies of skeleton connected with mechanical joints. Rigid body-based modeling comprises a compact notation and representation of kinematic and dynamic models capable of real-time computation. The human body is not quite similar to robot mechanisms because human modeling aims to recognize how human body motion behaves, while the objective of robot modeling is to command how the robot must behave.

Modeling the functionality of the human musculoskeletal system requires a profound knowledge of human motion behavior. This behavior results from mechanisms and definitions of the biomechanical structure consisting of bones and tendons. There are different reasons: (1) The human body comprises variable length joint distances held by tendons, (2) The human body differs in motion constraints from human to human and over time, (3) High and low level control is not trivial and not completely understood. The connection of sequential bones with connective tissues at body joints has often been simplified and represented using kinematic models. Several software tools have been presented to analyze human motion behavior for

Table 1 The implications of digital human models for human–robot interaction

DHM types	Role of DHM in improving collaborative robot scenario in HRI
Kinematic (e.g., bio-vision hierarchy, skeleton tree)	– Ergonomic analysis in human–robot collaborative workspaces [4, 6] – Improving safety and reducing physical strain by optimizing human–robot interactions in shared workspaces [7] – To control digital human model movements based on immediate feedback from sensors, ensuring smooth and coordinated interaction [8] – Integrate with real-time data from sensors (e.g., motion capture systems or cameras) to provide continuous feedback on the human's position and orientation to allow the robot to refine its movements dynamically, enhancing the interaction's fluidity and responsiveness [9]
Dynamic (e.g., articulated rigid bodies)	– Predict motion trajectories of both humans and robots to avoid collision [10] – Simulate and analyze force and impact scenarios during physical interaction [11] – Design ergonomic interventions such as adjusting the human–robot interaction level [12] – Generate a natural human motion for designing robot motion behavior and enhance safety and user experience by enabling collaborative robots to predict and adapt to human actions in real-time [13]
Data-driven (e.g., deep learning, statistical)	– Generate a human model that can help to predict and replicate human actions more accurately (cf. [14, 15]) – Anticipate and synchronize with human actions, such as predicting when a human will need assistance or when they are about to move into the robot's workspace [16] – Enable robots to adapt or respond appropriately to various human actions, preferences, and even intentional states [17]
Mesh and morphable (e.g., skeletonized skin mesh)	– Human-like human body modeling [18] and motion keyframes for collaborative robot interaction-rich environments (e.g., assembly, healthcare, service robots) – Improves user acceptance and comfort in interactions by providing more natural and intuitive movements [19] – Work process design and optimization in manufacturing and assembly [20]

different purposes (e.g., Siemens® Jack/Jill [6]). Most DHM tools comprise algorithms that translate high-dimensional human motion behavior into graphical visualizations that complement analytical quantifications. In order to represent poses, rigid bone models with rotatory joint kinematization and skinned meshes are mapped using 3D rigs. In this aspect, 3D rigged skeletons with skinned meshes are the commonly adopted approaches in various works c. [27]. The skeleton rig helps to map and retarget joint motion data to animate or simulate a DHM. For sensor-driven simulations, direct kinematics is usually applied to map sensor data to individual joints of a DHM. When a DHM is desired to follow a given path, then inverse kinematics models are used for computing joint variables based on the target motion data. However, the way motion data is represented directly affects the performance of the kinematic model. In this regard, various approaches for pose representation are being discussed in the scientific community.

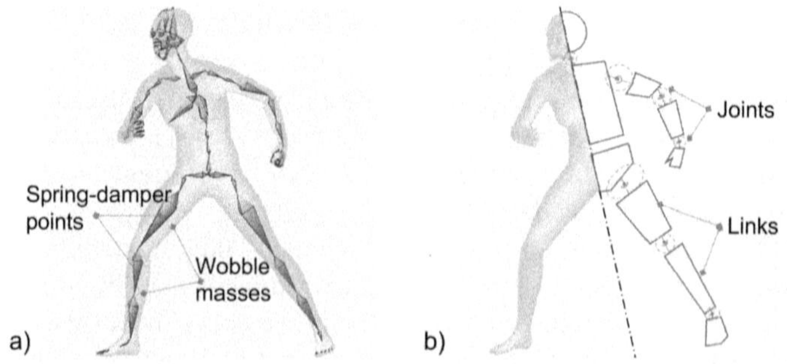

Fig. 2 **a** Human model with skin mesh, rig, and wobbling mass on a lower limb (generated with MakeHuman™ software); **b** Representation of human body kinematics with bone segment (links) and joints

According to the summary of Mourot et al. [27], orientation representation has been considered in various approaches such as Euler angles, Lie algebra, Rotation matrices, unit quaternions, Gram-Schmidt-like, and hierarchical limitations. The main difference between these approaches lies in the way they represent different notations and dimensionalities for solving invertibility, singularity, computational efficiency, non-linearity, and positional errors cf. [27]. The retargeting and rigged skin mesh modeling approaches are discussed in Sects. 2.1 and 2.2.

2.1 Skeleton Modeling and Re-targeting

Most DHM software provides skin modeling with automated skin-rig generation tools (see Fig. 2a). The generated rig is usually arranged in a tree-like hierarchical structure. The skeleton model comprises joint names, position and orientation offset, link length, and root joint. Skeleton parameters can be fixed, e.g., for wearable Inertial Measurement Unit (IMU) based systems, human body bone lengths may be fixed [28], or variable, e.g., optical sensor-based motion captures may or may not dynamically change the bone lengths based on the human posture [29, 30].

2.2 Mesh, Morphing Models

The need to create a realistic and natural-looking digital human model includes model parameterization and artistic visualization. This has led to studies of the human body morphology and its characteristics. An approach that has been considered for modeling the human body employs surface meshing and morphing for creating forms and shapes cf. [18, 31]. Morphing meshes are usually generated by linearly interpolating mesh vertices' position and triangulations' orientation. This linear mapping considers two extreme situations: male or female, tall or short, fat or thin, and long-limbed or short-limbed. According to [18], linear morphing is efficient for rapid human model creation using fuzzy widgets that can be easily selected by scrolling between the upper and lower limit values. In this aspect, translational morphing, rotational morphing, form factor, and mesh relaxing are vital parameters for manually optimizing the geometric models.

Unlike handcrafted and parameterized human body modeling, data-driven approaches have also proved efficient for completing human body shapes using morphable models. Techniques such as SCAPE [32] employ human body scanners to capture ranges from four directions to construct entire body meshes of the instances. The constructed mesh is later processed using feature registration to generate an automated articulated rigged model. Reshaping and automated rigging are crucial for developing realistic and deformable human motion simulations [31].

3 Dynamic Models

Today, many commercial and open-source physics engines are being used for simulating the characters of human and robot motion behavior. The most common physics libraries up to now include Open Dynamics Engine (ODE), PhysX™, Havok™, PyBullet™, and DART [33]. According to the report of Erez et al. [34], based on a simulation comparison of ODE, PhysX, Havok, PyBullet, and MuJoCo®, each engine has unique performance that makes them best for the types of systems they are designed for. Standard game engines such as Nvidia Omniverse®, Unity3D and Unreal have implemented physics libraries to simulate game object characters. In dynamic human motion modeling, the DHM is capable of responding to externally applied forces such as push or collision with objects. Controllers such as zero moment point control [35], and simple biped control (SIMBICON) [36] are shown to be feasible for bi-pedal locomotion dynamic control. These approaches can generate motion data based on a motion capture system or trajectory generator models. The controller consists of strategy and policy that transform the joint state into actions. These actions need a dynamic model that determines the required torque and force to simulate the DHM. Using modeling techniques such as Newton–Euler, Lagrange, or articulated body dynamics, the musculoskeletal dynamics can be modeled. The output of the musculoskeletal dynamics sometimes provides high dimensional information, which may require further simplification retargeting into the standard skeleton model, so that it fits standard animation frameworks [37].

3.1 Multi-body Dynamics Approach for DHM

Multi-body dynamics has been a well-established research focus in robotics and computer character animations for decades [38]. Even though it does not fully represent the reality of the human musculoskeletal system, the assumption of the human body system as a rigid multi-body system has been presented for DHM. A rigid body system comprises models for connecting graphs, links, joint geometries, and joint models [39]. These models can be simplified and applied to a group of particles forming a rigid body system. This group of particles must maintain relative position and orientation throughout the motion.

3.1.1 Dynamics of the Articulated Rigid Bodies

Considering the human body as a multi-body system, the body segments are hierarchically connected like tree structures from root to leaf. For such interconnected bodies, forward dynamics have been shown to be an efficient approach [38]. The articulated body systems have been represented with a tree structure that can be recursively computed from root to leaves using articulation body algorithms. The Articulation Body Algorithm (ABA) [39, 40] has been applied to simulate articulated body mechanisms given the joint position, velocity, and applied forces/torques. Based on these inputs, the ABA algorithm computes joint acceleration cf. [40]. The overall procedure includes velocity, acceleration, force, and torque computation.

3.1.2 Constrained Dynamics

In reality, human body joints are limited to certain motions, and in ergonomically feasible operations, there has been a further suggestion for limiting joint space to avoid muscle strains or workspace freedom. In this aspect, it is necessary to define constraints for generating realistic human motion or sequence of postures. In dynamic models, constraints arise due to link length, joint orientation, range of motion, collision or contact (e.g., due to ground, workspace object, or self-collision), and model parameter constraints (e.g., unit norm quaternions).

In forward dynamics models, externally applied forces and defined constraints are important to enable human body joints to achieve their goal. The review article presented in [3] describes two approaches for enforcing constraints. The first approach is to apply equal but opposite forces so that the body will remain in its position. The second approach involves reducing dimensionality or degree of freedom. Reduced dimensionality ensures enhanced simulation performance while compromising accuracy and process stability.

Various basic types of research on dynamic constraints have been presented. For instance, a Gaussian process latent variable has been implemented in [41] that learns contextual ergonomics from musculoskeletal model simulation, which ultimately helps to reduce muscle activation. When humans and collaborative robots physically interact, a holonomic constraint is one of the proposed approaches in [42] for establishing the same velocity for both end effecter frames of a robot and human. Dynamic noises resulting from sensor models such as inertial measurement unit (IMU) have been resolved using frequency constraints (e.g., Kalman filter [28]). Furthermore, constraint models such as exponential maps [30], fuzzy joint constraints [43], and path constraints can be differentiated. There exist pose dynamic constraints which are applied to one or all frames of the skeleton model, while joint space constraints are enforced to fit a specific condition.

3.2 Physics-Based Motion Models (Sense—Predict)

Human motion modeling using dynamic physical models has been successfully employed in domains such as generating realistic human motions, mainly in graphic animations, virtual reality, and computer-aided designs [44]. This motion generation can be implemented using a spring mass damper, oscillator and wobble mass models.

3.2.1 Mass-Spring-Damper (MSD) Models

Mass-Spring-Damper (MSD) modeling involves the multi-segmentation of human body structure into mechanical body segments. The MSD model considers these bodies as discrete mass nodes distributed throughout the human body with spring and dampers network connection. In this approach, the human body can be considered as a single degree of freedom body consisting of a cumulative mass, damper, and spring. Such an approach has been used to analyze a mechanical power output, for example, while hitting the ground during running cf. [45].

For collaborative robotic applications, MSD methods have been applied to determine the biomechanical forces and motion limits to establish safe and efficient human–robot physical interaction cf. [47, 48]. It is highly simplified to consider the human model as a single degree of freedom. Instead, various works have represented the human model into multiple segments depending on the target application. For example, Jenny and Jenny [46] described the human body with two masses (upper with mass-spring and lower with mass-spring-damper), segmenting the body into two regions, as shown in Fig. 3b. Such an approach requires a multi-degree of freedom MSD system model representation consisting of mass (M), damping (C), and spring (K) matrices.

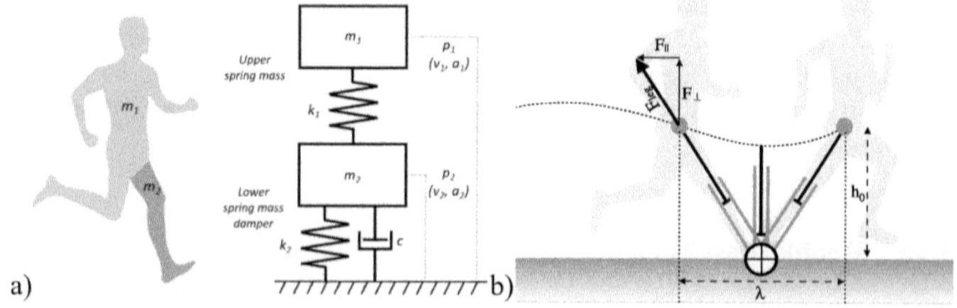

Fig. 3 Classical mass-spring-damper model; **a** Illustration of Mass-spring-damper model for running motion by Jenny and Jenny [46], CC BY 4.0; **b** mechanical output power estimation during running by Durikovic and Numata [45], CC BY 4.0

3.2.2 Oscillator Models

The initial concept that assumes human behaviors as complex oscillatory processes has led to the application of pendulum physics in human motion studies [49]. Physical observation of human motions, such as walking, is considered to fulfill the definition of an oscillator due to its periodicity around a dynamic equilibrium. The motion results from supplied power for maintaining the temporal motion frequency, which is a driving force.

In physics, repeating motions, e.g., back and forth, against the restoring force causes a harmonic motion. Oscillating mass objects generate a harmonic motion, in which the restoring force is proportional to the object's displacement. For example, releasing a swinging pendulum from some height will cause the pendulum to oscillate about its mean position. Oscillating motions can be represented with Newton's second law of equation of motion in time state where F(t) denotes external force that drives the harmonic motion. Harmonic motion can be simple, damped, or force-driven oscillations that are based on model properties (e.g., spring, mass, and damping).

In actual practice, the velocity of harmonic motion decreases if a joint is affected by a friction acting along the motion. Similarly, damping forces can cause an additional force that resists the motion. In this aspect, the simple harmonic motion model has to consider the damp effect, called the *damped harmonic oscillator* model. Oscillator models generate a harmonic trajectory that can be governed analytically using sinusoidal patterns (see Fig. 3);

The fundamental concept of the oscillator model has been extended into human motion modeling for generating harmonic motions for various aspects. Humans sometimes produce repetitive motions that are statistically constant. Studies show a strong correlation between frequency and motion path for a freely selected speed. Considering walk motion, the stride rate and length may be approximated by employing harmonic oscillators, ultimately predicting the walking pace and distance. A stride comprises two consecutive steps from the right and left instance. In [50], the study has shown successful prediction of stride rate for backward and forward walk motion using pendular force-driven harmonic oscillators. The force-driven harmonic oscillation has been implemented for comparing the predicted frequency (0.56 Hz) and preferred frequency (0.79 Hz) by Refs. [49, 51] for walking at a comfortable cadence. Harmonic motion models have been used with the inverted pendulum concept for designing wearable robotic systems that aim to assist human body movement.

3.2.3 Wobble Mass Models

In reality, the human body is composed of a complex structure comprising bones, joints, tendons, and soft tissues or muscles. In order to reduce the complexity of the digital human model, the rigid body approach from the robotics concept has been adapted for modeling and simulating human motions. Moreover, the approach has proven applicable for combining rigid body and soft tissue modeling cf. [52, 53]. Motions that are parameterized by frequency, e.g., running, carrying an object while bending down, walking on an inclined surface, or consisting of impact contact, may affect the naturalness. Specifically, wobbling masses from soft tissues may affect motion dynamics more than the rigid body model. The study presented in [52] investigates metabolic expenditure through soft tissue deformation by analyzing mechanical energy dissipation during running activity. On top of that, the effect of soft tissue on rigid body dynamics has been shown to be crucial for dynamic motions such as running [54].

Wobble mass modeling is vital for maintaining the balance of the human body model so that it remains stable and steady while performing actions. Motions such as walking, carrying an object, and reaching a target demonstrate this, as the human model is required to control its balance.

4 Data-Driven Motion Models (Sense—Learn—Predict)

Data-driven motion models refer to sets of captured motions that can be generalized. From these models, skeleton trajectories that are similar to many of the captured motions can be derived. This section focuses on models of motions mapped to a kinematic skeleton model with fixed bone lengths, resulting in 6 DOF for the root joint and a set of parameters (such as Euler angles or quaternions to avoid singularities) for 3 DOF of each rotatory joint. When considering the representation of motion data in the time domain, three types of motion models are distinguished:

Key frame motion modeling—motion data can be represented frame-wise, i.e., postures are represented for specific points in time. This type of representation is common for real-time applications [14]. Because of the large amounts of data per frame, pre-processing often comprises a step that maps the frame data onto a latent space. The frame data can be reduced to a vector with a fixed number of elements in size, e.g., using principal component analysis (PCA). When setting up the data-driven model, each frame represents one data point.

Trajectory-based motion modeling—motion data can also be represented as short trajectories, which are small motion snippets that may be semantically annotated (s. [55]). However, semantically reasonable trajectories such as "reach" actions are likely to exhibit a different duration for each repetition, which may not be desirable for setting up data-driven models. Therefore, the trajectories are projected onto a canonical timeline, for example using the dynamic time warping algorithm (s. [56]). After this, the time mapping is pre-processed separately from the spatial trajectory. B-Spline representations have been proposed that may be mapped onto latent spaces with functional principal component analysis (fPCA) [57, 58]. The trajectory-based approach is well suited for planning motions and motion corridors ahead of time c. [59]. Each trajectory represents one data point in the desired latent space when setting up the data-driven model. The Gaussian Mixtures and Gaussian Processes are also investigated in several works (e.g. [60, 61]).

Deep-learning based motion modeling—motion models generated by training using exemplary recorded motions apply common algorithms such as Convolutional Neural Network (CNN) [15, 62], Long Short Term Memory (LSTM) [63, 64], Gaussian Adversarial Network (GAN) [65], or hybrid models (e.g., LSTM + CNN) [66, 67]. These motion models synthesize stylistic human motion types, e.g., locomotion c. [68]. With these models, trajectories can be generated that have not been recorded before.

Therefore, data-driven motion models generalize recorded motions. These approaches can be fast, allowing sampling to generate motions that meet constraints. Even though data-driven approaches allow for generating visually impressive motions, there is a risk of obviously unrealistic motions.

5 Analytical Models (Sense—Reason—Predict)

5.1 Planning Based Models

Modeling motions with long-term goal requires a computational policy or path hypothesis that enable the digital human model (DHM) to accomplish the desired task. This requires the DHM to sense the surroundings to identify the goal target and explicitly reason the long-term motion plans while predicting the impact of future motion on the current action. This can be achieved either through forward or backward planning methods. The forward planning approach employs a predefined cost function, whereas the backward planning infers the cost function from human behavior observations to predict human motion [69–71].

The generation of goal-oriented motion that requires the process of defining problems for generating a sequence of actions is called planning [72]. The interconnection of segmented motions into low-level actions can lead to a complex problem, e.g. with multiple potential motion candidate making it challenging for the planners. Therefore, sub-tasks have been decomposed into basic activities, and the network of the action has been represented using AND/OR graphs [73], heuristic visualization of behavior trees [72, 74], or state machines [75]. Moreover, semantic planning languages such as Planning Domain Definition Language (PDDL) have been applied in various works in both robot and human task planning [75, 76]. The PDDL based planning is a human readable task description that can be implemented in behavior trees or state machines with forward or backward planning approaches.

Applied planning models become difficult to use when the complexity of the process increases, yielding large motion variations and involving body joints. This requires special consideration for optimizing the approach's usability, expressibility, and applicability. Several motion planning frameworks, such as MoveIt [77], ComFlow [78], MOSIM [19], ROSPlan [79], and OPML [80] exist. In research, planning has been considered at the higher level for describing task sequences represented with semantics [81, 82]. At the lower level, the planning comprises frame-by-frame motion in time or latent space [83–85]. Some approaches combine high- and low-level planning techniques to execute motion sequences for collaborative tasks. For instance, using AJAN [74] helps plan task sequences using descriptive languages and enables control of motion patterns, e.g., using motion model units. Further investigations shown in [86] highlight that process planning of human and collaborative robot activities should be explainable and expressible so that engineers can effectively use them.

6 Industrial Case Studies

The application of digital human modeling in collaborative robotics environment can be differentiated based on the maturity and technological readiness. The first stage comprises basic research to prove the feasibility of collaboration between humans and robots. In this regard, various scientific activities have been presented to recognize human motion patterns, actions, and sequences of activities. In this aspect, data-driven and physics-based motion classification and generation techniques have been employed, (e.g., [66]). Moreover, one of the emerging research questions is how humans and robots share autonomy, jointly handle objects and anticipate mutual care, e.g., [87–90]. The second stage comprises applied research projects that aim to develop methods, tools, or models that are suitable for being transferred to industrial applications. In this aspect, several frameworks and tools have been demonstrated by various projects for workflow planning (e.g., AJAN [91]), ergonomics analysis, and modular motion simulation (e.g., [92]). At the third stage, higher technological readiness and maturity of applications of DHM in an operational environment are being used for value-adding in the production process by reducing production time, unnecessary walk paths, and high ergonomic strains (e.g., EMA™ [93]).

6.1 Application of Musculoskeletal Modeling for Manual Operations and Exoskeleton Design

Manual material handling is a typical task in many industries, including construction, manufacturing, and retailing. Such tasks comprise actions such as pushing an object from one position to another for the purpose of relocation, lifting an object from a lower position to a higher position, or transferring from one higher position to another higher position by carrying the object. Similarly, lowering from a higher position to a lower position as well as locomotion while carrying an object are commonly required.

Various researchers have applied the musculoskeletal modeling approach to investigate the effects of dynamics on the digital human model using motion capture data (e.g., [37, 94]). In [37] the effects of load mass and position on the dynamics of the body joints, such as knees, shoulders, and lumbar spines were investigated. The motion capture based on a camera system was applied to acquire the human posture on a frame basis, and AnyBody™ modeling system software was employed to analyze the forces and motions of human activities. The illustration in Fig. 4 depicts the workflow of musculoskeletal modeling using digital human models.

Applying musculoskeletal modeling in exoskeletal design is becoming interesting for assessing the ergonomic feasibility of activities in shop floor environments [93]. In this aspect, the output of the assessment is crucial to design exoskeleton systems for avoiding musculoskeletal disorders cf. [95] that may arise due to the effects of

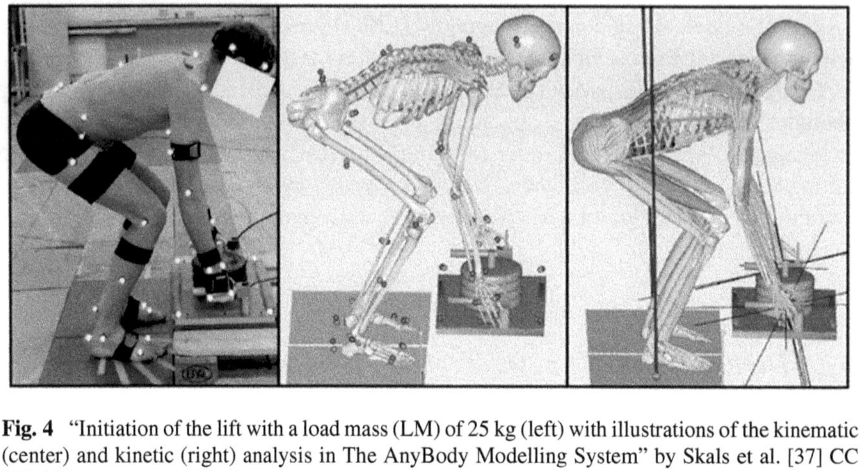

Fig. 4 "Initiation of the lift with a load mass (LM) of 25 kg (left) with illustrations of the kinematic (center) and kinetic (right) analysis in The AnyBody Modelling System" by Skals et al. [37] CC BY 4.0

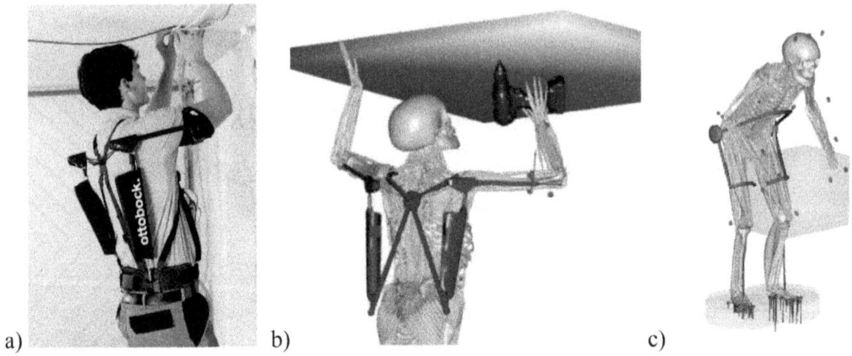

Fig. 5 **a** "Paexo shoulder exoskeleton used in the construction industry" and **b** "AnyBody Model with Paexo shoulder exoskeleton in the overhead task" by Gaisbauer [93] CC BY 4.0, and **c** AnyBody Model with back support exoskeleton for lower (used with permission)

workloads on the human worker during repetitive manual and collaborative operations with robots. Since musculoskeletal models can approximate the forces generated by muscles and tendons of the human body and holistically determine the muscle activities, they offer the potential to compute the body's expended energy given motion data capture [95] from physical activities. Furthermore, in the case of exoskeletal systems, the musculoskeletal models can help to analyze the effects of the interaction forces between the human and exoskeletal systems. However, musculoskeletal models face limitations due to their intensive computational requirements. At the same time oversimplification of the human body system representation can lead to unrealistic results. Moreover, the approach is reported promising for designing exoskeletal systems to investigate its ergonomic and functional feasibility with exoskeletal and without skeletal. The works presented in [94] analyze musculoskeletal models using electromyography (EMG) sensors-based data to determine whether wearing a exoskeleton is required or not for tasks such as drilling operation.

Furthermore, assisting devices have been used for overhead activities, e.g., using shoulder exoskeletons (see Fig. 5a, b) or material handling using hip support exoskeletons (see Fig. 5c) in order to minimize physical strains that can be caused during industrial tasks. Ultimately, such approaches are crucial for the design optimization of exoskeleton or assistive products, movement analysis, and orthopedics rehabilitation.

6.2 *Digital Twins of an Automotive End Assembly Operation*

Automotive end assembly is a domain in which human models have been proposed for applications such as time planning [96], workplace design and process validation [19], or ergonomic analysis [97]. However, the level of granularity of the models varies for different applications. In order to generalize the models in digital twins, a system for a comprehensive simulation of manual assembly with digital human motion models has been proposed in [92]. The concept is based on the idea of "varying granularity based on limited information." The core idea is that initially, information on the process and resources is limited, and the detail that the planner requires is low. The level of detail is then increased for later planning phases.

In order to achieve this, the following models for automated generation of scenes, motion constraints, and motions have been proposed by Antakli et al. [92]:

- *Automated grasp point estimation*: Reachability models are employed as a basis to generate feasible and ergonomically viable grasp points. This generation is presented for differently parameterized human skeleton models.
- *Automated material zone population*: Considering avatars that are located in predefined work positions, grasp points form the basis for optimizing the material locations.
- *Automated task sequence generation*: Task sequences are automatically derived from a planned process sequence. The process steps are split up into a set of predefined building blocks. However, the process is not completely automated, so manual adjustments remain necessary.
- *Automated walk point estimation*: Deriving adequate walk points is crucial for the automated setup of end assembly simulations because these points define the locations, at which the avatar is situated when performing tasks. In order to determine adequate walk points in real-time, two-dimensional matrices, so-called "heat maps" ([92], p. 208), are created before simulation. Matrix elements are rated for their suitability, i.e., a combination of reachability, collision freeness (with a box cast algorithm), and visibility assessment (using a ray-cast algorithm) (c. [98]).
- *Assembly path generation*: The algorithm rapidly-exploring random trees (RRT) [99] is used for a sampling based approach to estimate collision free assembly paths. Manual adjustments of keyframes supplement it.
- *Modular motion units*: The actual human trajectory can be derived from different algorithms. These are stored in Modular Motion Units (MMUs) cf. [19]. Each MMU may incorporate a different approach—from blending between poses to deep learning-based algorithms.

Real-time capability is considered crucial for practical acceptance by Antakli et al. [92]. To gain sufficient performance, a concept of motion reuse is implemented. Two practical applications in the automotive industry are presented in [92] (see also Fig. 6): Cleaning cylinder assembly and headlight unit pre-assembly. It shall be noted that both processes are presented in a form that deviates from the production line. An expert workshop is presented, in which the system is described as usable for planning and assessing processes for which no historical data exists, for walk path planning, for digital workshops for assembly preparation, for logistics and commissioning planning, and for layout planning. Shortcomings such as unrealistic hand poses, lack of support for tasks other than pick and place, and the lack of simulating the behavior of flexible parts are stated in [92] p. 223. These limitations are also reported for later planning stages with a more significant impact.

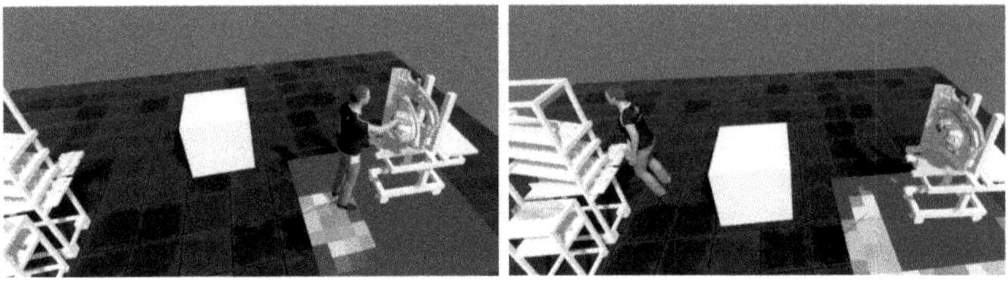

(a) (b)

Fig. 6 Modular human simulation [92]: **a** positioning process of a headlight module and **b** pick-up motion of a nut [92]

Overall, the presented approach of automatically synthesizing human motions in digital twins from task descriptions has been presented as a proof of concept in laboratory and demonstration environments for automotive end assembly.

6.3 Digital Planning and Simulation for Human–Robot Interaction

Recently, collaborative robotic workplace design is becoming a crucial element of human-centric production systems. In this respect, the human-centric production system is required to be resilient, sustainable, and reconfigurable [100]. This may involve human factor evaluation, ergonomic quantification, process planning, and control [101]. The outcome helps to increase the productivity and flexibility of existing work systems or to design a robust future generation of workplace systems. Furthermore, the interaction strategy of human–robot or human–machine is a hot research field.

Designing a collaborative robotic work system for an existing industrial scenario can be analyzed using several tools, including EMA Work Designer. This section considers the industrial use case scenario for manual object handling. In the actual scenario, the worker performs the handling operation, which is analyzed to be ergonomically poor. The worker repetitively lifts object parts and loads them into a CNC machine for processing. Since the process is physically demanding, a collaborative robot integration into the workplace has been proposed. In this aspect, a mobile robot platform handles parts up to 6 kg payload with omni-directional navigation capability in the shop floor environment (s. [5]). To investigate the feasibility of the proposed solution, EMA Work Designer software for digital planning and virtual validation is employed. The explicit visualization of the entire work process facilitates communication with all stakeholders (e.g., engineers, workers, work safety, and management) and facilitates a common understanding during the planning phase.

The digital simulation is used to validate desired task distribution between the human and robot system and its feasibility, safety, time synchronization, sequence

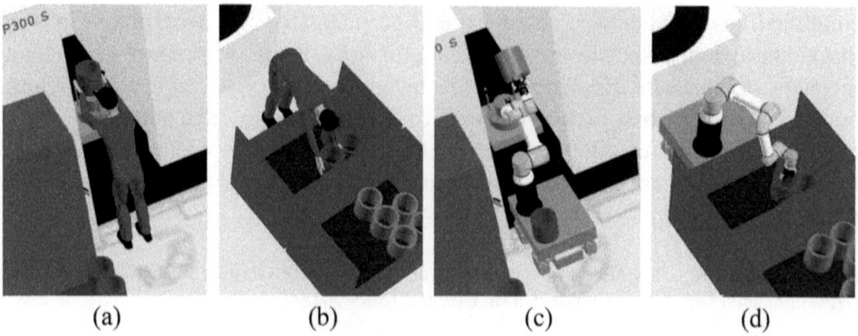

Fig. 7 Collaborative robot application for improving ergonomics and productivity; **a** Manual part loading on gear shaping machine, **b** manual placing of a finished part in the container, **c** robotic part loading, and **d** robotic part placing in the container using EMA software suite—used with permission [5]

planning, and navigation or walk paths. Finally, the results are optimized to ensure they fit standard industrial requirements such as speed, height, and distance. The qualitative results in [97] show a reduction in used active worker time reduction by 32% and ergonomic work strain by 59%, since the collaborative robot is set to overtake heavy lifting activities. Additionally, unnecessary walk ways are reduced by 83%, since the collaborative robot is set to overtake the motions between material supply and CNC machine. The workflow processes before and after digital planning are depicted in Fig. 7.

7 Conclusion

In general, the adoption of the digital human motion model in human-centered work designs in production processes is increasing. There is an active and growing research trend in the domain. According to Gartner®, digital human is one of the five technologies that will transform the digital future [102]. In this aspect, digital humans are designed to replicate human-like interactions and actions with machine systems like robots or actual humans as a human digital twin. With the advances in generative AI, humanoid robots, virtual reality, and augmented reality, the global market for DHM is projected to grow at a compound annual growth rate (CAGR) of more than 40% by 2028 [103]. Similarly, to address human and robot collaboration tasks, various tools, software, models, and methods have been presented to address the integration of digital humans.

Real-time capability, accuracy, process, and effort requirements can be crucial factors in human–robot interaction. Human safety and model accuracy may arise due to various model complexities. The applicability of digital human models for collaborative robotic applications requires further investigation as existing solutions employ simplified models. Besides, robots should be integrated into workplaces in ways that enhance, rather than diminish, meaningful work. Future work should investigate methods to balance the efficiency gains from humans and robots, preserving essential human aspects of work, such as purpose, social relationships, and autonomy. Moreover, further studies could focus on designing robots that complement human skills, fostering environments where robots enhance self-development and recognition in the workplace [104].

Acknowledgements The authors would like to acknowledge the German Research Foundation's (DFG) financial support within the HiSMoT project (500490184).

References

1. Jonek, M., Tuli, T.B., Manns, M.: Constraints for motion generation in work planning with digital human simulations. In: Towards Sustainable Customization: Bridging Smart Products and Manufacturing Systems, pp. 567–574. Cham (2022)
2. Mourtzis, D.: Simulation in the design and operation of manufacturing systems: state of the art and new trends. Int. J. Prod. Res. **58**(7), 1927–1949 (2020)
3. Geijtenbeek, T., Pronost, N.: Interactive character animation using simulated physics: a state-of-the-art review. Comput. Graph. Forum **31**(8), 2492–2515 (2012)
4. Chang, S.-W., Wang, M.-J.J.: Digital human modeling and workplace evaluation: using an automobile assembly task as an example. Hum. Factors Ergon. Manuf. Serv. Ind. **17**(5), 445–455 (2007)
5. Fritzsche, L., Funk, M., Rosen, P., Spitzhirn, M., Wischniewski, S.: Simulation and participatory design of HRI work systems in the SOPHIA project (2021)
6. Raschke, U., Cort, C.: Chapter 3—Siemens Jack. In: Scataglini, S., Paul, G. (eds.) DHM and Posturography, pp. 35–48. Academic Press (2019)
7. Kang, S., Kim, M., Kim, K.: Safety monitoring for human robot collaborative workspaces. In: International Conference on Control, Automation and Systems, vol. 2019 (2019)
8. Tuli, T.B., Manns, M.: Real-time motion tracking for humans and robots in a collaborative assembly task. Proceedings **42**(1), 48 (2019)
9. Popescu, M., Mronga, D., Bergonzani, I., Kumar, S., Kirchner, F.: Experimental investigations into using motion capture state feedback for real-time control of a humanoid robot. Sensors **22**(24), 9853 (2022)
10. Mronga, D., Knobloch, T., de Gea Fernández, J., Kirchner, F.: A constraint-based approach for human–robot collision avoidance. Adv. Robot. **34**(5), 265–281 (2020)
11. Liu, S.B., Althoff, M.: Online verification of impact-force-limiting control for physical human-robot interaction. In: IEEE/RSJ International Conference on Intelligent Robots and Systems (IROS), pp. 777–783 (2021)
12. Busch, B., Toussaint, M., Lopes, M.: Planning ergonomic sequences of actions in human-robot interaction. In: IEEE International Conference on Robotics and Automation (ICRA), pp. 1916–1923 (2018)
13. Latella, C., Lorenzini, M., Lazzaroni, M., Romano, F., Traversaro, S., Akhras, M.A., Pucci, D., Nori, F.: Towards real-time whole-body human dynamics estimation through probabilistic sensor fusion algorithms. Auton. Robot. **43**(6), 1591–1603 (2019)
14. Herrmann, E., Manns, M., Du, H., Hosseini, S., Fischer, K.: Accelerating statistical human motion synthesis using space partitioning data structures. Comput. Animation Virtual Worlds **28**(3–4), e1780

15. Holden, D., Saito, J., Komura, T.: A deep learning framework for character motion synthesis and editing. ACM Trans. Graph. (TOG) **35**(4), 138 (2016)
16. Hawkins, K.P., Bansal, S., Vo, N.N., Bobick, A.F.: Anticipating human actions for collaboration in the presence of task and sensor uncertainty. In: Proceedings—EEE International Conference on Robotics and Automation (2014)
17. Tuli, T.B., Manns, M.: Explainable human activity recognition based on probabilistic spatial partitions for symbiotic workplaces. Int. J. Comput. Integr. Manuf. **36**(12), 1783–1800 (2023)
18. Bastioni, M., Re, S., Misra, S.: Ideas and methods for modeling 3D human figures: the principal algorithms used by MakeHuman and their implementation in a new approach to parametric modeling. In: Proceedings of the 1st Bangalore Annual Compute Conference, New York, NY, USA, pp. 1–6 (2008)
19. Gaisbauer, F., Agethen, P., Otto, M., Bär, T., Sues, J., Rukzio, E.: Presenting a modular framework for a holistic simulation of manual assembly tasks. Procedia CIRP **72**, 768–773 (2018)
20. Bauer, S., Sylaja, V.J., Fritzsche, L., Ullmann, S.: Chapter 5—Task-based digital human simulation with editor for manual work activities—basic functionalities, applications, and future works. In: Scataglini, S., Paul, G. (eds.) DHM and Posturography. Academic Press, pp. 57–62 (2019)
21. Huck, T.P., Münch, N., Hornung, L., Ledermann, C., Wurll, C.: Risk assessment tools for industrial human-robot collaboration: novel approaches and practical needs. Saf. Sci. **141**, 105288 (2021)
22. Robla-Gomez, S., Becerra, V.M., Llata, J.R., Gonzalez-Sarabia, E., Torre-Ferrero, C., Perez-Oria, J.: Working together: a review on safe human-robot collaboration in industrial environments. IEEE Access **5**, 26754–26773 (2017)
23. Yigit, T., Kose, U., Sengoz, N.: Robotics rights and ethics rules. J. Multidiscip. Dev. **3**(1), 30–37 (2018)
24. Makrini, I.E., Merckaert, K., De Winter, J., Lefeber, D., Vanderborght, B.: Task allocation for improved ergonomics in human-robot collaborative assembly. Interact. Stud. **20**(1), 102–133 (2019)
25. Harari, Y., Bechar, A., Raschke, U., Riemer, R.: Automated simulation-based workplace design that considers ergonomics and productivity. Int. J. Simul. Model. **16**(1), 5–18 (2017)
26. Yamane, K., Nakamura, Y.: Robot kinematics and dynamics for modeling the human body. In: Robotics Research, Berlin, Heidelberg, pp. 49–60 (2011)
27. Mourot, L., Hoyet, L., Le Clerc, F., Schnitzler, F., Hellier, P.: A survey on deep learning for skeleton-based human animation. Comput. Graphics Forum **41**(1), 122–157 (2022)
28. Roetenberg, D., Luinge, H., Slycke, P.: Xsens MVN: full 6DOF human motion tracking using miniature inertial sensors. p. 7 (2009)
29. Mikołajczyk, T., Kłodowski, A., Mrozinski, A.: Camera-based automatic system for tool measurements and recognition. Procedia Technol. **22**, 1035–1042 (2016)
30. Murthy, P., Butt, H.T., Hiremath, S., Khoshhal, A., Stricker, D.: Learning 3D joint constraints from vision-based motion capture datasets. IPSJ T Comput. Vis. Appl. **11**(1), 5 (2019)
31. Feng, A., Casas, D., Shapiro, A.: Avatar reshaping and automatic rigging using a deformable model. In: Proceedings of the 8th ACM SIGGRAPH Conference on Motion in Games, New York, NY, USA, pp. 57–64 (2015)
32. Anguelov, D., Srinivasan, P., Koller, D., Thrun, S., Rodgers, J., Davis, J.: SCAPE: shape completion and animation of people. ACM Trans. Graph. **24**(3), 408–416 (2005)
33. Lee, J., Grey, M.X., Ha, S., Kunz, T., Jain, S., Ye, Y., Srinivasa, S.S., Stilman, M., Liu, C.K.: DART: dynamic animation and robotics toolkit. J. Open Source Software **3**(22), 500 (2018)
34. Erez, T., Tassa, Y., Todorov, E.: Simulation tools for model-based robotics: comparison of Bullet, Havok, MuJoCo, ODE and PhysX. In: IEEE International Conference on Robotics and Automation (ICRA), pp. 4397–4404 (2015)
35. Vukobratović, M., Borovac, B.: Zero-moment point—thirty five years of its life. Int. J. Human. Robot. **01**(01), 157–173 (2004)

36. Yin, K., Loken, K., van de Panne, M.: SIMBICON: simple biped locomotion control. ACM Trans. Graph. **26**(3), 105-es (2007)
37. Skals, S., Bláfoss, R., de Zee, M., Andersen, L.L., Andersen, M.S.: Effects of load mass and position on the dynamic loading of the knees, shoulders and lumbar spine during lifting: a musculoskeletal modelling approach. Appl. Ergon. **96**, 103491 (2021)
38. Redon, S., Galoppo, N., Lin, M.C.: Adaptive Dynamics of Articulated Bodies, p. 10
39. Featherstone, R., Orin, D.E.: Dynamics. In: Siciliano, B., Khatib, O. (eds.) Springer Handbook of Robotics, pp. 37–66. Springer International Publishing, Cham (2016)
40. Featherstone, R.: Forward dynamics—the articulated-body method. In: Featherstone, R. (ed.) Robot Dynamics Algorithms. Springer US, Boston, MA, pp. 89–105 (1987)
41. Marin, A.G., Shourijeh, M.S., Galibarov, P.E., Damsgaard, M., Fritzsch, L., Stulp, F.: Optimizing contextual ergonomics models in human-robot interaction. In: IEEE/RSJ International Conference on Intelligent Robots and Systems (IROS), pp. 1–9 (2018)
42. Tirupachuri, Y., Nava, G., Latella, C., Ferigo, D., Rapetti, L., Tagliapietra, L., Nori, F., Pucci, D.: Towards partner-aware humanoid robot control under physical interactions. In: Intelligent Systems and Applications, Cham, pp. 1073–1092 (2020)
43. Liu, H., Ju, Z., Ji, X., Chan, C.S., Khoury, M.: A unified fuzzy framework for human hand motion recognition. In: Liu, H., Ju, Z., Ji, X., Chan, C.S., Khoury, M. (eds.) Human Motion Sensing and Recognition: A Fuzzy Qualitative Approach pp. 147–170. Springer Berlin Heidelberg, Berlin, Heidelberg (2017)
44. Durikovic, R., Numata, K.: Human hand model based on rigid body dynamics. In: Proceedings. Eighth International Conference on Information Visualisation, 2004. IV 2004, pp. 853–857 (2004)
45. Jenny, D.F., Jenny, P.: On the mechanical power output required for human running—insight from an analytical model. J. Biomech. **110**, 109948 (2020)
46. Nedergaard, N.J., Verheul, J., Drust, B., Etchells, T., Lisboa, P., Robinson, M.A., Vanrenterghem, J.: The feasibility of predicting ground reaction forces during running from a trunk accelerometry driven mass-spring-damper model. PeerJ **6**, e6105 (2018)
47. Dong, S., Pan, X., Wang, H.: An improved equivalent impact model of human thorax for human-robot collaboration. Int. J. Intell. Robot. Appl. **6**(3), 426–436 (2022)
48. Maurice, P., Padois, V., Measson, Y., Bidaud, P.: Human-oriented design of collaborative robots. Int. J. Ind. Ergon. **57**, 88–102 (2017)
49. Holt, K.G., Hamill, J., Andres, R.O.: The force-driven harmonic oscillator as a model for human locomotion. Hum. Mov. Sci. **9**(1), 55–68 (1990)
50. Schot, P.K., Decker, M.J.: The force driven harmonic oscillator model accurately predicts the preferred stride frequency for backward walking. Hum. Mov. Sci. **17**(1), 67–76 (1998)
51. Holt, K.G., Hamill, J., Andres, R.O.: The force driven harmonic oscillator as a model for human locomotion: 576. Med. Sci. Sports Exerc. **21**(2), S97 (1980)
52. Riddick, R.C., Kuo, A.D.: Soft tissues store and return mechanical energy in human running. J. Biomech. **49**(3), 436–441 (2016)
53. Nguyen, H.T.P., Woo, Y., Huynh, N.N., Jeong, H.: Scoring of human body-balance ability on wobble board based on the geometric solution. Appl. Sci. **12**(12), 5967 (2022)
54. Masters, S.E., Challis, J.H.: The effects of wobbling mass components on joint dynamics during running. J. Appl. Biomech. **38**(2), 69–75 (2022)
55. Min, J., Chai, J.: Motion graphs++: a compact generative model for semantic motion analysis and synthesis. ACM Trans. Graph. (TOG) **31**(6), 153 (2012)
56. Johnen, B., Kuhlenkoetter, B.: A dynamic time warping algorithm for industrial robot motion analysis. In: Annual Conference on Information Science and Systems (CISS), pp. 18–23 (2016)
57. Manns, M., Martín, N.A.A.: Automated DHM modeling for integrated alpha-numeric and geometric assembly planning. In: Smart Product Engineering, pp. 325–334 (2013)
58. Manns, M., Martin, N.A.A.: Improving A* walk trajectories with B-splines and motion capture for manual assembly verification. Procedia CIRP **33**, 364–369 (2015)

59. Manns, M., Fischer, K., Du, H., Slusallek, P., Alexopoulos, K.: A new approach to plan manual assembly. Int. J. Comput. Integr. Manuf. 1–14 (2018)
60. Li, Q., Zhang, Z., You, Y., Mu, Y., Feng, C.: Data driven models for human motion prediction in human-robot collaboration. IEEE Access (2020)
61. Wang, J.M., Fleet, D.J., Hertzmann, A.: Gaussian process dynamical models for human motion. IEEE Trans. Pattern Anal. Mach. Intell. **30**(2), 283–298 (2008)
62. Petruck, H., Mertens, A.: Using Convolutional Neural Networks for Assembly Activity Recognition in Robot Assisted Manual Production, vol. 10902, p. 397. Springer Verlag, LNCS (2018)
63. Carrara, F., Elias, P., Sedmidubsky, J., Zezula, P.: LSTM-based real-time action detection and prediction in human motion streams. Multimed. Tools Appl. **78**(19), 27309–27331 (2019)
64. Zebin, T., Sperrin, M., Peek, N., Casson, A.J.: Human activity recognition from inertial sensor time-series using batch normalized deep LSTM recurrent networks. In: 40th Annual International Conference of the IEEE Engineering in Medicine and Biology Society (EMBC), pp. 1–4 (2018)
65. Härkönen, E., Hertzmann, A., Lehtinen, J., Paris, S.: GANSpace: discovering interpretable GAN controls. Adv. Neural. Inf. Process. Syst. **33**, 9841–9850 (2020)
66. Khan, I.U., Afzal, S., Lee, J.W.: Human activity recognition via hybrid deep learning based model. Sensors **22**(1), 323 (2022)
67. Tuli, T.B., Patel, V.M., Manns, M.: Industrial Human Activity Prediction and Detection Using Sequential Memory Networks (2022)
68. Du, H., Herrmann, E., Sprenger, J., Cheema, N., Hosseini, S., Fischer, K., Slusallek, P.: Stylistic locomotion modeling with conditional variational autoencoder. In: Eurographics 2019—Short Papers, p. 4 (2019)
69. Helmert, M.: The fast downward planning system. Jair **26**, 191–246 (2006)
70. Hoffmann, J.: FF: the fast-forward planning system. AIMag **22**(3), 57–57 (2001)
71. Rudenko, A., Palmieri, L., Herman, M., Kitani, K.M., Gavrila, D.M., Arras, K.O.: Human motion trajectory prediction: a survey. Int. J. Robot. Res. **39**(8), 895–935 (2020)
72. Magnaguagno, M.C., Pereira, R.F., More, M.D., Meneguzzi, F.: Web planner: a tool to develop classical planning domains and visualize heuristic state-space search, p. 7.
73. Lamon, E., Franco, A.D., Peternel, L., Ajoudani, A.: A capability-aware role allocation approach to industrial assembly tasks. IEEE Robot. Autom. Lett. **4**(4), 3378–3385 (2019)
74. Antakli, A., Spieldenner, T., Rubinstein, D., Spieldenner, D., Herrmann, E., Sprenger, J., Zinnikus, I.: Agent-based web supported simulation of human-robot collaboration. In: Proceedings of the 15th International Conference on Web Information Systems and Technologies, Vienna, Austria, pp. 88–99 (2019)
75. Brunner, S.G., Steinmetz, F., Belder, R., Domel, A.: RAFCON: a graphical tool for engineering complex, robotic tasks. In: IEEE/RSJ International Conference on Intelligent Robots and Systems (IROS), Daejeon, South Korea, pp. 3283–3290 (2016)
76. Jiang, Y., Zhang, S., Khandelwal, P., Stone, P.: Task planning in robotics: an empirical comparison of PDDL- and ASP-based systems. Front. Inf. Technol. Electron. Eng. **20**(3), 363–373 (2019)
77. Coleman, D.T., Sucan, I.A., Chitta, S., Correll, N.: Reducing the barrier to entry of complex robotic software: a moveit! case study (2014)
78. Niermann, D., Petzoldt, C., Dörnbach, T., Isken, M., Freitag, M.: Towards a novel software framework for the intuitive generation of process flows for multiple robotic systems. Procedia CIRP **107**, 137–142 (2022)
79. Cashmore, M., Fox, M., Long, D., Magazzeni, D., Ridder, B., Carrera, A., Palomeras, N., Hurtos, N., Carreras, M.: ROSPlan: planning in the robot operating system. In: Proceedings of the International Conference on Automated Planning and Scheduling, vol. 25, pp. 333–341 (2015)
80. Sucan, I.A., Moll, M., Kavraki, L.E.: The open motion planning library. IEEE Robot. Automat. Mag. **19**(4), 72–82 (2012)

81. Köster, N., Wrede, S., Cimiano, P.: An ontology for modelling human machine interaction in smart environments. In: Proceedings of SAI Intelligent Systems Conference (IntelliSys) 2016, pp. 338–350. Cham (2018)
82. Tuli, T.B., Kohl, L., Chala, S.A., Manns, M., Ansari, F.: Knowledge-based digital twin for predicting interactions in human-robot collaboration. In: 26th IEEE International Conference on Emerging Technologies and Factory Automation (ETFA), pp. 1–8 (2021)
83. Bitzer, S., Vijayakumar, S.: Latent spaces for dynamic movement primitives. In: 9th IEEE-RAS International Conference on Humanoid Robots, Paris, France, pp. 574–581 (2009)
84. Luck, K.S., Neumann, G., Berger, E., Peters, J., Amor, H.B.: Latent space policy search for robotics. In: IEEE/RSJ International Conference on Intelligent Robots and Systems, Chicago, IL, USA, pp. 1434–1440 (2014)
85. Tuli, T.B., Henkel, M., Manns, M.: Latent Space based collaborative motion modeling from motion capture data for human robot collaboration. Procedia CIRP **107**, 1180–1185 (2022)
86. Tuli, T.B., Manns, M.: Comparison of AI-based task planning approaches for simulating human-robot collaboration. In: Towards Sustainable Customization: Bridging Smart Products and Manufacturing Systems, pp. 158–165. Cham (2022)
87. DelPreto, J., Rus, D.: Sharing the load: human-robot team lifting using muscle activity. In: International Conference on Robotics and Automation (ICRA), pp. 7906–7912 (2019)
88. Amirshirzad, N., Kumru, A., Oztop, E.: Human adaptation to human-robot shared control. IEEE Trans. Human Mach. Syst. **49**(2), 126–136 (2019)
89. Jain, S., Argall, B.: Probabilistic human intent recognition for shared autonomy in assistive robotics. J. Hum.-Robot. Interact. **9**(1), 1–23 (2020)
90. Tuli, T.B., Manns, M., Jonek, M.: Understanding shared autonomy of collaborative humans using motion capture system for simulating team assembly. In: Towards Sustainable Customization: Bridging Smart Products and Manufacturing Systems, pp. 527–534. Cham (2022)
91. Antakli, A., Spieldenner, T., Köster, M., Groß, J., Herrmann, E., Rubinstein, D., Spieldenner, D., Zinnikus, I.: Optimized coordination and simulation for industrial human robot collaborations. In: Web Information Systems and Technologies, pp. 44–68. Cham (2020)
92. Gaisbauer, F.: Development of a Modular Digital Human Simulation Framework for Planning of Manual Assembly Processes. University of Ulm (2022)
93. Fritzsche, L., Galibarov, P.E., Gärtner, C., Bornmann, J., Damsgaard, M., Wall, R., Schirrmeister, B., Gonzalez-Vargas, J., Pucci, D., Maurice, P., Ivaldi, S., Babič, J.: Assessing the efficiency of exoskeletons in physical strain reduction by biomechanical simulation with AnyBody Modeling System. Wearable Technologies **2**, e6 (2021)
94. Kong, Y.-K., Choi, K.-H., Cho, M.-U., Kim, S.-Y., Kim, M.-J., Shim, J.-W., Park, S.-S., Kim, K.-R., Seo, M.-T., Chae, H.-S., Shim, H.-H.: Ergonomic assessment of a lower-limb exoskeleton through electromyography and anybody modeling system. Int. J. Environ. Res. Public Health **19**(13), 8088 (2022)
95. Tröster, M., Budde, S., Maufroy, C., Andersen, M.S., Rasmussen, J., Schneider, U., Bauernhansl, T.: Biomechanical analysis of stoop and free-style squat lifting and lowering with a generic back-support exoskeleton model. Int. J. Environ. Res. Public Health **19**(15), 9040 (2022)
96. Manns, M., Mengel, S., Mauer, M.: Experimental effort of data driven human motion simulation in automotive assembly. Procedia CIRP **44**, 114–119 (2016)
97. Wirsching, H.-J.: Chapter 4—Human Solutions RAMSIS. In: Scataglini, S., Paul, G. (eds.) DHM and Posturography, pp. 49–55. Academic Press (2019)
98. Schlick, C., Bruder, R., Luczak, H.: Arbeitswissenschaft. Springer Berlin Heidelberg, Berlin, Heidelberg (2018)
99. Lavalle, S.M.: Rapidly-exploring random trees: a new tool for path planning. Iowa State University, USA (1998)
100. Directorate-General for Research and Innovation (European Commission), Breque, M., De Nul, L., Petridis, A.: Industry 5.0: towards a sustainable, human centric and resilient European industry. Publications Office of the European Union, LU (2021)

101. ElMaraghy, H., Monostori, L., Schuh, G., ElMaraghy, W.: Evolution and future of manufacturing systems. CIRP Ann. **70**(2), 635–658 (2021)
102. Gartner identifies five technologies that will transform the digital future of enterprises. Gartner. [Online]. Available: https://www.gartner.com/en/newsroom/press-releases/2023-09-13-gartner-identifies-five-technologies-that-will-transform-the-digital-future-of-enterprises. Accessed 26 Aug 2024
103. R. and M. ltd: Digital Human Global Market Report 2024—Research and Markets. [Online]. Available: https://www.researchandmarkets.com/reports/5980619/digital-human-global-market-report. Accessed 26 Aug 2024
104. Smids, J., Nyholm, S., Berkers, H.: Robots in the Workplace: a threat to—or opportunity for—meaningful work? Philos. Technol. **33**(3), 503–522 (2020)

Occupational Wearable Robotics

Jesús Ortiz, Luigi Monica, Sara Anastasi, and Darwin G. Caldwell

Abstract Injuries and occupational diseases represent an important economic factor in different industrial sectors. In Italy alone, there have been over 60000 confirmed cases of occupational diseases in 2022, where musculoskeletal disorders (MSD) represent 70% of the total. This fact highlights the importance of protecting workers against physical efforts. While current ergonomic tools can mitigate most of the risk factors, there is still a non-negligible number of tasks that cannot be completely compensated for. This is due to the high complexity and costs of the workplace reorganization, or simply strict requirements of the target task. In some of these cases, exoskeletons are a viable solution. This chapter introduces different kinds of existing exoskeletons, with particular focus on occupational exoskeletons, considering commercial systems already deployed in industry, and some of the most promising technologies which will arrive in the next years. The deployment of occupational exoskeletons is conditioned by the regulatory frameworks and by the required reorganization of the workplace. Both concepts and the latest developments are discussed in this chapter, including their integration into Industry 4.0. Finally, this chapter describes the most common evaluation methods and discusses current efforts toward standardized benchmarking methods which will allow evaluating the benefit of using exoskeletons in real applications.

Keywords Exoskeletons · Wearable robots · Ergonomics · Regulations

1 Introduction

MSDs, including back pain, shoulder tendonitis, and carpal tunnel syndrome, are the leading cause of work-related injuries and absenteeism, costing employers billions of dollars annually. NIOSH estimates that WMSDs cost US employers more than

J. Ortiz (✉) · D. G. Caldwell
Advanced Robotics, Istituto Italiano di Tecnologia, Genova, Italy
e-mail: jesus.ortiz@iit.it

L. Monica · S. Anastasi
Department of Technological Innovation and Safety Equipment, INAIL, Rome, Italy

G. Barresi and S. Wischniewski (eds.), *Ergonomics in Robotics: Advances and Innovations*, Springer Tracts in Advanced Robotics 162,
https://doi.org/10.1007/978-3-032-19374-2_12

\$100 billion annually in direct medical expenses, lost productivity, and workers' compensation costs [1]. According to INAIL Open Data [2], in Italy there have been over 60000 confirmed cases of occupational diseases in 2022, where MSDs represent 68% of the total. Traditional ergonomic interventions, such as workstation modifications, job rotation, and lifting techniques, have proven effective in reducing MSD risk. However, these methods may not be sufficient to address ergonomic hazards associated with the most physically demanding tasks or workplaces with dynamic and unpredictable environments.

Occupational exoskeletons are wearable robotic devices that enhance human movement and capabilities. They can provide assistance to workers by reducing the load on their joints, improving postures, and enabling the execution of tasks that would otherwise be hazardous or impossible [3]. The integration of exoskeletons into the workplace requires a holistic approach to ergonomics. Exoskeletons can directly improve ergonomics in the workplace by reducing physical strain and facilitating safer postures. However, exoskeletons must be designed with the overall ergonomics of the workplace in mind. This includes ensuring that workplaces are designed to accommodate the capabilities and limitations of exoskeletons, such as the clearance required for the movements of the worker wearing the exoskeleton. While exoskeletons hold immense potential for improving ergonomics, their widespread adoption is contingent on several factors, including user acceptance, regulatory frameworks, and economic considerations [4]. User acceptance is crucial, as workers must find exoskeletons comfortable, convenient, and non-disruptive to their tasks and workflows. As exoskeletons move closer to commercialisation, regulatory bodies worldwide are establishing guidelines for their safety and performance. These regulations aim to ensure that exoskeletons meet minimum safety standards, including requirements for structural integrity, power system safety, and user interaction. Testing and validation are essential to demonstrate that exoskeletons meet these standards and deliver the intended ergonomic benefits without compromising worker safety. Additionally, the cost of exoskeletons remains a barrier to widespread adoption, particularly for small and medium-sized enterprises.

Passive exoskeletons, which rely on mechanical linkages or springs to provide assistance, have already entered the market and are being tested in various industries [5]. Active exoskeletons, which incorporate motors and sensors to provide dynamic and adaptive assistance, are still predominantly in the research stage. Active exoskeletons can provide dynamic assistance and real-time support tailored to specific tasks and user needs, offering a higher level of customisation and adaptability [6]. However, active exoskeletons are more complex and expensive to develop and manufacture, which has slowed their progress compared to passive exoskeletons.

This book chapter presents in Sect. 2 a general taxonomy of the exoskeletons. This classification is not limited to industrial scenarios, but it includes any kind of application. Section 3 focuses on occupational exoskeletons, explaining why current commercial exoskeletons correspond with certain design criteria. This section includes a description of the occupational exoskeletons from the point of view of the Industry 4.0. These two sections do not intend to be a detailed review of the state-of-the-art, but to present the current landscape with some relevant examples.

Section 4 describes the design process of occupational exoskeletons and the different testing and validation phases. Section 5 gives a general overview of the regulatory framework from the point of view of the exoskeleton manufacturers and users. To conclude, Sect. 6 wraps up the chapter with a short discussion.

2 Exoskeleton Taxonomy

The development of exoskeletons is a very active field at the moment, not only for industrial applications, but also in many other fields. Consequently, a wide variety of types and technologies exist. In this section, we will classify the existing exoskeletons that we can find in research and in the market according to different criteria, as shown in Fig. 1: (i) field of application, (ii) technology level, (iii) morphology, (iv) body area, (v) actuation system and (vi) structure.

This section does not intend to be a review of all the research and commercial exoskeletons and their classification, but just to give an overview with the most relevant examples up to date.

2.1 Field of Application

The design and development of an exoskeleton is very dependent on the requirements of the application. For this reason, the field of application is a very critical aspect that significantly affects the characteristics of the device. While first exoskeleton concepts were developed for industrial applications [7], it was in the military sector where the first functional prototypes were developed [8]. The medical field is also benefiting from these devices and has been traditionally the area with a wider application of exoskeletons [9]. However, exoskeletons for industrial applications, also named *occupational exoskeletons*, is currently the field with the highest development rate [10, 11]. This is due to the benefits to the health of the workers and their potential positive impact in the economy [12]. In sports, exoskeletons can provide different types of assistance too, but most of the devices are still in very early stages of application or they are not fully validated. Furthermore, the potential application fields of the exoskeletons is growing, and we can find a few other examples such as space exploration.

Military exoskeletons The usage of exoskeletons in military applications can provide benefits in two different kinds of scenarios. The first one is focused on tactical advantage, providing the soldiers with higher capabilities in terms of endurance and resilience. This allows soldiers to carry higher loads or run across longer distances. The best example is the Harvard Exosuit [13], a lower limb soft exoskeleton for walking/running assistance.

The second scenario is in logistics applications. In this case, the target is to provide superior strength to carry out logistic tasks, such as carrying or manipulating heavy

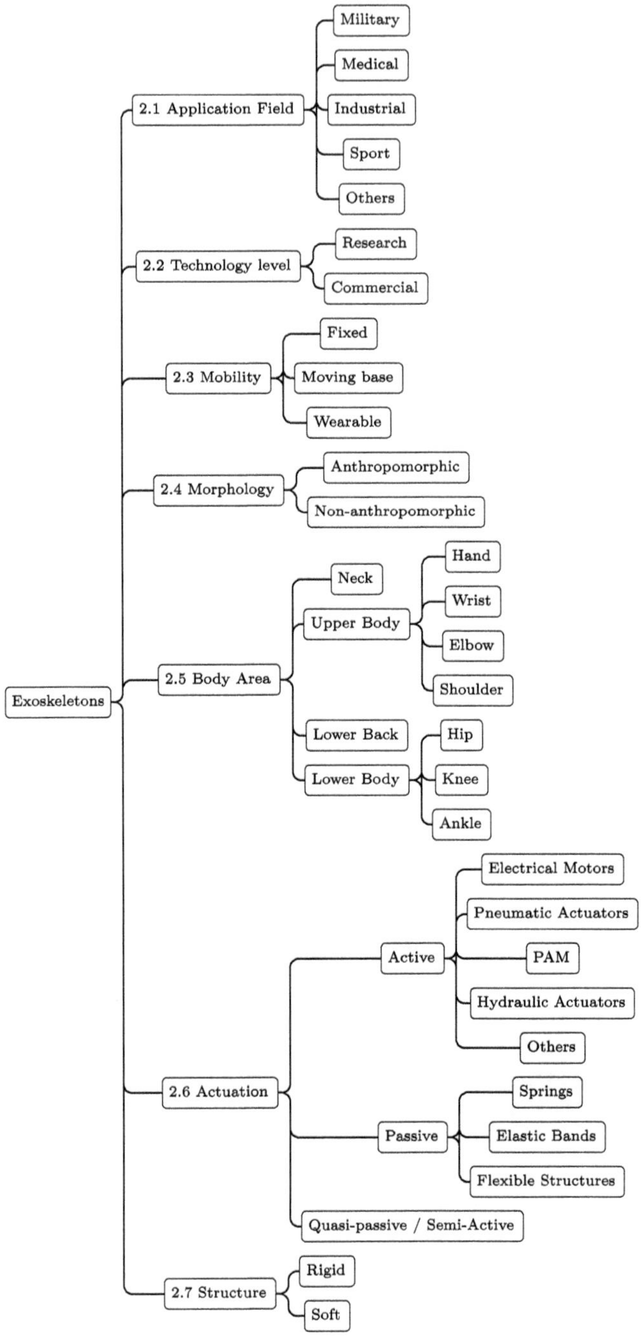

Fig. 1 Exoskeleton taxonomy

objects. These kinds of exoskeletons are full-body wearable active devices that unload the external load to the ground, where the wearer of the exoskeleton is fully carried by the exoskeleton. The best example of these kinds of device is Guardian XOS [14].

While the ergonomics is not the main focus of military exoskeletons, the comfort and acceptability of these devices play an important role in the maximisation of their benefits. For example, a lower-limb military exoskeleton that can be worn comfortably for longer periods of time, could potentially provide more tactical advantages. Secondly, the ergonomics of these devices has to be taken into account from the point of view of the side effects. The use of a wearable device in these applications should not create new ergonomic issues, which in turn would limit their use.

Medical exoskeletons Exoskeletons can be used in medical scenarios in different ways, but in this section we will refer to their use for the support of patients in rehabilitation scenarios. The usage of exoskeleton for the support of healthcare workers is included into the category of *occupational exoskeletons*, since they cope with ergonomics in the workplace and not medical care.

The main purpose of the rehabilitation exoskeletons is to improve the recovery of the patients. First examples were designed to substitute the therapy manually offered by a physiotherapist. Under this scenario, it was not required a mobile and untethered exoskeleton, i.e. the exoskeleton is fixed to the ground [15, 16]. These kinds of devices allow to perform a high variety of exercises and a close monitoring of the rehabilitation process. Their efficiency has been already proven in several publications and several scenarios [17]. There are efforts to improve the mobility of the rehabilitation exoskeletons combining them with moving bases, such as the Float exoskeleton [18].

A second group of medical exoskeletons are focused on the support of fully paralysed patients for daily life activities, such as walking [19–21]. While these devices are costly and not widely spread nowadays, they offer a matchless solution for subjects with paraplegia or hemiplegia, restoring to some degree the capacity of walking.

Finally, subjects with a low to moderate level of disability are also able to benefit from the support provided by exoskeletons. Under this group, soft exoskeletons are showing a good balance between performance and weight/bulkiness [22, 23].

Similarly to the military exoskeletons, ergonomics plays a secondary role in medical exoskeletons. The ergonomics of the exoskeleton is important for the comfort and acceptability of the devices, enabling longer training sessions without incurring into secondary issues caused by the exoskeleton itself.

Occupational exoskeletons As explained in detail in the introduction, occupational exoskeletons are focused on the reduction of the efforts exerted by the operators during manual handling work, with the objective to reduce the number of musculoskeletal disorders in industry and consequently their social and economical impact. Most of the existing devices provide from low to medium level of assistance, and they do not usually overcome the strength and capabilities of the operators. This complies with the objective of the reduction of the physical efforts, and at the same time do not pose any safety issue to the operators since the wearer is always able to compensate the forces produced by the exoskeleton. On the other hand, the exoskeleton has to compensate for the own weight of the operator plus the external load, which should stay always within the ergonomic safety limits.

Sport exoskeletons We can find a few examples of exoskeletons for sports. They can be used in two main scenarios: (i) offer support, or (ii) provide training to the wearer. In the first case, the exoskeleton supports the person wearing the exoskeleton in some specific way. For example, the knee exoskeleton [24] supports the lower limbs of the person during ski activities. Note that the use of these devices is still very scarce and it is difficult to find systematic studies about the efficiency of these devices. Furthermore, they are not yet regulated in professional sport competitions.

The second kind of devices, would provide training to the wearer, in the same way as wearing weights in the wrists, waist or ankles to improve the strength and speed. In this case, an exoskeleton can provide a focused support with monitoring capabilities [25].

Others There are other fields where the use of exoskeletons has been barely explored. For example, in space exploration, exoskeletons could provide support and assistance in many different scenarios. We can find an example of the use of a soft exoskeleton in microgravity conditions to provide support to the astronauts during long periods in the international space station [26]. This study is a feasibility study of the current technologies for this application, and a first step towards the development of a specialised device. However, there are plenty of unexplored scenarios not only in space applications, but also in other scenarios, where exoskeletons could provide assistance.

2.2 Technology Level

We can divide the exoskeletons into two main categories regarding their technological level: (i) research devices, and (ii) commercial products. Research devices are prototypes at different stages of development going from proof of concept of novel technologies to validated devices in industrial scenarios. While commercial products are certified devices that are sold and used for different applications. The availability of these commercial devices is in most of the cases limited to some geographic areas (Europe, America, Japan, etc.). Only a few multi-national companies, usually coming from the rehabilitation and prosthesis sector, are producing these devices, and most of the current existing companies are relatively small spin-off companies coming from research groups [27]. As a rapidly evolving technology, research is still very active in this area and novel concepts and technologies are still appearing.

2.3 Mobility

Exoskeletons can be classified according to their mobility as: (i) fixed base, (ii) moving base, and (iii) wearable. Rehabilitation exoskeletons commonly follow this approach, since they substitute rehabilitation equipment in a clinical setup, and thus do not require mobility [15, 16]. The second group, *moving base*, represents an improvement on the first group since they offer some degree of mobility. In this case, the exoskeleton is mounted on a moving base, usually passive, that follows the patient [18]. This allows the subjects to perform a higher variety of daily life activities. The last category are the fully wearable devices that are mounted on a person who can move around freely without restrictions.

2.4 Morphology

According to the similarity with the human body, exoskeletons can be classified as: (i) anthropomorphic, and (ii) non-anthropomorphic [28]. The first ones are the ones that approximately follow the morphology of the human body and are closely attached to the body. The joints of the exoskeleton are placed approximately aligned with the ones of the body of the wearer. This requires the inclusion of several mechanical adjustments that allow to configure the device for a specific user. Given the inherent difficulty of achieving perfect alignment, anthropomorphic exoskeletons usually include compensation mechanisms in order to avoid residual forces that can create discomfort and/or limit the range of motion.

On the other hand, non-anthropomorphic exoskeletons do not follow the morphology of the human body and actuate directly the body part through an end-effector. The exoskeleton connects a fixed base or a body part directly with the target joint or limb. These kinds of exoskeletons do not present misalignment issues, since the exoskeleton does not follow the human joints, but still could present range of motion limitations.

2.5 Body Area

Exoskeletons can be designed to assist the full body, certain body areas (such as the lower back, or hand), a limb (upper or lower) or a single joint (shoulder, elbow, knee, etc.) [28]. This aspect mostly depends on the target application. Rehabilitation exoskeletons always target a specific problem, and they mostly target the upper body for arm rehabilitation or lower body for gait training. Most of the commercial industrial exoskeletons target the lower-back, shoulder or knees. Military exoskeletons target the lower body for the assistance of carrying, or the full body to heavy object manipulation, which requires not only assistance to the upper body, but it requires to unload the weight to the ground too [29].

2.6 Actuation System

Exoskeletons can be divided into two main groups regarding the actuation system: (i) active, and (ii) passive. In the first group, the assistance is provided by an active element, such as a motor, while in the second group the assistant is provided by a passive element, such as a spring. Active exoskeletons are usually heavier, bulkier and more complex than the passive exoskeletons. Apart from the size and weight of the actuator, they require some extra elements such as electronics and batteries. On the other hand, passive exoskeletons do not require anything else than the passive element itself, except for some manual adjustment mechanisms. While it is difficult to overcome the deficits of the active exoskeletons, they have some potential advantages with respect to the passive ones, mainly related to their versatility due to their control system. They can potentially adapt dynamically to different tasks and conditions [30] without requiring manual adjustment. They can even support tasks that passive exoskeletons cannot [31]. Generally speaking, passive exoskeletons offer a better level of support and assistance in static tasks, while active exoskeletons have superior capabilities in dynamic tasks [6].

Active exoskeletons can make use of different kinds of actuation mechanisms, including but not limited to electrical actuators, hydraulics and pneumatics (including artificial muscles). Other actuation systems have been explored, such as Shape Memory Alloys (SMA) or electroactive polymers (EAP), but they are used mostly in proof of concept systems.

Passive exoskeletons commonly use coil springs, elastic rubber bands or carbon fiber rods. However, many other structures and materials have been explored.

Active and passive elements can be combined in other ways to improve the system efficiency, such as parallel elastic actuators (PEA) [32] or to improve the interaction with the user, such as serial elastic actuators (SEA) [33].

There is another group of exoskeletons that combine some characteristics from the passive and active exoskeletons: quasi-passive or semi-active exoskeletons. In these systems, the assistance is provided by a passive element, but it is controlled by an active system. The active system is low-power, and it just modulates or activates/deactivates the actuation [34, 35].

2.7 Structure

The structure of the exoskeleton is the part of the system in charge of transmitting the assistive forces to the wearer. We can divide the exoskeletons into two main groups according to this characteristic: (i) rigid exoskeletons, and (ii) soft exoskeletons. Rigid exoskeletons, also called hard exoskeletons, have a rigid structure and benefit from traditional sensing, actuation and control technologies used in robotics. These devices have been the first ones being developed, and can provide high levels of support.

Soft exoskeletons or exosuits do not have a rigid structure to transmit the forces to the wearer, but are based on textiles. They might have rigid components, but they are not used for the transmission of the assistive forces (and they are not usually in direct contact with the person). These devices can use traditional actuation systems with a remote force/torque transmission system (such as a cable driven mechanism), pneumatic actuation (such as pneumatic muscles) or other technologies. Since the structure of the exoskeleton cannot transmit forces in all directions, the assistance level is limited with respect to rigid exoskeletons.

There is an intermediate category which includes some characteristics of the two previous classes: flexible exoskeletons. These devices use a flexible structure which is able to transmit higher forces, but at the same time they can adapt to the shape and movements of the wearer to a certain degree. An example of this is FleXo [36], a fully flexible exoskeleton for the support of the lower back. There are other exoskeletons that make use of flexible elements, but in a limited way [37].

3 Occupational Exoskeletons

As with most of the robotic technologies, there is a gap between the devices on the market and the devices in research stages. The number of commercial devices is growing at an unprecedented rate, showing an increasing interest from industry and market opportunities. We could expect that the gap between the research and the market becomes closer in the next years.

Most of the commercial occupational exoskeletons commonly used currently in industry are (following the taxonomy described in Sect. 2) wearable, anthropomorphic, passive and rigid. The main body target areas are: (i) shoulder, and (ii) lower-back. The main reason for this distribution is the following:

Wearable Having a fixed or moving base exoskeleton would not be generally suitable for use in industry. This would limit the mobility of the workers and will show a low acceptance rate. For this reason, wearable exoskeletons are the first choice.

Anthropomorphic Anthropomorphic exoskeletons require less working space since they are closer to the human body. The bulkiness of an exoskeleton is an important factor, thus this kind of device is the preferred choice.

Passive The preference of passive exoskeletons is conditioned by several factors, for example low-cost, low-weight, simplicity and clearer regulations.

Rigid Rigid structures are usually easier to understand and characterise than soft structures, for example the transmission of the assistive forces or the connection to the actuation system. For this reason, current commercial exoskeletons are mostly based on rigid structures. However, there is an important coming trend of passive soft exoskeletons too.

With respect to the body area, lower-back exoskeletons have greater chances to be implemented on a wide scale due to the incidence of lower-back pain among the workers, accounting with almost the 50% of the Work Related Musculoskeletal Disorders [38]. Some examples of passive lower-back exoskeletons in the market are

Laevo [39], IX BACK [40] and PaeXo Back [41]. One of the few examples of active lower-back exoskeletons is Cray X [42]. Some soft passive lower-back exoskeletons can be found in the market, such as Apex 2 [43] and LiftSuit 2 [44].

The second group, shoulder exoskeletons, are mostly encouraged by the automotive industry. The main purpose of shoulder exoskeletons is to support overhead work, and this is a very common scenario in the automotive industry. Other ergonomic tools are unfeasible or ineffective in these scenarios, but shoulder exoskeletons can provide a solution in most of the cases [45]. Most notable examples of shoulder exoskeletons are IX SHOULDER AIR [46], Paexo Shoulder [47], Mate Xt 4.0 [48], Skelex 360 XFR [49] and Agadexo Shoulder [50].

3.1 Occupational Exoskeletons and Industry 4.0

Industry 4.0 is a broader term for the current industrial revolution which includes several robotic technologies, including occupational exoskeletons. To understand how exoskeletons play an important role in the factory of the future, we can refer to [51] and the four Design Principles: (i) technical assistance (virtual and physical), (ii) interconnection (collaborations, standards and security), (iii) information transparency (data analysis and information provision), and (iv) decentralised decisions. The way exoskeletons can address all these four Design Principles is explained as follows:

Technical assistance The main scope of an industrial exoskeleton is to provide physical assistance to the operator, reducing the muscular efforts [52–54] and the risk of injuries. As a side effect, the operators' physical stress would be reduced and the performance would be maintained more easily over time, increasing the quality of the work.

At the same time, a wearable device can be equipped with augmented capabilities in order to provide the operator with high-level information. For example, giving feedback to the operator about the process and performance, or providing global or semi-global information (i.e. not only information about the operator himself, but about other co-workers too). This exchange of information can be done using state-of-the-art techniques of virtual reality, augmented reality and haptics [55, 56].

Interconnection Exoskeletons can be easily equipped with an onboard computer with wireless connection capabilities. This allows the communication between each exoskeleton with a central computer, creating a network of exoskeletons. Interconnected exoskeletons can provide important advantages from the planning point of view, since it allows to monitor each *exo-worker* (exoskeleton powered worker).

Information transparency In the same way, the system status can be used to monitor single units and create a virtual replica of the system. This allows the management of a network of *exo-workers* in a very intuitive way.

Decentralised decisions This principle can be implemented on exoskeleton devices, for example grouping *exo-workers* in operative units with localised targets and independent organisation.

3.2 Implementation Challenges

Generally speaking, several challenges have been identified for the implementation of the Industry 4.0 [57]. Some of them are of special importance in the case of exoskeletons:

System malfunction The malfunction of an exoskeleton could cause the outage of a worker, causing loss of production. More importantly, a failure of an exoskeleton should not cause any injury to the operator by design. The robustness is consequently a very important requirement in the exoskeleton specifications [58].

Usability The introduction of new robotic platforms in the industrial environment requires the workers to adapt to the new systems. Special efforts should be put in the exoskeleton design in order to make them intuitive to control, limiting the required training and familiarisation time, and increasing the usability.

Unclear legal issues and data security Since one of the previously cited advantages of the exoskeletons is the worker monitoring, the use of the collected data is a critical aspect that needs to be managed appropriately. The data has to be used exclusively to improve the quality of work of the operators and industry, but in no case to discriminate the operators by their performance.

Lack of regulation, standard and forms of certifications The regulations around the use of exoskeletons in industry are not yet clearly defined, and several working groups are progressing towards a framework based on previous documents, for example on industrial robots [59], personal care robots [60], collaborative robots [61] or rehabilitation robots [62].

Unclear economic benefits or excessive investment While the cost of an exoskeleton is estimated to be equivalent to any basic robotic installation, the transition from well established robotic equipment to state-of-the-art exoskeletons is still a major problem. Recent research [52–54] have demonstrated the benefits for the workers, which translates into economic benefits due to the reduction of refunds to workers because of injuries at work and occupational diseases. However, since the cases of real implementations of exoskeleton systems in industry are still scarce, it is difficult to find a solid justification for the change.

General reluctance to change by stakeholders (user acceptance) Even if the potential benefits of the exoskeleton systems were clear the stakeholders are not willing to do the change to these new technologies. This is mainly because of ignorance of last developments on exoskeletons, or simply mistrust in new technologies [63].

Loss of jobs The broad introduction of robotics in industry could be seen as a potential source of loss of jobs. While this aspect is not fully clear yet, it seems plausible that there will be a shift in the workers' profile, towards more qualified operators. This is one of the critical issues in the Industry 4.0. However, in the case of exoskeletons, this is a possible positive outcome. The exoskeleton combines the flexibility of the human worker with the resistance of a robotic platform. This combination is competitive in many tasks against a traditional worker or an automated system. Consequently, an exoskeleton would not cause loss of jobs, and potentially would allow minorities to access new jobs.

4 Design and Development: From the Concept to the Validation

The definition of the user and industrial requirements are the basis and the starting point of the design process of an industrial exoskeleton. During the system design and development cycle, the industry and user feedback are critical for the proper alignment between the system characteristics and the final implementation of the device. In this process, we can differentiate between the different steps, which will be further described in the next sections:

Scenarios definition In Sect. 4.1 we analyse a number of generic tasks which could benefit from the use of an exoskeleton. It is however important to select some representative real scenarios to focus the definition of the system specifications and validation setup. The selection of the scenarios has to be done in a way that other similar tasks could benefit from the same requirements, but specific enough to be able to restrict all the system characteristics. Some scenarios could add specific requirements not strictly related to the function of the system but with environment conditions, such as presence of dust, humidity, encumbrance restrictions, etc.

User needs Sect. 4.2 describes general user needs, mainly linked with the usability and wearability of the system. The combination of the definition of the scenarios and the user needs is the basis for the definition of the system specifications. Section 4.3 gives some indications on important aspects to take into account while defining the specifications of an exoskeleton.

Design and Development During this phase, the system is designed in tandem with a continuous ergonomic assessment. Rather than a linear design-development process, early prototypes (potentially created using rapid prototyping techniques) are tested on subjects to assess comfort, usability, wearability, and other ergonomic factors. This iterative approach allows for the progressive refinement of prototypes, ensuring optimal integration with user needs. Section 4.4 outlines the structure of laboratory testing conducted during this stage.

Validation Once the prototypes achieve satisfactory performance and robustness, they undergo real-world testing. In this stage, the system is validated in controlled scenarios, similar to those defined in the initial phase. Compared to laboratory testing during system development, the number of monitored variables is reduced. Section 4.5 outlines the structure of this testing phase.

Deployment If the prototypes demonstrate significant benefits for the target tasks, from both user and industrial perspectives, they can be gradually deployed in real-world applications. Although there are few examples of the implementation of exoskeletons in industry, their deployment would not differ substantially from other equipment, aside from the increased level of operator involvement and interaction.

4.1 Scenarios Definition

The Health, Safety, and Environment (HSE) Manager typically identifies critical tasks, including those that could potentially benefit from exoskeleton use. The technical report [64] is a fundamental tool for workplace ergonomic analysis. This procedure involves a quick assessment to categorize tasks into three levels: a) no action required, b) critical (requiring immediate workplace or task review), and c) further analysis needed. Tasks are divided into four main categories, each assessed according to specific standards. A more detailed analysis is necessary if the assessed task meets any of the following criteria.
Lifting and carrying [65]: Handling of objects with a mass of 3 kg or more.
Pulling and pushing [66]: whole-body force exertions (i.e. while standing/walking); actions performed by one person; forces applied by two hands; forces used to move or restrain an object; forces applied in a smooth and controlled way; forces applied without the use of external support(s); forces applied on objects located in front of the operator; forces applied in an upright position (not sitting).
Handling of low loads at high frequency [67]: Repetitive work tasks (total duration of at least 1 h per shift) involving the manual handling of low loads at high frequency.
Static working postures [68]: Static or awkward posture of neck, trunk, arms or legs for more than 4 consecutive seconds and repeated during a significant part of the work.

Furthermore, it is important to define other kinds of information about the tasks related to the surrounding and environment conditions, such as working temperature, existence of narrow spaces, etc. These are critical factors in the implementation of an exoskeleton device in a real scenario.

4.2 User Needs

To ensure the successful adoption of exoskeletons in industrial settings, it is crucial to understand the needs and expectations of primary users, the individuals who will directly interact with the device. By identifying and addressing these needs, we can design exoskeletons that are both effective and user-friendly.
Key User Needs The following key user needs have been identified as critical for the successful implementation of exoskeletons:

- **Freedom of Movement** The system should minimally interfere with the operator's movements, both during operation and general wear. Any hindrance should be justified by significant benefits [69].
- **Comfort** The system should be as unobtrusive as possible, minimizing weight and perceived pressure. While increased assistance can improve user perception, it also leads to greater force exertion on the body. Balancing assistance level with force exertion is essential, often achieved through improved body attachments.

- **Wearability** The system should be easy to don, doff, and adjust independently by the operator.
- **Usability** Intuitive operation is key for widespread adoption. The system should seamlessly adapt to the operator's movements, providing appropriate assistance at the right time. Minimal user intervention, such as mode or assistance level changes, should suffice for regular use.

Secondary Considerations Several secondary factors can influence user acceptance. Aesthetic appeal can boost user confidence, while noise levels can either deter or enhance perception.

4.3 Specifications

A comprehensive set of specifications is crucial for the successful design and development of an exoskeleton. It is important not only to correctly categorize physical variables such as range of motion, speed, and torque, but also to carefully consider how these variables are expressed. For instance, simply defining the maximum torque and speed of a human joint during a specific task is insufficient. Understanding the joint's movement over time and the interdependencies between variables is essential to accurately characterize its behavior.

To ensure a well-defined exoskeleton system, the specifications for each target task should include the following information:

Task definition To gain a comprehensive understanding of the task, a detailed description, including visual aids like pictures and videos, is crucial. Secondary information, such as the number of repetitions per shift and the frequency of specific actions, is equally important for accurate task analysis.

Task context Any interactions or interruptions during the task should be clearly identified. For example, if the operator needs to perform secondary tasks between repetitions of the primary task, the exoskeleton should not hinder their ability to do so.

Joint classification Based on the task analysis and operator behavior, it is essential to identify the specific human joints that require assistance and the desired directions of motion. This includes defining the relevant axes of motion, such as flexion, extension, abduction, adduction, and rotation. The use of relative angles (angles between body segments) or absolute angles (angles relative to gravity) should be carefully considered. In the context of industrial exoskeletons, most assistance is typically provided against gravity, making absolute angles the primary focus. However, in certain cases, translating to relative joint angles may be feasible.

Joint movement profile The specifications should focus on the assisted joints, including their movement and torque profiles during the task. From this information, other essential parameters, such as speed, acceleration, power, and energy consumption, can be derived. Speed and acceleration are crucial for determining the maximum actuation requirements, while power indicates the necessary size of the actuation sys-

tem. Energy consumption, on the other hand, provides insights into potential energy recovery and harvesting strategies.
In some cases, visualizing joint information in torque-speed or torque-angle plots can provide valuable insights. A more systematic approach involves using the joint movement profile to optimize the actuation system, as demonstrated by [70].

The environmental conditions, such as dust, humidity, obstacles, surrounding spaces, and floor characteristics, must be considered in the specifications. This information is essential for defining other system characteristics, such as overall dimensions and protection measures.

From the user's perspective, the primary specifications relate to wearability and usability:

Maximum exerted force/pressure The user's body cannot withstand the same level of force or pressure in all areas. It is crucial to consider the maximum interaction forces between the exoskeleton and the user, taking into account the frequency and duration of the task.

System weight The distribution of the system's weight across the user's body is a critical factor. The connections between the exoskeleton's components and the internal force transfer mechanisms will influence the options for weight distribution.

Weight distribution For lightweight exoskeletons, weight distribution is less of a concern, as the system's mass is typically a small fraction of the user's body weight. However, for heavier systems, careful consideration must be given to weight distribution to avoid shifting the user's center of gravity and causing instability.

System dimensions From a user perspective, the exoskeleton's dimensions must be carefully considered to avoid accidental contact during normal operation. This often leads to positioning the bulkier components on the user's back. However, this approach can increase system complexity and weight, necessitating a trade-off between size and performance.

The specifications should include the type of energy source and its capacity. In scenarios where the operator remains relatively static, an external power supply can be considered to reduce system weight and improve autonomy. However, this approach limits the operator's mobility. Therefore, an on-board energy storage system, such as batteries, is often the preferred solution. Battery charging or replacement should be performed without disrupting the operator's workflow.

4.4 Laboratory Testing

This phase of testing is ideal for monitoring critical system and user variables within a controlled laboratory environment. Three primary types of testing can be distinguished:

Sub-system testing Individual components of the system can be tested independently on a test bench, without user involvement. This approach is common in robotic platforms and allows for focused evaluation of specific functionalities.

Functionality testing This type of testing involves user interaction to verify specific system functionalities. While it may not fully replicate the target task, it can involve simplified or synthetic tasks. For example, testing a kinematic chain might involve basic movements to assess the system's ability to follow the user's motions. This approach is similar to the methodology employed by [71] for assistive technology.
Task testing In this final stage, the entire system is evaluated during simulated industrial tasks defined in the specifications. The primary goal is to assess the overall system performance and potential user benefits. This involves collecting various biological measurements, including body motion, interaction forces, and muscular effort. By combining body motion and interaction force data with accurate biomechanical models, joint torques and muscular efforts can be calculated. Additionally, direct measurement of muscular effort using electromyography (EMG) is possible. Other relevant variables, such as contact pressure [54] and energy expenditure [72], can also be collected.
Qualitative User Feedback To gain valuable insights into the user experience, qualitative information can be collected through questionnaires [73–75].
User Selection For early-stage testing, users with familiarity with technology are often preferred, as they may adapt more easily to the system's interface. However, as testing progresses, it is advisable to involve users with limited experience in robotics to obtain more unbiased feedback and identify potential usability issues before industrial deployment.

4.5 Industrial Validation

Given the inherent variability of real-world environments, it is more challenging to control all system and user variables compared to laboratory testing. This phase primarily focuses on gathering subjective user feedback on the system's performance in its intended application. While similar to qualitative laboratory evaluations, this phase emphasizes the system's final use case. Additionally, it offers opportunities for measurements not feasible in laboratory settings:
End user evaluation Involving the target user group in this testing phase provides invaluable insights into the system's performance. Key aspects to evaluate include comfort, perceived level of assistance, and wearability (donning and doffing).
Interaction with the environment The interaction between the exoskeleton and the industrial environment is a crucial factor that cannot be fully replicated in a laboratory. This includes the working space, interactions with tools and machines, and potential interferences.
Integration in production Assessing the system's integration into the production process is essential. This involves evaluating potential impacts on preparation time, task timing (e.g., slowing down the operator or adding downtime for battery changes), and overall worker fatigue and productivity.

5 Regulations and Standards

5.1 Exoskeleton Manufacturers

Due to their versatility across rehabilitation, industry, and military sectors, and their diverse construction, exoskeletons lack uniform regulation. To address this, it is crucial to consider their functional design and intended purpose.

Rehabilitation exoskeletons can be classified as medical devices under European regulations (93/42/EEC). As such, they must adhere to stringent safety and performance standards. However, clinical evaluation of medical efficacy remains a challenging aspect.

Occupational exoskeletons can be categorized as technical devices that assist workers in performing their tasks. According to the European Union Machinery Directive (2006/42/EC), they are considered machinery and must comply with the relevant regulations.

It's important to note that occupational exoskeletons are not classified as Personal Protective Equipment (PPE) under Regulation (EU) 2016/425. This is because there is currently insufficient evidence to support their effectiveness in preventing musculoskeletal disorders (MSDs).

The European Union's new Machinery Regulation (EU) 2023/1230, which will replace the 2006/42/EC directive on January 20, 2027, introduces stricter requirements for human-machine interaction, particularly relevant for occupational exoskeletons.

Essential Health and Safety Requirement (EHSR) 1.3.7 specifically addresses the need to prevent contact risks and psychological stress associated with human-machine interaction. This highlights the importance of designing exoskeletons that seamlessly integrate with the user's workspace and minimize potential risks.

Currently, ISO 18646-4 provides performance criteria and test methods for lower-back support robots. While this standard is applicable to various types of lower-back support robots, it does not cover medical robots. However, the specified test methods can be adapted for medical applications.

Generally, for the active occupational exoskeleton there are useful general standards such as:

- EN ISO 10218-1:2011—Robots and robotic devices—Safety requirements for industrial robots—Part 1: Robots that specifies requirements and guidelines for the inherent safe design, protective measures and information for use of industrial robots. It describes basic hazards associated with robots and provides requirements to eliminate, or adequately reduce, the risks associated with these hazards.
- ISO 13482:2014—Robots and robotic devices—Safety requirements for personal care robots that specifies requirements and guidelines for the inherently safe design, protective measures, and information for use of personal care robots, in particular: mobile servant robot, physical assistant robot, person carrier robot.

Table 1 Standards

Category	Area	Standard
Operator's variability	Physical dimension	EN 547-3:1996+A1:2008—Safety of machinery—Human body measurements—Part 3: Anthropometric data EN 614-1:2006+A1:2009—Safety of machinery—Ergonomic design principles—Part 1: Terminology and general principles EN 894-1:1997+A1:2008—Safety of machinery—Ergonomics requirements for the design of displays and control actuators—Part 1: General principles for human interactions with displays and control actuators EN 1005-1:2001+A1:2008—Safety of machinery—Human physical performance—Part 1: Terms and definitions EN 1005-4:2005+A1:2008—Safety of machinery—Human physical performance—Part 4: Evaluation of working postures and movements in relation to machinery access into machinery EN ISO 14738:2008—Safety of machinery—Anthropometric requirements for the design of workstations at machinery EN ISO 15536-1:2008—Ergonomics—Computer manikins and body templates—Part 1: General requirements EN ISO 7250-1:2017—Basic human body measurements for technological design—Part 1: Body measurement definitions CEN ISO/TR 7250-2:2011+A1:2013—Basic human body measurements for technological design: Statistical summaries of body measurements from national populations
	Strength	EN 614-1:2006+A1:2009 EN 1005-1:2001+A1:2008 EN 1005-2:2003+A1:2008—Safety of machinery—Human physical performance—Part 2: Manual handling of machinery and component parts of machinery EN 1005-3:2002+A1:2008—Safety of machinery—Human physical performance—Part 3: Recommended force limits for machinery operation EN ISO 15536-1:2008

(continued)

- ISO/TS 15066:2016—Robots and robotic devices—Collaborative robots that specifies safety requirements for collaborative industrial robot systems and the work environment, and supplements the requirements and guidance on collaborative industrial robot operation.

Table 1 reports other useful ergonomic technical standards for the design characteristics of occupational exoskeleton.

Table 1 (continued)

Category	Area	Standard
Space for movement	Posture	EN 614-1:2006+A1:2009 EN 1005-1:2001+A1:2008 EN 1005-2:2003+A1:2008 EN 1005-3:2002+A1:2008 EN 1005-4:2005+A1:2008—Safety of machinery—Human physical performance—Part 4: Evaluation of working postures and movements in relation to machinery access into machinery EN ISO 14738:2008 EN ISO 15536-1:2008
	Dynamics	EN 614-1:2006+A1:2009 EN 1005-1:2001+A1:2008 EN 1005-4:2005+A1:2008 EN ISO 14738:2008 EN ISO 15536-1:2008
Work rate	Pace	EN 1005-3:2002+A1:2008
	Speed	EN 894-1:1997+A1:2008
Concentration	Mental operation	EN 614-1:2006+A1:2009 EN ISO 10075-1:2017—Ergonomic principles related to mental work-load—Part 1: General terms and definitions EN ISO 10075-2:2000—Ergonomic principles related to mental workload—Part 2: Design principles EN ISO 10075-3:2004—Ergonomic principles related to mental workload—Part 3: Principles and requirements concerning methods for measuring and assessing mental workload
	Cognitive performance	EN 614-1:2006+A1:2009 EN ISO 10075-1:2017 EN ISO 10075-2:2000 EN ISO 10075-3:2004
Human/machinery interface	Interaction	ISO 6385:2004—Ergonomic principles in the design of work systems EN ISO 9241-910:2011—Ergonomics of human-system interaction—Part 910: Framework for tactile and haptic interaction
	Visual	EN 614-1:2006+A1:2009 EN ISO 14738:2008
	Auditory	EN 614-1:2006+A1:2009
	Sensitivity	EN ISO 7730:2005—Ergonomics of the thermal environment—Analytical determination and interpretation of thermal comfort using calculation of the PMV and PPD indices and local thermal comfort criteria ISO 7933: 2004—Ergonomics of the thermal environment—Analytical determination and interpretation of heat stress using calculation of the predicted heat strain EN ISO 8996:2004—Ergonomics of the thermal environment—Determination of metabolic rate

5.2 Exoskeleton Users

Employers have a general duty to provide a safe and healthy working environment while minimizing potential risks during work. Workplace risk assessments, which account for all potential occupational hazards, are mandatory for all employers in Europe. The European Framework Directive (89/391/EEC) outlines specific measures to address these responsibilities. These measures include the prevention of occupational risks, the provision of information and training for workers and organizations, and the implementation of necessary safety protocols. Based on these regulations, potential risks associated with the use of exoskeletons in specific workplaces must also be considered.

The risks of using exoskeletons in working environments are varied and largely depend on their design and functionality. Active systems, for instance, may be prone to mechanical and technical defects. Malfunctions can lead to injuries, as the drive mechanisms of active exoskeletons are capable of exerting additional forces on the worker's body. Currently, it remains challenging to classify the forces exerted by on-body devices and their connection to injury risks. As a reference, the biomechanical threshold values outlined in the standards for collaborative robots (ISO/TS 15066:2016) may be considered. Exoskeletons might also heighten injury risks during slip, trip, or fall incidents. However, when upper-body exoskeletons are worn during level walking, their impact on such risks is currently assessed as low.

Nevertheless, depending on their construction and weight, exoskeletons could restrict natural freedom of movement, impairing a worker's ability to restore balance during a fall. The resulting injuries could be more severe than those experienced without an exoskeleton. Additionally, potential collisions between exoskeletons and work equipment, robots, or construction machinery must be accounted for, which may necessitate workplace redesign. In emergencies, rapid evacuation of buildings is crucial to ensure the safety of all employees. Therefore, exoskeletons must be designed for quick and easy removal. Designers should also account for scenarios where workers are operating alone.

Occupational exoskeletons are classified as work equipment and must comply with the requirements of Directive 2009/104/EC, which outlines minimum safety and health standards for workplace equipment. Employers are responsible for ensuring the safety of all equipment provided to workers. During the selection process, employers must consider specific workplace conditions, particularly those affecting the safety and health of workers. When risks associated with equipment cannot be entirely eliminated, employers must implement appropriate measures to minimize them. Furthermore, the equipment should adhere to relevant Community directives and meet the minimum requirements specified in Annex I.

Throughout its operational life, work equipment must be maintained in compliance with safety standards. Employers are responsible for ensuring that the equipment is properly installed and functions correctly through inspections and testing. These evaluations, which may include initial, post-assembly, periodic, and special inspections, must be conducted by competent personnel. Inspection results must be documented and retained.

If the use of work equipment poses specific risks, employers must restrict its access and allow modifications only by qualified personnel. Ergonomic and occupational health considerations must be fully addressed.

Employers must provide workers with clear, comprehensible information, such as written instructions, on the proper use of the equipment. These instructions should cover conditions of use, potential abnormal situations, and lessons learned from prior experiences. Workers must be made aware of relevant hazards, and employers are required to offer adequate training, including sessions focused on the risks and proper use of high-risk equipment.

6 Conclusions and Discussion

In this chapter we have described the different kinds of exoskeletons to understand what are the main differences between the different types and technologies. Occupational exoskeletons are a relatively new application area of exoskeleton, but with a great expansion. They fit within the factory of the future and the Industry 4.0 concept, and they represent a viable solution for some common ergonomic issues in today's industry. The first exoskeletons that are being evaluated in real scenarios are passive devices, due to their simplicity, weight and cost. Active exoskeletons on the other hand are being introduced at a slower pace due to the existing technological barriers and higher costs. Furthermore, the unclear regulatory framework makes even more difficult their integration. However, important efforts are being made for the framing of these devices with new specific standards. It is expected that active exoskeletons will have a higher importance in the next years.

Occupational exoskeletons represent a promising technological advancement in ergonomics, offering the potential to enhance worker safety, reduce MSD risk, and improve productivity. Despite these potential benefits, it is crucial to consider the long-term physical impact of exoskeleton use. While they can alleviate strain on specific muscle groups, prolonged use might introduce new ergonomic challenges or exacerbate existing conditions. Further research is needed to understand the potential long-term effects and to develop strategies for mitigating any negative consequences.

Beyond the physical benefits, exoskeletons can also impact psychosocial factors in the workplace. Proper training and education are essential to ensure that workers understand the benefits, limitations, and proper use of exoskeletons. This can help to increase user acceptance and minimize resistance to change.

Moreover, organizations need to consider the broader organizational implications of exoskeleton adoption. This includes factors such as infrastructure, maintenance, and safety protocols. A well-planned implementation strategy can help to mitigate potential challenges and maximize the return on investment.

References

1. National Research Council: Musculoskeletal disorders and the workplace: low back and upper extremities, (2001)
2. Inail open data. Last accessed Oct 11 2024. https://dati.inail.it/
3. Theurel, J., Desbrosses, K.: Occupational exoskeletons: overview of their benefits and limitations in preventing work-related musculoskeletal disorders. IISE Trans. Occup. Ergon. Hum. Factors **7**(3–4), 264–280 (2019)
4. Elprama, S.A., Vanderborght, B., Jacobs, A.: An industrial exoskeleton user acceptance framework based on a literature review of empirical studies. Appl. Ergon. **100**, 103615 (2022)
5. McFarland, T., Fischer, S.: Considerations for industrial use: a systematic review of the impact of active and passive upper limb exoskeletons on physical exposures. IISE Trans. Occup. Ergon. Hum. Factors **7**(3–4), 322–347 (2019)
6. Poliero, T., Fanti, V., Sposito, M., Caldwell, D.G., Di Natali, C.: Active and passive back-support exoskeletons: a comparison in static and dynamic tasks. IEEE Robot. Autom. Lett. **7**(3), 8463–8470 (2022)
7. Mosher, R.S.: Handyman to hardiman. Sae Transactions, 588–597 (1968)
8. Kopp, C.: Exoskeletons for warriors of the future. Def. Today **9**(2), 38–40 (2011)
9. Plaza, A., Hernandez, M., Puyuelo, G., Garces, E., Garcia, E.: Lower-limb medical and rehabilitation exoskeletons: a review of the current designs. IEEE Rev. Biomed. Eng. **16**, 278–291 (2021)
10. Voilqué, A., Masood, J., Fauroux, J. C., Sabourin, L., Guezet, O.: Industrial exoskeleton technology: classification, structural analysis, and structural complexity indicator. In: 2019 Wearable robotics association conference (wearracon), pp. 13–20. IEEE (2019)
11. De Looze, M.P., Bosch, T., Krause, F., Stadler, K.S., O'sullivan, L.W.: Exoskeletons for industrial application and their potential effects on physical work load. Ergonomics **59**(5), 671–681 (2016)
12. Fournier, D.E., Yung, M., Somasundram, K.G., Du, B.B., Rezvani, S., Yazdani, A.: Quality, productivity, and economic implications of exoskeletons for occupational use: a systematic review. PLoS ONE **18**(6), e0287742 (2023)
13. Asbeck, A. T., Dyer, R. J., Larusson, A. F., Walsh, C. J.: Biologically-inspired soft exosuit. In: 2013 IEEE 13th International Conference on Rehabilitation Robotics (ICORR), pp. 1–8. IEEE (2013)
14. Kim, S., Srinivasan, D., Nussbaum, M.A., Leonessa, A.: Human gait during level walking with an occupational whole-body powered exoskeleton: not yet a walk in the park. IEEE Access **9**, 47901–47911 (2021)
15. Meuleman, J., Van Asseldonk, E., Van Oort, G., Rietman, H., Van Der Kooij, H.: Lopes ii-design and evaluation of an admittance controlled gait training robot with shadow-leg approach. IEEE Trans. Neural Syst. Rehabil. Eng. **24**(3), 352–363 (2015)
16. Riener, R., Lünenburger, L., Maier, I.C., Colombo, G., Dietz, V.: Locomotor training in subjects with sensori-motor deficits: an overview of the robotic gait orthosis lokomat. J. Healthc. Eng. **1**(2), 197–216 (2010)
17. Bayon, C., Raya, R., Lerma Lara, S., Ramirez, O., Serrano, J., Rocon, E.: Robotic therapies for children with cerebral palsy: a systematic review. Transl. Biomed. **7**(1), 44 (2016)
18. Buccelli, S., Tessari, F., Fanin, F., De Guglielmo, L., Capitta, G., Piezzo, C., Bruschi, A., Van Son, F., Scarpetta, S., Succi, A., et al.: A gravity-compensated upper-limb exoskeleton for functional rehabilitation of the shoulder complex. Appl. Sci. **12**(7), 3364 (2022)
19. Esquenazi, A., Talaty, M., Packel, A., Saulino, M.: The rewalk powered exoskeleton to restore ambulatory function to individuals with thoracic-level motor-complete spinal cord injury. Am. J. Phys. Med. Rehabil. **91**(11), 911–921 (2012)
20. Vassallo, C., De Giuseppe, S., Piezzo, C., Maludrottu, S., Cerruti, G., D'Angelo, M. L., De Michieli, L.: Gait patterns generation based on basis functions interpolation for the twin lower-limb exoskeleton. In: 2020 IEEE International Conference on Robotics and Automation (ICRA), pp. 1778–1784. IEEE (2020)

21. Wang, S., Wang, L., Meijneke, C., Van Asseldonk, E., Hoellinger, T., Cheron, G., Ivanenko, Y., La Scaleia, V., Sylos-Labini, F., Molinari, M., et al.: Design and control of the mindwalker exoskeleton. IEEE Trans. Neural Syst. Rehabil. Eng. **23**(2), 277–286 (2014)
22. Awad, L.N., Esquenazi, A., Francisco, G.E., Nolan, K.J., Jayaraman, A.: The rewalk restore™-soft robotic exosuit: a multi-site clinical trial of the safety, reliability, and feasibility of exosuit-augmented post-stroke gait rehabilitation. J. Neuroeng. Rehabil. **17**, 1–11 (2020)
23. Ortiz, J., Di Natali, C., Caldwell, D.G.: Xosoft: design of a novel soft modular exoskeleton. In: Soft Robotics in Rehabilitation, pp. 165–198. (2021)
24. Petrone, N., De Bettio, G.: Evaluation of muscular emg activity variations wearing ski-mojo on a skiing treadmill. University of Padova, Italy, Department of Mechanical Engineering (2009)
25. Iervolino, R., Bonavolontà, F., Cavallari, A.: A wearable device for sport performance analysis and monitoring. In: 2017 IEEE International Workshop on Measurement and Networking (M&N)
26. Di Natali, C., Chini, G., Totaro, M., Lora-Millan, J.S., Rocon, E., Beccai, L., Ortiz, J.: Quasi-passive resistive exosuit for space activities: proof of concept. Appl. Sci. **11**(8), 3576 (2021)
27. Exoskeleton report. Last accessed, 11 Oct 2024. https://exoskeletonreport.com/
28. Crea, S., Beckerle, P., De Looze, M., De Pauw, K., Grazi, L., Kermavnar, T., Masood, J., O'Sullivan, L.W., Pacifico, I., Rodriguez-Guerrero, C., et al.: Occupational exoskeletons: a roadmap toward large-scale adoption. Methodology and challenges of bringing exoskeletons to workplaces. Wearable Technol. **2**, e11 (2021)
29. Jia-Yong, Z., Liu, Y., Mo, X.M., Han, C.W., Meng, X.J., Li, Q., Wang, Y.J., Zhang, A.: A preliminary study of the military applications and future of individual exoskeletons. J. Phys: Conf. Ser **1507**, 102044 (2020) IOP Publishing
30. Poliero, T., Lazzaroni, M., Toxiri, S., Natali, C. D., Caldwell, D. G., Ortiz, J.: Lifting and carrying: do we need back-support exoskeleton versatility? In: Wearable Robotics: Challenges and Trends: Proceedings of the 5th International Symposium on Wearable Robotics, WeRob2020, and of WearRAcon Europe 2020, October 13–16, 2020, pp. 499–503. Springer (2022)
31. Lazzaroni, M., Poliero, T., Sposito, M., Toxiri, S., Caldwell, D.G., Di Natali, C., Ortiz, J.: Back-support exoskeleton control strategy for pulling activities: design and preliminary evaluation. Designs **5**(3), 39 (2021)
32. Lazzaroni, M., Poliero, T., Sposito, M., Toxiri, S., Caldwell, D.G., Di Natali, C., Ortiz, J.: Mechanical design and analysis of light weight hip joint parallel elastic actuator for industrial exoskeleton. In: 6th IEEE International Conference on Biomedical Robotics and Biomechatronics (BioRob), pp. 631–636. IEEE (2016)
33. Ragonesi, D., Agrawal, S., Sample, W., Rahman, T.: Series elastic actuator control of a powered exoskeleton. In: 2011 Annual International Conference of the IEEE Engineering in Medicine and Biology Society, pp. 3515–3518. IEEE (2011)
34. Di Natali, C., Poliero, T., Sposito, M., Graf, E., Bauer, C., Pauli, C., Ortiz, J., et al.: Design and evaluation of a soft assistive lower limb exoskeleton. Robotica **37**(12), 2014–2034 (2019)
35. Di Natali, C., Sadeghi, A., Mondini, A., Bottenberg, E., Hartigan, B., De Eyto, A., O'Sullivan, L., Rocon, E., Stadler, K., Mazzolai, B., et al.: Pneumatic quasi-passive actuation for soft assistive lower limbs exoskeleton. Front. Neurorobot. **14**, 31 (2020)
36. Ortiz, J., Fernández, J., Poliero, T., Monica, L., Anastasi, S., Draicchio, F., Caldwell, D.G.: Flexo–modular flexible back-support passive exoskeleton. In: Wearable Robotics: Challenges and Trends: Proceedings of the 5th International Symposium on Wearable Robotics, WeRob2020, and of WearRAcon Europe 2020, October 13–16, 2020, pp. 333–337. Springer (2022)
37. Baltrusch, S.J., Van Dieën, J. H., Koopman, A.S., Näf, C Rodriguez-Guerrero, M. B., Babič, J., Houdijk, H.: Spexor passive spinal exoskeleton decreases metabolic cost during symmetric repetitive lifting. Eur. J. Appl. Physiol. **120**(2), 401–412 (2020)
38. Bauaerwerbstätigenbefragung 2012 männer/frauen in vollzeit mit den schwerpunkten arbeitsbedingungen, arbeitsbelastungen und gesundheitliche beschwerden [basic analysis of the 2012 bibb/baua employment survey for men/women in full-time jobs and focusing on working conditions, workloads and health complaints]. Last accessed 11 Oct 2024. https://www.baua.de/DE/Angebote/Publikationen/Berichte/F2417-2.html

39. Laevo exoskeletons. Last accessed 11 Oct 2024. https://www.laevo-exoskeletons.com/
40. XuitX. Ix back. Last accessed 11 Oct 2024. https://www.suitx.com/en/products/ix-back-exoskeleton
41. Orthexo. Paexo back. Last accessed 11 Oct 2024. https://orthexo.de/en/exoskeletons/industrial-exoskeletons/paexo-back/
42. German Bionic. Crayx. Last accessed, 11 Oct 2024. https://germanbionic.com/en/solutions/exoskeletons/crayx/
43. Herowear. Apex 2. Last accessed 11 Oct 2024. https://herowearexo.com/the-apex-2-back-assist-exosuit/
44. Auxivo. Liftsuit. Last accessed 11 Oct 2024. https://www.auxivo.com/liftsuit
45. De Vries, A., De Looze, M.: The effect of arm support exoskeletons in realistic work activities: a review study. J. Ergon **9**(4), 1–9 (2019)
46. SuitX. Ix shoulder. Last accessed 11 Oct 2024. https://www.suitx.com/en/products/ix-shoulder-air-exoskeleton
47. Orthexo. Paexo shoulder. Last accessed 11 Oct 2024. https://orthexo.de/en/exoskeletons/industrial-exoskeletons/paexo-shoulder/
48. Comau. Mate-xt 4.0. Last accessed 11 Oct 2024. https://mate.comau.com/mate-xt-4-0/
49. Skelex. Skelex 360-xfr. Last accessed 11 Oct 2024. https://www.skelex.com/products
50. Agade. Agade shoulder. Last accessed 11 Oct 2024. https://agade-exoskeletons.com/en/product/
51. Hermann, M., Pentek, T., Otto, B.: Design principles for industrie 4.0 scenarios. In: 49th Hawaii International Conference on System Sciences (HICSS), pp. 3928–3937. (2016)
52. De Looze, M.P., Bosch, T., Krause, F., Stadler, K.S., O'sullivan, L.W.: Exoskeletons for industrial application and their potential effects on physical work load. Ergonomics **59**(5), 671–681 (2016)
53. Bosch, T., van Eck, J., Knitel, K., de Looze, M.: The effects of a passive exoskeleton on muscle activity, discomfort and endurance time in forward bending work. Appl. Ergon. **54**, 212–217 (2016)
54. Huysamen, K., de Looze, M., Bosch, T., Ortiz, J., Toxiri, S., O'Sullivan, L.W.: Assessment of an active industrial exoskeleton to aid dynamic lifting and lowering manual handling tasks. Appl Ergonomics **68**, 125–131 (2018)
55. Billinghurst, M., Clark, A., Lee, G., et al.: A survey of augmented reality. Found. Trends Hum. Comput. Interact. **8**(2–3), 73–272 (2015)
56. Paelke, V.: Augmented reality in the smart factory: supporting workers in an industry 4.0. environment. In: Emerging Technology and Factory Automation (ETFA), 2014 IEEE, pp. 1–4. IEEE (2014)
57. Wolter, M.I., Mönnig, A., Hummel, M., Schneemann, C., Weber, E., Zika, G., Helmrich, R., Maier, T., Neuber-Pohl, C.: Industry 4.0 and the consequences for labour market and economy: scenario calculations in line with the bibb-iab qualifications and occupational field projections (industrie 4.0 und die folgen für arbeitsmarkt und wirtschaft: Szenario-rechnungen im rahmen der bibb-iab-qualifikations-und berufsfeldprojektionen), Nuremberg, Germany (2015). Technical report
58. Masood, J., Mateos, L. A., Ortiz, J., Toxiri, S., O'Sullivan, L., Caldwell, D.: Active safety functions for industrial lower body exoskeletons: concept and assessment. In: Wearable Robotics: Challenges and Trends: Proceedings of the 2nd International Symposium on Wearable Robotics, WeRob2016, October 18-21, 2016, Segovia, Spain, pp. 299–303. Springer (2017)
59. Robots and robotic devices—Safety requirements for industrial robots—Part 1: Robots. Standard, International Organization for Standardization, Geneva, CH, July 2011
60. Robots and robotic devices—Safety requirements for personal care robots. Standard, International Organization for Standardization, Geneva, CH, February 2014
61. Robots and robotic devices—Collaborative robots. Standard, International Organization for Standardization, Geneva, CH, February 2016

62. Medical electrical equipment—Part 2-78: Particular requirements for the basic safety and essential performance of medical robots for rehabilitation, assessment, compensation or alleviation (Under development). Standard, International Organization for Standardization, Geneva, CH, 2017
63. Mahmud, D., Bennett, S.T., Zhu, Z., Adamczyk, P.G., Wehner, M., Veeramani, D., Dai, F.: Identifying facilitators, barriers, and potential solutions of adopting exoskeletons and exosuits in construction workplaces. Sensors **22**(24), 9987 (2022)
64. Ergonomics—Application document for International Standards on manual handling (ISO 11228-1, ISO 11228-2 and ISO 11228-3) and evaluation of static working postures (ISO 11226). Standard, International Organization for Standardization, Geneva, CH, March 2014
65. Ergonomics—Manual handling—Part 1: Lifting and carrying. Standard, International Organization for Standardization, Geneva, CH, May 2003
66. Ergonomics – Manual handling—Part 2: Pushing and pulling. Standard, International Organization for Standardization, Geneva, CH, March 2007
67. Ergonomics—Manual handling—Part 3: Handling of low loads at high frequency. Standard, International Organization for Standardization, Geneva, CH, April 2007
68. Ergonomics-Evaluation of static working postures: Standard. International Organization for Standardization, Geneva, CH (2000)
69. Amandels, S., Eyndt, H. O. H., Daenen, L., Hermans, V.: Introduction and testing of a passive exoskeleton in an industrial working environment. In: Proceedings of the 20th Congress of the International Ergonomics Association (IEA 2018) Volume III: Musculoskeletal Disorders 20, pp. 387–392. Springer (2019)
70. Ortiz, J., Poliero, T., Cairoli, G., Graf, E., Caldwell, D.G.: Energy efficiency analysis and design optimization of an actuation system in a soft modular lower limb exoskeleton. IEEE Robot. Autom. Lett. **3**(1), 484–491 (2018)
71. Toxiri, S., Ortiz, J., Caldwell, D.G.: Assistive strategies for a back support exoskeleton: experimental evaluation. In: International Conference on Robotics in Alpe-Adria Danube Region, pp. 805–812. Springer (2017)
72. Levine, J.A.: Measurement of energy expenditure. Public Health Nutr. **8**(7a), 1123–1132 (2005)
73. Corlett, E.N., Bishop, R.P.: A technique for assessing postural discomfort. Ergonomics **19**(2), 175–182 (1976)
74. Van der Grinten, M.P., Smitt, P.: Development of a practical method for measuring body part discomfort. Adv. Ind. Ergon. Saf. **4**, 311–18 (1992)
75. Bangor, A., Kortum, P., Miller, J.: Determining what individual sus scores mean: Adding an adjective rating scale. J. Usability Stud. **4**(3), 114–123 (2009)

Intuitive Robot Programming

Charly Blanc, Julius Jankowski, Andreas Sonderegger, Sylvain Calinon, and Sarah Dégallier Rochat

Abstract The introduction of robots in SMEs requires a drastic reconsideration of existing programming techniques and interfaces. The growing versatility of the market essentially demands the robots to be re-programmed frequently, calling for interfaces that are intuitive enough to require no robotic expertise, yet flexible enough to meet industrial needs. Rather than limiting the intervention of the workers, the technology should empower them to take control over the machine. Only then, the human-machine complementarity can be leveraged, combining the efficiency of automation with the flexibility of human work. Ergonomics and human factors are critical to the development of these new programming interfaces, which should typically be developed and tested in a cyclic fashion with the end-users. In this chapter, we highlight the differences between easy-to-learn and easy-to-use interfaces, as well as the differences between interfaces targeting naive/casual users and interfaces targeting expert/professional workers. We discuss the design of robot programming interfaces based on a user-centric perspective. The research developments in this project target different methods to collect demonstrations (including tracked devices and kinesthetic teaching), different graphical user interfaces (including block-based and timeline representations), and different approaches to augmented reality (including tablets/smartphones, headsets or actuated projection devices). Through this multimodular approach, we aim at providing workers with the most natural way possible to interact with the robot depending on the task or their preferences.

Keywords Human-machine interfaces · Robotic programming · Learning from demonstration · Anthropocentric production systems

This work was supported by the Swiss National Science Foundation (SNSF) through the CODIMAN project.

C. Blanc · A. Sonderegger · S. D. Rochat
Bern University of Applied Sciences, Bern, Switzerland
e-mail: charly.blanc@bfh.ch

J. Jankowski · S. Calinon (✉)
Idiap Research Institute, Martigny, Switzerland
e-mail: sylvain.calinon@idiap.ch

Ecole Polytechnique Fédérale de Lausanne, Lausanne, Switzerland

G. Barresi and S. Wischniewski (eds.), *Ergonomics in Robotics: Advances and Innovations*, Springer Tracts in Advanced Robotics 162,
https://doi.org/10.1007/978-3-032-19374-2_13

1 Introduction

New technologies such as artificial intelligence, augmented reality and robotics—and the associated 4th industrial revolution—are expected to drastically change the nature of work in the manufacturing industry [12, 26, 50]. Industrial production systems shall become leaner, but also more integrated, traceable, and adaptable. The past visions of fully computer-controlled factories where no worker is needed are now challenged by visions of agile companies where flexible production systems can be leveraged by human creativity to foster continuous improvement and innovation [25].

To cope with the versatility of the market and the rapid evolution of technology, agile automation solutions that can be rapidly adapted to new demands are needed. New technologies are seen as enablers for "anthropocentric production systems", that are systems that allow the close interaction of humans and machines on the shop floor through the enhancement of the worker's capabilities [48]. For this transformation to happen, a shift in paradigm in industrial interface design is required, like the democratization of computers 40 years ago: rather than expert systems, intuitable solutions that can leverage the worker's task expertise are needed [59]. Collaborative robots have opened the path to the democratization of automation: since they allow for physical interactions with the device, non-experts should be able to develop ad hoc solutions by interacting directly with the robot, without having to code. Two main modalities have been used to simplify the programming of these robots: visual programming interfaces and learning by demonstration.

Visual programming interfaces have been used since decades in the industry to simplify programming, as for instance the icon-based flow chart offered by systems such as MORPHA or the Engel Programming Language (EPL) [38]. Such visualization are also used by robot manufacturers such as FANUC (see Fig. 1a) and Franka Emika. Recently, a new approach to visual programming based on text blocks has been proposed with systems such as Blockly [27] or Scratch [47]. Each block represents a statement with open parameters and the blocks can be combined to create more complex programs. Several studies have shown the relevance of this approach in terms of efficiency and usability in pure programming tasks [62], but also when applied to robotic systems [61]. Companies such as ABB and Omron are now proposing robotic systems with block-based programming interfaces.

An important research effort was deployed recently to simplify the programming of robots by exploiting imitation mechanisms [8]. The goal is to transfer skills to robots by letting them learn new tasks from human demonstrations (learning from demonstration, LfD). Research in this field encompasses three aspects: (1) the development of models and algorithms that can extract relevant patterns from multiple demonstrations, in order to compress the data and keep only the essential components of the demonstrated skills; (2) the development of sensory integration capability to record the demonstrated skills; and (3) the development of intuitive interfaces to simplify the re-programming of robots and to collect multimodal sensory data characterizing the task to be transferred [19]. These interfaces can take various forms, including smartphones with augmented reality (AR) capability, virtual reality (VR)

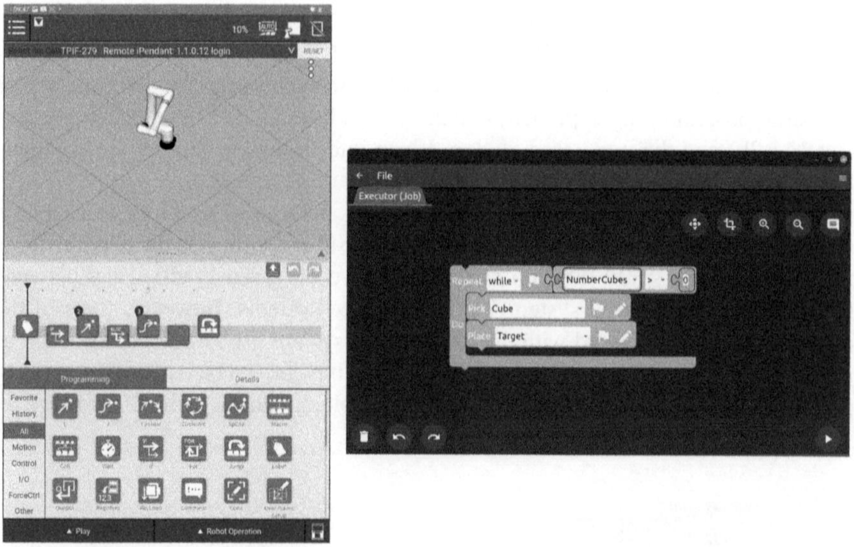

(a) Fanuc CRX interface (b) Prograblock interface

Fig. 1 Icon-based and block-based programming interfaces

headsets, teleoperation devices with force feedback, and robots endowed with torque sensing capability, allowing the user to guide the robot through the different steps of a task while the robot records movement information by proprioception (kinesthetic teaching).

From the user point of view, these new approaches to robot programming aim to enable interaction modalities with the machine that relies much more on the understanding of the task to be performed than of the machine itself. Rather than requiring robotics experts, the system shall need task experts, that is, workers, to be efficiently put to use. In this chapter, we present implementations of a block-based programming (BBP) interface and learning from demonstration (LfD) for industrial applications.

2 Block-Based Programming Languages

Block-based interfaces are visual programming tools that are based on blocks representing statements with open parameters. The blocks can be combined to create complex programs, as illustrated in Fig. 1b. This approach has been proven to efficiently ease programming by removing syntax-related obstacles [63] and by reducing cognitive load, but also because the user does not need to remember the language specific vocabulary since it is sufficient to recognize the correct block [6]. Previous work successfully combined Blockly with industrial robots [61] for a task where adult novices had to program a pick-and-place task. Other studies investigated robot programming teaching for adult novices [60] using Blockly (see Fig. 2a) with a framework called CoBlox. They compared the block-based approach with two widely used industrial programming approaches, namely a traditional, text-based teach pendant (see Fig. 2b) and the tree-based, dialog-driven interface providing by the company Universal Robot (see Fig. 2c).

The block-based system obtained better scores in terms of usability, learnability, and overall satisfaction. The participants were also able to program the required tasks faster. In a recent study [38], no differences were observed in efficiency between the traditional icon-based system and block-based systems. However, some of the obstacles that were noticed with the block-based system used for the study, such as the quantity of blocks and the lack of higher level blocks, can be easily overcome. In particular, different levels of abstractions for the blocks can be defined, for instance,

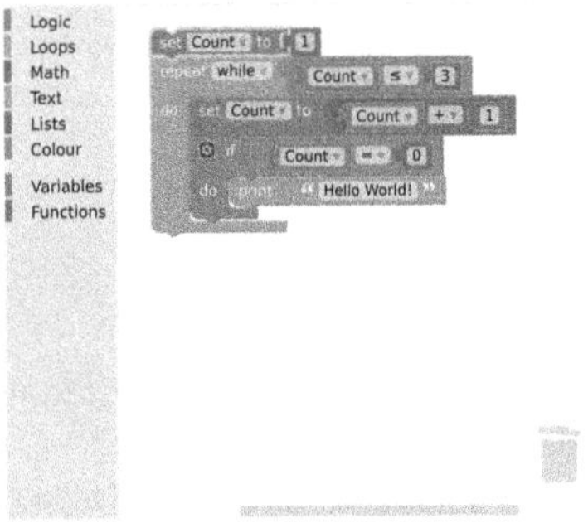

(a) Blockly interface

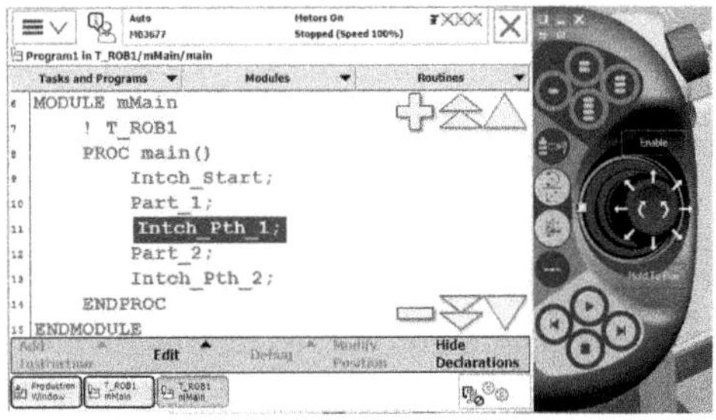

(b) ABB FlexPendant interface

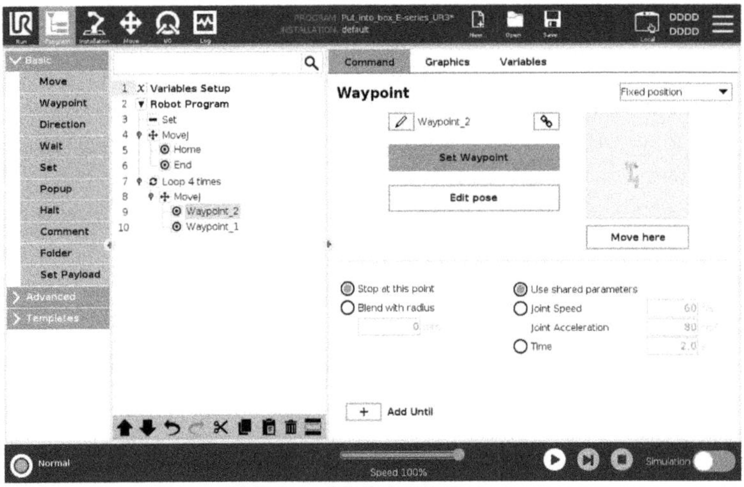

(c) Universal Robots interface

Fig. 2 Different types of robot programming interfaces

through a three level architecture based on the concept of *skill* [49]. In this model, the higher level (*tasks*) corresponds to the abstract description of the task (such as "build motor"), while the lower level (*device primitives*) reflects the machine API (such as "move in Cartesian space to this position"). The intermediate level is based on the concept of *skills* and corresponds to the steps needed to achieve the task (such as "take rotor"). The higher level is extremely intuitive for the workers, with a limited choice of actions, similarly to the typical operator interfaces, while the lower level represents the complete flexibility offer by the interfaced device. The intermediate level is object-oriented and still meaningful for the workers, although some robotic and programming concepts (such as *frames* or *loops*) are needed for the worker to be able to use the system to its full potential. Since *tasks* are a combination of *skills* and *skills* a combination of *devices primitives*, the user can learn these concepts by imitation by simply adapting existing programs to their production needs, for instance if the working environment changes or a new type of pieces needs to be produced (see Fig. 3). In this illustration, the user has the capability to choose an object from the dropdown menu (see Fig. 3b) for the robot to grasp. An object is defined by a unique identifier enabling references in the program and a frame.

However, for the workers to develop a feeling of ownership over the machine, they need to be able to predict, to a certain extent, the behavior of the robot. For this, a minimal understanding of the system is required [30]. It is therefore important to teach the worker the needed expertise in programming and robotics to allow them to develop a sense of autonomy and control over the system. In the context of robotic automation, contrarily to most leisure applications, the users are typically willing to spend time and energy to learn about the system to be able to leverage its full potential. They acknowledge that a deeper understanding of the technology enables them to make the most of its capabilities, improving productivity and efficiency in various tasks through the utilization of these robotic systems. Tutorials were therefore developed for the users to learn the required competencies to achieve more complex tasks. They are based on typical industrial use cases and are meant to be done by the

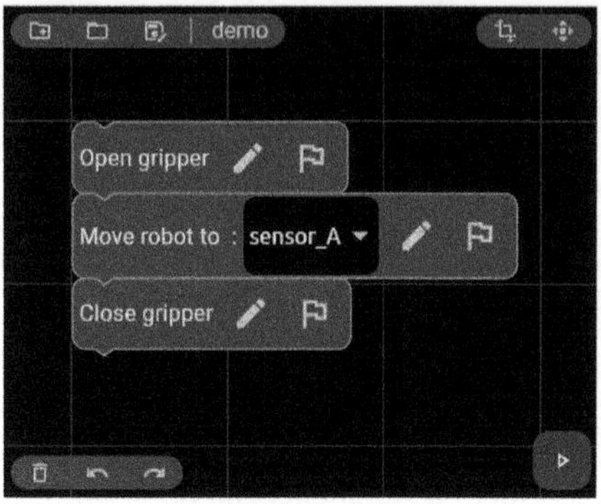

(a) Example of the task *Pick*

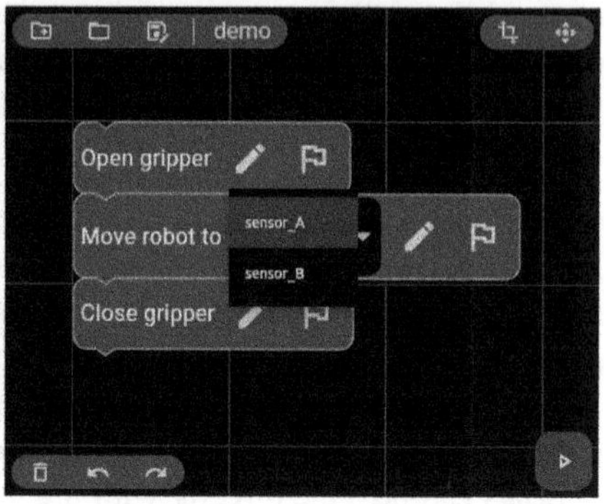

(b) Parameterization of the task *Pick*

Fig. 3 Example of task reprogrammed for a new type of components users when needed. Since interacting directly with the robot can be a source of stress if the user fears to get hurt by it or to break it [36], a simulation of the system was developed for the tutorials that can be used on a computer or tablet or with augmented reality (AR) goggles. In the following paragraphs, we describe more in details the different aspect of the interface.

2.1 A Flexible yet Intuitive Block-Based Programming Language: Prograblock

No-code programming. A common obstacle with programming is the learning of the syntax [24]. We therefore designed a no-code, block-based interface, *Prograblock*, illustrated in Fig. 1b. The blocks are parameterizable and can be edited. New blocks can be created by combining existing blocks, allowing for a greater flexibility. A similar system can be used for controlling the robot behavior and the vision algorithms, but also to control other ancillary devices, such as a vibrating bowl or the machine that need to be tended. The different blocks are classified in three different categories corresponding to different levels of abstraction as explained in the next paragraph.

Skill-based three-layer architecture. The programming interface utilizes three different levels of abstraction, going from common and intuitive human language (tasks) to the flexible but complex machine API (device primitives), with an intermediate language based on skills. Skills are object-oriented and serve as a communication bridge: they offer an intermediate language that can be used to describe the process steps in relation to the object being handled [49]. Skills are sensorimotor behaviors (such as pick and place) and require the coupling of a sensory information (for instance, the detection of the position of a component) and an action (for instance, moving to the location to be reached). New tasks can be created by simply combining skills, and new skills by combining device primitives. The user can create a new task (or skill) by simply editing an existing one. The idea behind the three-level architecture is to encourage the user to gain control over the system by progressively learning the lower-level languages, while maintaining a feeling of safety and control.

Tutorials. Tutorials were also developed for the users to learn the needed competencies to switch between the levels of abstraction and complete more complex tasks. In the case of robot programming, different concepts need to be understood, such as, for instance, loops or conditions in programming, as well as robot kinematic singularities. The tutorials are thus based on typical industrial use cases, with a small description of the concepts that will be learned. In the Experiments section below, we present the six first tutorials that have been developed and tested with students. The tutorials can be done directly with the robots since the elements can be 3D-printed or with a simulator (see Fig. 4).

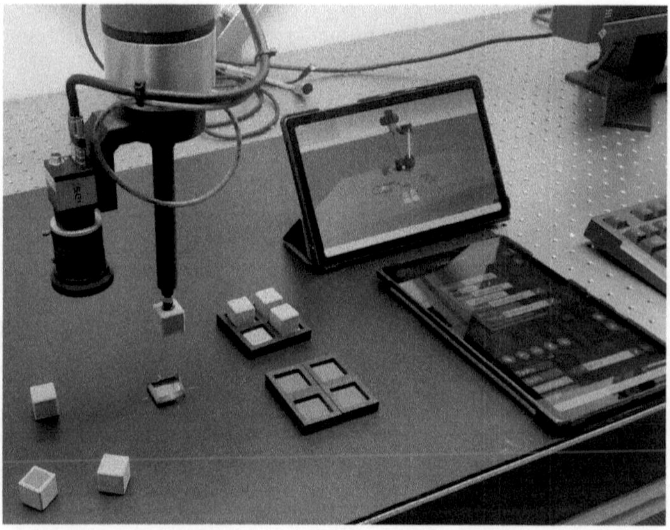

Fig. 4 Experimental setup

Simulator. A simulation of the system was developed for the user to be able to check the programmed behavior before executing it on the robot, following the concept of digital twin (see Fig. 5). We use it for the tutorials since the environment is known and can be modelled in advance. In the cases where the physical set-up is used, the simulation is automatically adapted to match the real set-up. It means that, since the environment is flexible (the positions of the elements are not predefined), the position of the objects in the simulation will be adapted to match the real set-up based on the camera feedback. The system can be used on a tablet, a computer or with AR goggles.

2.2 Experiments

Changes brought by the introduction of technology are usually accompanied by resistance and mistrust from the existing workforce [11]. This is especially true in the case of change linked with the implementation of technology that carries out one's own tasks and thus could pose a potential threat to one's own job [39]. Resistance in the workforce to the introduction of new technologies can have a negative impact on satisfaction, productivity and can jeopardize the success of the technology introduction [9, 53]. Therefore, it is important to make sure that the introduction of new technology will have a positive impact on the workers. To achieve this, we suggest adopting a user-centered approach, in which the development and introduction of the new technology is done in close collaboration with the people

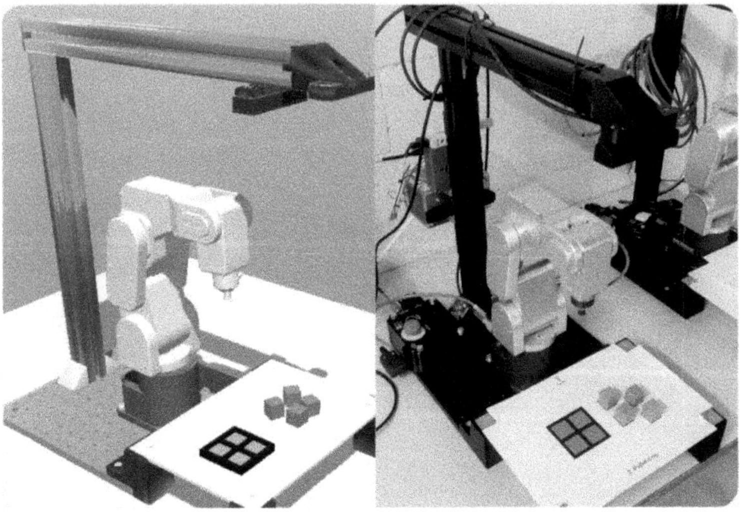

Fig. 5 Digital twin (simulation on the left, real system on the right)

concerned. Observations and interviews with the target operators have therefore been performed with the workers and a series of workshops is planned in the future to test the interface with them and adapt it to their needs once the system is robust enough. A preliminary evaluation of the whole system was performed with students through 2 different studies, as described below.

Description of the tutorials. In a first attempt to test this block-based, skill-oriented interface, two different experiments were performed with students. In all the experiments, the students had to perform six different exercises under different conditions. The four first ones were guided tutorials, while the two last ones were for testing the acquired competencies. The scenarios for the exercises were based on industrial applications such as pick and place or palletizing. The exercises were done with a simulation of the robot. The goal of the exercises were to help the user better understand the different functionalities of the interface, as well as some programming concepts, such as loops and conditions. More details about the design of the tutorials can be found in our previously published study [2].

Study 1: Effect of the Interface on Self-efficacy and Programming Anxiety. We aim at helping lay people develop a feeling of control over the robot through human-centered design approaches as well as the development of tailor-made tutorials. In order to test whether this was achieved with the proposed solution, a first study involving 58 student participants aged 16 to 39 years, with varying programming experience was conducted. The measures included programming anxiety [22], programming self-efficacy [22], attitudes towards robots [42], robot anxiety [5], using established questionnaires such as the Computer Programming Anxiety Questionnaire [22] and the Robot Attitude Questionnaire [42]. Assessments were performed before and after participants interacted with a robotic interface and also examined user experience [37] and task performance within a tutorial and test framework. We hypothesized a more positive attitude towards robots and a lower anxiety after interacting with the interface compared to the baseline measurement.

Results: Analysis of data revealed a significant decrease of **programming anxiety** after using the robot programming interface. Furthermore, participants reported significantly more positive **attitudes towards robots** after interacting with the programming tool compared to the baseline measurements. Participants' evaluation of their **robot anxiety** decreased also after using the robot programming tool, the effect did however not reach significance level. **Programming self-efficacy** did not change significantly after using the robot programming tool. In terms of user experience evaluation, most participants (79%) were able to complete the tasks during the 30 min given while only a minority of the participants (29%) completed the tasks without help. After completing the tasks, participants rated their user experience positively in terms of attractiveness, perspicuity, dependability, and stimulation. However, they gave lower ratings for efficiency and novelty.

Discussion: The purpose of our intuitive robotic interface is to make the programming of robots as easy as possible for laypeople, in such a way that their fears and reservations about the technology are reduced, but also increasing their positive attitudes toward robots and their deployment.

Considering the results of study 1, our objectives have been partially achieved: the initially rather high ratings of anxiety towards programming have been considerably reduced after interacting with the interface. Also, participants' ratings of their attitudes towards robots significantly increased after using the interface. In a similar direction, the evaluation of robot anxiety also decreased after using the interface. This indicates (small sample) that the use of intuitive and easy-to-use programming interfaces can positively influence the initially existing fears regarding programming and the attitude toward the use of robots. User experience evaluation unveiled that not all participants were capable to complete all the tasks, requesting to some extent help from the test facilitators. This indicates that, despite our efforts, the interface might not be intuitive enough yet and needs to be further improved so that everyone can build up self-efficacy and the necessary knowledge to program robots in the future.

Interestingly, however, the interaction with the interface did not positively influence participants' evaluation of their self-efficacy regarding their programming abilities. This could be related to the fact that the interaction with the tool lasted only a relatively short time (about 60 min), a duration which may not be sufficient to influence this important component of self-assessment of personal competences. This could be expected according to Yildiz et al. [65], who tested the effect of robot programming on computational thinking skills toward self-efficacy perception during seven weeks, two lessons per week. Their results showed a significant increase of self-efficacy.

Furthermore, a study by Hsu and Hwang [28] on a similar block-based interface found reduced learning anxiety among young students, aligning with our findings on programming and robot anxiety. Their study also indicates that block-based programming enhances playfulness and enjoyment in learning computational thinking.

Reich-Stiebert and Eyssel's research indicates positive attitudes toward robots in STEM (science, technology, engineering and mathematics) education contexts [46], consistent with our findings. Nomura et al. also found that personal robot experiences significantly influence these attitudes [43], suggesting that positive experiences with a block-based interface may enhance users' attitudes towards robots.

The implementation of an intuitive robotic interface in industrial settings offers several significant benefits. It can enhance workforce training by enabling production workers without prior programming knowledge to quickly learn and reprogram robots, reducing downtime and increasing production efficiency. Additionally, the interface can lower technological anxiety and improve attitudes toward robots, fostering greater acceptance and utilization of robotic technologies. This user-friendly approach also supports continuous learning and skill development, promoting innovation as workers conceive new ways to utilize robots. Moreover, by improving safety and ergonomics and enabling rapid adaptation to market changes, companies can gain competitively and better respond to evolving customer demands.

Despite the demonstrably positive effects of using the proposed interface, the user experience evaluation indicated that not all test participants managed to solve the test tasks independently after completing the tutorial. The analysis of their interactions with the interface as well as their feedback suggest that this might be related to reduced situational understanding linked with the use of the 3D simulation. A possible remediation would be to present a robot to the participants. Whether this actually influences the results was the subject of the second study conducted.

Study 2: Influence of Experiencing a Real Robot vs. Mere Interaction with a 3D-Simulation . In study 1, participants learned and tested their programming skills on a 3D-simulation of a robotic arm. The results of this study showed that not all participants were able to independently solve the tasks set after completing the tutorial. A possible reason may be due to the lack of situational understanding imposed by the mere use of a virtual environment. Seeing ones own programming efforts performed by a real robot might instead provide such situational understanding and hence improve programming performance. In order to address this research question, we conducted a second study to investigate the consequences of being introduced to a real robot on programming performances. We expected that participants attending such an introductory session would perform better in the programming tasks using the 3D-simulation compared to participants not attending the session.

This study involved 58 master students from a business school, aged between 23 to 46 years with no prior programming experience. Participants were divided into two groups to observe the effect of different introductory sessions on learning robotic programming. One group witnessed a live demonstration of a robotic cell, which included a collaborative robot with a gripper and a camera performing a sorting task, emphasizing that the robot operates solely on pre-programmed instructions. The other group proceeded directly to the programming tutorial without this demonstration. Participants worked in pairs throughout the tutorial and subsequent test session, fostering knowledge sharing as they interacted with the application used in the study 1.

Performance indicators were assessed using task completion time, the number of trials, and the number of program modifications. Participant experience was evaluated with the System Usability Scale (SUS [10]) for interface usability and the Animated Visualization of Affective States (AniSAM [51]) for emotional states.

Results: No differences were observed between the two groups in the tutorial phase regarding time spent on the tutorial, total number of trials and total number of modifications. For the simple task in the subsequent test session, no differences were observed between the two experimental groups neither. However, for the complex task in the test session, significant differences were observed, with participants who received an introductory session completing the task faster and making a lower number of modifications compared to the group not receiving the introductory session.

Analysis of participants' affective state revealed higher levels of **arousal** for participants in the introductory session group compared to participants not receiving an introduction. Effect of time and the interaction (introductory session x time) did not reach significance level. With regard to **valence**, data analysis revealed that the effect of the experimental manipulation was not significant while the effect of time was significant. Pairwise comparisons however indicated no significant differences between the three measurement points. Furthermore, the interaction (introductory session x time) did not reach significance level. The usability assessment of the system resulted in a SUS-score [10] of 71.7 which can be considered to be between good and excellent in terms of adjective ratings [4].

Discussion: The key findings of study 2 relate to the impact of the experimental manipulation on performance data in the test tasks. Although there was no discernible effect of the introductory session on the initial, less challenging task, participants demonstrated improved performance in the more difficult second task when they engaged in the introductory session.

In the first test task, the objective was to stack various cubes on different pallets. In the second test task, the robot arm had to navigate around an obstacle while stacking cubes. Most participants successfully programmed the robot to avoid the obstacle during the pickup and placement phases using a waypoint. However, a noteworthy proportion encountered difficulties in realizing that this avoidance procedure was also necessary when the robot arm returned to the pickup area due to the repetitive nature of the pickup and placement actions inherent to the stacking task. The observed improvement in performance among those who participated in the introductory session might be attributed to a better comprehension of how robots operate. Robots, being machines, strictly follow a predefined list of actions programmed by humans and lack the ability to interpret contextual information. Viewing the robot as adhering to a set of instructions without contextual understanding led to the correct resolution of the task. This suggests that attending the introductory session shapes participants' mental models (as described by Kieras [33]), enabling them to predict and explain the system's behavior.

In this second study, participants who attended the introductory session exhibited a higher level of arousal compared to the control group. While the exact cause of this effect remains speculative, it is conceivable that the heightened arousal is linked to witnessing a real robot arm in operation, potentially associated with increased stress. It is plausible that any negative consequences of this stressful experience were offset by the positive effects of gaining a deeper understanding of how robots function. In fact, Tisza et al. have found that arousal positively correlates with students' learning gains [57]. Further research is necessary to replicate this finding and ascertain its underlying cause.

Interpreting valence changes is difficult without significant post-hoc comparisons, suggesting further studies with increased statistical power are needed for definitive conclusions. The limitations of this study include its small sample size and paired programming, which might have distorted results due to group dynamics. Plans for a replication study with greater statistical power are underway to confirm these findings.

The results from study 2 indicate that an introductory session can significantly improve performance on complex tasks by helping participants understand how robots operate. This approach could be highly beneficial in industrial settings, allowing production workers with no prior programming experience to reprogram industrial tasks independently. This capability would enable quick adaptation to market changes. The introductory sessions help shape workers' mental models, making them better equipped to predict and explain robot behavior.

This method could be applied in industrial settings, empowering production workers with no prior programming experience to independently reprogram industrial tasks, thus facilitating quick adaptation to market changes. However, due to the unique characteristics of this target group, additional studies are required to assess feasibility. We are currently collaborating with a company to implement our system on their shop floor and plan to initiate trials with employees soon. The project aims to gradually improve the interface based on their feedback.

3 Learning from Sparse Human Demonstrations

Robot learning from demonstration (LfD) refers to the process used to transfer new skills to a robot by relying on demonstrations from a user [3, 8]. It is inspired by the imitation capability developed by humans and animals to acquire new skills. LfD aims at making programming accessible to novice users by providing them with an intuitive interface they are familiar with, as humans already exchange knowledge in this way. The goal is to provide user-friendly interfaces that do not require knowledge in computer programming or robotics. LfD can be considered at various levels, from the transfer of low-level motor control to the transfer of high-level symbolic reasoning capabilities. For a given skill to be acquired, several learning strategies can be considered, from the copying of the demonstrated actions to more elaborated abstractions, such as the extraction of the underlying objectives of the actions [21, 45]. The terms behavioral cloning and inverse optimal control can be used to refer to these two broad learning strategies, respectively. They have connections with imitation mechanisms studied in ethology and developmental psychology, where one can distinguish action-level mimicry to goal-level emulation (see e.g. [64]).

LfD is motivated by many applications and can be applied to various robots. In an industrial context, it is motivated by the lowered barrier to entry the field and a democratized access to robotics. These changes require quick and cost-effective adaptations of existing assembly lines, as well as the handling of small volumes such as personalized products. It typically requires the sharing of a common workspace between the workers and the robots. In practice, LfD enables robots to be reprogrammed by the persons who know the tasks to achieve, but who do not necessary have expertise in robot programming. In this context, LfD removes the costly and timely step of soliciting external expertise each time the robots need to be reprogrammed.

In service robotics, LfD aims at providing personalized assistance and services that could not be preprogrammed in advance due to the broad variety of tasks, persons, or environments that a robot can encounter. In some of these applications, LfD can rely on interactive social cues and on the natural human propensity to teach news skills to others—a communication behavior we already use to transfer knowledge.

Learning by Bilateral Interaction

In addition to the skill transfer perspective, a key aspect of LfD is the exploitation of the social learning mechanisms behind LfD (see [41] for an overview). In a classical machine learning problem it is often assumed that expert datasets are available (typically, assuming a preliminary access to the samples that are relevant solutions to the problem). Similarly, the learning strategy to be used is also often explicitly specified. These developments do not account for the way in which data are collected. In contrast to many machine learning applications in which the learning systems are independent of the acquired data, a remarkable characteristic of LfD is that the iterative interaction with the user and the robot can be exploited to influence the quality and nature of the collected data. It was observed in robotics that several learning strategies need to be combined to acquire skills efficiently (see, e.g., [15]).

Demonstration Modalities

In LfD, the demonstrations can take various forms, requiring the design of interfaces, devices, and control techniques that can be used for efficient collection of data. Some of these can be linked to social learning processes such as *observational learning* and *kinesthetic teaching*. The first refers to the visual observation of a demonstrator achieving a task with the goal of reproducing it. In robotics, observational learning will typically exploit vision systems, but it is often conceded that the perception system can extend to marked-based systems (mocap) or other forms of motion recording devices (e.g., gyroscopes and accelerometers).

Kinesthetic teaching, also called direct teaching, refers to the process of moving the robot manually, by haptic interaction, while the robot records the demonstration through its sensors (proprioception). In the case of articulated robots, the user can demonstrate the task in the workspace of the robot by constraining the movement to the robot capability. Compared to observational learning, kinesthetic teaching simplifies the correspondence problem [41] (also called motion retargeting or body mapping problem). A potential downside is that the user does not execute the task on his/her own and can control only a limited number of articulations simultaneously. In practice, such limitation often restricts kinesthetic teaching to tasks that do not involve highly dynamic movements. At a technical level, kinesthetic teaching is often implemented by having the robot actively compensate for the effect of gravity (typically implemented with torque-controlled robots), sometimes also complemented by compensation of inertia, friction and Coriolis effects.

LfD also considers demonstration modalities in the form of specialized devices used to teleoperate the robot. This includes commands from a graphical user interface as discussed in the previous sections, but also various forms of joysticks and exoskeletons. These interfaces either act in a passive way, or they can be provided with feedback mechanisms. The most sophisticated interfaces exploit recent developments in telepresence to allow the user to demonstrate tasks by feeling as if she/he was executing it on her/his own.

Movement Primitives

A key research aspect underlying LfD is the design of compact and adaptive movement representations that can be used for both analysis and synthesis. The term *movement primitives* is often employed in this context. The goal of such encoding strategy is to represent movements as a set of adaptive building blocks that can be (re)organized in parallel and in series to create more complex behaviors, see [17] for a review.

Exploitation of (Co)variations

LfD algorithms typically need to cope with several forms of variations, which make the problem harder than a simple *record and play* process. First, variations can arise from the constraints of the task to be transferred (e.g., dropping a sugar cube in a small cup requires more precision than dropping a bouillon cube in a big pot). Then, variations can arise from the kinematic structure of the robot (a redundant arm can achieve the same task in different manners). Most often, these variations are better described by taking into account the covariations (instead of each dimension separately). From a statistical perspective, this corresponds to the use of full covariance matrices instead of diagonal covariance matrices. In high-dimensional problems, the full covariance matrices will often be considered with a low rank structure (e.g., PCA decomposition in joint angle space). From a motor control perspective, such approach allows the encoding of coordination patterns and synergies that are of primary importance in many different skills, and at diverse levels [32].

One of the challenges of LfD is to exploit the detected covariations in the demonstrations to retrieve an adaptive and robust controller for the reproduction of the task. This can, for example, take the form of competing constraints through the weighting of movement primitives activated in parallel, or through a hierarchical organization of the task [35]. This can also be incorporated more directly within a control strategy by retrieving a controller following a minimal intervention control strategy [18, 58].

Task-Adaptive Movements
An important aspect of LfD is to enable robots to acquire skills that can be adapted to new situations. A common approach to achieve such generalization capability is to associate *task variables* (what describes the situation) with *movement variables* (what describes the skill) and then use a regression technique to retrieve new *movement variables* from new *task variables* (see, e.g., [44]). Such a regression approach is generic since the task variables can represent a wide range of context features organized in a vector form.

An alternative approach is to encode demonstrations from the perspective of multiple coordinate systems [16, 56]. This is achieved by providing a list of "observers" or frames of reference that can potentially be relevant for the movement or the task to transfer. Such an approach is motivated by the observation that skillful movement planning and control often require the orchestration of multiple coordinate systems that can have varying levels of importance along the task (see, e.g., [7]). Typical examples are movements in object-, body-, head-, or gaze-centered frames of reference that can collaborate in various manners for the different phases of a task. Invariance and coordination extraction in movements are also closely related to the coordinate systems in which the analysis takes place [52]. Task-parameterized movement primitives [16, 56] take inspiration from these lines of research to encode movements from the perspective of multiple coordinate systems, where a statistical analysis is simultaneously conducted in each coordinate system. Such a model encapsulates the variations and coordinations in each frame, enabling the robot to learn from demonstration how to handle perturbations and how to switch between coordinate systems, resulting in improved extrapolation capability (e.g., for the adaptation of movements to new positions of objects involved in a manipulation task).

3.1 Learning from Demonstrations of Key Positions

Inspired by a teacher showing a student the correct tennis swing by moving the student's body into static key poses (e.g. *take back*, *impact*, *follow through*), advanced kinesthetic teaching techniques require the user to demonstrate only the skeleton of the task in the form of key positions of the robot as snapshots of desired behavior [29], which can also be called viapoints or waypoints. A learning and control algorithm can then be employed to optimize the low-level motor control to replicate the demonstrated skeleton while filling the information gaps by dynamic and efficient control commands. These techniques allow the user to accurately control all articulations of the robot, enabling an LfD framework for both accurate and highly dynamic movements.

During the learning phase, the user input is processed into a distribution of optimal trajectories. The user provides a set of K demonstrations $\mathcal{D} = \{\mathcal{D}_k\}_{k=1}^{K}$, with N key positions per demonstration, i.e., $\mathcal{D}_k = \{\{^k\boldsymbol{y}_n\}_{n=1}^{N}, \boldsymbol{s}_k\}$. The approach can be applied in the robot joint space as well as in the robot's end-effector space. $\boldsymbol{s}_k$ reflects the state of the environment during the kth demonstration. While the K demonstrations show different solutions of one task, we assume that the number of key positions is constant among all demonstrations and that the order of the key positions is given by the user. In contrast to learning from full trajectories, there is no information about the timing of the demonstrated task (except for the order in which key positions are given). We recover this information by means of optimal control by exploiting the velocity and acceleration limits that we want to impose. These limits are given (e.g., by the user) in the form of element-wise intervals, e.g., $\dot{y}^i \in [-\dot{y}^i_{\text{lim}}, \dot{y}^i_{\text{lim}}]$ and $\ddot{y}^i \in [-\ddot{y}^i_{\text{lim}}, \ddot{y}^i_{\text{lim}}]$ for the ith DoF.

The learning phase consists of: (1) constructing optimal basis functions; and (2) learning a distribution of optimal trajectories by computing a probability density function of the basis function weights and the correlations with the task context.

Optimal Basis Functions

In order to construct optimal basis functions, we formulate a cascaded time-optimal control problem. The block diagram in Fig. 6 (*right*) gives an overview of the cascaded optimization scheme.

The inner optimal control problem considers a linear second-order system that is constrained to pass through key positions $\boldsymbol{y}_n$ at a given time t_n with minimal effort, i.e.,

$$
\begin{aligned}
\boldsymbol{y}^*(t) = \underset{\boldsymbol{y}(t)}{\arg\min} \; & \int_0^T \ddot{\boldsymbol{y}}(t)^\top \ddot{\boldsymbol{y}}(t) dt, \\
\text{s.t.} \; & \boldsymbol{y}(t_n) = \boldsymbol{y}_n, \quad n = 1, \dots, N, \\
& \ddot{\boldsymbol{y}}(0) = \ddot{\boldsymbol{y}}_1, \quad \ddot{\boldsymbol{y}}(T) = \ddot{\boldsymbol{y}}_N.
\end{aligned} \tag{1}
$$

A corresponding timing parameter $\boldsymbol{h} \in \mathbb{R}^{N-1}$ that encodes the time passed between two consecutive key positions, i.e., $h_n = t_{n+1} - t_n$. Consequently, the total duration

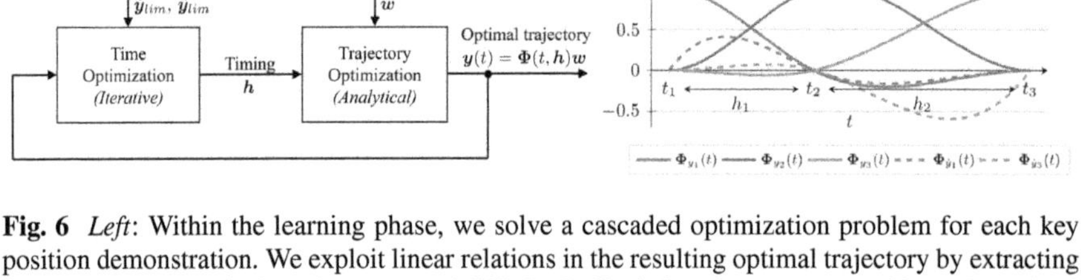

Fig. 6 *Left*: Within the learning phase, we solve a cascaded optimization problem for each key position demonstration. We exploit linear relations in the resulting optimal trajectory by extracting basis functions $\boldsymbol{\Phi}(t, \boldsymbol{h})$ for a demonstrated skill. *Right*: Basis functions constructed from the optimal control problem in (1) for a single DoF and $N = 3$ key positions

of the motion is given by $T = \sum_{n=1}^{N-1} h_n$. In [66], it has been shown that the solution of the linear optimal control problem in (1) is a polynomial spline in time. We find that the optimal trajectory can be expressed as a linear function of the key points that may come from a human demonstration. The analytical solution to the optimization problem in (1) allows us to derive *optimal basis functions* that map the key positions to an optimal trajectory. Figure 6 (*right*) shows a set of basis functions for a task instance with three key positions with a general timing parameter $\boldsymbol{h}$. Each colored signal represents a basis function that corresponds either to one of the key positions or to one of the boundary velocities.

Time Optimization

Within the outer optimization problem, the timing parameter $\boldsymbol{h}$ can be optimized. This is required to achieve efficient and smooth robot motions despite the fact that the human demonstrations do not provide information about timing and velocities of the robot. We find a timing parameter $\boldsymbol{h}^*$ that locally minimizes the movement duration T while satisfying the constraints through the sequential least squares quadratic programming algorithm [34]. For each demonstration of key positions $\mathcal{D}_k$ (setting $\dot{y}_1 = \dot{y}_N = \mathbf{0}$), we compute the optimal timing $\boldsymbol{h}_k^*$ and select a conservative timing parameter $\bar{\boldsymbol{h}}$ by selecting each element (i.e., the individual duration between consecutive key positions) as the maximum that appeared among all demonstrations. This conservative timing parameter ensures that the robot does not violate velocity and acceleration constraints when reproducing motions demonstrated by a human through key positions.

Probability Distribution of Optimal Trajectories

The linear mapping from human-demonstrated key positions and optimal trajectories can be exploited when multiple demonstrations of the same task are available. In this case, the set of demonstrated key positions can be interpreted as samples from a probability distribution over the space of key positions. The optimal basis functions then allow us to map not only single demonstrations to optimal robot trajectories but also Gaussian distributions of key positions to Gaussian distributions of robot trajectories. By also providing a context variable $\boldsymbol{s}$ that describes the state of the workspace for a given demonstration, correlations between the workspace state and the robot's motion can be learned.

Learning in Multiple Frames

Such correlations can be learned in a variety of coordinate systems. If the robot's motion is coupled with the location of object, for example if the robot is supposed to pick up this object, correlations are easier to detect and to learn in an object-centric coordinate system. In [18], poses of objects in the environment are specifically handled by learning the movement statistics in a coordinate system that is attached to the corresponding object. The same can be applied to the statistics of key positions in various frames. The robot's motion can then be generated by mapping the statistics in all coordinate systems into a world-fixed frame and subsequently fused by computing the product of Gaussians.

3.2 User Study

We here present the results of a user study comparing the effectiveness of teaching through key positions (Method **KP**) against full trajectories (Method **FT**), alongside other experiments showcasing the effectiveness of the proposed approach.

We conducted a study where 16 novice robot users (mean age 28.8, SD 3.2) provided demonstrations to solve a pick-and-place task with a 7-axis Franka Emika manipulator.

Experimental Setup: The task consisted of picking a toy drill (from the YCB dataset [20]) and placing it inside one of two boxes, as shown in Fig. 7. The participants provided 8 demonstrations by displacing the gravity-compensated robot arm. For each demonstration, a drill pose was randomly selected from a pool of predefined demonstration scenarios, along with a corresponding robot's starting configuration.

Conditions and Protocol: Participants first filled a brief questionnaire consisting of three 7-point Likert scale statements, aimed at assessing their robotics expertise. An average score of 1.95 was observed (Cronbach's $\alpha = 0.9$), indicating novice users. Each method was used by 8 participants.

Results: We compared the participants' effectiveness of teaching with Method **KP** and Method **FT** and with different numbers of demonstrations, in terms of success of the task reproductions and of quality of the reproduced trajectory.

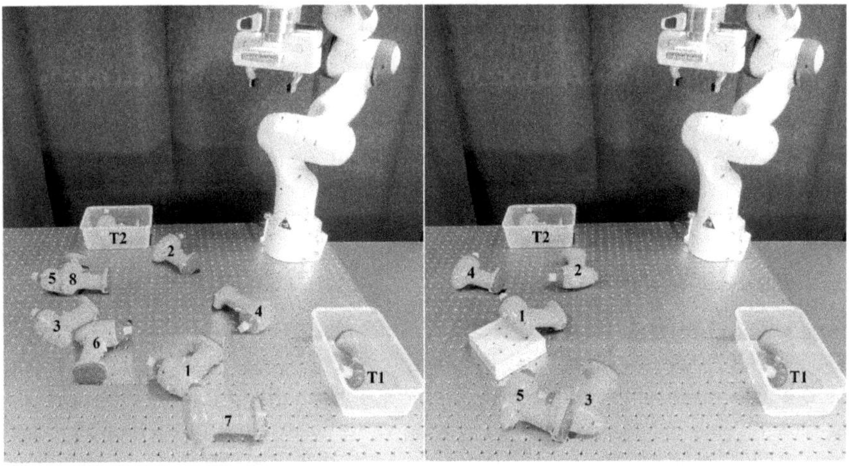

Fig. 7 Experimental setup of the user study: drill pick-up (1–8, left figure) and drop locations (T1 and T2) used for the demonstration collection, and test locations used in the evaluation phase (1–5, right figure)

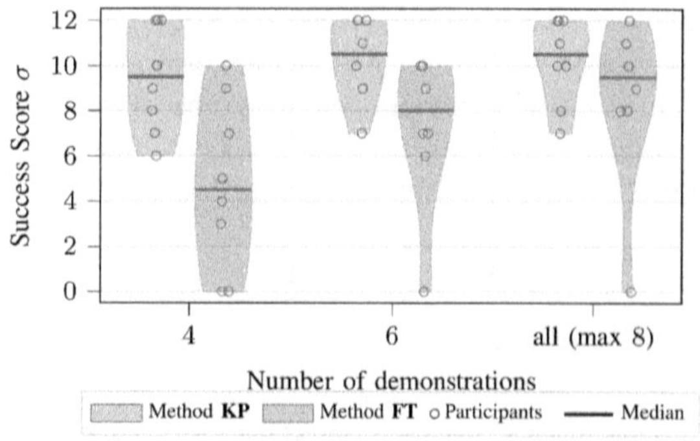

Fig. 8 Distribution of cumulative success score σ, separated by method and by number of demonstrations utilized for training

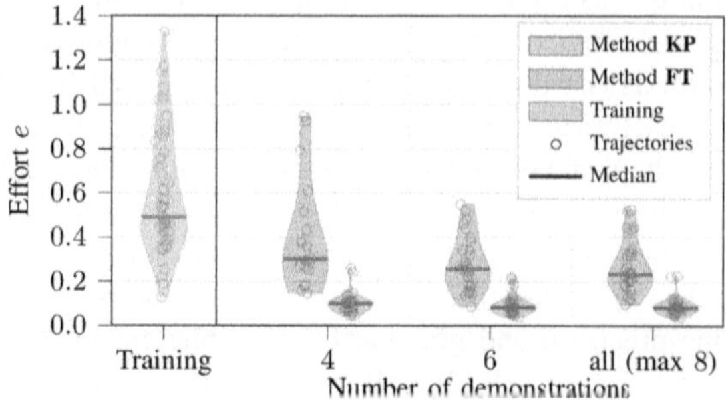

Fig. 9 Distribution of acceleration effort e of trajectory reproductions, separated by method and by number of demonstrations utilized for training. The leftmost violin instead shows e of the collected demonstrations

Figure 8 compares the distribution of cumulative score σ for each teaching method. We observe a statistically significant score difference between methods when trained with 4 and 6 demonstrations (Mann–Whitney U test,[1] $p < .01^{**}$), with a median score of 9.5 for Method **KP** vs 4.5 for Method **FT** when trained with 4 demonstrations. This difference can be explained by the fact that Method **KP** needs to learn in a lower-dimensional space than Method **FT**; a relevant difference when learning from a few demonstrations. As expected, σ improves for both methods with more than 6 demonstrations, with negligible differences between methods.

Figure 9 compares the distribution of acceleration effort e for each teaching method and for the participants' demonstrations. The acceleration effort of the robot is a metric for measuring the efficiency of the generated robot motion. A low-acceleration robot motion may require less energy to execute and in higher predictability for the human co-worker. Method **KP** produces trajectories with significantly lower e compared to Method **FT** (Mann–Whitney U test, $p < .01^{**}$ for any number of demonstrations used in the training). This is a direct consequence of the underlying optimization process.

As also observed in previous work [1], participants are almost twice as fast when providing full trajectory demonstrations (median 18 s) compared to key position demonstrations (median 28 s). However, the success score achieved by Method

[1] Data was tested for normality, always rejecting the null hypothesis (Shapiro-Wilk test, $p < .01^{**}$). We, therefore, used non-parametric tests.

KP trained with 4 demonstrations ($\sigma^{4}_{KP,\text{median}} = 9.5$) is comparable with the one of Method **FT** with twice the demonstrations ($\sigma^{8}_{\text{FT, median}} = 9.5$). It can be argued that, in a real scenario, operations other than the demonstration collection, such as setting the environment or the robot arm's initial configuration, would dominate the process, making the aforementioned time difference negligible.

4 Conclusion

In this chapter, we have introduced two approaches to simplifying robot programming: block-based programming (BBP) and learning from demonstration (LfD). These methods offer complementary strengths for intuitive robotic control. BBP, with its visual drag-and-drop interface, is well-suited for the programming of trajectories, providing precise control over task logic and structure. However, it can become cumbersome when handling complex behaviors, particularly for tasks that cannot be easily defined by a trajectory, such as polishing or other nuanced processes. LfD, on the other hand, allows robots to learn tasks through human demonstration, making it more intuitive for continuous or motor-driven behaviors. However, LfD lacks the precision and granular control inherent in BBP. Combining these approaches creates a powerful hybrid system: BBP can define high-level task structures, such as conditions and sequences, while LfD can handle complex, dynamic actions like object manipulation. This hybrid system allows users to refine LfD-learned behaviors within BBP or convert demonstrated tasks into reusable blocks for more sophisticated workflows. The result is an iterative improvement process in which users begin with basic programming in BBP and enhance it with LfD, creating a flexible and adaptable system that balances precise control with intuitive learning.

The two implementation of BBP and LfD that we have presented allow non-expert users to learn to autonomously program robots and develop a feeling of control over the machines. However, tools must be designed specifically for the target users, taking into account their unique backgrounds, motivations, and interests, to ensure software dependability [14]. If further testings with target users is still needed to fully guarantee this dependability, our results corroborate the relevance of the proposed approaches. On the long term, the integration of empowering tools for production work might enhance work conditions and, hopefully, give back the control over their work to the workers, turning them into makers rather than passive users [54]. Indeed, two visions regarding the future of partial automation can be identified: Human Emulation and Human Complementary [55]. On the one hand, in the emulation scenario, the machine will become more and more intelligent and autonomous to finally be able to replace humans. The human intervention is a temporary solution to help the robot perform the work until it can do it autonomously. The tasks left to the worker are typically extremely monotonous, possible leading to deskilling [23] and a feeling of mechanization [13]. On the other hand, in the complementary scenario, it is argued that humans and machines have complementary capabilities that can be optimally combined through collaboration. Indeed, Human-Robot Collaboration allows for automation productivity with human flexibility. Technology is a tool for humans to render work more efficient. If robots as human emulators represent a danger for workers, robots as human complements (or tools) can on the contrary help improve working conditions and even re-humanize work [40]. Indeed, procedural, repetitive tasks that requires the human to act like a machine can be automated, while tasks requiring common sense, creativity and dexterity can be left to the human. However, in order to achieve this empowerment through technology, programming solutions that foster worker reskilling are needed [31]. In this chapter, we have presented two possible approaches that allow to cover a wide range of tasks. Our approach is particularly interesting for small to medium enterprises (SMEs) that typical produce smaller volumes. Flexibility is key since the systems need to be usable for different tasks to reach the volume needed to make the system profitable, but, even more importantly, for the company to be able to re-task the robot rapidly. Indeed, if cycle time is crucial for mass production and requires expert, specialized solutions that are optimized for speed, the critical factor for smaller production volumes is the time needed to switch from one application to another, the cycle time being less important. In addition, because of the market versatility and the associated rapid changes in the market demand, reactivity is a key issue: since most SMEs cannot afford to hire an expert for their production, having to outsource the reprogramming of the system for new tasks also implies a delay in production launch. Solutions that allow non-robotic experts to re-task production systems are thus needed not only to reduce costs, but also to save time. In this chapter, we have presented two promising approaches to empower production workers and give them the needed autonomy to easily program robots on their own, contributing to the development of a Human Complementarity approach to automation.

References

1. Akgun, B., Cakmak, M., Yoo, J.W., Thomaz, A.L.: Trajectories and keyframes for kinesthetic teaching: a human-robot interaction perspective. In: Proceedings of the Seventh Annual ACM/IEEE International Conference on Human-Robot Interaction, pp. 391–398. (2012)
2. Blanc, C., Boudry, L., Sonderegger, A., Nembrini, J., Dégallier-Rochat, S.: Empowering production workers to program robots: a no-code, skill-based approach. In: International Conference on Computer-Human Interaction Research and Applications, pp. 21–39. Springer (2023)
3. Argall, B.D., Chernova, S., Veloso, M., Browning, B.: A survey of robot learning from demonstration. Robot. Auton. Syst. **57**(5), 469–483 (2009)
4. Bangor, A., Kortum, P., Miller, J.: Determining what individual SUS scores mean: adding an adjective rating scale, (2009)
5. Barbeite, F.G., Weiss, E.M.: Computer self-efficacy and anxiety scales for an Internet sample: testing measurement equivalence of existing measures and development of new scales. Comput. Hum. Behav. **20**(1), 1–15 (2004)
6. Bau, D., Gray, J., Kelleher, C., Sheldon, J., Turbak, F.: Learnable programming. Commun. ACM **60**(6), 72–80 (2017)

7. Bennequin, D., Fuchs, R., Berthoz, A., Flash, T.: Movement timing and invariance arise from several geometries. PLoS Comput. Biol. **5**(7), 1–27 (2009)
8. Billard, A.G., Calinon, S., Dillmann, R.: Learning from humans. In: Siciliano, B., Khatib, O. (eds.) Handbook of Robotics, pp. 1995–2014. Springer, Secaucus, NJ, USA (2016) 2nd edn., chap. 74
9. Bitonneau, D.: Conception de systèmes cobotiques industriels: approche robotique avec prise en compte des facteurs humains: application à l'industrie manufacturière au sein de Safran et ArianeGroup, Université de Bordeaux (2018). Ph.D. thesis
10. Brooke, J.: SUS: A "Quick and DirtyâŁž Usability Scale, pp. 207–212. CRC Press (1996)
11. Bruckman, J.C.: Overcoming resistance to change: causal factors, interventions, and critical values. Psychol.-Manag. J. **11**(2), 211–219 (2008). Routledge. Eprint: https://www.tandfonline.com/doi/pdf/10.1080/10887150802371708
12. Brynjolfsson, E., McAfee, A.: The Second Machine Age: Work, Progress, and Prosperity in a Time of Brilliant Technologies. W. W. Norton & Company (2014). Google-Books-ID: WiKwAgAAQBAJ
13. Bui, L.: Asian roboticism: connecting mechanized labor to the automation of work. Perspect. Glob. Dev. Technol. **19**(1–2), 110–126 (2020)
14. Burnett, M., Cook, C., Rothermel, G.: End-user software engineering. Commun. ACM **47**(9), 53–58 (2004)
15. Cakmak, M., DePalma, N., Arriaga, R.I., Thomaz, A.L.: Exploiting social partners in robot learning. Auton. Robot. **29**(3–4), 309–329 (2010)
16. Calinon, S.: A tutorial on task-parameterized movement learning and retrieval. Intell. Serv. Robot. **9**(1), 1–29 (2016)
17. Calinon, S.: Mixture models for the analysis, edition, and synthesis of continuous time series. In: Bouguila, N., Fan, W. (eds.) Mixture Models and Applications, pp. 39–57. Springer (2019)
18. Calinon, S., Bruno, D., Caldwell, D.G.: A task-parameterized probabilistic model with minimal intervention control. In: Proceedings of the IEEE International Conference on Robotics and Automation (ICRA), pp. 3339–3344. , Hong Kong, China (2014)
19. Calinon, S., Lee, D.: Learning control. In: Vadakkepat, P., Goswami, A. (eds.) Humanoid Robotics: A Reference, pp. 1261–1312. Springer (2019). https://doi.org/10.1007/978-94-007-6046-2_68
20. Calli, B., Walsman, A., Singh, A., Srinivasa, S., Abbeel, P., Dollar, A.M.: Benchmarking in manipulation research: using the Yale-CMU-Berkeley object and model set. IEEE Robot. Autom. Mag. **22**(3), 36–52 (2015)
21. Coates, A., Abbeel, P., Ng, A.Y.: Apprenticeship learning for helicopter control. Commun. ACM **52**(7), 97–105 (2009)
22. Connolly, C., Murphy, E., Moore, S.: Programming anxiety amongst computing students–a key in the retention debate? IEEE Trans. Educ. **52**(1), 52–56 (2009). Conference Name: IEEE Transactions on Education
23. de Assis Dornelles, J., Ayala, N.F., Frank, A.G.: Collaborative or substitutive robots? Effects on workers' skills in manufacturing activities. Int. J. Prod. Res. 1–34 (2023)
24. Derus, S.R.M., Ali, M., A.Z.: Difficulties in learning programming: views of students. In: 1st International Conference on Current Issues in Education (ICCIE 2012), pp. 74–79. (2012)
25. Dimirovski, G.M., Talha Dinibütün, A., Kile, F., Neck, R., Stahre, J., Vlacic, L.B.: Control system approaches for sustainable development and instability management in the globalization age. Annu. Rev. Control. **30**(1), 103–115 (2006)
26. Ford, M.: Rise of the Robots: Technology and the Threat of a Jobless Future, Basic Books (2015)
27. Fraser, N.: Ten things we've learned from Blockly. In: IEEE Blocks and Beyond Workshop (Blocks and Beyond), pp. 49–50. (2015)
28. Hsu, T.-C., Hwang, G.-J.: Interaction of visual interface and academic levels with young students' anxiety, playfulness, and enjoyment in programming for robot control. Univ. Access Inf. Soc. **22**(1), 213–225 (2023)

29. Jankowski, J., Racca, M., Calinon, S.: From key positions to optimal basis functions for probabilistic adaptive control. IEEE Robot. Autom. Lett. **7**(2), 3242–3249 (2022)
30. Kaasinen, E., Schmalfuß, F., Özturk, C., Aromaa, S., Boubekeur, M., Heilala, J., Heikkilä, P., Kuula, T., Liinasuo, M., Mach, S., Mehta, R., Petäjä, E., Walter, T.: Empowering and engaging industrial workers with Operator 4.0 solutions. Comput. Ind. Eng. **139**, 105678 (2020)
31. Kadir, B.A., Broberg, O.: Human well-being and system performance in the. Int. J. Ind. Ergon. **76**, 102936 (2020). transition to industry 4.0
32. Kelso, J.A.S.: Synergies: atoms of brain and behavior. In: Sternad, D. (ed.) Progress in Motor Control. Advances in Experimental Medicine and Biology, vol. 629, pp. 83–91. Springer, US (2009)
33. Kieras, D.E., Bovair, S.: The role of a mental model in learning to operate a device. Cogn. Sci. **8**(3), 255–273 (1984)
34. Kraft, D.: A software package for sequential quadratic programming. Forschungsbericht. Wiss. Berichtswesen d. DFVLR, Deutsche Forschungs- und Versuchsanstalt für Luft- und Raumfahrt Köln (1988)
35. Kulic, D., Takano, W., Nakamura, Y.: Incremental learning, clustering and hierarchy formation of whole body motion patterns using adaptive hidden Markov chains. Int. J. Robot. Res. **27**(7), 761–784 (2008)
36. Lasota, P.A., Fong, T., Shah, J.A.: A survey of methods for safe human-robot interaction. Found. Trends Robot. **5**(4), 261–349 (2017)
37. Laugwitz, B., Held, T., Schrepp, M.: Construction and evaluation of a user experience questionnaire. In: Holzinger, A. (ed.) HCI and Usability for Education and Work. Lecture Notes in Computer Science, pp. 63–76. Springer, Berlin, Heidelberg (2008)
38. Mayr-Dorn, C., Winterer, M., Salomon, C., Hohensinger, D., Fürschuss, H.: Assessing industrial end-user programming of robotic production cells: a controlled experiment. J. Syst. Softw. **195**, 111547 (2023)
39. McClure, P.K.: "You're fired", says the robot: the rise of automation in the workplace, technophobes, and fears of unemployment. Soc. Sci. Comput. Rev. **36**(2), 139–156 (2018)
40. Mormann, H., Endrissat, N.: Humanizing organizations: the professional role of HRM in organizational transformation. Die Unternehmung **76**(4), 418–429 (2022)
41. Nehaniv, C.L., Dautenhahn, K. (eds.): Imitation and Social Learning in Robots, Humans, and Animals: Behavioural. Social and Communicative Dimensions. Cambridge University Press, Cambridge, UK (2007)
42. Ninomiya, T., Fujita, A., Suzuki, D., Umemuro, H.: Development of the multi-dimensional robot attitude scale: constructs of people's attitudes towards domestic robots. In: Tapus, A., André, E., Martin, J.-C., Ferland, F., Ammi, M. (eds.) Social Robotics. Lecture Notes in Computer Science, pp. 482–491. Springer, Cham (2015)
43. Nomura, T., Kanda, T., Suzuki, T.: Experimental investigation into influence of negative attitudes toward robots on human-robot interaction. AI Soc. **20**, 138–150 (2006)
44. Paraschos, A., Daniel, C., Peters, J., Neumann, G., Burges, C.J.C., Bottou, L., Welling, M., Ghahramani, Z.: Probabilistic movement primitives. In: Weinberger, K.Q. (ed.) Advances in Neural Information Processing Systems (NIPS), pp. 2616–2624. Curran Associates, Inc (2013)
45. Ratliff, N., Ziebart, B.D., Peterson, K., Bagnell, J.A., Hebert, M., Dey, A., Srinivasa, S.: Inverse optimal heuristic control for imitation learning. In: International Conference on Artificial Intelligence and Statistics (AIStats), pp. 424–431. (2009)
46. Reich-Stiebert, N., Eyssel, F.: Learning with educational companion robots? Toward attitudes on education robots, predictors of attitudes, and application potentials for education robots. Int. J. Soc. Robot. **7**(5), 875–888 (2015)
47. Resnick, M., Maloney, J., Monroy-Hernández, A., Rusk, N., Eastmond, E., Brennan, K., Millner, A., Rosenbaum, E., Silver, J., Silverman, B., Kafai, Y.: Scratch: programming for all. Commun. ACM **52**(11), 60–67 (2009)
48. Romero, D., Stahre, J., Taisch, M.: (2020). The operator 4.0: towards socially sustainable factories of the future

49. Schou, C., Andersen, R.S., Chrysostomou, D., Bøgh, S., Madsen, O.: Skill-based instruction of collaborative robots in industrial settings. Robot. Comput.-Integr. Manuf. **53**, 72–80 (2018)
50. Schwab, K.: Shaping the Future of the Fourth Industrial Revolution. Currency , (2018)
51. Sonderegger, A., Heyden, K., Chavaillaz, A., Sauer, J.: AniSAM & AniAvatar: animated visualizations of affective states. In: Proceedings of the 2016 CHI Conference on Human Factors in Computing Systems, p. 4837 (2016)
52. Sternad, D., Park, S.-W., Mueller, H., Hogan, N.: Coordinate dependence of variability analysis. PLoS Comput. Biol. **6**(4), 1–16 (2010)
53. Sætren, G.B., Hogenboom, S., Laumann, K.: A study of a technological development process: human factors-the forgotten factors? Cognit. Technol. Work **18**(3), 595–611 (2016)
54. Taylor, M.P., Boxall, P., Chen, J.J.J., Xu, X., Liew, A., Adeniji, A.: Operator 4.0 or Maker 1.0? Exploring the implications of Industrie 4.0 for innovation, safety and quality of work in small economies and enterprises. Comput. Ind. Eng. **139**, 105486 (2020)
55. Terveen, L.G.: Overview of human-computer collaboration. Knowl.-Based Syst. **8**(2), 67–81 (1995)
56. Ti, B., Razmjoo, A., Gao, Y., Zhao, J., Calinon, S.: A geometric optimal control approach for imitation and generalization of manipulation skills. Robot. Auton. Syst. **164**, 104413 (2023)
57. Tisza, G., Sharma, K., Papavlasopoulou, S., Markopoulos, P., Giannakos, M.: Understanding fun in learning to code: a multi-modal data approach. In: Proceedings of the 21st Annual ACM Interaction Design and Children Conference, IDC '22, pp. 274–287. Association for Computing Machinery, New York, NY, USA (2022)
58. Todorov, E., Jordan, M.I.: A minimal intervention principle for coordinated movement. In: Advances in Neural Information Processing Systems (NIPS), pp. 27–34. (2002)
59. Villani, V., Pini, F., Leali, F., Secchi, C.: Survey on human-robot collaboration in industrial settings: safety, intuitive interfaces and applications. Mechatronics **55**, 248–266 (2018)
60. Weintrop, D., Afzal, A., Salac, J., Francis, P., Li, B., Shepherd, D.C., Franklin, D.: Evaluating CoBlox: a comparative study of robotics programming environments for adult novices. In: Proceedings of the 2018 CHI Conference on Human Factors in Computing Systems, pp. 1–12. ACM, Montreal, QC, Canada (2018)
61. Weintrop, D., Shepherd, D.C., Francis, P., Franklin, D.: Blockly goes to work: block-based programming for industrial robots. In: 2017 IEEE Blocks and Beyond Workshop (B&B), pp. 29–36 (2017)
62. Weintrop, D., Wilensky, U.: To block or not to block, that is the question: students' perceptions of blocks-based programming. In: Proceedings of the 14th International Conference on Interaction Design and Children, pp. 199–208. (2015)
63. Weintrop, D., Wilensky, U.: Comparing block-based and text-based programming in high school computer science classrooms. ACM Trans. Comput. Educ. **18**(1), 3:1–3:25 (2017)
64. Whiten, A., McGuigan, N., Marshall-Pescini, S., Hopper, L.M.: Emulation, imitation, over-imitation and the scope of culture for child and chimpanzee. Philos. Trans. R. Soc. B **364**(1528), 2417–2428 (2009)
65. Talha Yildiz, T., Seferoglu, S.S.: The effect of robotic programming on coding attitude and computational thinking skills toward self-efficacy perception. J. Learn. Teach. Digit. Age **6**(2), 101–116 (2021)
66. Zhang, Z., Tomlinson, J., Martin, C.: Splines and linear control theory. Acta Math. Appl **49**, 1–34 (1997)

Shared Control for Assistive and Collaborative Robotics

Claudio Pacchierotti, Divyaksh Subhash Chander, Louise Devigne, Marie Babel, and Charles Pontonnier

Abstract Assistive and collaborative robots offer a promising solution for reducing human physical stress and exposure across a broad range of applications. The design and control of such robotic systems should account for humans needs, so as to assist their motion and activity as well as optimize their efficiency and performance throughout the considered tasks. In this respect, designing *around* the human means accounting for their musculoskeletal constraints, features, and limitations during the development of robotic mechanical systems and control algorithms, such as powered exoskeletons or cobots. This Chapter presents an overview of human-centered design and shared control techniques in robotics, including supporting examples and applications in the field of industrial teleoperation and mobility assistance. We also discuss the perspectives of the field and how such human-centered approaches can be beneficial to a broader range of applications than those where it is implemented nowadays.

Keywords Human-centered robotics · Shared control · Assistive robotics · Collaborative robotics · Ergonomics · Haptics · Disability

1 Introduction and Scope

Assistive and collaborative robots are robotic systems that work with and for humans to help or support them in a physical task. Such systems are increasingly popular [1]

C. Pacchierotti (✉)
CNRS, Univ Rennes, Inria, IRISA, Rennes, France
e-mail: claudio.pacchierotti@irisa.fr

D. S. Chander
AnyBody Technology A/S, Aalborg, Denmark

L. Devigne · M. Babel
Univ Rennes, INSA Rennes, Inria, IRISA, Rennes, France

C. Pontonnier
Univ Rennes, ENS Rennes, Inria, IRISA, Rennes, France

G. Barresi and S. Wischniewski (eds.), *Ergonomics in Robotics: Advances and Innovations*, Springer Tracts in Advanced Robotics 162,
https://doi.org/10.1007/978-3-032-19374-2_14

and their broad spectrum of application spans from industrial teleoperation to surgery, from industrial training to tele-working and elderly care [2]. However, their design and control often do not take into account how humans use and interact with them, leading to generic designs that do not fit well the purpose of these systems. Indeed, differently from standard general-purpose industrial robots, assistive and collaborative robotic systems need to be designed accounting for the specificity of the considered task, as well as the user's anthropometrics and/or pathologies, depending on the area of application. This reasoning has motivated the concept of "human-centered design," in which the knowledge of the task to accomplish is combined with that of the human to assist, towards the development of effective and safe human-centered robotic systems.

This Chapter discusses current techniques that aim at including the "human" in the design and control of robotic systems, focusing on two representative activities that are at the forefront of human-centered assistive and collaborative robotics: industrial teleoperation and mobility assistance for people affected by sensory disabilities.

Robotic teleoperation. Teleoperation is a tool to extend human sensing and manipulation capabilities to remote environments. Teleoperated robotic systems are composed of a remote robot, which acts on the target environment, and of a local control system, which is commonly operated by a human user. In the most common implementations, the remote robot reproduces the movements of the human operator who, in turn, receives suitable information about the environment and the task to accomplish, in the form of audio, visual, and haptic feedback. Standard human-in-the-loop teleoperation systems enable a human operator to fully control the motion of the robot. However, in some cases, it is beneficial to *share* the control of the remote robot between the human operator and an autonomous algorithm, helping operators in carrying out increasingly challenging tasks. Shared control offers significant advantages, allowing for the combined strengths of human operators and autonomous algorithms. Despite the broad use of shared-control algorithms in robotic teleoperation [3–6], researchers have rarely focused their attention on accounting for the users' needs during robotic telemanipulation, e.g., their comfort, as well as for their specific capabilities, e.g., their experience. Nonetheless, this is a very important issue, as operators in many high-impact applications use robotic teleoperation system for long uninterrupted time periods (e.g., a robotic prostatectomy generally takes 2–4 h), making these considerations relevant.

Assisted mobility systems. Robotic systems for assisting the mobility of users exist in many forms, including power wheelchairs, robot-assisted walkers, and robotic white canes. They are used to overcome mobility impairments resulting from orthopaedic or neurological disabilities. Unfortunately, many individuals who need these mobility aids are unable to use them. Some people with cognitive and/or visual impairments are in fact not able to safely use them [7]. Navigation assistance is crucial for individuals with visual and/or cognitive impairments, allowing them to safely operate power wheelchairs or walkers This can be enforced by directly controlling the movement of the assistive device either in assist-as-needed (shared control) or autonomous ways [8, 9]. However, it is proven that doing a task yourself rather than letting a machine do it for you is more rewarding [10, 11], suggesting shared control approaches as the most promising. In this context, similarly to what we said for robotic teleoperation, researchers have rarely focused their attention on accounting for the users' needs during the use of such robots for assisted mobility, as well as for the users' specific capabilities, e.g., their pathology and preferences.

The remaining of this Chapter discusses how robotic control algorithms can be designed to include the needs of the human in the design and the decision loop of the control algorithm, through two representative examples of shared control.

2 Human in Control: Shared Control Techniques for the Control of Human-Centered Robotic Systems

2.1 Shared Control in Robotics

The evolution of robotic technologies toward addressing more challenging tasks requires an increasing degree of training, skills, and concentration from the human operator. For this reason, researchers have begun to design innovative techniques making the control of robotic systems systems more effective and intuitive, with the objective of increasing the performance of the task and reducing the cognitive and physical loads of the human operator. In fact, standard human-in-the loop robotic control, where the human user is in charge of all aspects of the robotic movements, has a number of drawbacks: human operators may not be fully aware of the robot capabilities and constraints (e.g., joint limits, singularities, limited workspace, collision avoidance) and end up spending large amounts of cognitive mental load for dealing with issues that can (and should), instead, be directly tackled by the robot's autonomy. In this respect, shared control algorithms have been investigated as one of the main tools for designing complex yet intuitive robotic systems, helping operators carrying out increasingly challenging tasks. This approach enables to *share* the available degrees of freedom of the robotic system between the operator and an autonomous controller.

Haptic feedback is a good mean for dynamically guiding and/or constraining the movements of an operated robotic system [12]. It informs the user about suggestions or warnings the controller considers beneficial for the task, such as the presence of dangerous regions of the workspace (e.g., using a repulsive force field pushing away from the forbidden area) or preferred directions of motion (e.g., using an attractive force field toward the target path). This assistance can be (i) unconstrained when the information is provided via ungrounded cutaneous haptics, allowing the user to choose whether to comply with the given suggestion, (ii) semi-constrained, when the information is provided via grounded kinesthetic haptics but the user can still deviate a bit from the suggested action, or (iii) fully constrained, when the information is provided again via grounded kinesthetic haptics but the user is forced to follow the given suggestion.

Unconstrained assistance (also known as guidance) has been used for providing navigation information to the visually-impaired [13], when driving a motorbike [14], during robotic surgery [15, 16], needle insertion [17, 18], telemanipulation [19, 20], as well as control of drones [21–23] and microrobots [24]. To leave the user free whether to comply or not to the given suggestion, unconstrained assistance provides feedback through ungrounded cutaneous devices, usually providing either vibrations [13, 15, 16, 18, 22, 24] or skin stretch [19, 22, 25], to convey information on a suggested path to follow [14, 18, 19, 22], an area to avoid [16], or the distance from a target pose of the robot [15, 22]. Thanks to using ungrounded cutaneous devices, this technique is very safe, as it is not affected by instability issues and leaves the operator free to deviate from the proposed suggestion, if needed. However, for the very same reasons, unconstrained approaches are also unfit to render important constraints (e.g., approaching of dangerous areas of the environment), as they are too easy to disregard. Moreover, the scarcity of suitable ungrounded cutaneous devices limits the quality and type of feedback available. Most unconstrained approaches use vibrotactile guidance through Eccentric Rotating Mass displays, which cannot provide rich information and can only be used for a limited amount of time, as sustained vibrations become quickly uncomfortable.

Semi-constrained approaches (also known as active constraints, virtual fixture, or haptic shared control) have been used in robot-assisted surgery [26, 27], palpation [28], cooperative manipulation [29], assembly [30], suturing [31], microrobotics [24], mobile robotics [32, 33], and grasping [34–37]. Most of these approaches have one of two purposes: *Forbidden-region constraints* [29] prevent the tool from entering certain poses and configurations, offering benefits such as task simplification [30], reduced risk of tools damaging protected regions [28, 31, 38], and avoidance of kinematic singularities [36, 39]; *navigational constraints* encourage the operator to move the tool along a specific pathway or toward a specific target [24, 27, 31, 34, 35]. Semi-constrained techniques enable constraints of variable strength, compliant when the user's deviation from the suggested action is acceptable and stiff otherwise. However, of course, to achieve this effect, the system must use grounded kinesthetic feedback, which might lead the teleoperation system to an unsafe behaviour. Moreover, when employing a *single-point* kinesthetic interface to render *both* assistance feedback and contact sensations, users might get confused regarding the origin of the feedback (am I feeling a contact with the environment or an active constraint?). This ambiguity can be avoided by distributing multiple interfaces throughout the body (multi-point interfaces), which is easy to do with ungrounded cutaneous devices, or by using sensory substitution techniques, where force information is substituted by audio or visual feedback.

Finally, constrained assistance has been used in prosthetics [40], mobile robotics [41], robotic cutting [42], car driving [43], and telemanipulation [44–46]. As the allowed user's motion reduces, constrained approaches become more and more similar to fully autonomous techniques. Constrained assistance is used in those scenarios where the division of work between the autonomous controller and the user is well defined. However, as teleoperation systems become more sophisticated and flexible, the environments in which they are employed become less structured and predictable. For this reason, there is an increasing need for approaches combining the knowledge and capabilities of autonomous systems with those of a human user. Moreover, in some applications, for reasons of safety and acceptance, it is necessary to favour techniques enabling a human to directly control the robotic tools (e.g., in medical robotics).

Despite the broad adoption of (haptic) shared control techniques, systems rarely focus on the user's comfort, leading to very few shared control techniques designed with this specific objective in mind. This is however a relevant issue, especially in those situations where the human operator controls the robotic system for a long uninterrupted period of time [47] (e.g., in medical surgery, consider that a robotic prostatectomy takes 2–4 h).

To deal with such issues, understanding and formalizing the tasks to be assisted as well as the human capabilities are key to the design of devices able to share the task realization between the robot and the human, especially for wearable devices such as exoskeletons. From the specification of the systems [48] to their control [49], their design benefits greatly from this knowledge.

From this perspective, modeling and simulating the human-robot system is a powerful and efficient way to test the robot functionalities and a potential shared control. Modeling and simulating means using experimental or simulated motion and force data to emulate or estimate the human biomechanical behavior in interaction with the assistive device. The control itself may use in real time a biomechanical model representative of the user to adapt its assistance with regard to the capabilities of the user at a given time. Biomechanical models can remain simple robotic representations of the human or being much more detailed such as in muscuskeletal simulation. Musculoskeletal simulation involves solving the kinematic and dynamic equations of motion to estimate the forces required for movement. Generally, in MS, a musculoskeletal model (MSM) relies on an arborescent structure of rigid segments linked by joints and actuated through muscles. This model leads to a large set of parameters influencing such simulation [50] and the estimated biomechanical variables. Musculoskeletal simulation (MS) can offer valuable insights into human movement and motor control at kinematic, dynamic, and muscle levels. Many available libraries tackle these simulation with a high degree of biomechanical fidelity [51–55].

A few examples including human modeling are available in the field of human robot interaction (HRI). For example, Busch et al. [56] used a simplified human model to calculate the human operator's body pose during HRI. A continuous cost function was derived using the Rapid Entire Body Assessment score (REBA). Then, it was used to position the robot so as to optimize the ergonomic comfort during interaction. Marin et al. [57] aimed at optimizing the ergonomics of an HRI task where the human operator has to drill on a board held by a robot. Musculoskeletal simulation was employed to train a Contextual Ergonomics Model through a probabilistic supervised learning Gaussian process. Results showed reduced muscular activations for the optimized drilling position. Chen et al. [58] modeled the human arm as a 7-degrees-of-freedom (7-DoF) robotic manipulator. The muscular effort was evaluated from the estimated value of the joint torques. Also Peternel et al. [59] estimated muscular effort from torques during co-manipulation tasks. The robot interacting with the human uses a whole-body dynamic model of the human to optimize for the position of the co-manipulation task in the workspace.

In all these cases, the robot pose during the interaction is optimized to help the user perform the interactive task with the least effort or the highest comfort possible. Such an approach is of course also beneficial during robotic telemanipulation. Through shared control, it is possible to guide the teleoperation along directions that maximize the human operator's comfort, taking into account ergonomics-related measures.

2.2 *Human-Centered Shared Control: An Example for Robotic Telemanipulation*

Human-centered design as well as ergonomics and biomechanics insights can drive the design of a shared control algorithm aiming at increasing the operator's comfort and other ergonomics-related metrics during robotic telemanipulation and cooperation. This Section summarizes a haptic-enabled shared-control approach aimed at minimizing the user's workload during a teleoperated pick-and-place task, originally presented in [60, 61]. Using an inverse kinematic model of the human arm and the rapid upper limb assessment (RULA) metric, the proposed approach estimates the current user's comfort at runtime. From this measure and an a priori knowledge of the task, it then generates dynamic active constraints guiding the users toward a successful completion of the task, along directions that improve their posture and increase their comfort.

Apparatus and reference frames. The operator's side consists of a Haption Virtuose 6-DoF haptic interface, while the remote side is composed of a Franka Panda 7-DoF manipulator equipped with a parallel gripper (see Fig. 1). We consider four reference frames, shown in Fig. 1.

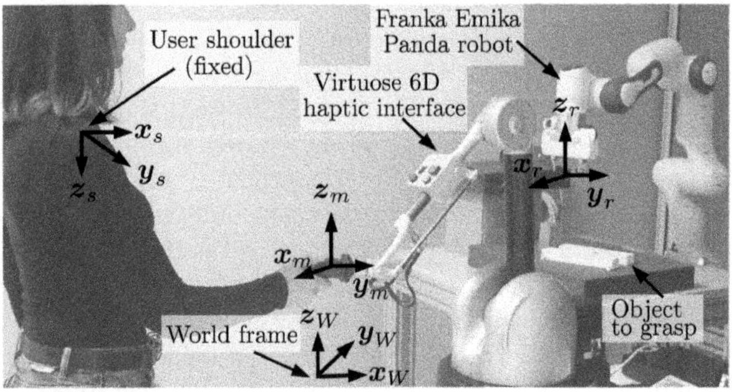

Fig. 1 The experimental setup for the pick and place task. On the user's side, a Haption Virtuose 6-DoF haptic device, on the robot side a 7-DoF Franka Emika Panda robot. The user shoulder is assumed to be fixed. Picture adapted from [60]

Estimating the user workload. As mentioned at the beginning of this Chapter, human posture is often used to estimate and prevent work-related musculoskeletal risks, as posture is indeed one of the risk factors for workers. There are many ways to assess the user's workload during a manual task. One of the them is the Rapid Upper Limb Assessment (RULA) method [62]. It is a discrete metric in which a score is given for each upper limb configuration. The sum of these scores represents the overall workload experienced by the user, i.e., the higher the RULA score, the higher the workload. Since this work aims to estimate the user's load or discomfort continuously throughout the task, a RULA-inspired continuous metric is employed. It is designed to increase as the deviation from the resting and most comfortable position of each joint angle in the arm increases. At each time step, the following quantities are calculated: the shoulder angles $\boldsymbol{q}_s = [q_1, q_2, q_3]^T$, the elbow angle $q_e = q_4$, and the wrist angles $\boldsymbol{q}_w = [q_5, q_6, q_7]^T$. Finally, $\boldsymbol{q}_{\text{arm}} = [\boldsymbol{q}_s^T, q_e, \boldsymbol{q}_w^T]^T$ denote the representation of the arm angles. From these quantities, it is possible to derive the workload $\mathcal{W}$ as the sum of the squared differences between the angles and their rest positions ($\pi/2$ for the elbow angle, and 0 for the others),

$$\mathcal{W} = \boldsymbol{q}_s^T \boldsymbol{q}_s + (q_e - \pi/2)^2 + \boldsymbol{q}_w^T \boldsymbol{q}_w. \tag{1}$$

Inverse kinematics. Estimating the workload $\mathcal{W}$ assumes that the joint angles $\boldsymbol{q}_{\text{arm}}$ of the user arm are known at each time step. However, given that the arm is represented with a 7-DoF kinematic model, the inverse kinematics is not straightforward. Let us consider the angle $\psi \in [0, \pi]$, representing the swivel angle of the arm around a virtual line connecting the shoulder to the wrist. If ψ is known, the redundancy is resolved. In this respect, we assume that the user naturally chooses the most comfortable configuration *of the elbow*, choosing the ψ leading to the least discomfort between all possible solutions (lowest $\mathcal{W}$).

Shared control architecture. The objective of the shared control should be to maximize the user's comfort while, of course, taking into account the requirements and objectives of the task, e.g., grasp that object, move along that trajectory, stay away from robot's singularities. To do so, the haptic feedback is divided into two components: a human-centered component $\boldsymbol{f}_{\mathcal{W}}$, which focuses on maximizing the user's comfort, and a task-centered one $\boldsymbol{f}_r$, which takes the task needs into account.
Human-centered cost function. To minimize the muscular discomfort experienced during the task execution, we minimize the cost function $\mathcal{H}_{\mathcal{W}} = \mathcal{W}$, i.e., the workload itself. The user is in a comfortable position when $\mathcal{W}$ is low.
Task-centered cost function. It is of course also important to consider what the task aims at achieving. A second cost function $\mathcal{H}_r$ represents this task-centered metric. This cost function is general and should be adapted to one's specific task. In this work, we minimize the Euclidean distance to the pose of the object to grasp, i.e.,

$$\mathcal{H}_r = \begin{bmatrix} \boldsymbol{p}_{r,g} - \boldsymbol{p}_r \\ {}^r(\theta \boldsymbol{u})_{r,g} \end{bmatrix}^T \begin{bmatrix} \boldsymbol{p}_{r,g} - \boldsymbol{p}_r \\ {}^r(\theta \boldsymbol{u})_{r,g} \end{bmatrix}, \tag{2}$$

where $\boldsymbol{p}_r$ and $\boldsymbol{p}_{r,g}$ are the current and target positions of the robot, and ${}^r(\theta \boldsymbol{u})_{r,g}$ is the angle-axis representation of the relative rotation between the current and target robot poses.

Haptic feedback. Haptic cues guide the operator such that both task- and human-centered costs are minimized, providing

$$\boldsymbol{f}_m = \alpha \boldsymbol{f}_{\mathcal{W}} + \beta \boldsymbol{f}_r, \tag{3}$$

to the human operator. $\boldsymbol{f}_{\mathcal{W}}$ is the force guiding the operator towards the position with the highest comfort, while $\boldsymbol{f}_r$ is the force minimizing the task-centered cost function. α and β are parameters that can be adapted to the importance of each force ($\alpha + \beta = 1$). The feedback component related to $\mathcal{H}_{\mathcal{W}}$ is designed to be proportional to a desired velocity of the haptic interface in a direction that minimizes the cost function,

$$\boldsymbol{f}_{\mathcal{W}} = K_{\mathcal{W}} \boldsymbol{v}_{m,d_{\mathcal{W}}}, \tag{4}$$

where $K_{\mathcal{W}}$ is a proportional constant, and $\boldsymbol{v}_{m,d_{\mathcal{W}}}$ is the desired velocity of the haptic device, based on the metric maximizing the comfort.

It is calculated from the desired joint velocities of the human arm, $\dot{\boldsymbol{q}}_{\text{arm},d}$, in this way: $\boldsymbol{v}_{m,d_{\mathcal{W}}} = \boldsymbol{T}\boldsymbol{J}\dot{\boldsymbol{q}}_{\text{arm},d}$, where $\boldsymbol{T}$ is a transformation matrix mapping the desired velocity from the shoulder to haptic interface frame, and $\boldsymbol{J}$ is the Jacobian of the human arm. Velocity $\dot{\boldsymbol{q}}_{\text{arm},d}$ is selected such that $\mathcal{H}_{\mathcal{W}}$ is minimized, or, in other words, $\dot{\mathcal{H}}_{\mathcal{W}}(\boldsymbol{q}_{\text{arm}}) = (\partial \mathcal{H}_{\mathcal{W}} / \partial \boldsymbol{q}_{\text{arm}}) \dot{\boldsymbol{q}}_{\text{arm}} \leq 0$. The desired angular velocity of the arm angles is chosen to be in the negative direction of the gradient of the human-related cost function,

$$\dot{\boldsymbol{q}}_{\text{arm},d} = -\frac{\partial \mathcal{H}_{\mathcal{W}}}{\partial \boldsymbol{q}_{\text{arm}}}. \tag{5}$$

The feedback component related to $\mathcal{H}_r$ is designed similarly,

$$\boldsymbol{f}_r = K_r \boldsymbol{v}_{m,d_r}, \tag{6}$$

where K_r is a proportional constant, and

$$\boldsymbol{v}_{m,d_r} = \begin{bmatrix} \dot{\boldsymbol{p}}_{m,d_r} \\ \boldsymbol{\omega}_{m,d_r} \end{bmatrix} = \begin{bmatrix} \boldsymbol{p}_{m,g} - \boldsymbol{p}_m \\ {}^m(\theta \boldsymbol{u})_{m,g} \end{bmatrix}, \tag{7}$$

where $\boldsymbol{p}_m$ and $\boldsymbol{p}_{m,g}$ are the current and target positions of the haptic device end-effector, and ${}^m(\theta \boldsymbol{u})_{m,g}$ is the angle-axis representation of the relative rotation between the current and the target poses of the haptic interface.

Evaluation in robotic telemanipulation. Figure 1 shows the experimental setup. The remote environment is composed of three different sets of objects to grasp, pick up, and place on a target location. Fifteen subjects participated in the study (10 males,

5 females). The pose of the user's shoulder was measured and kept fixed throughout the experiment.

Participants used the haptic interface to control the robotic manipulator and complete the considered pick-and-place manipulation task. First, they were asked to grasp the objects (first part of the task) and then move them to a target location marked on the table, where they were released (second part of the task).

As discussed before, we combine different cost functions to consider human-centered metrics as well as robot- and task-related ones. Here the estimated human workload is considered to be the human-related cost $\mathcal{H}_{\mathcal{W}}$, while the distance from a target pose of the robot, i.e., the grasping pose for the first part of the task and the put-down location in the second part, is considered to be our task-related cost $\mathcal{H}_r$. Of course, the proposed approach can be used for any other set of cost functions.

In this work, three different weighting schemes for the contribution of $\mathcal{H}_{\mathcal{W}}$ and $\mathcal{H}_r$ to the haptic guidance $\boldsymbol{f}_m$ have been considered:

(A0) $\alpha = 0, \beta = 1$ (the user is only guided toward the target, no human-centered cost);
(A20) $\alpha = 0.2, \beta = 0.8$ (the user is guided toward the target with a weak human-centered component);
(A40) $\alpha = 0.4, \beta = 0.6$ (the user is guided toward the target with a strong human-centered component).

Only conditions with $\beta \geq 0.6$ are considered, so as to ensure that the guidance feedback always brings the user towards the completion of the task. If β is too low, the haptic feedback would in fact simply take the user towards a comfortable arm position.

For each of the above conditions A0, A20, A40, users were asked to pick and place these sets of objects:

(B) a lightweight box made of cardboard of dimensions 14×4×4 cm;
(C) two cubes, each of dimensions 4.2×4.2×4.2 cm;
(L) a wooden piece, stylized like the letter "H", with outer dimensions 21×13×2.5 cm.

Each participant carried out eighteen repetitions of the considered manipulation task, two for each weighting condition and set of objects. Figure 2 shows a representative trial, highlighting the effects of employing different weighting schemes in a simple reaching movement between two points.

To evaluate the effectiveness of the proposed human-centered shared control approach, we recorded (i) the completion time, (ii) the error in placing the objects at the target, (iii-iv) the mean and maximum $\mathcal{H}_{\mathcal{W}}$ registered, and (v) the NASA Task Load Index (NASA-TLX). Results are shown in Fig. 3. The Bonferroni correction was used to reduce the chances of obtaining false-positive results when multiple pairwise tests were performed on a single set of data. Figure 4 shows a representative evolution of both cost functions versus time for the A0 and A40 weighting schemes.

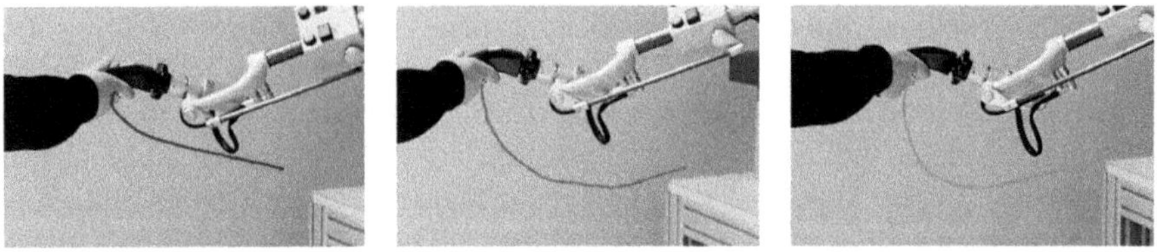

(a) A0 (b) A20 (c) A40

Fig. 2 Example of a representative reaching movement for the three weightings conditions A0 (blue), A20 (red), and A40 (green). In A0, $\mathcal{H}_r$ is the only cost considered. Therefore, the trajectory of the user is almost horizontal, i.e., the shortest towards the target. As the role of $\mathcal{H}_\mathcal{W}$ increases, the haptic feedback tries to minimize the user discomfort by making the arm move along a downward (more comfortable) trajectory

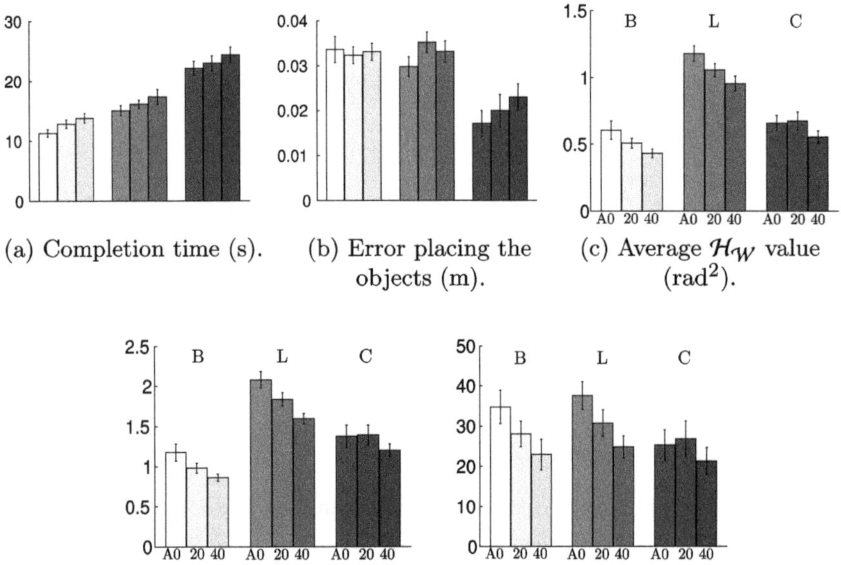

(a) Completion time (s). (b) Error placing the objects (m). (c) Average $\mathcal{H}_\mathcal{W}$ value (rad^2). (d) Maximum $\mathcal{H}_\mathcal{W}$ value (rad^2). (e) NASA TLX load index.

Fig. 3 Human subjects experiment. Mean and standard error of the mean of **a** completion time, **b** error in placing the objects, **c** average $\mathcal{H}_\mathcal{W}$, **d** maximum $\mathcal{H}_\mathcal{W}$, and **e** NASA TLX load index for the three control conditions (A0, A20, A40) and the three target objects (B, L, C)

Completion time showed a significant degradation when adding our human-centered guidance (A20, A40) versus standard task-centered guided teleoperation (A0). This result is quite expected, and it is the major drawback of our approach, as the additional guidance inevitably deviates the user from the shortest path. However, this (small) performance degradation is compensated by a significant reduction of the estimated muscular discomfort (mean and max $\mathcal{H}_\mathcal{W}$) and measured workload (NASA TLX). In fact, while the completion time and placing error degrade by 14%

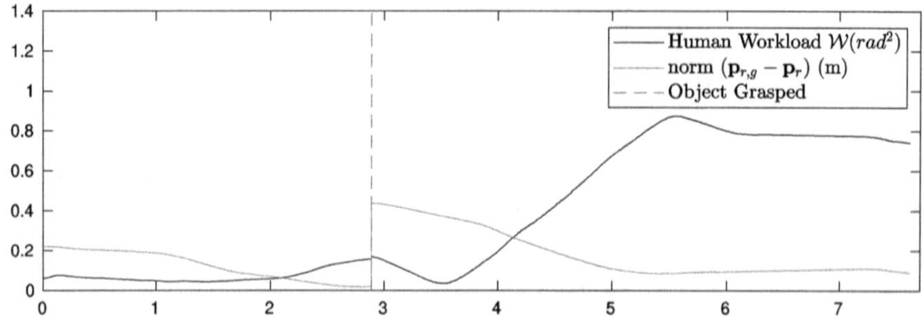

(a) Weighting condition A0 vs. time (s)

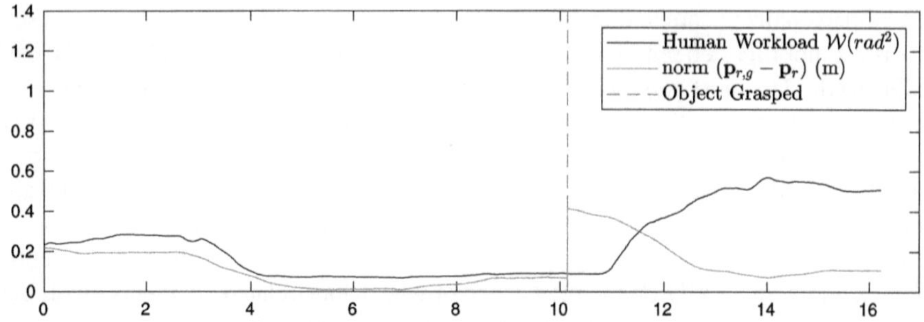

(b) Weighting A40 condition vs. time (s)

Fig. 4 Representative evolution of the human workload metric $\mathcal{W}$ and the norm of the translational error between the current and the desired robot pose, for the A0 and A40 weighing conditions. The user is faster in completing the task using A0; however, a lower $\mathcal{W}$ is recorded using A40, especially in the second part of the task (after grasping the object and moving towards the put-down location). The dashed line represents the moment the user grasps the object and the target position changes

and 10%, respectively, the NASA TLX value is improved by 30%. It is also important to highlight that we considered a rather short task (approx. 20 min in total), and we expect that the effect and usefulness of our approach increase with the duration of the task. Finally, we did not register any noticeable degradation of the performance due to implementing the haptic shared control at the user's side. Indeed, where to implement haptic active constraints in robotic teleoperation is still an open question. There are arguments for both. Implementing the active constraint at the user's side solves any issue related to the stability of the system, but it opens to the risk of badly controlling the remote robot. On the other hand, enforcing the active constraint at the robot side solves any issue related to the commanding of the robot, but it exposes to the risks of instabilities.

2.3 Human-Centered Shared Control: An Example for Power Wheelchair Navigation

This Section summarizes a shared-control framework aiming to provide adapted mobility assistance to people with disabilities, in the form of a power wheelchair driving assistance.

Assistive Technologies. People with disabilities have diverse needs and circumstances but share a common goal: maximizing independence in daily life despite mobility challenges. In this context, autonomy and social integration of people with disability can be augmented by several solutions such as the Assistive Technologies (AT), personalized care and inclusive design [63, 64]. Multiple devices can be used to assist mobility depending on the disability, mobility issues and individual's specific needs. The most common mobility aids include walking aids such as canes and crutches to transportation aids such as wheelchairs and stair lifts. Powered mobility devices can be prescribed for long-term use for people with severe or progressive disability or to address fatigue issues experienced while propelling a manual wheelchair. However, the ability to independently operate a mobility device safely can be altered by visual, visuo-spatial and/or cognitive impairments [65] thus leading to accidents. For example, in a study on the prevalence of wheelchair-related accidents involving 95 participants, 54.7% of them reported at least 1 accident in the past 3 years [66]. Therefore, when a person needs a power wheelchair, their physician must decide to write a prescription after a rigorous evaluation of their ability to drive safely [67, 68]. The consequence of this safety process is that some people are either not allowed to drive a power wheelchair, or if they do, they cannot maneuver it safely by themselves and need to be supervised and accompanied. In this context, robotic assistive solutions can increase wheelchair navigation safety. Indeed, navigation support can be supplied in the form of autonomous or semi-autonomous assistance [69–72].

User-driven design and use evaluation. The conception and evaluation of such ATs generally focuses on the functioning of the technology instead of the matching with user needs and self-concept [73]. To take up the autonomy and acceptability challenge, ATs have to be co-designed and meet users' needs which have to be addressed at different levels, namely interaction needs, functional needs and psychological needs. This needs to be conducted in conjunction with a design of an interface adapted to users' capacities. Moreover, the assessment of usability, acceptability and user experience all along the conception and use cycle has to be done through the the evaluation of the performance to achieve the task with the robotic assistive solution. In addition, the design of the assistive solution needs to take into account the involvement of user and professionals through Living Labs and following a methodological framework focusing on design for all, intuitiveness, etc. This objective of co-design can be reached thanks to multidisciplinary collaboration leading to transdisciplinary research following user-driven design methodologies. In particular, the Living Lab approach can be used to structure the development process of mobility AT. Indeed, Living Labs do not consider users as objects of investigation but as key actors in the innovation process [74]. Within a Living Lab, the collection of needs is carried out with user cohorts that are composed of people with disabilities, ageing people, family caregivers and healthcare professionals. The scientific and ethical evaluation of the clinical benefit are ensured through investigations in accordance with regulation on medical devices, allowing thereafter the AT's prescription by physicians, which is a necessary requirement for use by end users. The Living Lab evaluation methodology can be based on ecological experiments allowing technical choices validation and bringing out new needs based on users' feedback. The capacity of users to project themselves in the use throughout the development process is assessed thanks to methodologies relying on the psycho-social concept of intention of use based in particular on the model of Unified Theory of Acceptance and Use of Technology-UTAUT [75] and its derivative UTAUT2 [76]. Wheelchair navigation assistance is here detailed as an example of navigation support in the form of a semi-autonomous assistance relying on a sensor-based shared control solution. This solution is ergonomically designed to be adapted to user capacities while ensuring driving comfort.

Shared-control framework. Power wheelchair driving assistance requires the shared control to ensure user safety by taking over the wheelchair control when there is a danger for the driver. The strategy needs to be generic and directly compliant with any power wheelchair and control interface type, in any environment. It also needs to be customizable to each individual needs and difficulties. To ensure acceptability, the solution has to be transparent and intuitive to be accepted by the user. Such a strategy relies on a generic low complexity sensor-based control independent of the type of sensor.

Power wheelchair modeling. The power wheelchair is modeled as a rigid body composed of 2 actuated wheels; each of which being actuated by independent motors mounted on a common axis). The differential drive system induces a non-holonomic constraint on the wheelchair motion which is kinematically equivalent to a unicycle type robot from the class of type $(2, 0)$ robots as defined in [77]. In the following, let $\mathcal{F}_R(R, \vec{X}_R, \vec{Y}_R, \vec{Z}_R)$ be the frame attached to the robotic wheelchair (Fig. 5a). The control input $\mathbf{u}$ for such a 2-DoF system is given by $\mathbf{u} = [\upsilon, \omega]^T$, where the 2 variables υ and ω respectively denote the translational velocity along $\vec{X}_R$ and the rotational velocity around $\vec{Z}_R$ (Fig. 5b).
Sensor-based modeling. Let $\mathbf{x}$ be the set of data provided by a sensor s. The sensor jacobian $\mathbf{J}_s$ links the control input $\mathbf{u}$ to the sensor signal variation $\dot{\mathbf{x}}$ so that $\dot{\mathbf{x}} = \mathbf{J}_s\mathbf{u}$. Let $\mathbf{v}_s$ be the velocity of the sensor s. The sensor motion can then be characterized by its instantaneous velocity

$$\mathbf{v}_s = \begin{bmatrix} \boldsymbol{\upsilon}_s \\ \boldsymbol{\omega}_s \end{bmatrix}, \tag{8}$$

where the translational and rotational velocities are respectively given by $\boldsymbol{\upsilon}_s = [\upsilon_x \; \upsilon_y \; \upsilon_z]$ and $\boldsymbol{\omega}_s = [\omega_x \; \omega_y \; \omega_z]$. The sensor jacobian $\mathbf{J}_s$ is related to the robot jacobian ${}^r\mathbf{J}_r$ expressed in the robot frame by

$$\mathbf{J}_s = \mathbf{L}_s\, {}^s\mathbf{W}_r\, {}^r\mathbf{J}_r. \tag{9}$$

where ${}^s\mathbf{W}_r$ is the velocity screw transformation matrix from the sensor frame $\mathcal{F}_s$ to the robot frame $\mathcal{F}_R$ [78], and $\mathbf{L}_s$ is the sensor interaction matrix defined in [79] and [80] which links the variation of the sensor features $\dot{\mathbf{x}}$ with $\mathbf{v}_s$.

Sensor-based control. Sensor-based control only requires knowledge about the interaction matrix as well as the sensor location with respect to the robot frame. This method achieves low-computational cost allowing system embedding on a wheelchair. In particular, it achieves flexible and reactive control while reaching satisfying precision and speed performances. The goal of the sensor-based servoing for semi-autonomous navigation consists of regulating the wheelchair velocity control input in order to avoid collisions with obstacles while maintaining full control to the user when there is no risk of collision. To do this, a proportional corrector is designed following $\dot{\mathbf{e}} \geq -\lambda\mathbf{e}$, where $\mathbf{e} = \mathbf{s} - \mathbf{s}^*$ is defined as the difference between the set of data $\mathbf{s}$ and its desired value $\mathbf{s}^*$, with $\lambda \in \mathbb{R}^{+*}$ is a gain. Using this proportional corrector together with Eq. 9 and sensor data $\mathbf{x}$, we can compute a 2D speed map with an allowed risk free area and a forbidden area where collision would occur (Fig. 6a). For more details on the computation, see [81].

User-centered wheelchair control. Here, shared control aims to allow full control when there is no danger of collisions (allowed risk free area) while providing smooth velocity adaptation when the wheelchair gets close to an obstacle (forbidden area). Then, the proposed shared control is in the form of on a low complexity sensor-based servoing approach, where $\mathbf{u}$ is computed from distance sensor measurements constraints. This approach results in a trade-off between the robot velocity $\mathbf{u}$ and the user desired velocity $\mathbf{u}_{op}$. The wheelchair velocity is computed from user desired velocity under constraints by minimizing a cost function f that corresponds to the minimization and the user desired velocity $\mathbf{u}_{op}$, such that

$$f(\mathbf{u}_{op}, \mathbf{u}) = \alpha(\upsilon - \upsilon_{op})^2 + \beta(\omega - \omega_{op})^2, \tag{10}$$

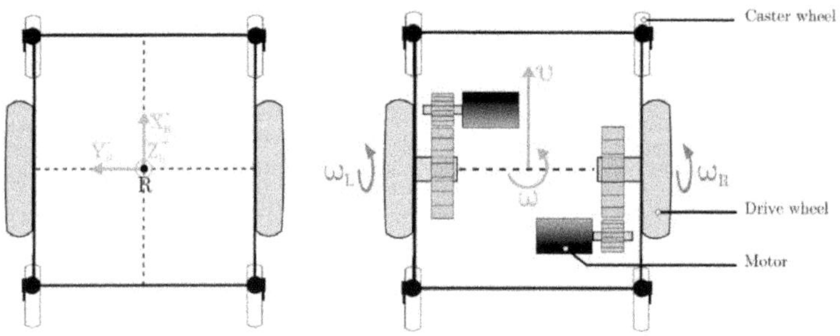

(a) Robot frame. (b) Differential control.

Fig. 5 Modeling of the robotic wheelchair

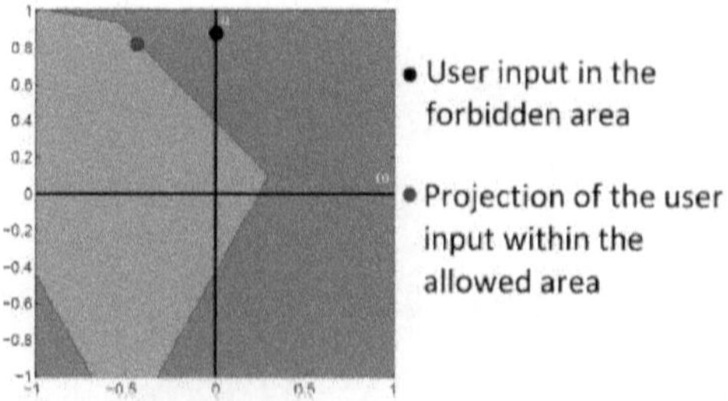

(a) The two areas deduced from the constraints computed from sensors measurements, in the wheelchair velocity domain [27].

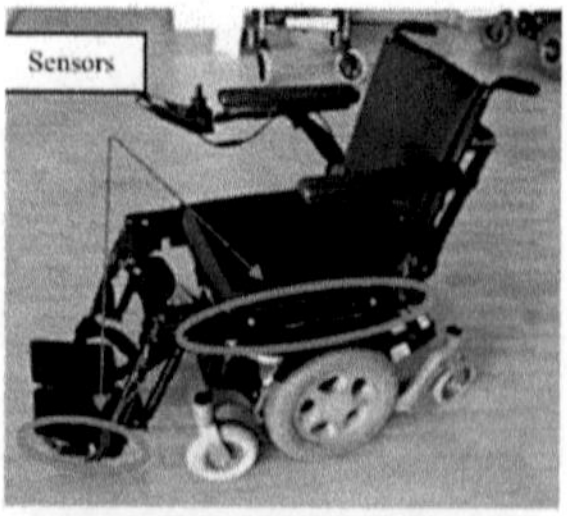

(b) Power wheelchair equipped with distance sensors.

Fig. 6 Wheelchair sensor-based shared control: constraints in the wheelchair velocity domain deduced from sensor measurements

with $\alpha \geq 0$ and $\beta \geq 0$ being weighting parameters which are set to model the intention of the user. The underlying idea is then to prioritize the main direction given by the user. The proposed shared control allows then the user to control the wheelchair while observing safety constraints (adaptive obstacle avoidance). The shared control does not need fine calibration and is independent of both number and type of sensors equipped on the wheelchair (Fig. 6a). The formulation of the problem leads to solve a simple quadratic system under constraints resulting in the definition of two areas in the wheelchair velocity domain in which the user velocity input is considered as a desired coordinate (Fig. 6b). For more details on the computation, see [82].

Ergonomic driving assistance. Wheelchair driving assistance has to be co-designed with users to take into account the ease of use and intuitiveness of the assistance strategy, but also to ensure that the strategy fits with the needs for social interaction. Thus, ergonomics must be addressed from a technological point of view, but also from a social point of view.

An ergonomic wheelchair use. If there is a danger of collision (i.e. user velocity input is within the forbidden area), user velocity input is progressively blended with its projection value within the allowed area thus smoothly regulating the wheelchair velocity to avoid the collision, while keeping it moving and respecting as much as possible user velocity input (Fig. 6a). The proposed shared control provides a smooth reactive solution without trajectory planning, robust towards dynamic obstacles, and taking into account the user intention and comfort to avoid obstacles. This approach allows customization to comply with user specific needs. For example, any type or number of sensors can be equipped on the wheelchair, and the blending between user input and robot control can be tuned to adapt the level of assistance to each individual. The value $s*$ represents the margin for a sensor s, this can be easily customized for each sensor, allowing for example to set asymmetrical margins for a person with an impairment such as unilateral neglect. In addition, the proposed approach only uses the user input so that it can be easily adapted with any input

(a) The SWADAPT2 assisted driving clinical investigation.

(b) The SWADAPT3 assisted driving clinical investigation.

Fig. 7 Wheelchair assisted driving clinical investigations involving regular users and people with driving difficulties

controller device (joystick, chin controller, etc.). Moreover, it can be adapted on any standard wheelchair as it has been designed to be interfaced with any power module (responsible for the set up of parameters such as velocity and acceleration).

Wheelchair assisted driving and social ergonomics. The proposed solution also takes into account social ergonomics as it is designed to be compatible with social activities during wheelchair motion [83]. The solution's compatibility with standard wheelchairs, power modules, and various input controllers ensures seamless integration and requires no additional device-specific learning beyond basic wheelchair operation. The intuitiveness and handling of dynamic obstacles facilitates the use of the proposed shared control for social activities, which have an impact on social participation, inclusion, and quality of life. Moreover, social experiments in cultural and urban places such as in museums have shown positive impact on quality of experience of participants as well as their relatives (Fig. 8a) . The social navigation constraints can be modeled and incorporated in the control strategy thanks to experiments allowing to analyze pedestrian behaviour around a wheelchair [84] (Fig. 8b).

Benefit evaluation and clinical investigations. The benefit of the proposed shared control framework has been assessed within clinical investigations with regular users and people with driving difficulties because of neurological or visual impairments. As part of the Interreg France (Channel) England project ADAPT, the obstacle avoidance has been tested with more than 100 participants on specific obstacle courses designed by occupational therapists consisting of "positive" obstacles such as walls, door passing, elevators, etc. (Clinical investigations SWADAPT2 under the registration number NCT04259151, Fig. 7a) and "negative" obstacles such as holes, curbs, ramps, etc. (Clinical investigations SWADAPT3 under the registration number NCT05459779, Fig. 7b).

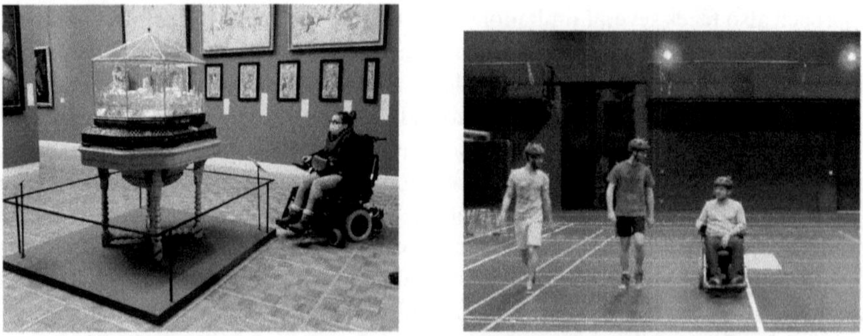

(a) Assisted wheelchair driving with shared control: experiment in a museum.

(b) Experiment to analyse pedestrian behaviour around a wheelchair.

Fig. 8 Wheelchair assisted driving experiments with social interactions

Moreover, shared control is also investigated during ecological situations such as a rehabilitation center corridor [82] or cultural places such as museums [85].

3 Conclusions and Perspectives

Attention to ergonomics and, in general, human needs and characteristics is considered increasingly important in the design of robotic systems as well as in their control. Designing *around* the human means considering its musculoskeletal constraints, features, and limitations during the development of robotic mechanical systems and control algorithms. This Chapter has presented an overview of human-centered design and human-centered shared control techniques for robotics, including supporting examples in the context of physical assistance at work, industrial teleoperation, and mobility assistance for people affected by sensory disabilities.

The potential of human-centered approaches in robotics is significant and far-reaching. By integrating human factors into the design and control of robotic systems, we can create more intuitive, efficient, and user-friendly interfaces that reduce cognitive and physical load on operators. This is particularly crucial in applications such as medical robotics, assistive technologies, and industrial automation, where prolonged human-robot interaction is common. The shared control paradigm, in particular, offers a promising avenue for balancing human expertise with robotic precision and capabilities. One of the key strengths of human-centered design in robotics lies in its ability to adapt to the specific needs and limitations of individual users. By considering factors such as ergonomics, cognitive load, and user comfort, these systems can potentially improve task performance, reduce operator fatigue, and enhance overall safety. This is especially valuable in fields like robotic surgery, where operator comfort and precision are paramount for successful outcomes. However, the approach also faces several limitations and challenges. One of the primary obstacles is the complexity of accurately modeling human behavior and preferences, which can vary significantly between individuals and even for the same individual over time. Developing robust and generalizable algorithms that can adapt to this variability remains a significant challenge. Additionally, there is often a trade-off between user comfort and task efficiency that needs to be carefully balanced. Another limitation is the current state of haptic technology. While haptic feedback has shown great promise in enhancing user experience and task performance, the available hardware for providing rich, nuanced feedback is still limited, particularly for ungrounded cutaneous devices. This can restrict the quality and types of information that can be conveyed to the user through haptic channels.

In terms of maturity, human-centered design in robotics has made significant advancements in recent years, but it is still an evolving field. While some applications, such as shared control in surgical robotics and assistive technologies, have reached a relatively high level of maturity, others are still in early stages of development. The integration of advanced sensing technologies, machine learning algorithms, and more sophisticated human modeling techniques is pushing the field forward, but there is still much work to be done to fully realize the potential of human-centered robotics. Looking ahead, several key areas require further research and development. These include: (i) Improving the accuracy and robustness of human modeling techniques to better predict and adapt to user behavior and preferences; (ii) Developing more advanced haptic interfaces capable of providing richer, more nuanced feedback to users; (iii) Exploring novel ways to balance user comfort with task efficiency and safety requirements; (iv) Investigating the long-term effects of human-centered robotic systems on user performance, comfort, and well-being; (v) Addressing ethical considerations related to shared control and autonomy in human-robot interaction.

As the field continues to mature, we can expect to see more widespread adoption of human-centered design principles in robotics across various domains. This shift towards more user-centric approaches has the potential to significantly enhance the effectiveness, safety, and acceptance of robotic systems in both professional and personal settings. However, realizing this potential will require ongoing collaboration between roboticists, human factors experts, and end-users to ensure that these systems truly meet the needs and expectations of the humans they are designed to assist and augment.

References

1. Wolfartsberger, J., Haslwanter, J.D.H., Lindorfer, R.: Perspectives on assistive systems for manual assembly tasks in industry. Technologies **7**(1), 12 (2019)
2. Maurice, P., Padois, V., Measson, Y., Bidaud, P.: Human-oriented design of collaborative robots. Int. J. Ind. Ergon. **57**, 88–102 (2017)
3. Arai, H., Takubo, T., Hayashibara, Y., Tanie, K.: Human-robot cooperative manipulation using a virtual nonholonomic constraint. In: Proceedings on IEEE International Conference on Robotics and Automation. Symposia Proceedings, vol. 4, pp. 4063–4069. (2000)

4. Bowyer, S.A., Davies, B.L., Baena, F.R.: Active constraints/virtual fixtures: a survey. IEEE Trans. Robot. **30**(1), 138–157 (2014)
5. Pacchierotti, C., Prattichizzo, D.: Cutaneous/tactile haptic feedback in robotic teleoperation: motivation, survey, and perspectives. IEEE Trans, Robotics (2023)
6. Pacchierotti, C., Scheggi, S., Prattichizzo, D., Misra, S.: Haptic feedback for microrobotics applications: a review. Front. Robot. AI **3**, 53 (2016)
7. Hardy, P.: Powered wheelchair mobility: an occupational performance evaluation perspective. Aust. Occup. Ther. J. **51**(1), 34–42 (2004)
8. An experimental assessment: Ahmetcan Erdogan and Brenna D Argall. The effect of robotic wheelchair control paradigm and interface on user performance, effort and preference. Robot. Auton. Syst. **94**, 282–297 (2017)
9. Narayanan, V.K., Spalanzani, A., Babel, M.: A semi-autonomous framework for human-aware and user intention driven wheelchair mobility assistance. In: Proceedings IEEE/RSJ International Conference on Intelligent Robots and Systems (IROS), pp. 4700–4707. (2016)
10. Hurst, A., Tobias, J.: Empowering individuals with do-it-yourself assistive technology. In: Proceedings ACM SIGACCESS Conference on Computers and Accessibility, pp. 11–18. (2011)
11. Meyerson, G., Dewettinck, B.: Effect of empowerment on employees performance. Adv. Res. Econ. Manage. Sci. **2**(1), 40–46 (2012)
12. Pacchierotti, C.: Cutaneous Haptic Feedback for Robotics and Virtual Reality. Habilitation à diriger des recherche (HDR), Rennes, France (2022)
13. Scheggi, S., Talarico, A., Prattichizzo, D.: A remote guidance system for blind and visually impaired people via vibrotactile haptic feedback. In: Proceedings Mediterranean Conference on Control and Automation, pp. 20–23. (2014)
14. Prasad, M., Taele, P., Goldberg, D., Hammond, T.A.: Haptimoto: turn-by-turn haptic route guidance interface for motorcyclists. In: Proceeding SIGCHI Conference on Human Factors in Computing Systems, pp. 3597–3606. (2014)
15. Howard, T., Szewczyk, J.: Visuo-haptic feedback for 1-d guidance in laparoscopic surgery. In: Proceedings IEEE RAS/EMBS International Conference on Biomedical Robotics and Biomechatronics, pp. 58–65. (2014)
16. Ramos, A., Prattichizzo, D.: Vibrotactile stimuli for distinction of virtual constraints and environment feedback. In: International Conference on Human Haptic Sensing and Touch Enabled Computer Applications, pp. 141–149. (2014)
17. Aggravi, M., Estima, D.A., Krupa, A., Misra, S., Pacchierotti, C.: Haptic teleoperation of flexible needles combining 3d ultrasound guidance and needle tip force feedback. IEEE Robot. Autom. Lett. **6**(3), 4859–4866 (2021)
18. Pacchierotti, C., Abayazid, M., Misra, S., Prattichizzo, D.: Teleoperation of steerable flexible needles by combining kinesthetic and vibratory feedback. IEEE Trans. Haptics **7**(4), 551–556 (2014)
19. Chinello, F., Pacchierotti, C., Bimbo, J., Tsagarakis, N.G., Prattichizzo, D.: Design and evaluation of a wearable skin stretch device for haptic guidance. IEEE Robot. Autom. Lett. **3**(1), 524–531 (2017)
20. Fleck, J.J., Zook, Z.A., Clark, J.P., Preston, D.J., Lipomi, D.J., Pacchierotti, C., O'Malley, M.K.: Wearable multi-sensory haptic devices. Nat. Rev, Bioeng (2025)
21. Aggravi, M., Elsherif, A.A.S., Giordano, P.R., Pacchierotti, C.: Haptic-enabled decentralized control of a heterogeneous human-robot team for search and rescue in partially-known environments. IEEE Robot. Autom. Lett. **6**(3), 4843–4850 (2021)
22. Aggravi, M., Pausé, F., Giordano, P.R., Pacchierotti, C.: Design and evaluation of a wearable haptic device for skin stretch, pressure, and vibrotactile stimuli. IEEE Robot. Autom. Lett. **3**(3), 2166–2173 (2018)
23. Aggravi, M., Sirignano, G., Giordano, P.R., Pacchierotti, C.: Decentralized control of a heterogeneous human-robot team for exploration and patrolling. IEEE Trans. Autom. Sci. Eng. **19**(4), 3109–3125 (2021)

24. Pacchierotti, C., Ongaro, F., Van den Brink, F., Yoon, C., Prattichizzo, D., Gracias, D.H., Misra, S.: Steering and control of miniaturized untethered soft magnetic grippers with haptic assistance. IEEE Trans. Autom. Sci. Eng. **15**(1), 290–306 (2017)
25. Pezent, E., Fani, S., Clark, J., Bianchi, M., O'Malley, M.K.: Spatially separating haptic guidance from task dynamics through wearable devices. IEEE Trans, Haptics (2019)
26. Bowyer, S.A., Davies, B.L., Baena, F.R.: Active constraints/virtual fixtures: a survey. IEEE Trans. Robot. **30**(1), 138–157 (2013)
27. Prada, R., Payandeh, S.: On study of design and implementation of virtual fixtures. Virtual Reality **13**(2), 117–129 (2009)
28. Meli, L., Pacchierotti, C., Prattichizzo, D.: Experimental evaluation of magnified haptic feedback for robot-assisted needle insertion and palpation. Int. J. Med. Robot. Comput. Assist. Surg 13(4), (2017)
29. Abbott, J.J., Marayong, P., Okamura, A.M.: Haptic virtual fixtures for robot-assisted manipulation. In: Robotics Research, pp. 49–64. (2007)
30. Rosenberg, L.B.: Virtual fixtures: perceptual tools for telerobotic manipulation. In: Virtual Reality Annual International Symposium, pp. 76–82. IEEE (1993)
31. Chen, Z., Malpani, A., Chalasani, P., Deguet, A., Vedula, S.S., Kazanzides, P., Taylor, R.H.: Virtual fixture assistance for needle passing and knot tying. In: Proceedings on IEEE/RSJ International Conference Intelligent Robots and Systems (IROS), pp. 2343–2350. (2016)
32. Cognetti, M., Aggravi, M., Pacchierotti, C., Salaris, P., Giordano, P.R.: Perception-aware human-assisted navigation of mobile robots on persistent trajectories. IEEE Robot. Autom. Lett. **5**(3), 4711–4718 (2020)
33. Cognetti, M., Aggravi, M., Pacchierotti, C., Salaris, P., Giordano, P.R.: Shared control active perception for human-assisted navigation. In: I-RIM 2020-2nd Italian Conference on Robotics and Intelligent Machines (2020)
34. Abi-Farraj, F., Pacchierotti, C., Arenz, O., Neumann, G., Giordano, P.R.: A haptic shared-control architecture for guided multi-target robotic grasping. IEEE Trans, Haptics (2019)
35. Rahal, R., Ghalamzan, A.M., Abi-Farraj, F., Pacchierotti, C., Giordano, P.R.: Haptic-Guided Grasping to Minimise Torque Effort During Robotic Telemanipulation. Auton. Robot. , (2023)
36. Selvaggio, M., Abi-Farraj, F., Pacchierotti, C., Giordano, P.R., Siciliano, B.: Haptic-based shared-control methods for a dual-arm system. IEEE Robot. Autom. Lett. **3**(4), 4249–4256 (2018)
37. Selvaggio, M., Cacace, J., Pacchierotti, C., Ruggiero, F., Giordano, P.R.: A shared-control teleoperation architecture for nonprehensile object transportation. IEEE Trans. Robot. **38**(1), 569–583 (2021)
38. Li, M., Kapoor, A., Taylor, R.H.: A constrained optimization approach to virtual fixtures. In: Proceedings IEEE/RSJ International Conference on Intelligent Robots and Systems, pp. 1408–1413. (2005)
39. Turro, N., Khatib, O.: Haptically augmented teleoperation. In: Experimental Robotics VII, pp. 1–10. (2001)
40. Cipriani, C., Zaccone, F., Micera, S., Chiara Carrozza, M.: On the shared control of an EMG-controlled prosthetic hand: analysis of user-prosthesis interaction. IEEE Trans. Robot. **24**(1), 170–184 (2008)
41. Yu, H., Spenko, M., Dubowsky, S.: An adaptive shared control system for an intelligent mobility aid for the elderly. Auton. Robot. **15**(1), 53–66 (2003)
42. Vozar, S., Chen, Z., Kazanzides, P., Whitcomb, L.L.: Preliminary study of virtual nonholonomic constraints for time-delayed teleoperation. In: Proceedings IEEE/RSJ International Conference on Intelligent Robots and Systems (IROS), pp. 4244–4250. (2015)
43. Li, M., Cao, H., Song, X., Huang, Y., Wang, J., Huang, Z.: Shared control driver assistance system based on driving intention and situation assessment. IEEE Trans. Ind. Inf. **14**(11), 4982–4994 (2018)
44. Abi-Farraj, F., Pacchierotti, C., Giordano, P.R.: User evaluation of a haptic-enabled shared-control approach for robotic telemanipulation. In: Proceedings on IEEE/RSJ International Conference on Intelligent Robots and Systems, pp. 1–9. (2018)

45. Abi-Farraj, F., Pedemonte, N., Giordano, P.R.: A visual-based shared control architecture for remote telemanipulation. In: Proceedings on IEEE/RSJ International Conference on Intelligent Robots and Systems, pp. 4266–4273. (2016)
46. Rahal, R., Abi-Farraj, F., Giordano, P.R., Pacchierotti, C.: Haptic shared-control methods for robotic cutting under nonholonomic constraints. In: Proceeding IEEE/RSJ International Conference on Intelligent Robots and Systems (IROS), pp. 8151–8157. (2019)
47. van't Hullenaar, C.D.P., Hermans, B., Broeders, I.A.M.J.: Ergonomic assessment of the da vinci console in robot-assisted surgery. Innov. Surg. Sci. **2**(2), 97–104 (2017)
48. Chander, D.S., Tomezzoli, A., Cavatorta, M.P., Gréau, M., Marie, S., Pontonnier, C.: Biomechanical requirements of meat cutting tasks: a pilot study quantifying tasks sustainability and assistance required at the wrist. Appl. Ergon. **116**, 104186 (2024)
49. Skuric, A., Padois, V., Daney, D.: On-line force capability evaluation based on efficient polytope vertex search. In: 2021 IEEE International Conference on Robotics and Automation (ICRA), pp. 1700–1706. IEEE (2021)
50. Erdemir, A., McLean, S., Herzog, W., van den Bogert, A.J.: Model-based estimation of muscle forces exerted during movements. Clin. Biomech. **22**(2), 131–154 (2007)
51. Damsgaard, M., Rasmussen, J., Christensen, S.T., Surma, E., Zee, M.D.: Analysis of musculoskeletal systems in the anybody modeling system. Simul. Model. Pract. Theory **14**(8), 1100–1111 (2006)
52. Delp, S.L., Anderson, F.C., Arnold, A.S., Loan, P., Habib, A., John, C.T., Guendelman, E., Thelen, D.G.: Opensim: open-source software to create and analyze dynamic simulations of movement. IEEE Trans. Biomed. Eng. **54**(11), 1940–1950 (2007)
53. Martinez, R., Michaud, B., Begon, M.: Pyomeca: an open-source framework for biomechanical analysis. J. Open Source Softw. **5**(53), 2431 (2020)
54. Michaud, B., Bailly, F., Charbonneau, E., Ceglia, A., Sanchez, L., Begon, M.: Bioptim, a python framework for musculoskeletal optimal control in biomechanics. IEEE Trans. Syst., Man, Cybernetics: Syst. **53**(1), 321–332 (2022)
55. Muller, A., Pontonnier, C., Puchaud, P., Dumont, G.: Custom: a matlab toolbox for musculoskeletal simulation. J. Open Source Softw. **4**(33), 1–3 (2019)
56. Busch, B., Maeda, G., Mollard, Y., Demangeat, M., Lopes, M.: Postural optimization for an ergonomic human-robot interaction. In: Proceedings on IEEE/RSJ International Conference on Intelligent Robots and Systems, pp. 2778–2785. (2017)
57. Marin, A.G., Shourijeh, M.S., Galibarov, P.E., Damsgaard, M., Fritzsch, L., Stulp, F.: Optimizing contextual ergonomics models in human-robot interaction. In: Proceedings IEEE/RSJ International Conference on Intelligent Robots and Systems, pp. 1–9. (2018)
58. Chen, L., Figueredo, L.F., Dogar, M.R.: Planning for muscular and peripersonal-space comfort during human-robot forceful collaboration. In: Proceedings on IEEE International Conference on Humanoid Robots, pp. 1–8. (2018)
59. Peternel, L., Kim, W., Babi, J., Ajoudani, A.: Towards ergonomic control of human-robot co-manipulation and handover. In: Proceeding IEEE-RAS International on Conference Humanoid Robotics, pp. 55–60. (2017)
60. Rahal, R., Matarese, G., Gabiccini, M., Artoni, A., Prattichizzo, D., Giordano, P.R., Pacchierotti, C.: Caring about the human operator: haptic shared control for enhanced user comfort in robotic telemanipulation. IEEE Trans. Haptics **13**(1), 197–203 (2020)
61. Rahal, R., Matarese, G., Gabiccini, M., Artoni, A., Prattichizzo, D., Giordano, P.R., Pacchierotti, C.: Haptic shared control for enhanced user comfort in robotic telemanipulation. In: SALC 2020-IEEE ICRA Workshop on Shared Autonomy: Learning and Control, pp. 1–2. (2020)
62. McAtamney, L., Corlett, E.N.: Rula: a survey method for the investigation of work-related upper limb disorders. Appl. Ergon. **24**(2), 91–99 (1993)
63. Divya, R., Dinesh Peter, J.: Smart healthcare system-a brain-like computing approach for analyzing the performance of detectron2 and posenet models for anomalous action detection in aged people with movement impairments. In: Complex and Intelligent Systems, pp. 1–20 (2021)

64. Perez, A.J., Siddiqui, F., Zeadally, S., Lane, D.: A review of IoT systems to enable independence for the elderly and disabled individuals. In: Internet of Things, p. 100653. (2022)
65. Cawood, J., Visagie, S., Mji, G.: Impact of post-stroke impairments on activities and participation as experienced by stroke survivors in a western cape setting. S. Afr. J. Occup. Ther. **46**(2), 10–15 (2016)
66. Chen, W.-Y., Jang, Y., Wang, J.-D., Huang, W.-N., Chang, C.-C., Mao, H.-F., Wang, Y.-H.: Wheelchair-related accidents: relationship with wheelchair-using behavior in active community wheelchair users. Arch. Phys. Med. Rehabil. **92**(6), 892–898 (2011)
67. Lukersmith, S., Radbron, L., Hopman, K.: Development of clinical guidelines for the prescription of a seated wheelchair or mobility scooter for people with traumatic brain injury or spinal cord injury. Aust. Occup. Ther. J. **60**(6), 378–386 (2013)
68. Routhier, F., Vincent, C., Desrosiers, J., Nadeau, S.: Mobility of wheelchair users: a proposed performance assessment framework. Disabil. Rehabil. **25**(1), 19–34 (2003)
69. Baklouti, E., Amor, N.B., Jallouli, M.: Autonomous wheelchair navigation with real time obstacle detection using 3d sensor. Automatika: časopis za automatiku, mjerenje, elektroniku, računarstvo i komunikacije **57**(3), 761–773 (2016)
70. Ezeh, C., Trautman, P., Devigne, L., Bureau, V., Babel, M., Carlson, T.: Probabilistic vs linear blending approaches to shared control for wheelchair driving. In: 2017 International Conference on Rehabilitation Robotics (ICORR), pp. 835–840. IEEE (2017)
71. Horn, O.: Smart wheelchairs: past and current trends. In: 2012 1st International Conference on Systems and Computer Science (ICSCS), pp. 1–6. IEEE (2012)
72. Kokosy, A., Floquet, T., Howells, G., Hu, H., Pepper, M., Sakel, M., Donze, C.: Sysiass? An Intelligent Powered Wheelchair, (2013)
73. van Dam, K., Gielissen, M., Bles, R., van der Poel, A., Boon, B.: The impact of assistive living technology on perceived independence of people with a physical disability in executing daily activities: a systematic literature review. In: Disability and Rehabilitation: Assistive Technology, pp. 1–10 (2023)
74. Dube, P., Sarrailh, J., Billebaud, C., Grillet, C., Zingraff, V., Kostecki, I.: Le livre blanc des living labs. Umvelt Service Design, Montréal, Canada (2014)
75. Venkatesh, V., Morris, M.G., Davis, G.B., Davis, F.D.: User acceptance of information technology: toward a unified view. In: MIS Quarterly, pp. 425–478 (2003)
76. Venkatesh, V., Thong, J.Y.L., Xu, X.: Consumer acceptance and use of information technology: extending the unified theory of acceptance and use of technology. In: MIS Quarterly, pp. 157–178. (2012)
77. de Wit, C.C., Siciliano, B., Bastin, G.: Theory of Robot Control. Springer Science & Business Media (2012)
78. Paul, R.P.: Robot Manipulators: Mathematics, Programming, and Control: The Computer Control of Robot Manipulators, Richard Paul (1981)
79. Samson, C., Espiau, B., Borgne, M.L.: Robot Control: The Task Function Approach. Oxford University Press, Inc (1991)
80. Espiau, B., Chaumette, F., Rives, P.: A new approach to visual servoing in robotics. IEEE Trans. Robot. Autom. **8**(3), 313–326 (1992)
81. Devigne, L.: Solutions robotiques bas coût pour l'aide à la navigation en fauteuil roulant électrique : vers une contribution dans le champ de la rééducation neurologique. PhD thesis, 2018. Thèse de doctorat dirigée par Babel, Marie et Gallien, Philippe Informatique Rennes, INSA (2018)
82. Devigne, L., Narayanan, V.K., Pasteau, F., Babel, M.: Low complex sensor-based shared control for power wheelchair navigation. In: IEEE/RSJ International Conference on Intelligent Robots and Systems (IROS), pp. 5434–5439. IEEE (2016)
83. Narayanan, V.K., Spalanzani, A., Luo, R.C., Babel, M.: Analysis of an adaptive strategy for equitably approaching and joining human interactions. In: 25th IEEE International Symposium on Robot and Human Interactive Communication (RO-MAN), pp. 341–346. IEEE (2016)

84. Olivier, A.-H., Le Borgne, N., Babel, M., Crétual, A., Pettré, J., Collision avoidance between a walker and a person on an electric powered wheelchair. In: ISPGR, Conference of the International Society for Posture & Gait Research, p. 2019. Edinburgh, United Kingdom, June (2019)
85. Leblong, E., Piette, P., Dandois, M., Ceze, E., Devigne, L., Pasteau, F., Babel, M., Fraudet, B., Mcfadyen, B.J., Olivier, A.-H.: Navigation dans les musées de la ville de rennes: approche écologique pour l'évaluation d'un module d'assistance à la conduite lors des interactions des usagers de fauteuil roulant électrique dans un environnement complexe. In: SOFPEL 2022-28ème congrès de la Société Francophone Posture Équilibre et Locomotion (2022)

Telepresence and Teleoperation

Nikhil Deshpande

Abstract This chapter elaborates on the handbook's premise of establishing the notions of ergonomics and human factors in robotics, here in relation to the technologies of telepresence and teleoperation. As implemented through robotic systems, telepresence and teleoperation are often interchangeably used terms. Both use the prefix term "tele", which pertains to communication over distance, and both refer to humans being able to do something—"being present" (telepresence) or "operating devices" (teleoperation)—at a distance. Semantically and technologically, telepresence and teleoperation exist by themselves, and one is not a necessary condition for the other. However, as technology has evolved, the interplay between the two has markedly increased. Nevertheless, humans, and their ability to interact with system elements, are central to both ideas. In that respect, ergonomics as a study of methods to optimise human well-being and improving overall system performance, must serve as an integral criterion in developing telepresence and teleoperation systems. The chapter details the idea of ergonomics in the domain of telepresence and teleoperation by tracing the evolution of telepresence and teleoperation, then illustrating the challenge of ergonomics and human factors engineering in this field, and further demonstrating how different aspects of this challenge can be addressed. The chapter concludes with a case study showing how the thread of ergonomics and human factors can be woven throughout the design and implementation of telepresence and teleoperation technologies, grounding them in human-centred arguments. The chapter aims to demonstrate how ergonomics principles fundamentally underpin the landscape of telepresence and teleoperation technologies. It lays the groundwork for advances in future telepresence and teleoperation systems that have ergonomics principles at their core.

Keywords Telepresence · Teleoperation · Robotics · Human-centred · Human factors · Interaction design · Cognitive ergonomics

---

N. Deshpande (✉)
Cyber-Physical Health and Assistive Robotics Technologies (CHART) Research Group, School of Computer Science, University of Nottingham, Nottingham, UK
e-mail: nikhil.deshpande@nottingham.ac.uk

G. Barresi and S. Wischniewski (eds.), *Ergonomics in Robotics: Advances and Innovations*, Springer Tracts in Advanced Robotics 162,
https://doi.org/10.1007/978-3-032-19374-2_15

1 Introduction

Wait, wait… it seems the feed is very glitchy! I am not able to see clearly there.

It looks like there's a lot of delay in the video! Let's wait till it syncs up.

Hello!? I think I lost you again. I'll just hang-up and try calling again.

Variations on these statements are common refrains of users when using the many video conferencing applications that are prevalent in the world today [1]. On the other hand, these complaints would not be out-of-place in the world of robotic teleoperation [2]. In the rapidly evolving technological landscape, telecommunication has emerged as an imperative that bridges physical distances and enables human interaction in previously unimaginable ways. The increased attention to it in recent years, no doubt accelerated by the pandemic, is focused on enabling humans to interact from a distance with other people, socially as well as professionally, while also allowing people to remotely work, interact with environments, and operate machines. Two specific instances of telecommunications technologies are the concepts of Telepresence and Teleoperation. Both use the prefix term "tele", which pertains to a form of remote operation or communication, and both refer to humans being able to do something—"feeling like they are present" (telepresence) or "acting in environments through machines" (teleoperation)—at a distance [3]. Semantically and technologically, telepresence and teleoperation exist by themselves, and one is not a necessary condition for the other. However, as telecommunications has evolved, the interplay between the two has markedly increased, and it has been observed in fact, that "feeling present" can improve teleoperation task performance, while being able to "act in environments remotely" enhances telepresence [3, 4]. Telepresence and teleoperation find applications in diverse domains, including social interaction, entertainment, tourism, and other professional fields ranging from healthcare and manufacturing to space exploration and disaster response [5, 6]. From virtual meetings and remote collaboration to telemedicine and tele-education, these applications have bridged geographical gaps and enabled seamless connections [1–6].

As these technologies seem to touch almost every aspect of the human lived experience, achieving seamless and efficient human interaction for telepresence and teleoperation hinges on the application of human factors and ergonomics principles in their design and implementation. Ergonomics, also known as human factors engineering, focuses on designing systems and equipment to fit the capabilities (and limitations) of human users [5]. In the context of telepresence and teleoperation, ergonomic considerations become crucial in designing devices, machines, interfaces, and indeed interactions, that simplify the human user experience. Centralising the developments around their human experience has become an important paradigm in the evolution of both these technological standards [7].

2 Understanding Telepresence and Teleoperation

In the broadest sense, telepresence refers to the feeling of being present in a remote environment. First defined as a neologism in 1980 by Minsky and Gunkel [8], the idea is to allow a person to feel as if they are at a place other than their true location. The "feeling of being present" being an abstract concept, has varying interpretations itself. Even so, a telepresence system needs to have certain elements to qualify as such—ability to visualise the remote environment for the user, and to speak and be heard there. In this sense, even simple video communications systems can qualify as basic forms of telepresence. Where the "feeling of being present" is considered, additional elements come in, e.g., more sophisticated stereo video and audio communication, ability to move around in the remote environment, tracking user gestures, etc., which provide specific stimuli to the user, and therefore have bidirectional information exchange between the user and the remote location [3]. The ability to move around brings the idea of robotic telepresence [9], or mobile robotic presence. This is basically a video conferencing system through which a user can move around in a remote environment and interact with the inhabitants there. In this sense, introducing the robotics element in telepresence melds it with the concept of teleoperation, to indirectly enhance the feeling of presence [9].

Teleoperation itself, on the other hand, is a much older concept, with regards to "remotely controlling" a device at a distance, remotely operated weapons being some of the first examples seen in the nineteenth century itself [10]. In the academic world, teleoperation tends to have a more mechanised, or robotic flavour, for the aspect of remote control. This began in the 1940s with the first "master-slave" manipulation system, which had a direct mechanical coupling between a master device—a mechanism with levers, clutches, and pedals for the operator to manipulate—and a slave system—a device with mechanical arms and grippers—allowing operators to work with radioactive material from behind shield barriers [11]. Unlike telepresence, the primary aim of teleoperation is to extend the human capability to "act" in an environment they are not physically present in, e.g., handling objects from a distance. The user imposes a motion and force on the device-based interface available, which in turn results in a displacement transmitted to a remote machine (or robotic manipulator) that mimics that movement. The key component here is the bidirectional force reflection between the operator and the remote machine.

Figure 1 gives a sense of what the general telepresence and teleoperation framework looks like. The articles in [3, 4, 12] give a comprehensive look at the evolution of telepresence and teleoperation systems, from the original direct mechanical linkages between the operator and the remote devices [4], to remote teleoperation over a communication network and over the internet [13], to the more recent incarnations in telexistence [14], prizes supported by the ANA Avatar XPrize [15], showing that the concept of teleoperation continues to be a fertile ground for innovation and progress. Although an exhaustive list of devices is beyond the scope of this chapter, Table 1 gives a peek into some of the devices and applications.

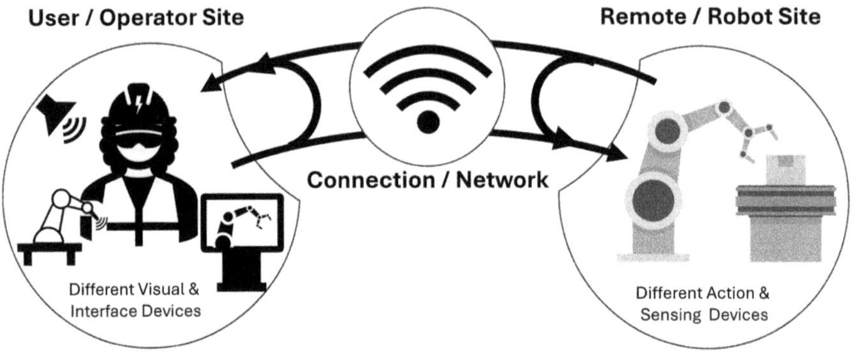

Fig. 1 General overview of a telepresence (or teleoperation) system

The foundation of successful telepresence and teleoperation lies in the ability to replicate sensory perceptions and motor actions accurately. In their modern incarnations, both telepresence and teleoperation have devices for the user and machines in the remote environment, with a dedicated communication channel in between. The user device acts as the interaction interface and usually involves presenting all the information coming from the remote side to the user, as well as capturing the commands from the user to send them to the remote side. On the other hand, the remote devices, usually robotic mechanisms with locomotion and manipulation capabilities of varying complexity, with means for multimodal sensing and display, allow users to perceive the remote environment and act in it as per their requirements. High-quality visual, auditory, and tactile feedback can help users' feeling of presence in the remote environment. In fact, telepresence and teleoperation are often conflated in state-of-the-art literature [31], and not without reason. To this point, it has been shown that interfaces that facilitate this sense of "presence" also enhance human task performance with teleoperation systems [32]. This shows the interdependency of telepresence and teleoperation owing to modern technological advances.

In order to bound the scope of this chapter, we reference both telepresence and teleoperation in terms of their robotic interaction. To that end, simple audio-video conferencing is not considered here. Humans, and their ability to interact with people and environments through the medium of robotics, are taken as central to both ideas. Research has made great progress in improving the capabilities of machines (and/or robots), improving autonomy in manipulation and navigation, as well as the tele-connectivity framework [33, 34]. Nevertheless, having precise and responsive control mechanisms, sufficient degrees-of-freedom (DOFs), and dexterity in the mechanisms, as well as communication channels with sufficient bandwidth and low latency, are critical challenges in ensuring a good quality interaction experience, and remain very much, a work-in-progress [2]. Through the course of this chapter, it will become evident that achieving seamless and efficient human-machine interaction in telepresence and teleoperation, grounded in ergonomics principles, requires consideration of multiple factors.

Table 1 Example domains, devices, and applications in telepresence and teleoperation

Domain	Device(s)	Sample features	Applications
Social and commercial telepresence	TEMI [16], DOUBLE 3/ PEOPLEBOT [17], OHMNI Labs [18]	Autonomous navigation; path planning; obstacle avoidance; remote call-in and motion control; settings for guided tours; integration with natural language and generative AI; face/voice/ emotion recognition; speaking avatars; hi-def/360 cameras; lidar; manipulation capabilities (sometimes)	Commercial telepresence (meetings, museum visits), hospital assistant, remote work, remote therapy/social care
Telesurgery	DA VINCI (intuitive surgical) [19], VERSIUS (CMR surgical) [20], LUNA (Asensus surgical) [21], HUGO (Medtronic) [22]	Multiple robotic arms for interventions; separate surgeon console with high-definition 3D display and finger- or touch-based control interface; high-def 3D cameras and instrumented arms (e.g., EndoWrist, TrueWrist) for motion and precision; integration with VR/AR for intuitive interaction	Commercial robotic surgery (Telesurgery)—All kinds of laparoscopic and soft tissue interventions
Underwater telerobotics	CURV [23], KAIKO/ADIOSMO [24], JASON/ ARGO [25]	Six degrees of motion with auto-control functions for depth, altitude, and heading; target identification, pinger detection, and on-board SONAR; hi-def cameras; customised tool packages for interventions	Remotely operated vehicle (ROV) for underwater exploration
Space telerobotics	ROBONAUT (NASA) [26], DEXTRE (NASA) [27], CURIOSITY/ SPIRIT (NASA) [28], INTERACT CENTAUR (ESA) [29], CANADARM2 (CSA) [30]	Telerobotics and tele-navigation in space; humanoid or dual-arm tele-manipulation; wheeled mobile base; multiple high-def cameras on board and in environment; multiple DOFs; on-board sensor suite; fully autonomous to fully teleoperated control	Remote operation on the International Space Station; moon or Mars rover; Space station-to planet telerobotics; on-board physics lab for material analysis

3 Role of Ergonomics in Telepresence and Teleoperation

As defined in the earlier chapters, the primary goal of ergonomics is to optimise the interaction between humans and their environment, thereby enhancing efficiency, safety, and well-being [35]. The key phrase "optimise the interaction between humans and their environment" is rather a short hop from the definitions of telepresence and teleoperation we have seen above, and in fact seems to overlap broadly. Ergonomics must play a central role in telepresence and teleoperation applications as they involve prolonged periods of interaction with digital and/or mechanical interfaces. In state-of-the-art systems, users engage with these technologies through various devices such as videoconferencing systems, mobile phones, mixed reality (MR) headsets and displays, TV monitors, joysticks, keyboards, haptic devices, gloves, etc. [31, 36]. The challenge lies in designing the interfaces and the interaction to enable users to perform their tasks more effectively and efficiently, while minimising the physical and cognitive strain on them, and without compromising their safety. Furthermore, as these systems become more prevalent in the workplace, their design needs to derive from the principles of organisational ergonomics, ensuring that devices are fit for purpose, allow workers to work effectively even when working remotely, encourage workers to maintain good posture, minimise their fatigue, and including proper training on equipment use, thereby ensuring organisational satisfaction and productivity [37]. In this respect, every aspect of the interaction needs to be looked at through the lens of ergonomics, e.g., kind, quality, and amount of sensory data being presented, the types and designs of the devices on the user and remote sides, the modes of interaction being asked of the users, etc. [38]

4 Understanding the Contours of the Challenge

Let us do a thought experiment. Imagine that you are called on to attend an art exhibition in the city, using their newly installed telepresence robot, one of the many that are currently available on the market [17]. You are a generally healthy person, without any known disabilities. What would be some of the first questions that would pop up in your mind? Some variations on—"Would I be able to control the robot myself, or someone else would control it for me?", "How fast/slow would it move around the exhibition area?", "How would I visualise the exhibits?", "What would be the amount of space I can see, how clearly, would it be 3D?", "What if I would like to speak to the artist/curator? Would they be able to see me or just hear me?", "Would I be able to shake hands with anyone, touch the exhibits, if allowed?", "Would I be able to feel such physical interactions?". Perhaps there are many more than these that may come to your mind, but you get where we are going with this. The fact of you not being present at the exhibition physically immediately brings up so many aspects related to the human action-perception-cognition loop, ones that we otherwise take for granted. To be sure, feeling present in a remote location, or performing an action through a robot at a distance, are tasks that are not natural to human beings. So, it is natural to have such apprehensions about it.

Then, before you embark on the visit itself, what would be among your very first actions? You would try to check if your video is coming through, whether and how much the robot really moves when you fiddle with the navigation controls, if the sound from the remote side is coming through, whether the video controls work, etc. Once you are comfortable with it, you will begin moving around the space, as you would once you enter the exhibition. You may then begin to notice a few things—(i) that looking at the video feed as the robot moves is quite discomfiting [39]; (ii) that once something is outside the camera's field of view, you are not able to see it [9]; (iii) it is quite a slow process to redirect the camera, or even to turn the robot around [40], etc. You do not get a complete sense of where you have been, and where you would like to go. For that you need additional information, e.g., a map of the area, and you are required to correlate the robot's position and the map yourself. Some things that are seemingly simple to us as humans, e.g., avoiding colliding with objects, surmounting gradients or small steps when navigating, become surprisingly difficult when the robot is to be commanded to do them. Any instability in the network connection while the robot is moving only exacerbates the problem. Often, being able to enjoy the visit takes a back seat, let alone connecting socially or emotionally with others in the remote environment. The situation in the above thought experiment inherently demands a physical and mental engagement from the user that is not natural to humans [41]. Figure 2 gives a glimpse of the factors involved. Just in those aspects, if not implemented well, visiting a place through telepresence can require a level of concentration and attention to the feedback and focus on the controls, for extended periods, that makes the experience very different to visiting a place in person [42].

An analogy for this level of continuous engagement and attention can be of a surgeon performing surgery [43]. Any distraction or missed information can result in increased mental workload and leave the subsequent actions susceptible to errors. If we then expand the above thought experiment into the world of teleoperation and the kind of difficult environments where humans may wish to act remotely, we realise that the aforementioned aspects may call for an even greater engagement from the

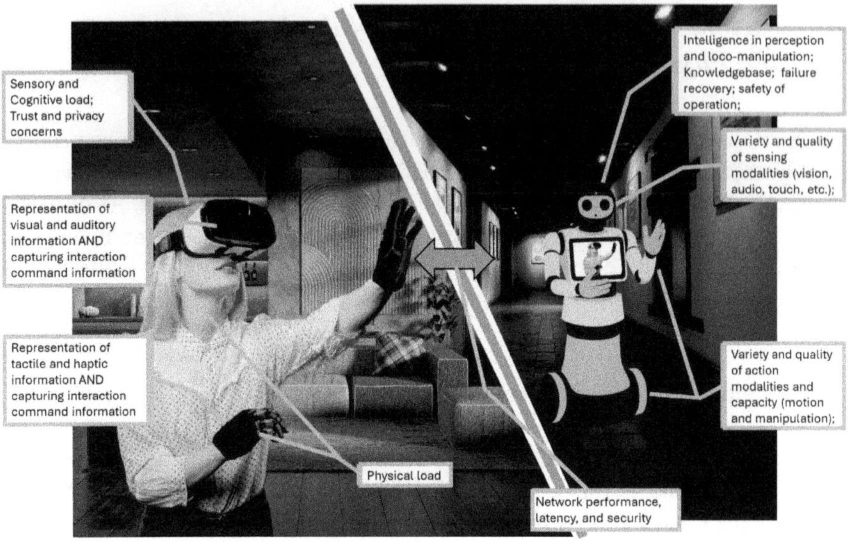

Fig. 2 An overview of the factors affecting ergonomics in telepresence and teleoperation, with the "visiting the museum" scenario

human user [44]. For instance, visiting an art exhibition remotely would not demand as high a level of concentration and mental processing of information on part of the user as would performing a rescue operation by teleoperating a robot.

Nevertheless, in the above thought experiment, if you are a disabled person, the nature of the discussion would indeed focus on the fact that telepresence affords you the ability to visit the museum at all, without leaving the comfort of your inhabited space [45, 46]. From the perspective of ergonomics and human factors therefore, different elements must play an important role in delivering an acceptable telepresence and teleoperation experience.

4.1 Factors Affecting Ergonomics in Telepresence and Teleoperation At-a-Glance

The challenge of considering ergonomics and human factors in telepresence and teleoperation applications is a multi-dimensional one, and consequently, requires addressing many different dimensions of the problem. Table 2 here gives a cursory glance across the different factors that need to be accounted for, giving an idea of how each of them plays a role in the ergonomics of the interaction in telepresence and teleoperation. In the subsequent subsections, we delve deeper to understand these factors.

4.2 Factors Regarding Motor Skills of the Human and Physical Ergonomics

Being able to understand how to maneuver a machine in a remote environment, how to quickly respond to interaction requirements through a robot, using only the available controls and interaction devices, is not an easy task for human users. As discussed below, the physical design, control capabilities, and overall interaction features are important factors for the devices afforded to the user in successfully establishing and executing telepresence and/or teleoperation tasks.

4.2.1 Remote-Side Mechanism Design and Control

Precise and responsive mechanisms are crucial for teleoperation tasks, especially those requiring fine motor skills. The capabilities, or the lack thereof, of the mechanisms in the remote environment can put a significant physical strain on the users. A significant hurdle arises from the constrained maneuverability, and at times insufficient DOFs in movement of various teleoperation robots, which often lack the

Table 2 Different factors affecting the ergonomics considerations in telepresence and teleoperation

Factors		User-side	Robot-side
Sensing modalities and communication design	Visual sense	Acuity; depth perception; resolution (image and motion); brightness; perspective; motion blur	Field-of-view; frame rate; resolution; data interface; physical design; colour + depth capture; pan-tilt-zoom option; placement (static or movable); perspective (ego- and/or exo-centric)
	Auditory sense	Direction; amplitude level; frequency; clarity; discerning different sounds; background noise suppression	Physical design; transducer principle; frequency response; directionality; placement (static or movable)
	Haptic sense	Feedback type—kinesthetic, vibration, texture, temperature, guidance; sensitivity; placement on the body; amplitude level	Sensing range; placement; hysteresis; sensing workspace
	Other sensors	Type of data representation or sensory substitution; data fusion	Modality—pressure, gases; sensing range; sensor fusion
	Network and communication	Latency; bandwidth; packet loss and stability; audio/video/data codecs	Data compression and packetisation; transport protocol; frame rate
Physical design and action modalities	Device design	Kinematics; DOFs; stability; type of control; force reflection capacity	Dimensions (shape and size); DOFs; kinematics; type of control; type of locomotion (wheels, legs, etc.) and manipulation (arm, hand, gripper, etc.); payload; power autonomy
	Physical interface	*Buttons*—types and quantity; *Headset*—weight, dimensions; *joystick*—Weight, physical design, button placements; *Wearable haptics*—weight, motion range, gesture resolution; force applied; *Haptic devices*—(see above)	Physical form of the device (bio-inspired, machine-like, human-looking, etc.); loco-manipulation capacity (structured or unstructured environments); precision and accuracy of actions, gestures
	Visual interface	Type and amount of visual information; balance of text and video; placement of information in field-of-view; visual cues; assistive overlays	

(continued)

Table 2 (continued)

Factors		User-side	Robot-side
	Other interfaces	*Free-hand gestures*—simplicity/complexity/diversity of gestures, accuracy and precision of recognition; *Speech*—speech rates, noise filtering, voice quality, speech-to-text processing algorithms; *Language*—generative AI models, model complexity (no. of parameters), perplexity, precision and recall, accuracy; *Eye-tracking*—sensitivity, accuracy, precision, noise reduction; *Brain-computer Interface*—electrode placement and density, sensor type, signal-to-noise ratio, accuracy, latency, robustness, noise reduction, real-time processing, privacy and security	*Language*—generative AI models, model complexity (no. of parameters), perplexity, precision and recall, accuracy; *Speech*—quality and modulation of machine speech; text-to-speech processing algorithms; noise filtering
Cognition and trust	Cognitive intelligence	Abstraction of control; information representation and relevance; interactive perception/action; mode switching; cues for decision-making	Granularity of execution autonomy; perception for navigation and manipulation; failure recovery modes; knowledgebase
	Privacy and security	Trust in system; data gathering and consent; access control; data storage and encryption	Safety of operation; failure recovery; network security and encryption; access control

natural loco-manipulative capabilities inherent in humans [2]. This limitation introduces complexities in precisely reproducing movements and gestures, as well as assessing force and pressure when interacting with remote objects and navigating remote environments. For example, in the thought experiment above, this may lead to difficulty for the user to adjust the body pose of the telepresence robot to help drive it around an obstacle and through an area.

4.2.2 User-Side Interaction and Control Interfaces

Similarly, the design and control capabilities of the user-side mechanisms, i.e., the interaction and control interfaces, play a crucial role in the ergonomics of the interaction. The traditional devices, e.g., control joysticks with 2D visual feedback can mean requiring users to often spend hours in front of screens or wear devices that may not be optimised for long-duration use. Prolonged sitting, improper screen height, and inadequate dexterity and positioning of devices can result in musculoskeletal problems. Telepresence applications often leverage virtual reality (VR) and augmented reality (AR) technologies to create immersive experiences [34]. The concept of these technologies is elaborated here:

Virtual/Augmented/Mixed Reality—The *Reality-Virtuality Continuum*

The reality-virtuality continuum, first introduced by Milgram and Kishino [47], describes the spectrum between the completely real and the completely virtual, and is the foundational concept in the fields of VR, AR, and mixed reality (MR). At one end of the continuum is the real environment, which consists entirely of real-world, physical objects, environments, and experiences. At the opposite end is the virtual environment, which is entirely digital and computer-generated. Between these two extremes, lie various combinations of real and virtual elements, creating a spectrum of experiences that blend the physical and digital worlds [48]. Moving from the reality end to the virtuality end, first comes AR, which overlays digital (or virtual) elements onto the real world. Examples include the Pokemon Go game, or AR glasses that show digital information in your field of view [48]. Then, comes augmented virtuality (AV), a less common term referring to virtual environments that incorporate real-world elements, e.g., a virtual meeting room that includes live video feeds of participants. At the virtuality end, also termed as VR, is a fully immersive digital environment, with virtual objects. Here users can interact with a completely virtual world, using dedicated displays, e.g., Meta Quest, HTC Vive, etc. Mixed Reality (MR), on the other hand, refers to the subset of the continuum that includes both AR and AV, seamlessly integrating real and virtual elements in real-time. This can range from simple overlays of digital information on the real world to complex environments where physical and digital objects coexist and interact.

In the context of telepresence and teleoperation, the immersive nature of VR can enhance the feeling of "being there," making users feel as if they are physically present in a remote location, e.g., virtual meetings, remote training, and virtual tourism. AR overlays can provide additional context and information without completely removing the user from their physical environment. For instance, in remote assistance scenarios, AR can allow remote therapists to guide at-home patients through their therapy, overlaying instructions and annotations on their field-of-view. This can be made more complete for participants with MR, where both real and virtual objects can be used, making the experience more dynamic and collaborative [49].

The reality-virtuality continuum enhances telepresence by providing a range of immersive and interactive experiences that bridge the gap between physical and virtual worlds [47]. While these technologies offer exciting possibilities, and indeed show improvement in teleoperation task performance, they also pose unique ergonomic challenges. Wearing VR headsets for extended periods can cause discomfort, motion sickness, and visual fatigue [50]. For the physical interfaces, although haptic control interfaces can provide a better perception of the forces involved, the delicate equilibrium between transparency and stability of the system can lead to untoward physical strains on users' hands and fingers [51]. Inaccurate force feedback can also cause users to apply unnecessarily high forces, which can be damaging both to the user and remote sides of the interaction [52]. Furthermore, generally the mismatch in the DOFs available to the user as against those of the robot, implies that the user is trying to mentally map the desired motions between their gestures and the outcomes at the robot. This can lead to significantly high mental workloads [40]. Other interfaces such as eye-tracking and brain-computer interfaces can be very useful in balancing the physical and cognitive loads, but rely on the sensitivity, resolution, and filtering of generated signals to be able to deliver satisfactory task performance [53]. The need for high-performance hardware and software can also be cost limiting. Social issues, such as privacy concerns and the potential for digital addiction, are key issues to be resolved before the technology becomes user ready [49].

4.3 Factors Regarding Information Representation and Decision-Making

Successful teleoperation relies heavily on the operator's ability to understand and reason about the remote environment where the robot is operating, as well as the environment around themselves, i.e., the command or pilot station. In telepresence and teleoperation, cognitive load can manifest itself not only in terms of the ability to perform a task taking into account the information provided, but also in terms of two other aspects: (i) the ability of the user to understand, quickly and intuitively, the requirements of the task based on the relevance, or indeed, irrelevance, of the information provided related to the task, or aspects of it; and (ii) the ability of the user to quickly learn how to perform the required task using the devices and interfaces available, based on their understanding and how they have performed the required task the initial few times [54]. Engaging in teleoperation tasks can impose a significant cognitive load on operators due to the complexity of the remote environment. Both aspects are related to the idea of situation awareness, which is not just perception of the environment and the objects in it, but also comprehending their meaning and correlations, and understanding their status in time [55]. A lack of situation awareness leads to users needing to use their imagination, and in many cases, prior knowledge, to process the information that is being relayed from the remote environment, and fill-in-the-gaps where all information is not available [56]. Furthermore, as noted previously, the need for the user to be constantly engaged with the system when performing the task, with the operator manually controlling each action, can lead to physical and mental fatigue [2].

4.3.1 Remote-Side Sensors for Capturing the Environment

Indeed, modern cameras, sensors, and embedded computers have dramatically enhanced the capturing, processing, and relaying of remote data, allowing the user to understand the remote scene and make informed decisions. Yet, one is always trying to build a mental 3D map of the remote environment and the task affordances from the data being shown—stereo video, point-cloud, ego- and exo-centric camera views, etc. [57] The point-of-view of the sensors, their sensing range, and their sensitivity are among the chief factors, which determine whether the information being provided back to the user is sufficient for the task [9]. Similarly, the acoustic sensors for auditory feedback and force/torque sensors for estimating force exertion, allow multi-modality in the perception of the environment [58, 59]. However, raw data from these sensors requires the user to mentally fuse the data spatio-temporally, and make sense of it, which becomes mentally demanding for the user. Adding more information via sensors may not always help with situation awareness, as this can stretch users' attention capabilities [60].

4.3.2 User-Side Interface for Visualisation and Sensory Feedback

The traditional interface of visualising raw sensory data on monitor screens provides limited spatial awareness and can lead to errors in tasks demanding intricate coordination between locomotion and manipulation. As discussed earlier, the modern interfaces with VR/AR are beneficial in terms of presenting data more intuitively and immersively [34]. However, these technologies are yet to be adopted at scale and most users remain unfamiliar with the hardware involved in the technology. Additionally, issues such as latency in visual feedback, discrepancies in motion perception, and inconsistencies in depth perception can lead to sensory confusion and cybersickness as noted earlier. Engaging in any kind of "tele"-tasks can impose varying levels of cognitive load on human users, depending on the complexity of the remote environments, the capabilities and design of the remote agent, the features of the interaction interface itself, the information presented, and in extreme cases, the need for rapid decision-making in difficult situations [56].

4.4 *Factors Regarding Social Interaction and Communication*

As seen earlier in the thought experiment, telepresence and teleoperation can extend beyond physical interaction, encompassing social and emotional aspects as well. In the domain of social interaction, telepresence aims to essentially mediate human-to-human interaction over distance. The quality of such interaction depends on how well the system can replicate face-to-face, in-person interaction and communication dynamics. Several technological challenges emerge as barriers to creating truly immersive and natural virtual connections. Apart from the quality of the audio-visual, one major challenge lies in replicating authentic nonverbal cues, such as facial expressions, gestures, and body language, with high fidelity [61]. The limitations of current hardware and software often result in a loss of nuance and subtlety in conveying emotions and intentions, hindering the establishment of genuine rapport between the users and the remote environment [62]. The controllability and stability of the telepresence platform also play a crucial role. For instance, aspects like maintaining eye-contact, building trust, being able to observe and react to body language, etc., are impacted adversely if the platform is not responsive enough, unstable in their control, or cannot adapt well to the environment. Additionally, achieving synchronised and real-time interactions across varying internet speeds and network latencies poses a significant hurdle. Delays and disruptions can lead to disjointed interactions in turn reducing the feeling of presence, immersion, as well as engagement [63]. Finally, accessibility and inclusivity of the telepresence systems are key to having meaningful interactions for diverse user groups in different cultural and social contexts [64].

4.5 *Factors Regarding Safety, Trust, and Privacy*

In the realm of telepresence and teleoperation, beyond the technical developments, three factors are crucial to their adoption: (i) ensuring safety of the human user, the system itself, and the surrounding environment; (ii) building trust between the human users and the technology itself; and (iii) ensuring the privacy of the interaction and related data. All three factors relate to the comfort, confidence, and satisfaction that users would feel in any human–machine (or, human–machine–human) interaction [65].

4.5.1 Safety

Safety concerns stem from the potential disconnection between the user and the remote environment, leading to a reduced ability to identify hazards or respond swiftly to unexpected events [66]. This challenge is particularly pronounced in scenarios involving remote operation of machinery, vehicles, or mission-critical environments, where a delay in decision-making can result in accidents. Various operational stresses and potential misuse can lead to mishaps, e.g., harming a patient in a telesurgery scenario. Without the fail-safes, redundancy, capacity to handle unexpected situations [66], and proper training of users of the system covering all aspects from basic controls to emergency procedures [67], users will not feel safe using the technology.

4.5.2 Trust

Moreover, establishing trust in the technology is intricate, as users must feel confident that the system will accurately convey information, respond predictably, and safeguard their interests [68]. System reliability is crucial here, and the lack of consistent performance under various conditions, and unexpected failures can lead to diminishing trust [68, 69]. This also touches on the transparency of the system—users need to understand how the system works, and the system's limitations and potential risks. Clear and comprehensive documentation, real-time feedback, and a well-designed user interface that is intuitive and easy to navigate can significantly enhance trust, helping users feel more in control and aware of the system's operations [67].

4.5.3 Privacy

Concerns about data security, privacy breaches, and unauthorised access are becoming increasingly relevant in today's world with the proliferation of social media and mass communication [67]. Not only is protecting user data paramount, helping users understand what data is collected and how it is used is just as important. Strong encryption protocols can secure data during transmission and storage, ensuring lower susceptibility to attacks [70]. Robust authentication and authorisation mechanisms are necessary so only authorised users can access sensitive data [67]. Compliance with privacy laws and regulations, e.g., GDPR, involves not only implementing technical measures but also establishing policies and procedures for data handling, user consent, and breach notification [70].

5 Addressing the Challenge

Ergonomics is the study of the relationship between humans and their working environment, emphasising the design and arrangement of tools, equipment, and systems to promote user well-being and productivity. Ergonomics principles focus on optimising the fit between humans and machines, considering factors such as user characteristics, device design, environmental conditions, information feedback, and, importantly, task requirements. Solving the challenge of ergonomics in telepresence and teleoperation applications requires a collaborative effort involving designers, engineers, human factors specialists, and end-users.

5.1 *Metrics to* Measure *Ergonomics and Human Factors*

A key aspect is to be able to measure the role of ergonomics and human factors, i.e., to be able to understand how the devices, interfaces, and interactions affect the physical and cognitive involvement of human users in task performance in telepresence and teleoperation. The multidisciplinary approach relies on subjective as well as objective assessments.

5.1.1 Subjective Assessments

Delving into the realm of human perception, cognition, and emotion, subjective assessments serve as a bridge between quantitative data and the often complex, multi-faceted nature of human experiences. Subjective assessments encompass a range of research methods and tools, including surveys, questionnaires, interviews, focus groups, and direct observation of user behaviour [71], carefully designed to elicit users' feelings, thoughts, and interactions within specific contexts. For example, in users' interaction with robots in a telepresence or teleoperation scenario, understanding people's attitudes and emotions prior to, during, and after the interactive study, e.g., through the NARS (Negative Attitude toward Robots) scale [72], is just as important as knowing their ease of comprehending the presented visual feedback, clarity of details, pain points, and overall satisfaction. This can provide designers with insights into the user's perception of the system's usability, as well as understanding human behaviour in relation to the tested interfaces [73–75]. In terms of situation awareness in telepresence and teleoperation, the SAGAT (Situation Awareness Global Assessment Technique) questionnaire is a useful model, involving intra-task queries concerning both the current and future states of the task itself [55]. More specifically, understanding users' attributes and propensities towards experiencing telepresence, through assessing their perceptual, cognitive, motor abilities, and prior experience becomes an important factor in adapting the system designs. An example of this is the Immersive Tendencies Questionnaire [76]. Proper methodology, well-constructed questionnaires, ensuring a representative sample of participants, addressing potential biases, and using standard models for data interpretation, enhances the validity and reliability of subjective assessments [71].

5.1.2 Objective Assessments

Unlike the qualitative focus of subjective assessments, objective assessments rely on empirical data, making them particularly valuable for obtaining precise and quantifiable performance metrics, physiological measurements, and behavioural observations. For example, in usability testing of telepresence interfaces, objective assessments might measure the time it takes for users to complete specific tasks, the number of errors made during the process, or the efficiency of navigation through the different parts of the interface [77]. Physiological measurements e.g., heart rate, skin conductance, eye-tracking data, electro-encephalography, brain imaging, etc. [78, 79], tie into the modern field of *neuroergonomics*—analysing neural processes of users as they interact with systems, to judge the ergonomics of the designs [80]. For instance, eye-tracking technology can reveal where a user's attention is focused on a teleoperation interface, helping designers identify areas of interest or distraction in a user interface [63]. Behavioural observations can range from tracking the movement of a user's hand while manipulating a device to recording verbal and non-verbal communication patterns during teleoperation tasks [71].

Combining objective assessments with subjective assessments can yield a more comprehensive understanding of the user experience. Statistical analyses of the data obtained from these assessments, e.g., Analysis-of-variance (ANOVA), Student's t-test, etc., are crucial to establishing whether, and by how much, the developed systems are better at satisfying ergonomics requirements [81]. By leveraging objective assessments alongside their subjective counterparts, researchers and designers can make informed decisions to optimise the usability, safety, and overall quality of the telepresence and teleoperation systems for users.

5.2 Grounding Telepresence and Teleoperation in Ergonomics Principles

Addressing the challenge of grounding the developments in telepresence and teleoperation within the principles of ergonomics, is easier said than done. It involves reevaluating the priorities of design and development of these technologies from the very beginning, ensuring the correct foundation is available to enable the human user to be at the centre of it all [82]. Towards achieving this goal, four independent axes are defined here that can help researchers and system designers to develop and evaluate the performance of their system in terms of ergonomics and human factors:

1. *To act (and interact) effectively*—any system must first be functionally effective to achieve the goals. For telepresence and teleoperation, this means that the human user is first able to reach their objective (visit the museum, conduct the surgery, diffuse the bomb, etc.), when they use the system. But then, the user is also able to do with it similar or better task performance than without an ergonomically designed system.

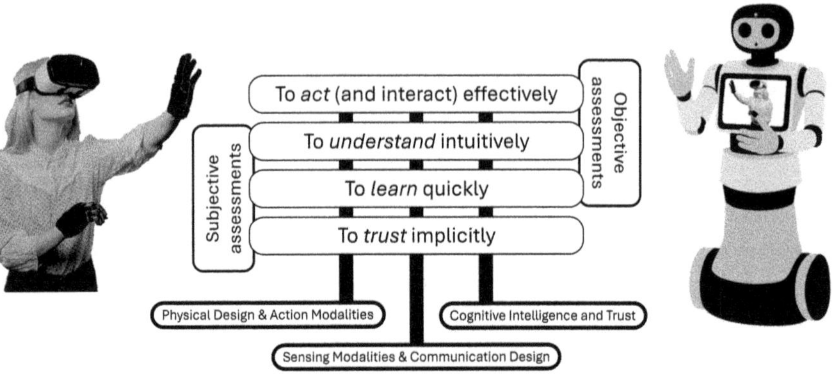

Fig. 3 Addressing telepresence and teleoperation systems using the principles of ergonomics and human factors; The four *axes* of act, understand, learn, and trust are built on the three *pillars* of *physical design*, *sensory and communication design*, and *cognition and trust design*. The four axes are influenced transversally by the subjective and objective assessment techniques of ergonomics

2. *To understand intuitively*—where the user comprehends the task being done and the information presented without the need for deliberate analysis, using the natural correlations between action and perception.
3. *To learn quickly*—where the user learns to use the physical interfaces and the information representations without specific training, utilising the intuitive understanding of the system.
4. *To trust implicitly*—where the user is confident in the system's behaviour and performance so as to trust it without having to explicitly check against its deliberately harmful or detrimental actions.

Any telepresence/teleoperation system may earn scores along each of these four axes. Figure 3 gives an overview of this scheme, and a deeper explanation of each of these axes follows here.

5.2.1 To Act Effectively

Mechanism design or, in general, device design and control is a complex and general problem. The key point is that there is no one-size-fits-all solution for the design of the device. Certainly, there are some attributes that it needs to have: sensors, cameras, actuators, mechanism to move and interact in a space, etc. The combination of these attributes into a physical form is where the application domain, the utility of the device, and the user themselves must play an important role. For instance, strategically placing a couple of motion-capable cameras on the front of a robot, or using a 360-degree camera, can both provide the operator sufficient situational awareness, but perhaps in different application scenarios, one may work better over the other [9, 39]. The physical design must satisfy the requirements of the task itself first. This functional purpose, above all the other aspects, is fundamental to successful ergonomics designs, which allows a consistency in task performance [83]. Similarly, having a level of autonomy in action, e.g., collision avoidance, failure recovery, scene understanding, object grasping, etc., can significantly enhance the robot's versatility and task performance, while improving the confidence of the user [84]. The level of this autonomy again comes down to the application and the requirements of the operator in the specific scenario. One can think of this "acting" also in terms of the "interacting" that the user will do through the physical devices at their end, and the same attributes apply. In general, the interaction scheme must aim to simplify the user's mode of commanding the system to complete the task [85]. This can be done through reducing the DOFs in the interface device [86, 87], mode switching to allow the same device to control multiple aspects of the system [88], etc. The recent advances in generative AI and language transformers can contribute to this simplification greatly by allowing the user to speak their desired action outcomes, and the system uses speech-to-text and generative AI to plan and execute the action [89]. This is especially key to helping the confidence of non-expert users.

Furthermore, efficient power management and battery life are crucial in minimising interruptions and downtime during task performance. In case of such interruptions or failures, the ability of the system to involve the user through timely and transparent communication of reasons and recovery modes, is crucial to improving the engagement of the user [90].

5.2.2 To Understand Intuitively

As noted earlier, workload, mental and physical, has a strong correlation with task performance in telepresence and teleoperation. A focus on ergonomics implies helping human users do the tasks and process the incoming information as naturally as possible. This means interfaces presenting processed data instead of raw data, and where possible, presenting knowledge of the situation, instead of simply information, leveraging the properties of the task, environments, or users [91]. Hiding/presenting detailed information as required can help declutter the visual field, reducing the workload on non-expert users. Situation awareness can be enhanced by combining such information and representing it so as to help the user think and act effectively [92]. For example, sensors can sometimes provide superhuman abilities, e.g., seeing in the dark with infrared depth sensors or ultrasonic sensors to detect obstacles [93, 94]. Cameras or directional sensors from different perspectives, e.g., ego- and exo-centric cameras, can provide a better detailed view. The system can also promote interactive understanding, by actively controlling sensor orientations, robot poses, etc., accounting for any latency bottlenecks in the feedback [95]. A related feature comes from the field of Explainable AI [96], where the concept of generative AI and language transformers is used by the robot to explain its actions in natural language in real-time, improving the quality of interactions [96]. Ultimately, user actions in telepresence and teleoperation are driven by the decision-making geared towards achieving the goal of the task, which in-turn relies not only on the quality of the information being received by the user, but the characteristics of the user as well, e.g., whether they are an expert or a novice operator. The aim of the ergonomics-centred approach is to facilitate easier decision-making, prioritising the relevance and modality of the information.

5.2.3 To Learn Quickly

Along with the ability to intuitively understand and make decisions for the task, an important aspect of the ergonomics-centred approach is for the user to be able to learn simply and quickly as they use the system over time. Though, as noted earlier, telepresence and teleoperation being unnatural experiences, some training is inevitable. The interface designs and the actions required of the user should focus on common denominators in everyday experiences of users, aiming for more natural and familiar designs rather than specialised interfaces, thereby minimising the need for upfront training [88]. The more consistent the relationship among perceptions of the user, their actions, and the outcomes achieved, the flatter and shorter the learning trajectory. The ability to learn does not then remain restricted to the user, but also implies the robot's ability to learn and adapt on-the-fly. In this case, the robot is seen as a coach of the user, in helping build their skill, and help maintain it over longer periods in their career [97]. Systems where the robot enables error correction and learning from mistakes for a user can be seen as aiding comprehension and training for the user, e.g., understanding where the mistake could have been prevented. Such interventions can help not only expert users, but more constructively, non-expert users can be brought into the domain and up-to-speed more quickly, opening up access to these technologies to different demographics e.g., aging workforce, disabled people, marginalised population, etc.

5.2.4 To Trust Implicitly

A final key tenet of any human-commanded or human-used system is the level to which the user is able to trust the system [65]. Without trust, the engagement, confidence, and comfort of the user can rapidly degrade. Trust can manifest itself in many ways, but two fundamental ones are: (i) that the system will perform according to the requirements, be consistent in its performance, and eventually achieve the goal; and (ii) that it will act safely without injuring the user or other humans while doing the task. Trust can be expedited where the user has been engaged with the system for a long time, or has been involved since its inception, not only when testing or using it. In the modern data-obsessed world, certain other dimensions get added to this—(i) ethical handling of the data being collected through all the sensors in the system, maintaining requisite levels of privacy and access control; (ii) authenticity in communication, i.e., restricting impersonations, or faking/concealing identities and emotions, etc., among others. Trust in telepresence and teleoperation systems therefore is built over time, after continued use of the system. This would only be enabled by the above-mentioned aspects of *acting, understanding,* and *learning* with the system. Therefore, once put into practice, through consistent experiences, and transparent interactions, the process can become self-perpetuating—better *acting, understanding,* and *learning*, leading to deeper trust in the system, in-turn leading to enhanced *acting, understanding,* and *learning.*

6 Case Study: Ergonomics in Telesurgery

This illustrative case-study highlights how keeping ergonomics, both physical and cognitive, and human factors at the centre of the design and research methodology can positively impact the developments in telepresence and teleoperation. Drawing on the research by authors in [43, 98, 99], the case-study focuses on telesurgery—the idea of surgeons performing surgery using robotic devices and appropriate interfaces—showing how the use of motorised devices, natural visual and command interfaces, and assistive features can help achieve significant enhancements in surgical outcomes, while keeping the surgeon's comfort and control at the highest priority.

6.1 The Case of Transoral Laser Microsurgery

In surgical laryngology, transoral laser microsurgery (TLM) refers to the procedure of using a free-beam surgical laser to treat abnormalities in the larynx, through the effective coordination of the following: (1) a manual micromanipulator with a small joystick to direct the beam of the laser (CO_2 is the prevalent variety) at the surgical site—an area about 20×20 mm^2, at a distance of 400–500 mm; (2) visualisation through a surgical microscope from 400–500 mm away; and (3) laser activation through a footswitch [43]. The challenges and limitations of this traditional setup with respect to controllability and ergonomics are noted in literature [86], prompting the introduction of computer-assisted surgical approaches in TLM, introducing motorised laser control mechanisms allowing higher quality incisions [98]. Yet, the problems of manual micromanipulation in a constrained space persist, with the limitations of poor operating ergonomics and discomfort [98]. The following subsections detail how the introduction of teleoperation technology, in collaboration with the surgeons at each step of the way, led to dramatic enhancements in all aspects of the surgical procedure, as noted in [98, 99]. Figure 4 demonstrates graphically the ergonomic improvements introduced.

6.2 The Physical Ergonomics

In the first instance, authors in [43, 98] focused their design on improving two key aspects: (i) simplifying the interface for the surgeon to control the laser motion; and

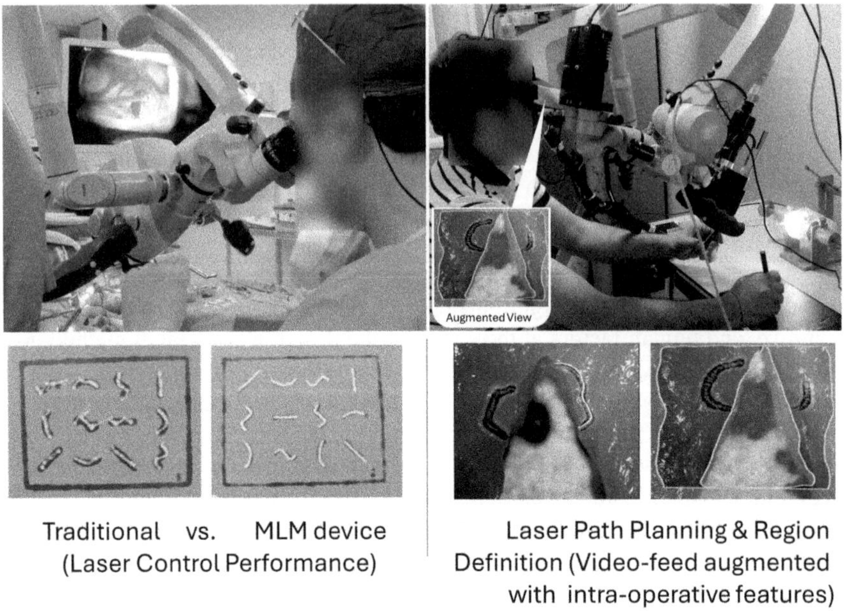

Fig. 4 The case of telesurgery: The left figure shows the traditional setup in transoral laser microsurgery, with the joystick laser manipulator and the microscope. The right figure shows the setup enhanced through the introduction of automation using a tablet device for laser manipulation, a head-mount display with stereoscopic video augmented with intra-operative assistive features. The clear benefits of applying physical and cognitive ergonomics principles to the setup are visible in the laser control performance shown [98, 99]

(ii) performing the surgery more accurately and efficiently. Both these aspects are interlinked. The surgical accuracy and efficiency were improved by replacing the manual micromanipulator with a motorised laser micromanipulator (MLM), which controls the motion of the beam-deflection mirror. The MLM provided high positioning accuracy (35 μm), a programmable working area (up to 40×40 mm^2 at 400-mm distance), and user-definable motion speeds (up to 0.1 m/s) [99]. Following this, for the surgeon to control the MLM device, a graphics tablet with stylus was introduced as the command interface [99]. Both these technological interventions aimed at satisfying not only the suitability criteria for the device design, i.e., that the device is physically able to perform the required surgical tasks, when guided appropriately by the surgeon, but also reduce the requirement of high psychomotor skills and difficult hand-eye coordination on the part of the surgeon. Replacing the micromanipulator joystick with a pen-like stylus for laser aiming immediately simplified the laser control for the surgeon, making it more familiar as an interface. As seen in Fig. 4, the joystick also required the surgeon to hold their hands in difficult poses for extended periods, resulting in hand fatigue and discomfort [100]. The pen-like stylus significantly improved surgeons' hand posture, and in-turn their body posture while performing the surgical task, thereby enhancing their comfort and confidence with the system. This design approach, therefore, touched upon the aspects of *acting* (achieving the laser incisions accurately and efficiently), *learning* (using a simple pen-like interface), and *trust* (consistency in action and visualisation) for the surgeon-machine telesurgery interface.

6.3 The Cognitive Ergonomics

To explore the enhancements possible in the cognitive ergonomics aspects of the design, two further elements were introduced [86]: (i) decoupling the visualisation of the surgical site from the microscope to a 3D stereoscopic display; and (ii) intra-operative assistive features, which combined the real-time stereo video capture with the automatic control capabilities of the MLM device. For the visualisation, two 1080p full HD cameras were attached to the surgical microscope, and the acquired video was displayed in real-time in stereoscopic mode through a modified 3D head-mount display. This modification meant that the surgeon now performed the surgery using a more comfortable display than the microscope, without disturbing the surgical workflow. The intra-operative assistive features were implemented in two ways, by programming the stylus to: (i) allow the surgeon to virtually plan laser paths and regions, which can then be automatically executed by the MLM (a type of shared autonomy); and (ii) define regions within the surgical area where the laser can be automatically turned on/off, as desired. The visualisation and intra-operative features together addressed the *acting* (as above), *learning* (using the same pen-like device for added features; automatic execution of laser incisions), and *trust* (added comfort due to features; no privacy concerns) aspects of the challenge. They also addressed the *understanding* aspect, where the surgeon was able to comprehend the visual information more easily (as against a microscope), and add-on visual information (incision paths, regions, etc.) to aid in improving surgical outcomes.

Although the design resulted in a dramatic change in the way the surgeon performed the surgery, it was important that the surgeon always felt in control of the surgical procedure, along with improving the ergonomics of the devices and the interaction. The impact of the change was evaluated in terms of objective and subjective assessments.

6.4 The Objective Assessment

A representative trajectory following task for laser tracing was designed for the comparative objective evaluation, where surgeon subjects were asked to perform it using the traditional interface and the stylus interface. The system accuracy was evaluated in terms of the root-mean-square-error (RMSE in mm) of the trajectory following, while the efficiency was evaluated in terms of the time taken (in seconds) to complete the task. Having both novice and expert subjects perform the task, it was observed that for both the metrics, the stylus-based interface performed significantly better than the traditional interface, showing a 57.3% improvement in RMSE error (0.35 mm vs. **0.15 mm**) and a 25.5% improvement in execution time (14.4 s vs. 10.7 s) [99]. These metrics are directly related to the operational safety and efficiency of the interfaces. A high value for the RMSE gives an account of how far the laser-traced shape is from the desired shape. This implies that the laser is active in non-desired areas. In real surgeries, this would mean that the laser is ablating more than the desired area of the tissue, perhaps healthy tissue, directly impacting the surgical outcomes, i.e., voice quality in this case.

6.5 The Subjective Assessment

A separate subjective assessment was carried out by the authors with the complete system, including the intra-operative assistive features, through a representative microsurgical setup [43]. With expert and non-expert surgeons asked to perform real TLM surgical tasks with ex-vivo pig larynxes, e.g., incising a layer off the vocal fold mucosa, the setup allowed the evaluation of the applicability of the developed stylus-based interface in real TLM scenarios. For the evaluations, a new questionnaire was devised based on the System Usability Scale and the NASA-TLX [74, 75]. The questionnaire accounted not only for the ergonomics and intuitiveness aspects of the system but also the suitability and perceived usefulness of the features of the system in relation to TLM. As noted in [43], the system scored favourably on all scales on 7-point Likert-type scores:

1. It scored 6.0 or better (out of 7.0) in terms of ease of use and learning, precision, satisfaction, safety, and improvements it offered in TLM.
2. It scored 5.0 or better in terms of avoiding: errors, hand fatigue, attention loss, and stress, while being appropriate for particular operations in TLM.
3. On two scales, the scores indicated that the system still needs to be improved further: (i) offering new laser operations; and (ii) covering all possible laser operations.

The high usability scores underline the comfort and adaptability of the system, where the pen-like stylus replaces the manual joystick and allows intuitive control, dramatically simplifying the surgeon-machine interface. The assistive feature of automatic laser scanning plays a major role in alleviating the control and fatigue problems associated with the manual micromanipulator. It frees up the surgeon's laser controlling hand, reducing the attendant hand fatigue and potentially allowing the hand to be used for other tasks, e.g., bimanual tissue manipulation—a distinct advantage when accessing difficult regions in the surgical site. The positive evaluations here demonstrate the potential safety, usefulness, and benefit of the newly designed system for future clinical use.

7 Summary and Conclusion

Telepresence and teleoperation research and commercial development has seen significant progress in recent years. However, despite the apparent ubiquity of cheaper and more capable robots, the anticipated proliferation of these in daily life seems a fair while away. The argument in this chapter is that this is, at least in part, due to the fundamentally non-natural nature of human interaction in telepresence and teleoperation. A sound approach to bring the interaction closer to the natural, human-like ways is to focus on applying ergonomics and human factors principles in all aspects of system design, especially the human-centred interface. This is critical to promoting user comfort, well-being, and overall satisfaction.

Addressing the challenge of physical and cognitive ergonomics requires a holistic approach that considers user diversity, collaborative design, customizable interfaces, and sensory feedback. This chapter proposes a four-axis scheme to bring ergonomic principles to the centre of the system development: *acting effectively*, *understanding intuitively*, *learning quickly*, and *trusting implicitly*. Focusing on these four axes can help prioritise human-centred approaches in the technologies leading to more productive and engaging system utilisation by a broader user base. Fundamental technologies for sensing, actuation, control, networking, cognition, etc. serve transversally to enhance the system along each of the four axes, both in terms of action and interaction, as well as subjective and objective assessments. As telepresence and teleoperation technologies continue to evolve, prioritizing ergonomic design principles will ensure that the outcomes and benefits are maximised for humans in everyday life, without compromising human health and well-being.

References

1. Karl, K.A., Peluchette, J.V., Aghakhani, N.: Virtual work meetings during the COVID-19 pandemic: the good, bad, and ugly. Small Group Res. **53**(3), 343–365 (2022)
2. Rea, D.J., Seo, S.H.: Still not solved: a call for renewed focus on user-centered teleoperation interfaces. Front Robot AI. **9** (2022)
3. Ijsselsteijn, W.A.: History of telepresence. In: Schreer, O., Kauff, P., Sikora, T. (eds). 3D Video Communication: Algorithms, Concepts and Real-Time Systems in Human Centred Communication, pp. 5–21 (2005)
4. Lichiardopol, S.: A survey on teleoperation. Technical Report, Technische Universiteit Eindhoven (2007)
5. Yu, S., Lee, J., Park, B., Kim, K., Cho, I.: Ergonomic analysis of a telemanipulation technique for a pyroprocess demonstration facility. Nucl. Eng. Technol. **46**(4), 489–500 (2014)
6. Delmerico, J., Mintchev, S., Giusti, A., Gromov, B., Melo, K., Horvat, T., Cadena, C., Hutter, M., Ijspeert, A., Floreano, D., Gambardella, L.M., Siegwart, R., Scaramuzza, D.: The current state and future outlook of rescue robotics. J. Field Robot. **36**(7), 1171–1191 (2019)
7. Lin, T.C., Krishnan, A.U., Li, Z.: Intuitive, efficient and ergonomic tele-nursing robot interfaces: design evaluation and evolution. ACM Trans. Hum. Robot. Interact. **11**(3), 23:1–23:41, (2022)
8. Minsky, M.: Telepresence. Omni, 45–51 (1980)

9. Johnson, S., Rae, I., Mutlu, B., Takayama, L.: Can you see me now? How field of view affects collaboration in robotic telepresence. In: Proceedings of the 33rd Annual ACM Conference on Human Factors in Computing Systems, pp. 2397–2406 (2015)
10. Teleoperation: Wikipedia. https://en.wikipedia.org/wiki/Teleoperation. Accessed 3 Mar 2024
11. Goertz, R.C.: Master-Slave manipulator. Technical Report, Argonne National Lab, USA (1949)
12. Niemeyer, G., Preusche, C., Stramigioli, S., Lee, D.: Telerobotics. In: Siciliano, B., Khatib, O. (eds.) Springer Handbook of Robotics, pp. 1085–1108 (2016)
13. Fiorini, P., Bejczy, A., Schenker, P.: Integrated interface for advanced teleoperation. In: Proceedings on IEEE International Conference Systems, Man and Cybernetics, pp. 1413–1418 (1992)
14. Telexistence Inc. https://tx-inc.com/en/home/. Accessed 16 Nov 2024
15. ANA Avatar XPrize, ANA. https://www.xprize.org/prizes/avatar. Accessed 16 Nov 2024
16. Temi—Robot as a service. https://www.robotemi.com/. Accessed 5 March 2024
17. Telepresence Robots.com. https://telepresencerobots.com/robots/. Accessed 16 Nov 2024
18. Ohmni Labs. https://ohmnilabs.com/. Accessed 16 Nov 2024
19. Da Vinci robotic surgical system. Intuitive Surgical Inc. https://www.intuitive.com/en-gb/. Accessed 16 Nov 2024
20. Versius surgical system. CMR Surgical. https://us.cmrsurgical.com/. Accessed 16 Nov 2024
21. Luna Surgical System. Asensus Surgical. https://www.asensus.com/. Accessed 16 Nov 2024
22. Hugo Surgical System. Medtronic. https://www.medtronic.com/covidien/en-gb/robotic-assisted-surgery/hugo-ras-system.html. Accessed 16 Nov 2024
23. CURV-21, USA Navy. https://www.navy.mil/Resources/Fact-Files/Display-FactFiles/Article/2173379/curv-21-remotely-operated-vehicle/. Accessed 16 Nov 2024
24. Kaiko/Abismo, JAMSTEC, Japan. https://www.jamstec.go.jp/e/about/press_release/archive/2007/20071218/. Accessed 16 Nov 2024
25. JASON/ARGO, WHOI, USA. https://ndsf.whoi.edu/jason/history-of-jason/. Accessed 16 Nov 2024
26. Robonaut, NASA, USA. https://www.nasa.gov/robonaut2/. Accessed 16 Nov 2024
27. Dextre, NASA, USA. https://www.nasa.gov/international-space-station/mobile-servicing-system/. Accessed 16 Nov 2024
28. Curiosity/Spirit Mars, ROVERS, NASA, USA. https://science.nasa.gov/mission/msl-curiosity/. Accessed 16 Nov 2024
29. Interact centaur Telerobotics platform, ESA, Europe. https://www.esa.int/ESA_Multimedia/Videos/2015/09/Meet_ESA_s_Interact_Rover. Accessed 16 Nov 2024
30. Canadarm2/3, CSA, Canada. https://www.asc-csa.gc.ca/eng/iss/canadarm2/about.asp. Accessed 16 Nov 2024
31. Sheridan, T.B.: Teleoperation, telerobotics and telepresence: a progress report. Control. Eng. Pract. **3**(2), 205–214 (1995)
32. Bystrom, K.-E., Barfield, W., Hendrix, C.: A conceptual model of the sense of presence in virtual environments. Presence Teleop. Virt. **8**(2), 241–244 (1999)
33. Deshpande, N., Ortiz, J., Sarakoglou, I., Semini, C., Tsagarakis, N., Brygo, A., Fernandez, J., Frigerio, M., Saccares, L., Toxiri, S., Caldwell, D.G.: Next-generation collaborative robotic systems for industrial safety and health. WIT Trans. Built Environ. **174**, 187–200 (2018)
34. Naceri, A., Mazzanti, D., Bimbo, J., Tefera, Y., Prattichizzo, D., Caldwell, D.G., Mattos, L.S., Deshpande, N.: The Vicarios virtual reality interface for remote robotic teleoperation: teleporting for intuitive tele-manipulation. J. Int. Robot. Syst. Theory Appl. **101**(4) (2021)
35. Mehta, R.K.: Integrating physical and cognitive ergonomics. IIE Trans. Occup. **4**(2–3), 83–87 (2016)
36. Frank, J. A., Moorhead, M., Kapila, V.: Mobile mixed-reality interfaces that enhance human–robot interaction in shared spaces. Front. Robot. AI **4** (2017)
37. Ipsen, C., Nardelli, G., Poulsen, S., Ronzoni, M.: Implementing tele presence robots in distance work: experiences and effects on work. In: Bagnara, S., Tartaglia, R., Albolino, S., Alexander, T., Fujita, Y. (eds) Advances in Intelligent Systems and Computing, vol. 821 (2018)

38. Goodrich, M.A., Crandall, J.W., Barakova, E.: Teleoperation and beyond for assistive humanoid robots. Rev. Hum. Factors Ergon. **9**(1), 175–226 (2013)
39. Litleskare, S., Calogiuri, G.: Camera stabilization in 360° videos and its impact on cyber sickness, environmental perceptions, and psychophysiological responses to a simulated nature walk: a single-blinded randomized trial. Front. Psychol. **10** (2019)
40. Deml, B.: Human factors issues on the design of telepresence systems. Presence Teleop. Virt. **16**(5), 471–487 (2007)
41. Boudouraki, A., Fischer, J.E., Reeves, S., Rintel, S.: 'Being in on the Action' in mobile robotic telepresence: rethinking presence in hybrid participation. In: Proceedings on 2023 ACM/IEEE International Conference on Human-Robot Interaction, pp. 63–71 (2023)
42. Rae, I., Mutlu, B., Takayama, L.: Bodies in motion: mobility, presence, and task awareness in telepresence. In: Proceedings SIGCHI Conference on Human Factors in Computing Systems, pp. 2153–2162 (2014)
43. Deshpande, N., Peretti, G., Mora, F., Guastini, L., Lee, J., Barresi, G., Caldwell, D.G., Mattos, L.S.: Design and study of a next-generation computer-assisted system for transoral laser microsurgery. OTO Open. **2**(2) (2018)
44. Trevelyan, J.P., Kang, S., Hamel, W.R.: Robotics in hazardous applications. In: Siciliano, B., Khatib, O. (eds.) Springer Handbook of Robotics, pp. 1101–1126 (2016)
45. Petrushin, A., Barresi, G., Mattos, L.S.: Gaze-controlled laser pointer platform for people with severe motor impairments: preliminary test in telepresence. In: 2018 40th Annual International Conference IEEE Engineering in Medicine and Biology Society (EMBC), pp. 1813–1816 (2018)
46. Zhang, G., Hansen, J.P.: Telepresence robots for people with special needs: a systematic review. Int. J. Hum. Comput. Interact. **38**(17), 1651–1667 (2022)
47. Milgram, P., Kishino, F.: A taxonomy of mixed reality visual displays. IEICE Trans. Inf. Syst. **E77-D**(12), 1321–1329 (1994)
48. Tremosa, L.: Virtuality continuum, interaction design foundation. https://www.interaction-design.org/literature/topics/virtuality-continuum. Accessed 16 Nov 2024
49. Woolley, S.I., Collins, T.: User experience and engagement in the reality–virtuality continuum: a special issue guest editorial. Presence Teleoper. Virtual Environ. **28**, 203–206 (2019)
50. Breves, P., Stein, J.-P.: Cognitive load in immersive media settings: the role of spatial presence and cybersickness. Virtual Real. **27**(2), 1077–1089 (2023)
51. Diaz, I., Juan, J., Huli, T.: Stability boundary and transparency for haptic rendering. In: Zadeh, M.H. (ed.) Advances in Haptics (2010)
52. Yin, M.S., Haddawy, P., Suebnukarn, S., Schultheis, H., Rhienmora, P.: Use of haptic feedback to train correct application of force in endodontic surgery. In: Proceedings on 22nd International Conference on Intelligent User Interfaces, pp. 451–455 (2017)
53. Escolano, C., Antelis, J.M., Minguez, J.: A telepresence mobile robot controlled with a noninvasive brain–computer interface. IEEE Trans. Syst. Man Cybern. Part B. **42**(3), 793–804 (2012)
54. Chen, J.Y.C., Haas, E.C., Barnes, M.J.: Human performance issues and user interface design for teleoperated robots. IEEE Trans. Syst. Man Cybern. Part C Appl. Rev. **37**(6), 1231–1245 (2007)
55. Riley, J.M., Kaber, D.B., Draper, J.V.: Situation awareness and attention allocation measures for quantifying telepresence experiences in teleoperation. Hum. Factors Ergon. Manuf. Serv. Indus. **14**(1), 51–67 (2004)
56. Zimmer, M., Al-Yacoub, A., Ferreira, P., Hubbard, E.-M., Lohse, N.: Experimental study to investigate mental workload of local vs remote operator in human-machine interaction. Prod Manuf Res. **10**(1), 410–427 (2022)
57. DeJong, B.P., Colgate, J.E., Peshkin, M.A.: Improving teleoperation: reducing mental rotations and translations. In: IEEE International Conference on Robotics Automation, pp. 3708–3714 (2004)
58. Petzold, B., Zaeh, M., Faerber, B., Deml, B., Egermeier, H., Schilp, J., Clarke, S.: A study on visual, auditory, and haptic feedback for assembly tasks. Presence Teleop. Virt. **13**(1), 16–21 (2004)

59. Rodríguez-Sedano, F.J., Conde, M.Á., Rodríguez-Lera, F.J., Chaparro-Peláez, J.: Measuring the impact of haptic feedback in collaborative robotic scenarios. Univ. Access Inf. Soc. (2023)
60. Endsley, M.R., Jones, D.G.: Designing for Situation Awareness: an Approach to User-Centered Design, 2nd edn. CRC Press (2004)
61. Almeida, L., Menezes, P., Dias, J.: Telepresence social robotics towards co-presence: a review. Appl. Sci. **12**(11), 5557 (2022)
62. Ma, S., Simon, T., Saragih, J., Wang, D., Li, Y., La Torre, F., Sheikh, Y.: Pixel codec avatars. In: 2021 IEEE/CVF Conference on Computer Vision and Pattern Recognition, pp. 64–73 (2021)
63. Tefera, Y.T., Mazzanti, D., Anastasi, S., Caldwell, D.G., Fiorini, P., Deshpande, N.: FoReCast: real-time foveated rendering and unicasting for immersive remote telepresence. In: Uchiyama, H., Normand, J.-M. (eds.) ICAT-EGVE 2022—International Conference on Artificial Reality and Telexistence and Eurographics Symposium on Virtual Environments. The Eurographics Association (2022)
64. Tuli, T.B., Terefe, T.O., Rashid, M.M.U.: Telepresence mobile robots design and control for social interaction. Int. J. Soc. Robot. **13**, 877–886 (2021)
65. Dassonville, I., Jolly, D., Desodt, A.M.: Trust between man and machine in a teleoperation system. Reliab. Eng. Syst. Saf. **53**(3), 319–325 (1996)
66. Rogers, H., Khasawneh, A., Bertrand, J., Madathil, K.C.: An investigation of the effect of latency on the operator's trust and performance for manual multi-robot teleoperated tasks. In: Proceedings on Human Factors and Ergonomics Society Annual Meeting, vol. 61, no. 1, pp. 390–394 (2017)
67. Youssef, K., Said, S., Al Kork, S., Beyrouthy, T.: Telepresence in the recent literature with a focus on robotic platforms, applications and challenges. Robotics. **12**(4) (2023)
68. Simpson, T.W.: Telepresence and trust: a speech-act theory of mediated communication. Philos. Technol. **30**(4), 443–459 (2017)
69. Melchiorri, C.: Robot teleoperation. In: Baillieul, J., Samad, T. (eds.) Encyclopedia of Systems and Control (2014)
70. Barua, H.B., Sau, A., Roychoudhury, R.D. A perspective on robotic telepresence and teleoperation using cognition: are we there yet?. arxiv.org. https://arxiv.org/abs/2203.02959 (2022)
71. Insko, B.E.: Measuring presence: Subjective, behavioral and physiological methods. In: Riva, G., Davide, F., IJsselsteijn, W.A. (eds.) Being there: concepts, effects and measurements of user presence in synthetic environments, pp. 109–119 (2003)
72. Nomura, T., Suzuki, T., Kanda, T., Kato, K.: Measurement of negative attitudes toward robots. Interact. Stud. Soc. Behav. Commun. Biol. Artif. Syst. **7**(3), 437–454 (2006)
73. Finstad, K.: The usability metric for user experience. Interact. Comput. **22**(5), 323–327 (2010)
74. Hart, S.G., Staveland, L.E.: Development of NASA-TLX (Task Load Index): results of empirical and theoretical research. In: Hancock, P.A., Meshkati, N. (eds.) Advances in Psychology, vol. 52, pp. 139–183 (1988)
75. Brooke, J.: SUS: a 'Quick and Dirty' usability scale, usability evaluation in industry. In: Jordan, P.W., Thomas, B., McClelland, I.L., Weerdmeester, B. (eds.) vol. 189 (1996)
76. Rózsa, S., Hargitai, R., Láng, A., Osváth, A., Hupuczi, E., Tamás, I., Kállai, J.: Measuring immersion, involvement, and attention focusing tendencies in the mediated environment: the applicability of the immersive tendencies questionnaire. Front. Psychol. **13** (2022)
77. Deshpande, N., Mattos, L.S., Barresi, G., Brogni, A., Dagnino, G., Guastini, L., Peretti, G., Caldwell, D.G.: Imaging based metrics for performance assessment in laser phonomicrosurgery. In: IEEE International Conference on Robot Autonomous, pp. 4850–4856 (2013)
78. Salatino, L., Deshpande, N., Demarzi, G., Berta, R., Zambotti, M., Boccardo, N., Freddolini, M., Laffranchi, M., Michieli, L., Barresi, G.: Spatial augmented respiratory cardiofeedback design for prosthetic embodiment training: a pilot study. In: Proceedings of IEEE International Conference on Systems, Man and Cybernetics, pp. 3390–3395 (2022)

79. Barresi, G., Olivieri, E., Caldwell, D.G., Mattos, L.S.: Brain-controlled AR feedback design for user's training in surgical HRI. In: Proceedings on IEEE International Conference on System, Man and Cybernetics, pp. 1116–1121 (2015)
80. Barresi, G., Nam, C.S., Esfahani, E.T., Balconi, M.: Editorial: neuroergonomics in human-robot interaction. Front. Neurorobot. **16** (2022)
81. Sauro, J. and Lewis, J.R.: Quantifying the user experience: practical statistics for user research (2012)
82. Mair, G.M.: Telepresence—the technology and its economic and social implications. In: International Symposium on Technology and Society at a Time of Sweeping Change, pp. 118–124 (1997)
83. Small, N., Lee, K., Mann, G.: An assigned responsibility system for robotic teleoperation control. Int. J. Intell. Robot. Appl. **2**(1), 81–97 (2018)
84. Brooks, C., Szafir, D.: Balanced information gathering and goal-oriented actions in shared autonomy. In: 14th ACM/IEEE International Conference Human-Robot Interaction (HRI), pp. 85–94 (2019)
85. Norman, D.A.: Cognitive engineering. In: Norman, D.A., Draper, S., (eds.) User Centered System Design, pp. 31–62 (1986)
86. Deshpande, N., Ortiz, J., Caldwell, D.G., Mattos, L.S.: Enhanced computer-assisted laser microsurgeries with a 'virtual microscope' based surgical system. In: Proceedings on IEEE International Conference Robotics and Automation, pp. 4194–4199 (2014)
87. Singh, A., Seo, S.H., Hashish, Y., Nakane, M., Young, J.E., Bunt, A.: An interface for remote robotic manipulator control that reduces task load and fatigue. In: IEEE International Symposium Robot and Human Interactive Communication (RO-MAN), pp. 738–743 (2013)
88. Rea, D.J.: Now you're teleoperating with power: learning from video games to improve teleoperation interfaces. PhD Dissertation, University of Manitoba (2020)
89. Du, G., Han, R., Yao, G., Ng, W.W.Y., Li, D.: A gesture- and speech-guided robot teleoperation method based on mobile interaction with unrestricted force feedback. IEEE/ASME Trans. Mechatron. **27**(1), 360–371 (2022)
90. Adamides, G., Christou, G., Katsanos, C., Xenos, M., Hadzilacos, T.: Usability guidelines for the Design of Robot Teleoperation: a taxonomy. IEEE Trans. Hum. Mach. Syst. **45**(2), 256–262 (2015)
91. Szafir, D., Szafir, D.A.: Connecting human-robot interaction and data visualization. In: ACM/IEEE International Conference on Human-Robot Interaction (HRI), pp. 281–292 (2021)
92. Seo, S.H., Rea, D.J., Wiebe, J., Young, J.E.: Monocle: interactive detail-in-context using two pan-and-tilt cameras to improve teleoperation effectiveness. In: 26th IEEE International Symposium on Robot and Human Interactive Communication (RO-MAN), pp. 962–967 (2017)
93. Mast, M., Materna, Z., Španěl, M., Weisshardt, F., Arbeiter, G., Burmester, M., Smrž, P., Graf, B.: Semi-autonomous domestic service robots: evaluation of a user Interface for remote manipulation and navigation with focus on effects of stereoscopic display. Int. J. Soc. Robot. **7**(2), 183–202 (2015)
94. Nielsen, C.W., Goodrich, M.A., Ricks, R.W.: Ecological interfaces for improving mobile robot teleoperation. IEEE Trans. Robot. **23**(5), 927–941 (2007)
95. Bajcsy, R., Aloimonos, Y., Tsotsos, J.K.: Revisiting active perception. Auton. Robot. **42**(2), 177–196 (2018)
96. Hunt, W., Ramchurn, S.D., Soorati, M.D. A survey of language-based communication in robotics. arxiv.org. https://arxiv.org/html/2406.04086v1 (2024)
97. Vinoski, J.: Robots as job creators? Upskilling, cobots and AI may prove job loss doomsayers wrong. Forbes Inc. (2019). https://www.forbes.com/sites/jimvinoski/2019/10/22/robots-as-job-creators-upskilling-cobots-and-ai-may-prove-job-loss-doomsayers-wrong/. Accessed 8 Mar 2024
98. Mattos, L.S., Deshpande, N., Barresi, G., Guastini, L., Peretti, G.: A novel computerized surgeon-machine interface for robot-assisted laser phonomicrosurgery. Laryngoscope. **124**(8) (2014)

99. Deshpande, N., Mattos, L.S., Caldwell, D.G.: New motorized micromanipulator for robot-assisted laser phonomicrosurgery. In: Proceedings on IEEE International Conference on Robotics and Automation, pp. 4755–4760 (2015)
100. Gholami, S., Lorenzini, M., De Momi, E., Ajoudani, A.: Quantitative physical ergonomics assessment of teleoperation interfaces. IEEE Trans. Hum. Mach. Syst. **52**(2), 169–180 (2022)

Developments and Opportunities

Quantifying the Subjective Experience in Human–Robot Interaction

Simone Borsci, Elisa Prati, Jule Landwehr, and Margherita Peruzzini

Abstract Assessing the quality of interaction between collaborative robots (i.e., cobots) and humans to achieve a (shared) goal is an essential part of supporting the diffusion of usable and safe collaborative robotic systems. The key issue here is that cobots within and between domains can vary substantially in terms of features even when these are meant to support similar tasks. Under this perspective, it seems very hard, if not impossible, to identify coherent ways to benchmark the quality of cobots. However, all the cobots share one element which is that they have to deal with humans and satisfy them. This chapter claims that focusing on how a system can satisfy the human counterpart can (indirectly) provide insights into the quality of cobots that can be a piece of valuable information for the designers. In this sense, identifying and validating a set of key aspects to assess the subjective experience in interaction with robots during collaborative tasks is a stepping stone for benchmarking cobots. We will discuss how, in other domains, researchers developed standardized scales to ensure comparable results regarding the user experience (UX), and to model the differences between systems. Moreover, based on a recent systematic literature review, an initial list of key aspects to evaluate UX with cobots is discussed as a step to build and validate a future tool for unifying how experts evaluate cobots.

Keywords Human–Robot collaboration · Artificial intelligence agents satisfaction · Usability · User experience · Assessment

S. Borsci (✉) · J. Landwehr
Faculty of Behavioural Management and Social Sciences, Department of LDT, Group of Cognition, Data and Education, University of Twente, Enschede, The Netherlands
e-mail: s.borsci@utwente.nl

S. Borsci
Department of Surgery and Cancer, Imperial College of London, NIHR HealthTech Research Center, London, UK

E. Prati
Deep Blue, Rome, Italy

M. Peruzzini
Department of Industrial Engineering, University of Bologna, Bologna, Italy

G. Barresi and S. Wischniewski (eds.), *Ergonomics in Robotics: Advances and Innovations*, Springer Tracts in Advanced Robotics 162,
https://doi.org/10.1007/978-3-032-19374-2_16

1 Why Validated Tools for Subjective Evaluation Are Important

In recent years, the artificial intelligence (AI) race between major players such as the USA, China, and the EU [1] pushed more and more researchers to seek ways to embed AI and robots in our society. Ideally, this development should enable symbiotic exchanges and collaborations between humans and systems that could benefit people and the overall humanity [1].

To make sure these systems benefit humans the value of interaction and collaboration with robots should be evaluated validly and comparably to determine effects on people and the society of cobots intended here as—robotics systems designed to collaborate with humans to perform actions and achieve tasks. Certainly, the value of cobots can be observed from different perspectives. In an industrial setting, cobots that improve the productivity of the workers (thus bring more economic value) can be considered a success from the business point of view. Nevertheless, these successful cobots might enhance the workload for the operators, require them to perform new and additional tasks, and/ or force them to be more vigilant during the task execution. In this sense, the cobots could be valuable for certain stakeholders, but not much or even be detrimental to the tasks of the main end-user (i.e., the operators). As highlighted by several researchers the lack of value for the end user can be due to several reasons e.g., a lack of trust [2], Fluency [3], safety [4], embodiment [5], etc. To prevent a lack of value and ensure successful collaboration, we want to account for these aspects while designing the cobots.

However, when designing and assessing in a formative way, usually, there is no time to control and account for all the possible aspects that might affect the collaboration in detail, nor is it always possible to test for and collect all the necessary objective metrics (e.g., time on task, correctness, etc.) and capture psycho-physiological aspects (such as eye movements, heart rate, etc.) to account for cognitive related aspects. Therefore, the quantity and quality of measurements that can be performed to support the development of cobots are dependent on the available resources (mainly budget and time). Certainly, broad and complete evaluation processes are expensive and usually performed in advanced phases of development. Hence, tools are needed for the evaluation process that supports lean and agile approaches to design by iteratively improving the symbiotic collaboration [1] between humans and robots. Such types of tools are usually standardized inventories or scales that could help designers keep track of human subjective reactions after the exchange with the robot; these subjective scales are usually adopted as ways to measure indirectly (through the mean of human experience) the objective aspects on top of the objective metrics of performance, like i.e. time of the task, correct execution, etc.

The subjective perspective captured by such types of scales is of course holistic and complementary to the objective one, nevertheless, it is not secondary. This is well known in the context of usability e.g., ISO 9241-11 [6]. Satisfaction (the subjective metrics) is often considered affected by the objective metrics of interaction, e.g., efficiency and effectiveness, suggesting that good objective performance with a system should result in good levels of satisfaction. Nevertheless, the relationship between objective and subjective is not always so direct. As Tractinsky noted in 2018 [7] several studies and meta-analyses reported a weak to moderate correlation between satisfaction (measured in various ways) and effectiveness and efficiency [8–11]. As suggested by Borsci et al. [12], this means that more attention should be placed on measuring usability in a replicable way because objective and subjective aspects affect each other in a way that can be measured more or less systematically and accurately. This distance between objective and subjective assessment suggests that aspects such as the quality in use, the decision to use, and, ultimately, the overall experience of people interacting with systems are not determined solely by the objective quality, characteristics, and functioning of a system.

It may seem contradictory that objectively usable systems could be subjectively perceived as unsatisfactory or unable to provide a positive user experience. However, people consider the objective characteristics (e.g., design, functionality, etc.) of a system as one aspect among a larger set of information that they weigh to decide whether a system is worth using or not. Certainly, the paradigm of the user experience (UX, ISO 9241-210, [13]) extends the one of usability. In the context of UX objective and subjective aspects of usability are complemented by additional criteria (e.g., value, usefulness, acquired skills, etc.) to obtain a better picture of the potential impact of a technology on humans. HRI researchers insist that when it comes to experience with robots, people are using multiple criteria to assess the interactive exchange with systems, for the specific case of collaboration tasks. For instance, elements such as how teaming up with a robot could be set up in line with human needs and preferences [14], and the possibility for people to interact with an agent to perform a task by minimizing expensive cognitive processes (e.g., attention, memory [15], etc.) could be considered some of the relevant aspects that should be considered when designing a system to be used in collaboration with humans. Therefore, all of these elements must be included in the evaluation of the experience of people with cobots. Albeit there are some indications from the literature about potential aspects that should be considered in assessing UX in the HR context, as recently pointed out by Prati et al. [16] it is not yet clear how to assess the concept of UX in a comparable way.

Associated with that the overall scientific perspective on how to assess the impact of technology on people is changing, as proposed by Donald Norman [17] we are moving from a human-centered approach to a "humanity-centered design" in which the quality of systems is not only designed around the human needs but the quality is also assessed in terms of a ratio between resources utilized and outcomes/benefits for the users. Under this perspective designing reliable robotic systems able to objectively improve people's performance e.g., in terms of time, accuracy, and minimizing human effort, does not translate automatically into having designed a system that is subjectively considered usable (e.g., satisfactory) or into a system that could maximize people experience i.e., high level of UX.

Identifying the most relevant set of aspects to subjectively evaluate the HRI, and then measuring these aspects in a reliable and replicable way is an essential step to ensure replicability and benchmarking of UX assessment in the domain of robotics. This could potentially facilitate the embedding of robotics in human society by making systems more in line with users' mental models [18, 19] and therefore easy to learn, satisfactory, and more usable and acceptable. To achieve this goal it is necessary to understand how it is possible to quantify and somehow objectify something that is quite subjective like people's common understanding of a concept. This is what we will discuss in Sect. 2. In this section we are summarizing, without the intent to be exhaustive, the process necessary to develop a standardized scale with the precise scope of quantifying and comparing people's subjective assessment after the interaction with a product. In line with the process of developing an evaluation scale, in Sect. 3 we first describe the ongoing work carried out by researchers from the University of Twente (the Netherlands) and the University of Modena and Reggio Emilia (Italy) to develop and define an initial conceptual framework for composing a scale to assess UX in the context of HRI. We then discuss the first review of the framework, which was carried out by a large panel of international experts in order to redefine the key dimensions for assessing the experience of humans interacting with robots. Section 4 concludes the chapter by discussing future steps to develop a definitive scale for assessing UX with cobots, and provides a summary in terms of takeaways for HRI practitioners who want to assess the user experience with robots.

2 The Process of Quantifying People's Subjective Experience

When people are asked to judge the quality of a service or to rate their experiences with products, they are usually asked to answer a series of questions related to aspects or attributes of the system under evaluation. If the aspects used to survey potential users are empirically validated, then, and only then, does the subjective evaluation provide reliable information about a specific framework or construct. This is a theoretical or empirically validated construction of people's minds regarding a particular argument, and it is often composed of an inferred set of attributes related to a larger concept. A framework is essentially the conceptualization of knowledge about a specific concept, such as user experience, satisfaction, acceptance, trust, etc. This abstract concept becomes measurable and quantifiable when it is broken down into concrete attributes [20, 21]. For instance, people might implicitly share a similar idea of what entails "satisfaction" with technology, and someone could be mistakenly persuaded that simply asking one or two questions along the line of how satisfied you are with this system, could be enough to capture the concept of satisfaction.

Nevertheless, even if humans share roughly the same idea of what it means to "have a positive experience" with a system, people could associate the idea of "positive experience" also with very different attributes and therefore weigh the importance of these attributes differently. It should be considered that when we are talking about interacting and collaborating with systems, the type of technology is not a secondary aspect. The attributes by which we measure the construct might, indeed, vary substantially based on the types of technology under assessment—e.g., think of the different attributes that can determine the quality of interaction with a website or with a robot.

It should be clear that the definition of a measurable construct and the assessment of such a concept by a standardized scale requires a rigorous and non-trivial process. Of course, standardized scales are already available to assess, for instance, satisfaction or experience with interactive systems e.g., The System Usability Scale [22] etc. However, these scales even if reliable cannot necessarily be translated into the field of HRI without the risk of only partially measuring relevant aspects of interaction. Thus, to maximize the usefulness of evaluation scales it is important to define a construct specifically for the domain of application or accounting for the specificity of the technology of interest, and to design and validate the inventory tool for capturing all the relevant information to assess the specific experience of people with such technological set.

The main advantages of using such specifically designed tools of assessment, usually presented under the form of scales, are that: (i) this standardized scale offers more insights to developers about key aspects suggesting what should be redesigned to maximize the perceived (positive) impact on the users, and (ii) such statistically reliable scales enable replicability of the assessment and enable comparability and benchmarking of systems [12, 23, 24]

Experts suggested [20, 21, 25] multiple valid processes that practitioners can follow to design and validate a reliable scale. We originally summarized the process of developing a scale in five main phases, as follows (see Fig. 1):

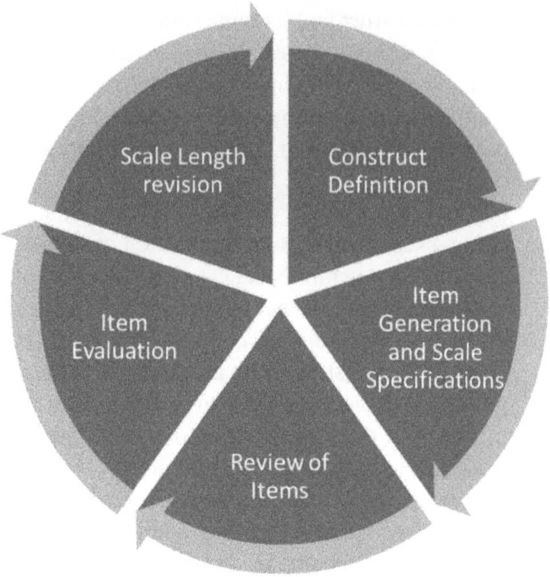

Fig. 1 The 5 main steps to develop a standardized scale: (1) construct definition, (2) item generation and scale specifications, (3) review of items, (4) item evaluation; (5) scale length revision

1. **Construct (Framework) Definition**. So how can we ensure we have a good overview of the aspects and dimensions of the construct we want to measure? There are several ways to go about this, but the most used strategies are using: (i) previous theory or theoretical model; (ii) systematic literature review, and, (iii) expert review and agreement. Each strategy has a different purpose, and all these together help you to get a complete picture of your construct. Theories can be used and reviewed to research different definitions of the construct and its attributes. By consulting different theories, it is possible to collect preliminary information on the content of a construct. Often a construct has different meanings in different contexts and domains. So, theory can also help you to find the right definition for your purpose. This in turn will help to operationalize the construct. Operationalizing in this context means transforming abstract concepts into measurable aspects. To further expand and specify the construct, a systematic literature review can help to identify different attributes (also called factors, aspects, or levels) of the construct based on previous empirical studies. These aspects could also be grouped or clustered together based for instance on similarity criteria and composed of macro aspects or sets of aspects, i.e., dimensions. Once an initial theoretical construct is defined in terms of a list of dimensions and aspects, a group of experts can be involved to review and decide which aspects listed in the constructs are important to measure it. Experts are usually asked to assess the construct by using focus groups, interviews, or the Delphi technique [26]. The agreement and the insights from the experts create a concrete and clear definition of the construct and its dimensions. One pitfall while defining your construct is to make it too general or too specific.
2. **Scale Specifications and Item Generation**. Once we have achieved consensus on our construct and its dimensions the type and quantity of questions per each dimension should be established, usually this is also part of the review process done by experts. In terms of scale specification, there are several options when it comes to scale formats and types of close-ended questions. Often formats are categorical (yes/no) or continuous (strongly agree–strongly disagree). The most used continuous scale format is the Likert, albeit other formats can be used e.g., pairwise, etc. [20, 21]. Besides deciding on a scale a fundamental step in developing a scale includes the generation of items associated with each dimension. There are two main steps: (1) create a list of items to assess the dimensions of the construct; and (2) select a representative sample of items [27]. The list created during the first step can be even four times larger than the number of dimensions of the construct. The purpose of this step is to generate as many items as possible to make it possible to select the best items later [21]. Several strategies can be used to create potential items. These strategies can be divided into two main methods deductive and inductive. Deductive methods include conducting a literature review and using pre-existing scales to find relevant items. Inductive methods are based on gaining qualitative information regarding your construct based on the opinions of a target population. This might include focus groups, expert panels, and interviews [25]. You can use one or more strategies of either both methods or one of them to create your item pool. Generally, you should

consider that the language is clear, that the content is relevant, and you include both positively and negatively formulated items [21]. It is also important to only integrate one statement per item and that the item fits the dimension you want to measure.

3. **Review of items**. Once your item pool is created you want to test whether the items reflect your construct. This is also referred to as content validity. To do so you can use expert judges and or pilot testing [20, 21, 25] your item pool. After testing your content validity, you want to end up with an initial list of items that are associated with each dimension. In this step, you also want to reduce your initial item pool to the first final version of your scale. This phase can be performed for instance by card sorting techniques [28], asking potential users to inspect the clarity of the formulation of the item but also to group items and dimensions randomly presented. This is a useful way to observe the distance between the mental model of potential respondents of the scale and the construct. If participants struggle to associate items and attributes that the items should measure, it is possible that the formulation of the question is not clear.
4. **Item Evaluation**. After the review of items, you want to conduct a validation study. Here you want to test what the scale you created is measuring (construct validity) and you also want to estimate the scoring consistency (reliability) of the scale [25]. To do so we can use exploratory factor analysis (EFA) and confirmatory factor analysis (CFA). It is necessary to have a representative sample and a large amount of data, to test the reliability. As a rule of thumb, we need at least 10 participants for each item on a scale [25]. The EFA is performed when there are no expectations regarding the association between the attributes of the construct. You want to explore the underlying model and which items belong to which factors. The EFA results might provide you with indications regarding which dimensions are really necessary to measure the construct and the association between items and aspects of the construct. The final EFA model could be also significantly different in terms of aspects and items from the model that emerged from the phase of construct definition. In order to confirm the EFA model a CFA is usually performed.
5. **Scale length revision**. This is the last step. After you confirmed your model and validated your scale you can decide, based on the data of the CFA analysis in step 4, to keep or drop items based on how much each item provides useful information about a dimension of the construct—i.e., how much variation in the data the item can explain or factor loadings. Generally, it is often recommended to reformulate an item that is performing badly instead of dropping it. Of course, the trade-off between the number of items in a scale and the exhaustiveness of information that can be provided by the scale should be carefully considered by the designers. This is a trade-off between efficiency and effectiveness; you may decide to reduce the scale to the minimum number of the most reliable items per dimension in order to have a tool that can be used for quick assessment during formative testing, or you may decide to have a long list of items that try to capture as many aspects of all the dimensions as possible.

3 Definition and Validation of a Construct for the User eXperience (UX) in HRI: A Work in Progress

3.1 Theoretical Construct

In line with the process of construct definition presented above, a systematic literature review was performed to identify the aspects that are considered important to characterize and measure the concept of interactive experience with a cobot. An initial attempt to systematize aspects deemed important by experts to assess UX with cobots was recently published by Prati et al. [29]. In this review, a list of 19 international articles was used by the authors to extrapolate an initial list of 40 aspects that emerged as commonly associated with the theory or the practice of designing and assessing UX with cobot. A team of three experts on HRI, industrial design, and human factors during a series of brainstorming sessions extended the previous review by identifying eight additional articles [30–37] that mentioned relevant aspects by (i) looking at and re-examining the reference list of the items identified by Prati and colleagues [29], and (ii) searching for articles that were published after 2021 mentioning in the title or abstract one or more aspects relevant to the design or assessment of UX with cobots. This team identified a final list of 85 issues reported by experts in the literature as relevant to the assessment of HRI (see Table 1).

By convenience, the 85 aspects reported in Table 1 were grouped in 15 dimensions intended here as clusters of similar aspects that can be used to assess HRI. Moreover, as the dimensions could be also oriented toward certain aspects of the collaboration between humans and robots, the experts also proposed to split the dimensions into three groups. The first group contains all those dimensions that can inform mainly the Robot side of the interaction e.g., "Robot dimensions". Specifically, this group contains seven dimensions from easiness to regulation to robot task performance emerged from bringing together several aspects deemed important in the literature to assess the quality of interaction (see Table 2). The second group contains the so-called "Human dimensions" (Table 3). This group contains aspects aggregated to assess the human side of HCI and it is divided into 4 main dimensions: (1) Human opinion about the robot (Human judgment before the interaction with the robot, Human judgment of the performance with a robot), (2) Human factors self-judgment (Personality-based and Ability-based). The final group is composed of four dimensions that we grouped under the umbrella of "context of interaction" between people and robots (Table 4). These dimensions include aspects pertaining: (1) the task, (2) the environment of the interaction, (3) The team in interaction, and (4) the interaction function.

With the goal of further simplifying the construct, by maintaining the 15 dimensions presented above (Tables 2, 3, and 4), the experts tried to summarize the main aspects that characterized each dimension using descriptors, as follows:

1. **Easiness of robot regulation. This dimension is described as**: the easiness of the robot's physical regulation e.g., robot's components positioning.

Table 1 List of aspects associated with the experience with cobots that emerged from the review of the literature and brainstorming activities of experts

1	Easiness of regulation	30	Mutual care	59	Self-efficacy
2	Physical appearance/ features	31	Sociability	60	Ability-based
3	Anthropomorphism	32	Privacy, control, initiative	61	Ethics
4	Dimension/size	33	Friendly relationship and company	62	Task
5	Weight	34	Social relationship	63	(Perceived) usefulness
6	Type	35	Social norms	64	Type
7	Noise	36	Concern about the status	65	Physical effort
8	Form	37	Communication	66	Difficulty of performing tasks
9	Robustness	38	Efficiency	67	Criticality of tasks
10	Emotional appearance	39	Effectiveness	68	Operator training
11	Likeability	40	Utility	69	Situation awareness
12	Warmth	41	Perception and affect	70	Feedback
13	Disturbance	42	Anxiety	71	Functionality
14	Discomfort	43	Attitudes toward use	72	Easy of use
15	Attractiveness	44	Negative attitudes	73	Learnability
16	Competences	45	Expectation	74	Memorability
17	Flexibility	46	Acceptance	75	Physical-based interface
18	Modality	47	Control	76	Graphical-based interface
19	(Perceived) intelligence	48	Comfort	77	Vocal-based interface
20	Cybersecurity	49	(Perceived) safety	78	Gesture-based interface
21	Agency	50	Trust	79	Members role
22	Movement mode	51	Intention to use/(again)	80	Humans/robots number
23	Fluency	52	Enjoyment	81	Forms of grouping
24	Organic	53	Satisfaction	82	Workstation layout
25	Autonomy	54	Usability	83	Workstation elements
26	Adaptability	55	Frustration	84	Environment aspects
27	Animacy	56	Stress	85	Application context
28	Interactivity	57	Cognitive workload		
29	Companionship	58	Personality-based		

2. **Robot physical appearance. This dimension is described as**: how the physical features of the robot can affect the user's judgment. In particular, the dimension considers aspects such as e.g., level of Anthropomorphism etc.
3. **Robot's emotional appearance. This dimension is described as**: how the robot's physical and behavioral characteristics delineate the "robot's emotional

Table 2 List of aspects grouped under the Robot dimensions

Robot dimensions	Aspects per dimension
Easiness of regulation	Easiness of regulation
Robot's physical appearance	Anthropomorphism; dimension & size; weight; type; noise; form; robustness; emotional appearance
Robot's emotional appearance	Likeability; warmth; disturbance; discomfort; attractiveness
Robot's competence features	Competences; flexibility; modality (perceived); intelligence; cybersecurity; agency; feedback
Robot's physical behavior	Movement mode; fluency; organic; autonomy; adaptability; animacy; interactivity
Robot's social behavior	Companionship; mutual care; sociability; privacy, control, initiative; friendly relationship and company; social relationship; social norms; concern about the status; communication
Robot task performance	Efficiency; effectiveness; utility

Table 3 List of aspects grouped under the Human dimensions

Human dimensions	Aspects per dimension
Human judgment before the interaction with the robot	Perception and affect; anxiety; attitudes toward use; expectation; acceptance; (perceived) safety; trust; intention to use/(again)
Human judgment of the performance of a robot	Acceptance; control; comfort; (perceived) safety; trust; intention to use/(again); enjoyment; satisfaction; usability; frustration; stress; cognitive workload
Personality-based	Personality-based; ethics
Ability-based	Self-efficacy; ability-based; operator training

profile" and how it can affect the user's judgment. It considers e.g., robot's likeability etc.

4. **Robot's competence features. This dimension is described as**: the user's judgment of the robot's competencies (e.g., reliability, responsiveness) and perceived intelligence (e.g., knowledgeable, responsible) based on its behavior during the interaction.
5. **Robot's physical behavior. This dimension is described as**: the user's judgment of the robot's physical behavior during the interaction, considering parameters such as, e.g., movement etc.

Table 4 List of aspects grouped under the context of interaction dimensions

Context of interaction dimensions	Aspects per dimension
Task	Task; (perceived) usefulness; type; physical effort; difficulty; criticality; operator training; mutual care; utility
Environment	Workstation layout; workstation elements; environment aspects; application context
Team	Members' role; humans/robots number; forms of grouping
Interaction	Knowledge of the robot's status/situation awareness; feedback; functionality; easy of use; learnability; memorability; physical-based interface; graphical-based interface; vocal-based interface; gesture-based interface

6. **Robot's social behavior. This dimension is described as**: the user's judgment of the robot's social behavior considering parameters such as e.g., companionship etc.
7. **Robot task performance. This dimension is described as**: the user's judgment of the robot during a specific performance, considering the efficiency (e.g., time on task), effectiveness (e.g., task completeness, number of errors), and utility.
8. **Human judgment before the interaction with a cobot. This dimension is described as**: the user's perception of the robot before the interaction, based on e.g., perception and effect, anxiety etc.
9. **Human judgment of the performance with a cobot**. **This dimension is described as**: the user judgment of the robot during a specific performance task, considering. This includes e.g., Acceptance etc.
10. **Human-Factors personality-based. This dimension is described as**: the user's self-description regarding their own personality characteristics, like e.g., ethics aspects etc.
11. **Human-Factors ability-based. This dimension is described as**: the user's self-description regarding their own work characteristics, like e.g., self-efficacy etc.
12. **Task performed. This dimension is described as**: the characteristics of the specific performed task during the interaction e.g., type of task, etc.
13. **The environment of interaction. This dimension is described as**: the specific characteristics of the environment where the task was performed. This dimension refers to e.g., workstation layout etc.
14. **Team involved during the task performance. This dimension is described as**: the members involved in the specific task performed, considering e.g., the number of humans and robots etc.
15. **Interaction aspects. This dimension is described as**: the interaction aspects of the specific performed task, in terms of e.g., knowledge of the robot's status etc.

These 15 dimensions with descriptors represent the final theoretical construct agreed upon by three experts. This construct was divided into 7 robot-related dimensions (from 1 to 7), 4 human-related (from 8 to 11), and 4 dimensions related to context of interaction (from 12 to 15).

3.2 Empirical Revision of the Theoretical Construct with Experts Using a Delphi Approach

Once the theoretical construct was established the process of validation, as indicated in Sect. 2, started with a first cycle of a Delphi technique. A survey in Qualtrics ©2023 (https://www.qualtrics.com) was designed and used to collect the opinion of a large group of HRI experts. The survey was structured as follows: an initial part to explain the study objective; a central part with questions for the experts to rate their agreement toward the model. Specifically, the experts were asked to express their opinion about how much they agreed on the importance of each dimension to properly evaluate the UX during an HRI task, using a 9-point Likert scale ranging from 1 (Not important at all) to 9 (Extremely important), with 5 as the cut-off (Neither not important nor important). Moreover, each expert was asked to report aspects that were missing or that were unclear in the labels and the descriptions of the dimensions. Finally, additional insights about experts were collected by demographic questions e.g., sex, country, role, expertise, domain, etc.

3.2.1 Data Analysis

The consensus was checked based on descriptive statistics (mean, standard deviation, median and interquartile range). Specifically, in line with the literature [38–41] for the 9-point Likert scale adopted in this study we were seeking to establish agreement about the importance of the aspects for (i) a median score ≥ 6, and (ii) at least 75% of experts agreeing on the importance of each aspect with (iii) an IQR of 2 or less.

3.2.2 Characteristics of the Expert Panel

Out of 90 international experts who started the survey only 81 submitted a completed answer. 74.1% of experts were male, 21% declared to be experts but with one year of experience, the remaining participants declared to work as experts for more than a year but less than 5 (34.6%), and the remaining sample declared expertise from more than 5 years (25.9%) to more than 10 years (18.5%). It is worth highlighting that a panel of 81 international experts is considered a quite large panel for a Delphi study [38]. Experts from 14 countries were involved in the study, 90.2% of the participants were from EU countries, the largest group of participants was from Italy (69.1%),

Table 5 Percentage of panelists that declared expertise on HRI application domain: (i) industry and warehouse; (ii) health, surgery, and social robotics; (iii) education; (iv) others e.g., military, police, house, etc.

The domain of expertise declared	Frequency %
Industry and warehouse	88.9
Health, surgery, and social robotics	23.5
Education	16
Others e.g., military, police, house, etc.	27.2

and only 9.8% of the respondents were from non-EU countries. Table 5 summarized the domain of expertise declared by the panelists.

Because most experts declared a focus on industry and warehouse HRI applications, we decided to aggregate all the other applications in one group. Therefore, we divided our sample into two main groups of experts: (i) Industry oriented (IO), composed of the 88.9% who declared expertise in industry and warehouse applications of robotics, and (ii) commercially oriented (CO) application aggregating the answer of the 66.7% experts who declared to work also on HRI application in domains other than industry.

3.2.3 Agreement on the Theoretical Construct

A regression model was utilized to test if the domain of expertise (IO or CO) of the panelists significantly affected their answers concerning the importance of the different dimensions of the construct. No significant difference emerged among the aspects, apart from the following dimensions: **Easiness of robot regulation (Dimension 1)** and **Robot's social behavior (Dimension 6)**. Specifically, the IO experts tend on average to rate as significantly more relevant for assessing UX easiness of regulation **compared** to experts of CO ($B = 0.51444$, SE = 0.24561, Wald = 2.094; $p = 0.036$). Similarly, IO experts tend to rate **Robot's social behavior** as significantly less important compared to CO experts ($B = -0.61964$, SE = 0.22092, Wald = −2.805; $p = 0.005$). As summarized in Table 6 experts agreed with ten out of fifteen dimensions of the theoretical construct. Conversely, the panel did not manage to reach an agreement regarding the importance of five dimensions.

It is worth highlighting that the experts' consensus on the importance of D2 (Robot's physical appearance) as a set of aspects that should be assessed to quantify the user experience, but the level of agreement among the panelists appears to be borderline. Even if IQR is acceptable the percentage of experts who rated this dimension as important is particularly low, and the overall average of consensus that this dimension is relevant for assessing HRI is equal to 5.9 out of 9 with an SD of 1.87.

Table 6 Reports disagreement among the experts in the panel

Dimensions	Experts' opinions (all domains)					IO Experts	CO Experts	Issues
	Average	SD	Mdn	IQR	%	IQR	IQR	
D3	5.70	2.05	6	3	60	3	2	*
D6	5.79	2.05	6	3	62	3	3	**
D8	6.96	1.61	7	2	80	2	3	*
D10	5.52	2.37	6	3	53	3.25	3	**
D14	6.10	1.96	6	2	65	2	3	*

Average of the rating, Standard Deviation (SD) Median (Mdn) Interquartile range (IQR, irrespectively of the domains and divided in IO and CO), and percentage (%) of agreement. Issues are highlighted when the IRQ is higher than 2

*Problematic IQR agreement only for one of the domain groups (IO or CO); **Problematic IQR agreement irrespectively from the domain

3.2.4 Discussion

The panel of experts substantially agreed with 67% (ten out of fifteen dimensions) of the theoretical construct. These ten dimensions represent a set of important aspects to consider when assessing people's experience in HRI i.e., 1, 2, 4, 5, 7, 9, 11–13, and D15. Two dimensions, specifically D4 (**Robot competence features**) and 15 (**Interaction aspects**), obtained the highest level of consensus among experts regardless of their domain. Among the valid dimensions, we included also D2 (**Robot physical appearance**), **nevertheless**, the importance of this dimension for experts should be further clarified as the aspects included in this dimension seem relevant only for 65% of the experts in our panel. Previous analysis suggested that experts from CO and IO groups tend to rate significantly different D1 (**Easiness of robot regulation), however, the agreement analysis clarified that overall,** this dimension is considered important and should be retained. Conversely, D6 (**Robot's social behavior**) seems problematic in terms of consensus and further analysis should be performed to decide if this dimension should be retained or removed from the list of dimensions. D3 (**Robot's emotional appearance**) and D10 (**Human-Factors personality-based**) are considered less relevant by experts to assess the UX with cobots, and there is no consensus regarding the importance of the aspects included in these dimensions. Therefore, a new round of Delphi will be needed to ask experts if these dimensions should be completely removed or reworded or incorporated into other dimensions. Regarding D8 **(Human judgment before the interaction with a cobot)** and D14 **(Team involved during the task performance)** experts at least partially disagree about the importance of these aspects, and further investigation is needed to ascertain the importance of these factors.

The ten dimensions over which agreement was already achieved can be considered an initial list of aspects that designers and evaluators might want to review when designing a new system or reviewing a system that is already in use. The next steps of this work, now that we have a solid construct, will include: generating a list of items and testing the items with EFA and CFA to define an initial scale and to adjust the length of the scale for different purposes.

4 Conclusion

This chapter presented the initial empirical validation of a new framework to assess the UX with robots. The road is still long as future stages of validation with experts should be organized. Indeed, we started the phase of construct definition, and yet we have to start the phase of scale specifications and item generation. Nevertheless, the analysis suggests that ten out of the fifteen dimensions are already retainable. These dimensions can be already used by practitioners (with caution) for heuristic review in two ways: (i) during the testing phase to check if (and evaluate) how the aspects have been considered in the current design; and (ii) during the design phase as a reminder of all the necessary aspects that require to be considered to provide a good UX.

With the diffusion of artificial intelligence systems like ChatGPT and the growing presence of autonomous systems in our society [42] there is a growing demand for measuring the impact of such systems and maximizing the quality of humans' and robots' ability to act together [1, 43]. While the ethical and philosophical implications of the spreading of such systems are under discussion [44], the construct we proposed and partially validated is opening the possibility for experts to reliably compare systems and their quality from the perspective of the end-users. The possibility to measure, compare, and benchmark reliably the experience of the end users will pave the way to the possibility of quantifying the relationship between quality of design, UX, and other relevant socio-technical aspects such as trust toward systems, perceived reliability and safety, and objective measures of safety in usage and usability. The long-term goal is to provide a tool (e.g., an evaluation scale) that could help to anticipate, identify and solve issues that might affect people's interaction with Robots. Whether we will achieve such a goal this diagnostic tool will boost the market and the diffusion of highly usable systems able to maximize the experience of people co-acting with robots.

References

1. Borsci, S., et al.: Embedding Artificial Intelligence in Society: Looking Beyond the EU AI Master Plan Using the Culture Cycle, pp. 1–20. AI & society (2022)
2. Hancock, P.A., et al.: A meta-analysis of factors affecting trust in human-robot interaction. Hum. Factors **53**(5), 517–527 (2011)
3. Hoffman, G.: Evaluating fluency in human–robot collaboration. IEEE Trans. Hum.-Mach. Syst. **49**(3), 209–218 (2019)

4. Akalin, N., Kristoffersson, A., Loutfi, A.: Do you feel safe with your robot? Factors influencing perceived safety in human-robot interaction based on subjective and objective measures. Int. J. Hum. Comput. Stud. **158**, 102744 (2022)
5. Roesler, E., Manzey, D., Onnasch, L.: Embodiment matters in social HRI research: Effectiveness of anthropomorphism on subjective and objective outcomes. ACM Trans. Hum.-Robot Interact. **12**(1), 1–9 (2023)
6. ISO, ISO 9241-11 Ergonomic requirements for office work with visual display terminals—Part 11: Guidance on usability. CEN: Brussels, BE (2018)
7. Tractinsky, N.: The usability construct: a dead end? Human-Comput. Interact. **33**(2), 131–177 (2018)
8. Hornbæk, K.: Current practice in measuring usability: Challenges to usability studies and research. Int. J. Hum Comput. Stud. **64**(2), 79–102 (2006)
9. Hirsch, P.M., Levin, D.Z.: Umbrella advocates versus validity police: a life-cycle model. Organ. Sci. **10**(2), 199–212 (1999)
10. Hornbæk, K., Law, E.L.-C.: Meta-analysis of correlations among usability measures. In: Proceedings of the SIGCHI conference on Human factors in computing systems (2007)
11. Sauro, J., Lewis, J.R.: Correlations among prototypical usability metrics: evidence for the construct of usability. In: Proceedings of the SIGCHI Conference on Human Factors in Computing Systems (2009)
12. Borsci, S., et al.: Shaking the usability tree: why usability is not a dead end, and a constructive way forward. Behav. Inf. Technol. **38**(5), 519–532 (2019)
13. ISO, ISO 9241-210: 2010 Ergonomics of human-system interaction—Part 210: Human-centred design for interactive systems. CEN, Brussels, BE (2010)
14. Steinfeld, A., et al.: Common metrics for human-robot interaction. In: Proceedings of the 1st ACM SIGCHI/SIGART Conference on Human-Robot Interaction (2006).
15. Olsen, D.R., Goodrich, M.A.: Metrics for evaluating human-robot interactions. In: Proceedings of PERMIS. Citeseer (2003)
16. Prati, E., et al.: How to include user eXperience in the design of Human-Robot interaction. Robot. Comput.-Integr. Manuf. **68**, 102072 (2021)
17. Norman, D.A.: Design for a Better World: Meaningful, Sustainable, Humanity Centered. MIT Press (2023)
18. Rueben, M., et al.: Helping users develop accurate mental models of robots' perceptual capabilities: a first approach. In: Workshop on Trust, Acceptance and Social Cues in Robot Interaction (SCRITA) at RO-MAN (2019)
19. Bejarano, A., et al.: You had me at hello: The impact of robot group presentation strategies on mental model formation. In: 2022 17th ACM/IEEE International Conference on Human-Robot Interaction (HRI). IEEE (2022)
20. DeVellis, R.F.: Scale development: Theory and applications, vol. 26. Sage publications. Los Angeles, London, New Delhi, Singapore, Washington DC, Melbourne (2016)
21. Kyriazos, T.A., Stalikas, A.: Applied psychometrics: The steps of scale development and standardization process. Psychology **9**(11), 2531–2560 (2018)
22. Brooke, J.: SUS-A quick and dirty usability scale. Usability Eval. Ind. **189**(194), 4–7 (1996)
23. Borsci, S., et al.: Assessing user satisfaction in the era of user experience: comparison of the SUS, UMUX, and UMUX-LITE as a function of product experience. Int. J. Human-Comput. Inter. **31**(8), 484–495 (2015)
24. Federici, S., et al.: Inside pandora's box: a systematic review of the assessment of the perceived quality of chatbots for people with disabilities or special needs. Disabil. Rehabil. Assist. Technol. **15**(7), 832–837 (2020)
25. Morgado, F.F., et al.: Scale Development: Ten Main Limitations and Recommendations to Improve Future Research Practices, vol. 30. Reflexão e Crítica, Psicologia (2017)
26. Linstone, H.A., Turoff, M.: The Delphi Method. Addison-Wesley Reading, MA (1975)
27. Dimitrov, D.M.: Statistical Methods for Validation of Assessment Scale Data in Counseling and Related Fields. Wiley (2014)
28. Spencer, D., Warfel, T.: Card Sorting, vol. 7. Boxes and arrows (2004)

29. Prati, E., et al.: A systematic literature review of user experience evaluation scales for human-robot collaboration. In: Moser, B. (ed.) Advances in Transdisciplinary Engineering (2022)
30. Schmidtler, J., et al.: Black or White? Influence of robot arm contrast on distraction in Human-Robot interaction. In: Engineering Psychology and Cognitive Ergonomics: 12th International Conference, EPCE 2015, Held as Part of HCI International 2015, Los Angeles, CA, USA, August 2-7, 2015, Proceedings, vol 12. Springer (2015)
31. Bartneck, C., Forlizzi, J.: Shaping human-robot interaction: understanding the social aspects of intelligent robotic products. In: CHI'04 Extended Abstracts on Human Factors in Computing Systems (2004)
32. Weiss, A., et al.: The USUS evaluation framework for human-robot interaction. In: AISB2009: Proceedings of the Symposium on New Frontiers in Human-Robot Interaction (2009)
33. Gervasi, R., Mastrogiacomo, L., Franceschini, F.: A conceptual framework to evaluate human-robot collaboration. Int. J. Adv. Manuf. Technol. **108**, 841–865 (2020)
34. Gualtieri, L., Revolti, A., Dallasega, P.: A human-centered conceptual model for integrating augmented reality and dynamic digital models to reduce occupational risks in industrial contexts. Proced. Comput. Sci. **217**, 765–773 (2023)
35. Villani, V., et al.: A general methodology for adapting industrial HMIs to human operators. IEEE Trans. Autom. Sci. Eng. **18**(1), 164–175 (2019)
36. Scholtz, J.: Theory and evaluation of human robot interactions. In: Proceedings of the 36th Annual Hawaii International Conference on System Sciences. IEEE (2003)
37. Beauchamp, Y., Stobbe, T.J.: The effects of factors on human performance in the event of an unexpected robot motion. J. Safety Res. **21**(3), 83–96 (1990)
38. Birko, S., Dove, E.S., Özdemir, V.: Evaluation of nine consensus indices in Delphi foresight research and their dependency on Delphi survey characteristics: a simulation study and debate on Delphi design and interpretation. PLoS ONE **10**(8), e0135162 (2015)
39. Von der Gracht, H.A., Darkow, I.-L.: Scenarios for the logistics services industry: a Delphi-based analysis for 2025. Int. J. Prod. Econ. **127**(1), 46–59 (2010)
40. von der Gracht, H.A.: Consensus measurement in Delphi studies: review and implications for future quality assurance. Technol. Forecast. Soc. Chang. **79**(8), 1525–1536 (2012)
41. Polisena, J., et al.: Health technology assessment methods guidelines for medical devices: how can we address the gaps? The international federation of medical and biological engineering perspective. Int. J. Technol. Assess. Health Care **34**(3), 276–289 (2018)
42. Vemprala, S., et al.: Chatgpt for robotics: design principles and model abilities. Microsoft Auton. Syst. Robot. Res **2**, 20 (2023)
43. EU robotics and artificial intelligence team A1: White Paper on Artificial Intelligence: Public consultation towards a European approach for excellence and trust (2020)
44. Floridi, L., et al.: AI4People—an ethical framework for a good AI society: opportunities, risks, principles, and recommendations. Mind. Mach. **28**(4), 689–707 (2018)

User Preferences for Enhanced Robot Adaptation in Assistive Tasks

Gerard Canal, Guillem Alenyà, and Carme Torras

Abstract The adaptation of robot behaviour to each user can be seen as an ergonomic feature. Adaptation plays a key role in increasing human–robot interaction efficiency and safety, and this is especially relevant in assistive scenarios, where users may be unable to adapt to the robot or help in task execution. Robot adaptation may happen at different levels of abstraction, from low-level trajectory parameters to higher-level task concepts. These higher-level abstractions relate to human factors such as personal preferences on how the robot should perform the task. This chapter introduces the concept of user preferences to adapt the robot's behaviour and also help in the robot's decision-making, with a special focus on assistive robotic tasks such as dressing or feeding a user with mobility impairments. It also presents the challenges associated with adaptation in this kind of scenario, as well as some existing methods that may be used to adapt robot behaviour to high-level abstract preferences. Such adaptation makes it easier for the user to interact with the robot in a more efficient and comfortable manner, thus improving its ergonomics.

Keywords User preferences · Robot behaviour adaptation · Robot personalisation · Assistive robotics · Daily living assistance

1 Introduction

Robotic assistants have a great potential to improve the lives of dependent people[1] and to empower them by helping them perform activities of daily living more

[1] We define dependent people as those individuals who require help from other people to perform activities of daily living.

G. Canal (✉)
Department of Informatics, King's College London, London, UK
e-mail: gerard.canal@kcl.ac.uk

G. Alenyà · C. Torras
Institut de Robòtica i Informàtica Industrial (IRI), CSIC-UPC, Barcelona, Spain
e-mail: galenya@iri.upc.edu

C. Torras
e-mail: torras@iri.upc.edu

G. Barresi and S. Wischniewski (eds.), *Ergonomics in Robotics: Advances and Innovations*, Springer Tracts in Advanced Robotics 162,
https://doi.org/10.1007/978-3-032-19374-2_17

autonomously, effectively extending and increasing their independence. In turn, these robots may also help improve the psychological well-being of their users; as described by Williams et al. [1], dependent patients who suffered a loss of independence may have a great psychological impact due to their condition: loss of self-worth and self-identity, feelings of burden and guilt. Similarly, Boström et al. [2] found an association between depressive symptoms and the lack of independence in tasks such as transfer and dressing. Lee and Riek [3] define autonomy as how much users such as older adults feel in control over their lives and argue that to support these users' autonomy, they should have enough control over the robots. Therefore, in these cases the robot becomes a tool that a dependent person may use and control to achieve tasks such as eating, grooming, or dressing [4].

When considering such kinds of robotic assistants, ergonomics play a crucial role in the effectiveness of the assistance and the perceived usefulness by the users. Assistive robots that take into account human factors and user needs shall be able to better adapt to the specific needs of the assisted individuals, making the use of the robots easier [5, 6]. Moreover, the levels of autonomy [7] should also be adapted, so that the user still plays an active role in the task's achievement, depending on their needs and capabilities [8]. This makes the problem harder, given that the user is also acting, which calls for cross-task generalisation [9]. This chapter will cover some methods to adapt robot behaviours to individual users with examples in two domains: robots for feeding and dressing assistance. These tasks fall into the realm of physically assistive robots, where the robots physically interact with the user, and contacts between the robot and the user may be present. This makes these tasks particularly challenging as there is an increased risk of harm and discomfort. Ergonomics are, therefore, key to improving the user's well-being in such tasks.

Assistive tasks such as assisted feeding have great potential to improve people's lives. Some works have focused on trajectory planning [10], while others look at the safety considerations of the task [11], the adaptations that humans do when feeding [12], the bite acquisition [13, 14], bite transfer [15], and bite timing [16].

Regarding the dressing scenario, different kinds of garments and user interactions have been explored. When putting on a jacket, the user's movement space may be modelled [17–19], considering force information and the observed range of user movements. Similar approaches generate plans for the user and the robot to optimise trajectories to the user [20], as well as predicting garment forces that will be applied to the user [21], adapting to different user poses and garments [22]. Other examples in this area include [23–25], where combinations of user pose and clothing types are used to detect dressing errors such as cloth snagging, and the user is allowed to correct the robot's position using voice. When dressing lower limbs, Yamazaki et al. [26] developed a procedure to help disabled users to put on trousers, where visual information is used to recognise the trousers' state, while force sensing is also employed to detect failures and adapt to leg differences fitting the trajectories to the user. Shirt dressing was investigated by Tamei et al. [27], using reinforcement learning to put on a shirt while considering the arm motions. Collaborative dressing has also been explored. For instance, Klee et al. [28] tackle assistance in putting on a hat through repositioning requests to the user, adapting to the user constraints by monitoring the compliance to the requests. Other works have looked at personalised shoe fitting [29, 30], where smart low-level controllers may adapt to online user movements, while task planners may adapt to high-level preferences.

While low-level robot posture adjustments may be useful to adapt to the physical limitations of the user, these high-level preferences may focus on improving the experience while also considering the person's predilections. For example, preferences for hand-overs are analysed by Cakmak et al. [31], where robot and object configurations are adapted to the user preferences related to the position and orientation of the object. User preferences for robot motion planning are learned by Wilde et al. [32] in the form of user constraints regarding the task, while Hayes et al. learn preferences to improve navigation tasks [33]. Preferences may also be used to adapt the stopping distance of a robot [34] based on the user's personality. More examples of social robots include [35], where the robot interacts with post-stroke patients in a rehabilitation context, adapting the personality of the robot to favour a good match with the user, so as to improve the interaction experience.

These adaptations have the potential to increase the acceptance of these novel technologies by users who may be reluctant to learn how to use the robots. Society might still be reluctant to accept social robots for domestic purposes, whereby it is important to take into account the opinions of future users so as to satisfy their preferences [36]. All in all, potential users would be more open to these new technologies when there is a benefit in their covering a clear need, and openness to using these robots may be higher for instrumental tasks [37, 38].

This chapter will be structured as follows: in Sect. 2, we introduce physical adaptations of the robot behaviour, while the use of user preferences is explored in Sect. 3. Section 4 deepens in the use of such preferences along with symbolic planning techniques, and Sect. 5 explores different assistive robotics examples and the evaluation of the adaptation with users. Finally, Sect. 6 concludes the chapter with some pointers to future research.

2 Physical Adaptation to Ergonomic Needs of Individuals

Robotic assistance must be inherently ergonomic, not only by design but also regarding the robot's behaviour. Therefore, these robots have to be adaptive to the needs and capabilities of the user.

When it comes to robot behaviour adaptation, this should happen when the robot is set to perform the task (i.e., at the user's home or care facility). One way to perform such adaptation is through modifying the motion trajectories the robot performs while doing the task. These adaptations can be applied in a natural and intuitive way (e.g., through guiding) by a caregiver without any technical robotics knowledge. Adaptations like this are considered low-level and mainly modify the trajectory that is being executed by the robot.

Here, the FUTE framework [39] is described, which defines a strategy to tailor the robot behaviour to users; for instance, by means of kinesthetic adaptations of the robot movement.

2.1 *The FUTE Personalisation Framework*

The "FUTE framework" is a three-phase framework to design and develop, among others, the kind of adaptive assistive applications described in Sect. 1. The three phases are called "Factory setting", "User Tailoring" and "Execution tuning". They are depicted in Fig. 1 and described as:

1. **Factory setting**: the robot is provided with the skills needed to perform the assistive task in a generic way. This would suit either the design of a new robot or the enhancement of an existing platform to carry out a new task.
2. **User Tailoring**: this second phase takes place in the user's home. The robot performs a nominal skill, but personalisation is encouraged in order to adapt its behaviour to the user's needs. In this phase, the robot should acquire, as automatically as possible, information about how the task has to be done for the user at hand while it performs the task in the generic way learned in the previous phase. This personalisation may be done by the user or by an external agent (such as a carer), and it could be either explicit or implicit. In the feeding example, this could consist of the selection of the feeding point, it being either inside or outside the mouth. In the proposed example, the data include the proprioceptive robot perception as well as 3D images from a camera located at the hand of the robot (Fig. 2).
3. **Execution tuning**: in this last phase, the robot performs the task designed in the first phase but takes into account the personalisation introduced in the second one. In the feeding example, the 6D pose of the user can be computed using an RGBD camera and a face detection algorithm, and the robot trajectories adapted to the current pose of the user. If the user is not satisfied with the robot's behaviour, the User Tailoring phase can be repeated to further adapt the robot's behaviour.

The FUTE framework has been effectively used to personalise robot behaviours in a diverse set of assistive tasks, such as feeding and dressing [40]. Experimental results may be found in [39], where Probabilistic Movement Primitives (ProMPs) [41, 42]

Fig. 1 The FUTE process. Extracted from [40]

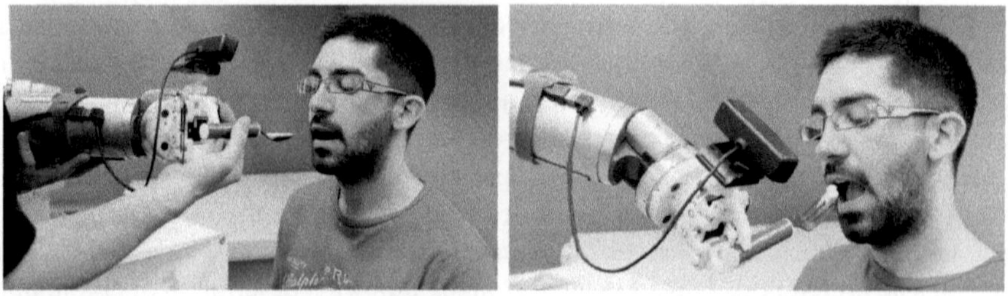

(a) Caregiver personalising a spoon feeding skill.

(b) A user eating from a fork.

Fig. 2 Assistive personalised feeding application example. Extracted from [39]

are used to encode trajectories that are then adapted through kinesthetic disturbances of the original (factory setting) trajectory while it is being executed. The results also show how these types of primitives may be interesting to extract contextual information about the task, such as the feeding point in a feeding trajectory.

3 Symbolic Preferences to Adapt Robot's Behaviour

Physical and low-level trajectory adaptations are able to provide a certain amount of flexibility to make the robots more ergonomic for the user they are assisting. This mainly handles the physiological aspects of the adaptation. However, for a robot to be more ergonomic, it shall also consider psychological factors that may influence how the robot should behave, reduce the errors of the system, and increase the comfort and satisfaction of the users.

Therefore, it is important to capture the user's psychological traits that are relevant to personalise robot behaviour so as to tailor it to the user's specific needs and capacities. One way to carry out such personalisation is to acquire user preferences to create a model of the user and then apply it to uniquely determine the behaviour, actions, and reactions of the robot so that the interaction becomes safer, more comfortable, and better accepted.

User preferences can be of many types and evolve over time, as well as depend on specific situations. Focusing on the ergonomics of assistive robotics tasks, a preference classification was developed in [43] to permit building user models that can drive robot behaviour personalisation.

3.1 A Taxonomy of Preferences for Assistive Robots

A hierarchical taxonomy of user preference categories designed for Human–Robot Interaction (HRI) applications was presented in [43]. This taxonomy focuses on assistive scenarios where the robot aids users facing difficulties in performing Activities of Daily Living (ADLs). The taxonomy does not distinguish between preferences and user constraints (such as mobility issues), as the latter can be expressed as a preference. For instance, a user who cannot move the right arm will "prefer" not to use this arm. Encoding impairments as preferences results in a more generic taxonomy without the need to separate similar preferences and impairments into different categories. The taxonomy's aim is to describe preferences that are useful for an autonomous system to make decisions in an assistive scenario, either physical or social. The preferences defined in the taxonomy allow the description of ergonomic adaptation of the robot movements and behaviour in a flexible way.

The taxonomy can be found in Fig. 3. It defines two types of preferences: the *primary* preferences and the *modifier* ones. The former are preferences that are directly applicable, while the latter are used to accompany the primary preferences and modify them, effectively conditioning their applicability.

The taxonomy divides the preferences into two main category groups: the decision-making and the configuration preferences. The *Decision-making* (DM) preferences are those that help the robot to choose between the different actions that it can execute at a given moment, provided that they all lead to the final assistive goal (see left branch of Fig. 3). DM preferences are divided into two categories, which are again subdivided into more fine-grained preference types:

- *Communication preferences* regulate the desired amount of different kinds of interaction with the robot. They are subdivided into:
  - Information providing: preferences stating whether the robot should inform regarding performed actions and events, or if it should omit unimportant information. It relates to the *verbosity* of the robot, namely the amount of speech that the robot should use.
  - Information obtaining: they define if the user prefers the robot to inquire about missing information or the robot should try to infer such information from other sources.

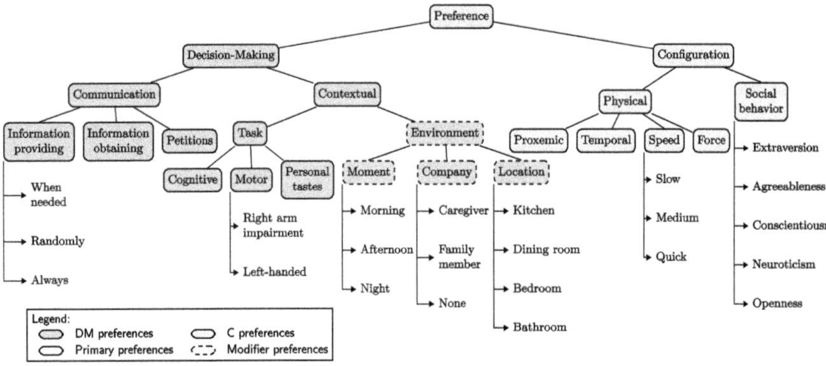

Fig. 3 Preference taxonomy for assistive physical human–robot interaction. Extracted from [43]

- Petitions: state if the robot should ask the user to perform some action (such as repositioning himself to ease the solution of a task) or if the robot should risk accomplishing the task without "bothering" the user, provided that safety is not in jeopardy.

• *Contextual preferences* specify how the robot's behaviour may change depending on the execution context, it being defined as the user's environment, place and time. Four subcategories are distinguished here:

- Task: preferences that may have implications about the task that is being performed. They may consist of general user constraints (such as limited right arm mobility) or simpler preferences such as how it is more comfortable to scoop the spoon when eating. They are subdivided into:

  Cognitive preferences related to cognitive disabilities of the user.
  Motor constraints of the user which may limit the task.
  Personal tastes (other personal needs and desires).

- Environment: preferences regarding the execution setting. They are mainly modifier preferences (see Fig. 3) that supplement primary preferences and limit their application range. These are subdivided into three categories:

  Moment: define the time of the day in which the task is executed. Preferences may vary depending on, for instance, whether the task is performed in the morning or at night, as the state of the person may be different (i.e., tiredness and mood may affect the user preferences).
  Company: preferences concerning the people that are in the environment. User preferences with the robot may be different when a caregiver is also assisting the user in contrast to when a family member is doing so. Preferences may also be different when the user is alone with the robot, given that more support may be needed in this case.
  Location: preferences related to the location where the user is situated. User preferences may change depending on where the task is being executed. For instance, it may not be the same to fit a shoe while sitting on the bed or to fit a slipper while resting on the couch.

The other type of preferences are the *configuration* (C) preferences (see the right branch of Fig. 3). Those are preferences that define how a chosen action is to be performed. They are used to tune the parameters of the actions rather than to choose which action sequence is going to be executed to solve the task. Configuration preferences are also divided into two categories:

- Physical preferences define the physical properties of the (physical) actions. These include:
  - Proxemic: preferences that relate to the spatial requirements of the user, the robot and the task.
  - Temporal: define temporal requirements of the user and the task. For instance, these could relate to the duration of actions, where a user may prefer to not have the foot lifted for more than one minute.
  - Speed: they specify the velocity of the robot movement (i.e., how quick or slow). This potentially relates to the feeling of safety, as the user may be uncomfortable if the robot makes sudden fast movements, but may also get impatient when the movements are too slow.
  - Force: some tasks, such as shoe fitting, require pressure against some body parts. The applied force may be limited based on user desires and abilities.
- Social behaviour preferences are used to shape the robot's personality. This is described following the definitions of personality from [44], through which the taxonomy is linked to the Big Five personality trait taxonomy [45], where personality is described in terms of five traits (extraversion, agreeableness, conscientiousness, neuroticism and openness). These preferences enable users to express their desired personality traits for the robot. Other elements that could be included as social behaviour preferences are the kind of voice, tone, formality level, prose and detail level of the interactive acts.

The proposed taxonomy allows us to define user preferences for HRI tasks, and more specifically, for physically assistive robots to help older adults and people with disabilities. However, there is a lot of redundancy when instantiating the preference taxonomy for every specific task. For instance, a user who prefers the robot to move slowly while fitting a shoe will probably prefer it to be slow also when dressing a jacket. A user with reduced mobility in the right arm will need special assistance to perform any activity involving this arm. To solve this redundancy, and to ease the description of the preferences, they are defined in two levels:

- *Global preferences*, are those that are applicable to most tasks. They define generic user preferences and personal constraints which may be used in any setup.
- *Specific preferences* define activity-related preferences. They only apply to certain cases and during the execution of specific tasks.

The two levels allow the specialization of preferences, including potentially conflicting or opposing elements. To solve potential conflicts, the executor should consider specific preferences to override existing global preferences, but use global preferences in the absence of a specific one.

4 Use of Symbolic Preferences to Adapt the Robot's Behaviour in the Short and Long Term

Preferences allow to model the user in terms of their needs and capacities. However, there are many challenges involved in their use to adapt the robot behaviour. For instance, the user may not be able to self-assess their preferences and needs concerning the robot's behaviour. And, even if they do, those preferences may be over or underestimated. This means that the robot not only needs to behave according to the preferences but also to re-adapt in case of an imprecise setting of the preference values. Therefore, acquiring and using such user preferences is a complex endeavour.

Some options to initialise the preference values focus on asking seemingly unrelated but easier questions to the user. For instance, in [29], the users are asked to rate their level of confidence in the robot's abilities and comfortability on a numeric rating scale. With this, an inference mechanism can derive the values of the preferences. Methods like this make it easier for a user to answer the questions, as directly deciding the value of their preferences may not be easy–given that users may not be accustomed to interact with a robot. However, this also restricts the amount of preferences obtained, and may lead to incorrect inferences. This also supports the need for long-term adaptations.

4.1 Preference-Adapted Task Planning

Automated planning methods compute a sequence of actions that bring a system from an initial state to a goal state, based on a model of the world dynamics. Therefore, these planning techniques are very useful for implementing robotic behaviours that adapt to the user and their needs, as the user preferences can be encoded in each action definition.

There are multiple approaches to integrate preferences into planning domains to aid decision-making and action selection to make choices that are aligned with the pre-defined user preferences.

More limiting approaches set the preferences as a precondition of the action. In this case, actions will only be executable or selectable by the planner when the corresponding preferences are present in the state. While this ensures no action is executed if it does not fulfil the user preferences, it may result in more restrictive approaches where dead-ends happen due to not having applicable preferences present. It also limits the chances for adaptability and exploration of different preference values.

An alternative approach, such as the one taken in [29], defines preferences as action costs in a cost-optimising planning scenario. This way, actions are "more expensive" to be selected by the planner when the preferences are not aligned with the action. This allows the exploration of other preference values and prevents potential dead-ends when there is no action that can be executed with the available preferences.

4.2 Execution-Led Adaptation to Preferences

Although the presented approaches are able to execute behaviours that adapt to preferences, these adaptations are mainly short-term and may need to be modified if the user behaviour changes. Therefore, to be fully adaptive, robots must keep on adapting their behaviours to changes in the person and their preferences which may happen in the longer term.

Towards this, the planning model may be expanded to a probabilistic planning approach, where actions' outcomes happen based on probability. Along with the preference-based costs, this allows an adaptation not only to preferences but also to the success of the different actions.

Having a model of an assistive task as a planning domain with action outcome probabilities and preference-based costs then permits the learning and adaptation of the costs and probabilities after each execution. To achieve this, the model may be first initialised to favour actions that satisfy the user preferences. Therefore, actions satisfying user preferences may have their success probabilities increased (whereas the success probabilities of the non-compliant actions are decreased), while the costs for not complying with user preferences are also increased.

Once the initial model has been refined, the task may be executed. Adaptation can then be performed in an ongoing manner: a satisfaction score from the user will be used to adapt the costs of the executed actions, and the individual success of the actions will be used to update the outcome probabilities.

Therefore, the costs of the preferences are adapted to promote those actions that were executed successfully and satisfied the user. To achieve this, the feedback value (i.e., a real number) will be used to increase the cost of the actions when the satisfaction is poor and lower their cost when the user's satisfaction is high. Consequently, after adaptation, the planner will be more prone to select the actions whose cost has been lowered, while the actions whose cost increased will be less appealing.

Given that user satisfaction may not be directly related to the success of the actions, probabilities have to be updated solely based on the observed outcome. Techniques such as the decreasing m-estimate [46] may be used to adapt the probabilities.

Using an approach like this, the planning model will converge towards the most successful and satisfactory set of costs and probabilities, which will make the planner more prone to choose actions which have been good for the user. Effectively, this means that the costs may have changed such that the preference values have diverged from the initial ones. For instance, if the initial preference was that the robot moved fast but the satisfaction and success of the user were low when the robot was moving fast, the increase in cost and decrease in probability will make the cost for moving fast too high, while the cost for moving slow may have decreased to a point where it is a better option for the planner to choose.

However, for this convergence to work well, the planner must be allowed to explore other preference values within a task. For instance, if an action keeps failing during execution due to a wrong set of preference values, the planner may insist on repeating the action when a replan is performed. To avoid this, exploration may be forced by disabling the offending actions in the planning model before the replanning step, forcing the planner to select a more costly action which may result in finding a more satisfactory combination. If that is the case, the adaptation method described above will decrease that action's cost, converging into a model that is better suited for the user. Alternatively, the cost refinement could be done before replanning rather than at the end of the task.

Example: adaptation in a shoe-fitting scenario As an example of these adaptation methods, take a robotic shoe-fitting task. The task may be split into three symbolic actions, namely approach, insert, and release. The approach action requires (as a precondition) that the shoe is grasped, and results in (as effects) the shoe being close to the foot. This is a precondition for the insertion, which has an effect on the shoe being fitted. Finally, release requires the shoe to be fitted and grasped, and as effect the shoe is not grasped anymore, completing the task. There may also be actions to inform the user and to perform verbal requests to the user.

To start the task, the user is seated near the robot and lifts the foot as a signal to start the shoe-fitting task. The robot already has the shoe in its end-effector. In the first execution, the initial preferences are obtained and the initial refinement of the action costs is performed. Then, the planner is used to obtain the action sequence to execute based on the perceived initial state, and the execution of the action is carried out. Once each action is executed, the symbolic state is recomputed and a new plan is obtained if the observed state does not match the expected one. This process is repeated until the shoe has been fit. Once the shoe-fitting has been completed, the user's satisfaction is obtained, and the costs and preferences of the domain are then refined. Figure 4 shows the robot executing the three shoe-fitting task actions.

When the observed state does not match the expected one after an action's execution the action is considered failed and a replan is triggered. In the new plan, the robot may start with a request for the user to move the foot closer or to stay still, with all these decisions taken by the symbolic planner.

If an action keeps failing with the current preferences, a behaviour change is forced by disabling some of the actions to encourage exploration; for instance, by executing the action at different speeds.

(a) *Approach* action, to get closer to the foot.

(b) *Insert shoe* action, using an elaborated robot wrist rotation.

(c) *Release* action, to complete the task.

Fig. 4 Example of the execution of the three shoe-fitting actions. Extracted from [29]

5 Ergonomic Adaptation Examples in Assistive Robotics

User preferences are a great tool to drive the adaptation of robot behaviours. These adaptations may be very useful to improve the assistance that robots provide in assistive scenarios. One reason for such improvement is that adaptations to preferences have the power to improve the ergonomics of the robotic system, making it more intuitive, comfortable, and eventually safer. In turn, the adaptations to user preferences may result in higher acceptance of the systems.

Taking comfortability as an example, [47] shows that users feel more pleased when their preferences are being used in an assistive robotics scenario. The study evaluates how preferences may change the robot behaviour, and whether users being assisted by a robot are able to distinguish different executions and to recognise when their own preferences are in use.

5.1 *Example Assistive Scenarios*

There are many potential assistive tasks that a robot could perform to help a user. Here, examples of three tasks are shown with a description of their ergonomic aspects and how preferences may help adapt the robot to the user. Examples of the setups for the different tasks are shown in Fig. 5.

- Assistive feeding: The robot feeds the user one spoonful. An example of task-related preferences is proximity, which defines how close the robot should be to the user's face. This may affect whether the robot inserts the spoon in the user's mouth, or waits in front of it for the user to take the foot.
- Shoe fitting: The robot approaches a shoe to the user, who prepares the foot in front of it, and the shoe is fitted. Then, the robot releases the shoe and the task is finished. Example preferences here may be the amount of movement of the foot, which might depend on the available mobility of the user. This may also be related to the amount of force exerted on the foot.
- Jacket (open-sleeved garment) dressing: The robot grasps the garment with two arms, and it dresses the user. Possible behaviours are starting with one arm or the other, or inserting both together. This could be done in each part of the task, be it the sleeve insertion, lower-arm dragging, or upper-arm dressing. Example preferences here are the amount of movement in a human arm, which may be limited due to injury or other user capabilities. This may hint the robot to start with the other arm, which might lower the amount of forced movement on the limited arm.

These tasks are widely addressed by the physically assistive robotics community, given that they have a potentially high impact on the dependent users' lives. They represent some of the most basic needs without which a human cannot live with dignity.

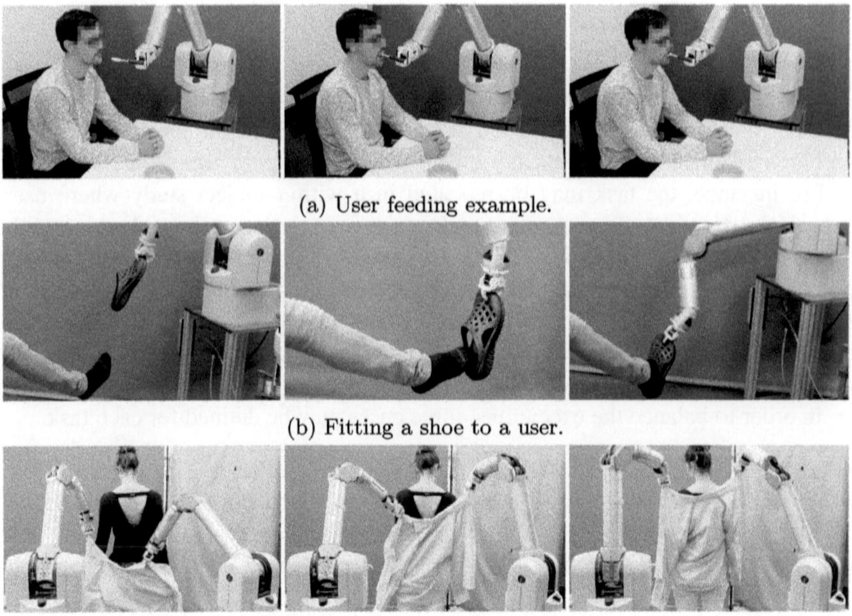

(a) User feeding example.

(b) Fitting a shoe to a user.

(c) Dressing of a sleeved garment.

Fig. 5 Example of the three assistive tasks. Extracted from [47]

Beyond individual task-related preferences or ergonomic adaptations, there are more generic preferences that may also be defined. These preferences would be applicable in a cross-task manner for which they would be defined in the global preferences level of the taxonomy of Subsect. 3.1, while the task-dependent ones may be defined as specific preferences. For instance:

- Robot speed: defines the speed at which the robot moves. This would be a physical configuration preference in the taxonomy of Subsect. 3.1.
- Robot verbosity: the amount of speech the robot should use. For example, the robot may inform the user when dressing a garment by saying "I am approaching your right hand". Possible values for the preference may be gradual levels of verbosity, such as speaking sometimes, always, or never. Following the taxonomy of preferences, this would be a decision-making preference of the information providing type in the communication category.

Note that preferences are often interrelated. The robot's speed may affect the amount of exerted force in a shoe-fitting task (or vice-versa, the amount of force may limit the speed). Similarly, the proximity in a feeding scenario may depend on the range of mobility the user has in the neck. Therefore, any method intending to adapt a robot to a user should consider that requirements and preferences may sometimes be conflicting or contradictory.

5.2 Evaluation of the Robotic Adaptation

The evaluation of the adaptation and ergonomic performance of assistive robots must necessarily be based on usage by the target end-users. However, the evaluation may still include objective elements along with subjective experiences from the users.

For instance, the task may be repeated in a within-subject study where users experience the task with and without adaptations, reporting at the end which of the executions was better suited for them. This kind of evaluation is less dependent on potential biases the users may have, and may lead to more objective conclusions. In such a case, even a control group may be included in which no adaptation is applied, and for which users may state they did not feel any different, or show doubt with respect to which of the two executions was better.

In order to balance the executions, three cases may be defined for each task.

- CASE A: The task is executed two times without adaptations.
- CASE B: The first execution uses the preferences/adaptations, while the second does not.
- CASE C: The first execution does not use the preferences/adaptations, the second execution does.

The three cases can then be experienced by all the users throughout different tasks or task repetitions. The three conditions may be permuted for different users so that tasks are evaluated in the different conditions. This overcomes any task-dependent element that may affect the choice of the users. This evaluation approach was used in [47].

Along with this and other objective and subjective measures to evaluate the specific aspects of the ergonomic adaptation of the robot, standardized questionnaires should also be used. For instance, the Almere model [48] to measure the acceptance of assistive agents by older adults, or the Godspeed questionnaire [49] to measure elements such as the likeability, perceived intelligence, or perceived safety of the assistive robotic system.

5.3 Pleasantness of Robotic Adaptations

Adaptation of robot behaviours to user preferences has been shown to improve the reported pleasantness of the interaction. For instance, [47] explored how the pleasantness was perceived in relation to when the user preferences were used.

The results showed that when no preferences were available, users tended to give either a low score or a middle score to express neutrality to questions regarding how pleasant was their interaction with the robot. When the preferences were used, the satisfaction increased, regardless of whether the users were aware of their preferences being used.

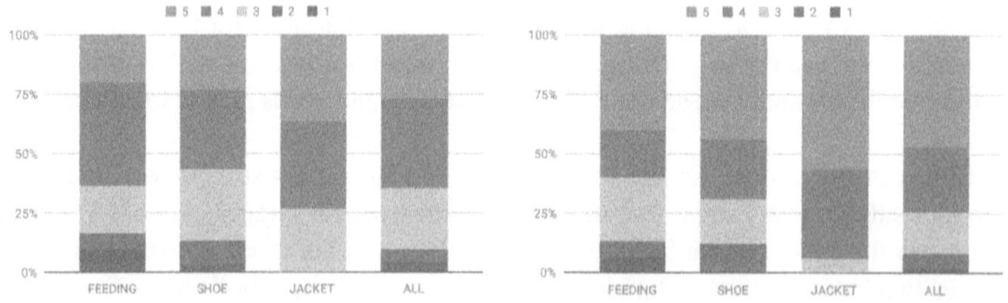

(a) Pleasantness answers

(b) Pleasantness answers considering only successfully guessed state (with preferences)

Fig. 6 Results of a 5-point Likert scale (5-agree, 1-disagree) regarding the pleasantness experienced by the users. Extracted from [47]

The results regarding the pleasantness of the interaction obtained in [47] are reflected in Fig. 6. Figure 6a includes the results considering all the answers, while Fig. 6b reflects only the answers in the cases where the users successfully identified their preferences were being used. It can be seen that the pleasantness levels increase when the preferences are correctly identified, as when taking into consideration all the answers the scores are lower, while considering only the guessed results scores are higher. This suggests that users feel more pleased with the interaction with the robot when they realise that the robot is adapting to their preferences, thus supporting the idea that ergonomic adaptation to user needs makes for better-interacting robots.

6 Conclusions and Future Research

Research has shown that adaptation to user preferences, behaviours, and needs has the potential to increase the acceptance of assistive robots, making them easier to use and more effective overall. These adaptations are one way to increase assistive robot ergonomics, making them more inclusive and practical.

While several elements of behaviour adaptation have been investigated, there are still many areas to work on in order to achieve fully adaptive assistive robots that have the potential to change users' lives. On the topic of preferences, elicitation and inference are some areas to be explored. Obtaining the preferences is expensive, and generalising them may also be complicated. Therefore, methods that allow inferring the preferences in a semi-automatic manner, without having to ask many questions to the user may be required to make the process less tedious for the user, which might decrease the acceptance of the system.

Another important element is objective evaluation and benchmarking. Most research relies on their own system evaluations, given the lack of an objective set of methods. User evaluations are extremely important, but often expensive to run, and not always easy when using prototypical robots that may not be up to a safety standard to be used with potential end-users. Therefore, starting with methods that automatically evaluate the adaptation to preferences and needs may help advance the field.

Safety is also crucial for the advancement of adaptive assistive robots. Assistive robotics safety is very important in general, with work on safe robot control needed to ensure the robot cannot harm the user in close-contact tasks. However, when it comes to adaptive robots, safety is even more crucial as the robot's behaviour is not preset anymore, which may make the system more unpredictable and, in turn, less safe.

Finally, thorough evaluation by potential users, both caregivers and end-users, is needed. This would require safe robots, as well as users able to use such robots for relatively extended periods of time. Such conditions would permit consistently evaluating the impact of these robots on the quality of life of users and their acceptance. In summary, it would provide a valuable assessment of the potential benefits and drawbacks that assistive robots and their adaptive behaviour would have for the healthcare ecosystem in the short and long terms.

Acknowledgements The authors would like to acknowledge the support received from project CHLOE-MAP funded by MCIU/AEI/10.13039/501100011033 and by ERDF, UE; by the European Union under the project ARISE (HORIZON-CL4-2023-DIGITAL-EMERGING-01101135959); by the Generalitat de Catalunya through the group RobIRI (2021 SGR 00514); and by the CHIST-ERA project COHERENT funded by EPSRC (EP/V062506/1).

References

1. Williams, A.M., Christopher, G., Jenkinson, E.: The psychological impact of dependency in adults with chronic fatigue syndrome/myalgic encephalomyelitis: a qualitative exploration. Health Psychol. (2016). https://doi.org/10.1177/1359105316643376
2. Boström, G., Conradsson, M., Rosendahl, E., Nordström, P., Gustafson, Y., Littbrand, H.: Functional capacity and dependency in transfer and dressing are associated with depressive symptoms in older people. Clin. Interv. Aging **9**, 249 (2014)
3. Lee, H.R., Riek, L.D.: Reframing assistive robots to promote successful aging. ACM Trans. Human–Robot Interact. **7**(1), 11:1–11:23 (2018). https://doi.org/10.1145/3203303
4. Chen, T.L., Ciocarlie, M., Cousins, S., Grice, P.M., Hawkins, K., Hsiao, K., Kemp, C.C., King, C.H., Lazewatsky, D.A., Leeper, A.E., Nguyen, H., Paepcke, A., Pantofaru, C., Smart, W.D., Takayama, L.: Robots for humanity: using assistive robotics to empower people with disabilities. IEEE Robot. Autom. Mag. **20**(1), 30–39 (2013). https://doi.org/10.1109/MRA.2012.2229950
5. Robinson, H., MacDonald, B., Broadbent, E.: The role of healthcare robots for older people at home: a review. Int. J. Soc. Robot. **6**(4), 575–591 (2014). https://doi.org/10.1007/s12369-014-0242-2
6. Xu, T., Zhao, T., Cruz-Garza, J.G., Bhattacharjee, T., Kalantari, S.: Evaluating human-in-the-loop assistive feeding robots under different levels of autonomy with VR simulation and physiological sensors. In: International Conference on Social Robotics, pp. 314–327. Springer (2022)

7. Beer, J.M., Fisk, A.D., Rogers, W.A.: Toward a framework for levels of robot autonomy in human-robot interaction. J. Human-Robot Interact. **3**(2), 74 (2014). https://doi.org/10.5898/JHRI.3.2.Beer
8. Bhattacharjee, T., Gordon, E.K., Scalise, R., Cabrera, M.E., Caspi, A., Cakmak, M., Srinivasa, S.S.: Is more autonomy always better? Exploring preferences of users with mobility impairments in robot-assisted feeding. In: Proceedings of the 2020 ACM/IEEE International Conference on Human-Robot Interaction, pp. 181–190 (2020)
9. He, J.Z.Y., Erickson, Z., Brown, D.S., Raghunathan, A., Dragan, A.: Learning representations that enable generalization in assistive tasks. In: 6th Annual Conference on Robot Learning (2022). https://openreview.net/forum?id=b88HF4vd_ej
10. Silva, C., Vongkulbhisal, J., Marques, M., Costeira, J.P., Veloso, M.: Feedbot—a robotic arm for autonomous assisted feeding. In: Progress in Artificial Intelligence, pp. 486–497. Springer International Publishing (2017)
11. Vila, M., Canal, G., Alenyà, G.: Towards safety in physically assistive robots: eating assistance. In: Robots for Assisted Living Workshop at the IEEE/RSJ International Conference on Intelligent Robots and Systems (IROS) (2018)
12. Bhattacharjee, T., Lee, G., Song, H., Srinivasa, S.S.: Towards robotic feeding: role of haptics in fork-based food manipulation. IEEE Robot. Autom. Lett. **4**(2), 1485–1492 (2019)
13. Banerjee, R., Dean, S., Bhattacharjee, T.: To ask or not to ask: robot-assisted bite acquisition with human-in-the-loop contextual bandits. In: First Workshop on Out-of-Distribution Generalization in Robotics at CoRL 2023 (2023)
14. Gallenberger, D., Bhattacharjee, T., Kim, Y., Srinivasa, S.S.: Transfer depends on acquisition: analyzing manipulation strategies for robotic feeding. In: 2019 14th ACM/IEEE International Conference on Human-Robot Interaction (HRI), pp. 267–276. IEEE (2019)
15. Belkhale, S., Gordon, E.K., Chen, Y., Srinivasa, S., Bhattacharjee, T., Sadigh, D.: Balancing efficiency and comfort in robot-assisted bite transfer. In: 2022 International Conference on Robotics and Automation (ICRA), pp. 4757–4763. IEEE (2022)
16. Ondras, J., Anwar, A., Wu, T., Bu, F., Jung, M., Ortiz, J.J., Bhattacharjee, T.: Human-robot commensality: bite timing prediction for robot-assisted feeding in groups. In: Conference on Robot Learning, pp. 921–933. PMLR (2023)
17. Zhang, F., Cully, A., Demiris, Y.: Probabilistic real-time user posture tracking for personalized robot-assisted dressing. IEEE Transactions on Robotics pp. 1–16 (2019). https://doi.org/10.1109/TRO.2019.2904461
18. Gao, Y., Chang, H.J., Demiris, Y.: Iterative path optimisation for personalised dressing assistance using vision and force information. In: IEEE/RSJ International Conference on Intelligent Robots and Systems (IROS), pp. 4398–4403 (2016). https://doi.org/10.1109/IROS.2016.7759647
19. Gao, Y., Chang, H.J., Demiris, Y.: User modelling for personalised dressing assistance by humanoid robots. In: IEEE/RSJ International Conference on Intelligent Robots and Systems (IROS), pp. 1840–1845 (2015). https://doi.org/10.1109/IROS.2015.7353617
20. Kapusta, A., Erickson, Z., Clever, H.M., Yu, W., Liu, C.K., Turk, G., Kemp, C.C.: Personalized collaborative plans for robot-assisted dressing via optimization and simulation. Auton. Robot. (2019). https://doi.org/10.1007/s10514-019-09865-0
21. Erickson, Z., Clever, H.M., Turk, G., Liu, C.K., Kemp, C.C.: Deep haptic model predictive control for robot-assisted dressing. In: 2018 IEEE International Conference on Robotics and Automation (ICRA), pp. 1–8 (2018). https://doi.org/10.1109/ICRA.2018.8460656
22. Wang, Y., Sun, Z., Erickson, Z., Held, D.: One policy to dress them all: learning to dress people with diverse poses and garments. In: Robotics: Science and Systems (RSS) (2023)
23. Chance, G., Camilleri, A., Winstone, B., Caleb-Solly, P., Dogramadzi, S.: An assistive robot to support dressing—strategies for planning and error handling. In: 6th IEEE International Conference on Biomedical Robotics and Biomechatronics (BioRob), pp. 774–780 (2016). https://doi.org/10.1109/BIOROB.2016.7523721
24. Chance, G., Jevtić, A., Caleb-Solly, P., Dogramadzi, S.: A quantitative analysis of dressing dynamics for robotic dressing assistance. Front. Robot. AI **4** (2017)

25. Chance, G., Caleb-Solly, P., Jevtić, A., Dogramadzi, S.: What's up?—Resolving interaction ambiguity through non-visual cues for a robotic dressing assistant. In: IEEE International Symposium on Robot and Human Interactive Communication (RO-MAN), pp. 284–291 (2017)
26. Yamazaki, K., Oya, R., Nagahama, K., Okada, K., Inaba, M.: Bottom dressing by a life-sized humanoid robot provided failure detection and recovery functions. In: IEEE/SICE International Symposium on System Integration, pp. 564–570 (2014). https://doi.org/10.1109/SII.2014.7028101
27. Tamei, T., Matsubara, T., Rai, A., Shibata, T.: Reinforcement learning of clothing assistance with a dual-arm robot. In: 11th IEEE-RAS International Conference on Humanoid Robots (Humanoids), pp. 733–738. IEEE (2011)
28. Klee, S.D., Ferreira, B.Q., Silva, R., Costeira, J.P., Melo, F.S., Veloso, M.: Personalized assistance for dressing users. In: Social Robotics, pp. 359–369. Springer International Publishing (2015)
29. Canal, G., Alenyà, G., Torras, C.: Adapting robot task planning to user preferences: an assistive shoe dressing example. Auton. Robot. **43**(6), 1343–1356 (2019). https://doi.org/10.1007/s10514-018-9737-2
30. Canal, G., Pignat, E., Alenyà, G., Calinon, S., Torras, C.: Joining high-level symbolic planning with low-level motion primitives in adaptive HRI: application to dressing assistance. In: IEEE International Conference on Robotics and Automation (ICRA), pp. 3273–3278 (2018). https://doi.org/10.1109/ICRA.2018.8460606
31. Cakmak, M., Srinivasa, S.S., Lee, M.K., Forlizzi, J., Kiesler, S.: Human preferences for robot-human hand-over configurations. In: 2011 IEEE/RSJ International Conference on Intelligent Robots and Systems, pp. 1986–1993 (2011). https://doi.org/10.1109/IROS.2011.6094735
32. Wilde, N., Kulić, D., Smith, S.L.: Learning user preferences in robot motion planning through interaction. In: 2018 IEEE International Conference on Robotics and Automation (ICRA), pp. 619–626 (2018). https://doi.org/10.1109/ICRA.2018.8460586
33. Hayes, C.J., Marge, M., Stump, E., Bonial, C., Voss, C., Hill, S.G.: Towards learning user preferences for remote robot navigation. In: Proceedings of the RSS 2018 Workshop on Models and Representations for Human-Robot Communication (2018)
34. Rossi, S., Staffa, M., Bove, L., Capasso, R., Ercolano, G.: User's personality and activity influence on HRI comfortable distances. In: Social Robotics, pp. 167–177. Springer International Publishing (2017)
35. Tapus, A., Ţăpuş, C., Matarić, M.J.: User–robot personality matching and assistive robot behavior adaptation for post-stroke rehabilitation therapy. Intel. Serv. Robot. **1**(2), 169 (2008). https://doi.org/10.1007/s11370-008-0017-4
36. de Graaf, M.M., Ben Allouch, S., van Dijk, J.A.: Why would I use this in my home? A model of domestic social robot acceptance. Human-Comput. Interact. **34**(2), 115–173 (2019)
37. Smarr, C.A., Prakash, A., Beer, J.M., Mitzner, T.L., Kemp, C.C., Rogers, W.A.: Older adults' preferences for and acceptance of robot assistance for everyday living tasks. Proc. Hum. Factors Ergon. Soc. Annu. Meet. **56**(1), 153–157 (2012). https://doi.org/10.1177/1071181312561009
38. Deutsch, I., Erel, H., Paz, M., Hoffman, G., Zuckerman, O.: Home robotic devices for older adults: opportunities and concerns. Comput. Hum. Behav. **98**, 122–133 (2019). https://doi.org/10.1016/j.chb.2019.04.002
39. Canal, G., Alenyà, G., Torras, C.: Personalization framework for adaptive robotic feeding assistance. In: International Conference on Social Robotics (ICSR), pp. 22–31. Springer International Publishing (2016). https://doi.org/10.1007/978-3-319-47437-3_3
40. Canal, G.: Adapting Robot Behavior to User Preferences in Assistive Scenarios. Ph.D. thesis, Universitat Politècnica de Catalunya · BarcelonaTech (UPC) (2020)
41. Paraschos, A., Daniel, C., Peters, J., Neumann, G.: Probabilistic movement primitives. In: Advances in Neural Information Processing Systems (NIPS) (2013)
42. Colomé, A., Neumann, G., Peters, J., Torras, C.: Dimensionality reduction for probabilistic movement primitives. In: IEEE-RAS International Conference on Humanoid Robots, pp. 794–800 (2014). https://doi.org/10.1109/HUMANOIDS.2014.7041454

43. Canal, G., Alenyà, G., Torras, C.: A taxonomy of preferences for physically assistive robots. In: IEEE International Symposium on Robot and Human Interactive Communication (RO-MAN), pp. 292–297 (2017). https://doi.org/10.1109/ROMAN.2017.8172316
44. Fong, T., Nourbakhsh, I., Dautenhahn, K.: A survey of socially interactive robots. Robot. Auton. Syst. **42**(3–4), 143–166 (2003). https://doi.org/10.1016/S0921-8890(02)00372-X
45. John, O.P., Srivastava, S.: The big five trait taxonomy: history, measurement, and theoretical perspectives. Handb. Pers.: Theory Res. **2**(1999), 102–138 (1999)
46. Martínez, D., Alenya, G., Torras, C.: Planning robot manipulation to clean planar surfaces. Eng. Appl. Artif. Intell. **39**, 23–32 (2015)
47. Canal, G., Torras, C., Alenyà, G.: Are preferences useful for better assistance?: A physically assistive robotics user study. ACM Trans. Hum.-Robot Interact. (THRI) **10**(4), 1–19 (2021). https://doi.org/10.1145/3472208
48. Heerink, M., Kröse, B., Evers, V., Wielinga, B.: Assessing acceptance of assistive social agent technology by older adults: the Almere model. Int. J. Soc. Robot. **2**(4), 361–375 (2010). https://doi.org/10.1007/s12369-010-0068-5
49. Bartneck, C., Kulić, D., Croft, E., Zoghbi, S.: Measurement instruments for the anthropomorphism, animacy, likeability, perceived intelligence, and perceived safety of robots. Int. J. Soc. Robot. **1**, 71–81 (2009)

Future of Anticipatory Robotics in Manufacturing

Marta Lorenzini, Marta Lagomarsino, Costanza Messeri, Arash Ajoudani, and Andrea Maria Zanchettin

Abstract Collaborative robots, designed to work alongside humans in shared environments, are transforming manufacturing production lines, enabling partial automation and boosting productivity. But their potential extends beyond these benefits. By implementing *anticipatory strategies* to prevent workers' risky behaviours and reduce excessive workloads, collaborative robots can proactively enhance working conditions. For example, they can take on repetitive or strenuous tasks to reduce physical effort or adjust the pace of operations to help alleviate stress levels. Indeed, Musculoskeletal Disorders (MSDs) and Common Mental Disorders (CMDs), which arise from uncontrolled physical and cognitive loads, are major work-related health problems in developed countries, with incommensurate social and economic costs. Collaborative robots can curb the effect of such disorders and help in preserving workers' health. This Chapter addresses how Human Factors and Ergonomics (HFE) can be monitored to provide robots with relevant inputs about the human state, and how robots can be employed to improve HFE. Novel strategies are proposed to adapt online robot behaviour to the workers' psycho-physical state and enhance their well-being and productivity.

Keywords Human-robot collaboration · Human-centred robotics · Physical and cognitive load estimation and assessment

Marta Lorenzini, Marta Lagomarsino and Costanza Messeri are co-first authors.

M. Lagomarsino · C. Messeri · A. M. Zanchettin
Dipartimento di Elettronica, Informazione e Bioingegneria, Politecnico di Milano, Milan, Italy
e-mail: marta.lagomarsino@iit.it

C. Messeri
e-mail: costanza.messeri@polimi.it

A. M. Zanchettin
e-mail: andreamaria.zanchettin@polimi.it

M. Lorenzini (✉) · M. Lagomarsino · A. Ajoudani
Human-Robot Interfaces and Interaction Laboratory,Istituto Italiano di Tecnologia, Genoa, Italy
e-mail: marta.lorenzini@iit.it

A. Ajoudani
e-mail: arash.ajoudani@iit.it

G. Barresi and S. Wischniewski (eds.), *Ergonomics in Robotics: Advances and Innovations*, Springer Tracts in Advanced Robotics 162,
https://doi.org/10.1007/978-3-032-19374-2_18

1 Introduction

Collaborative robots are increasingly integrated into industrial environments, where they operate directly alongside human workers in shared workspaces, opening new possibilities to enhance manufacturing productivity and efficiency. While ensuring safety remains paramount in this context, it is also critical to avoid that collaborative robots adversely impact the cognitive or physical well-being of their human counterparts [1]. This can be achieved by designing legible, predictable, and smooth robot movements, which may resemble human gestures, in addition to carefully planning and managing the space sharing and interaction between them. However, collaborative robots can do more than peacefully coexist with humans and partially automate their labour. They hold the potential to proactively enhance the working conditions through the implementation of *anticipatory strategies* [1], which prevent humans from harmful behaviours and reduce the risk of excessive workload. Robots can support human workers in performing physical (e.g., relieving them from part of the effort while handling heavy loads), cognitive (e.g., adapting the work pace to avoid increasing workers' stress levels), and hazardous (e.g., handling chemical material) operations. Besides the clear benefits for workers' well-being, i.e., the reduction in incidence of work-related Musculoskeletal Disorders (MSDs) and Common Mental Disorders (CMDs), this positively impacts workplace productivity and efficiency.

These premises make evident the strong interconnection between collaborative robotics and Human Factors/Ergonomics (HFE), particularly through the development of robotic *anticipatory strategies*. This Chapter intends to analyse how collaborative robotics can support manufacturing operations and optimise some HFE-related Key Performance Indicators (KPIs) in industrial Human-Robot Collaboration (HRC). Among the various HRC frameworks, we focus on those where collaborative robots work in direct cooperation with humans to complete shared tasks, anticipating human needs and adapting their behaviour accordingly to boost productivity and worker well-being. As such, teleoperation systems and exoskeletons are not covered. Given the innovative nature of these solutions, their Technology Readiness Level (TRL) remains relatively low, at around 4–5.

The remainder of this Chapter is organised as follows. An overview of methods and approaches for monitoring HFE online to provide robots with relevant inputs as well as validate their efficacy are described in Sect. 2. Some novel strategies to improve HFE by means of robotic intervention are then presented in Sect. 3. Finally, limitations and challenges are discussed, highlighting promising research trends to protect and promote psycho-physical health at work.

2 Ergonomics Assessment for Collaborative Robotics

The ultimate goal of HFE is to improve human well-being and maximise system efficiency. To fulfil such purposes, the careful monitoring of human physical and cognitive activity is essential. This monitoring underpins the first two domains of HFE (i.e., physical and cognitive ergonomics) and is also relevant to macro-ergonomics (i.e., organisational ergonomics). Indeed, a workplace logistical and organisational aspects indirectly affect the worker's individual dimension. By assessing human physical and cognitive workload, any job's potential ergonomic risk factors can be identified and quantified to lay the foundation of a risk prevention and reduction program. This is where collaborative robots come into play. Advanced human-robot shared workstations may offer a promising solution to enhance worker well-being and awareness. To this end, information about the human workload should be provided to the robot.

In the last decades, a considerable number of methods and protocols have been developed to quantify and evaluate both physical and cognitive ergonomics. The latter have been thoroughly reviewed in [1] and can be categorised into the following three groups:

- *subjective judgements*: self-questionnaires from workers or narrative interviews from experts that either way reflect the point of view of the former;
- *systematic observations*: procedures (e.g., worksheet to be filled in, parameters to be collected) carried out by experts based on simple observations of the examined subjects, collected on-site at the workplace or from video recordings;
- *direct measurements*: procedures performed on-site at the workplace or during simulations in laboratories using sensor systems, whose collected data are either used directly or integrated into ad-hoc models for ergonomic assessment.

In the current industrial scenario, most tools adopted to evaluate workers' ergonomics rely on *subjective judgements* and *systematic observations*. The first, among which the most widely used is the National Aeronautics and Space Administration - task load index (NASAâŁ"TLX) [2], are non-invasive and straightforward to use, applicable to a broad range of occupational situations, low-cost and time-saving. Hence, they represent a suitable tool to evaluate HRC frameworks a posteriori but, due to their format, cannot be converted to inputs for the robot. On the other hand, *systematic observations* may provide, still without disrupting the workers, a more specific analysis of the possible indicators of human workload. Concerning the physical domain, researchers proposed observational techniques that analyse a particular activity or a single risk factor (carrying and lifting, pushing and pulling, low loads at high frequency, posture and movements [3]) but also comprehensive methods to obtain a unique measure of workers' physical ergonomics [4]. Concerning the cognitive domain, performance measures (e.g., task time, run time, reaction time, accuracy, and error rate) on either the first or secondary task have been considered [5]. However, observational-based methods in their original form must be employed as offline procedures after collecting recordings on the analysed activities, which does not provide the immediate results that the robot would require to react timely. Moreover, they generally involve complex protocols that can only be completed by qualified personnel. Lastly, *direct measurements* of the workers, collected using various sensors, ensure an accurate and throughout evaluation of the human workload. By integrating motion-capture, interaction forces and/or electromyogram (EMG) data with musculoskeletal models of the human body, it is possible to obtain even factual physical indicators of the human biomechanical load (e.g., joint torques and muscle forces) that cannot be directly estimated. Nevertheless, such models are rather computationally expensive, affecting their online capabilities. On the other hand, psycho-physiological measurements such as electroencephalogram (EEG), electrocardiogram (ECG), and electrodermal (EDA) signals provide objective and quantitative information about the cognitive workload [6], but require expensive and impractical systems that are highly sensitive to human movements and often hinder users' activities. Due to their limitations, direct measurement methods have been exclusively employed in laboratory settings.

The tools to assess human workload within HRC frameworks should provide online quantitative indicators of HFE that can feed the robot control system and allow its adaptation to human intentions and actions. Fit-for-workplace sensor technologies (i.e. wearable and lightweight or camera-based, long-life, easy to set up, calibrate and use) that do not hinder human movements should be adopted, enabling them to perform their activities without any constraint or discomfort and without affecting the interaction with the robot.

To fulfil such requirements, thus allowing the online assessment of HFE demanded by robots, a possibility is to build on the available methods mentioned above. On the one hand, the laborious offline procedures underlying observational-based methods can be automated by combining motion-capture data with advanced expert-reviewed algorithms. On the other hand, the computational costs and complexity of direct measurement techniques can be reduced by designing simplified yet not simplistic models and employing easy-to-use sensors. In the following, after a brief overview of some state-of-the-art attempts in this direction, we present two studies that propose innovative indicators for the online monitoring of physical (Sect. 2.1) and cognitive (Sect. 2.2) HFE.

2.1 *Online Physical Load Assessment*

To meet contemporary industry demands, several reduced-complexity approaches for assessing physical load have been proposed, focusing on the online monitoring of human kinematics and dynamics. For instance, [7] proposed multiple ergonomic indicators that quantify workers' physical demands when executing various manual activities, addressing both postures/movements and forces/torques. Still chasing a reduced-complexity approach but adding accuracy for more sophisticated body districts, [8] developed a flexible model of the human spine mechanics to assess compressive loads. Finally, [9] presented a stochastic methodology for the simultaneous floating-base estimation of the human whole-body kinematics and dynamics toward online ergonomics assessment. Indeed, ergonomic targets such as muscular effort and body posture can be linked to human kinematic and dynamic quantities such as joint torque and joint angles/velocities that, in turn, can be modelled by employing the principles of humanoid robotics. This approach significantly lowers computational costs ensuring online capabilities and can provide instant data on the human physical load. On this basis, in [10] an online monitoring framework is introduced, which can estimate the required physical load just as the workers are carrying out their activities. In the following, the specifics of this strategy will be outlined. The defined set of indexes accounts for multiple ergonomic risk factors (e.g., mechanical overloading of the body, awkward postures, repetition frequency, etc.) enabling the assessment of diverse manual activities typically conducted in the manufacturing sector. Besides kinematic variables (i.e., joint displacement from a neutral/ergonomic position, velocities, and accelerations), a strong emphasis is placed on the dynamic aspects of the tasks (i.e., forces or torques acting on and inside the body). Indeed, some indicators that were investigated in previous studies [11–13], such as overloading torque, fatigue and power, the human centre of mass (CoM) potential energy, and compressive forces, are integrated into the framework to establish a thorough method for the evaluation of physical load. Additionally, to address workers' individual demands, the estimation of the proposed set of indexes is based on a subject-specific and fast re-identifiable human model.

An experimental analysis was conducted on twelve healthy subjects to evaluate the framework. Participants performed three different tasks, which represent typical manufacturing activities associated with various physical risk factors. For each task, experimental conditions with different physical loads thus risk levels were considered. The outcome of the proposed set of indexes was then investigated to establish their link with the risk factor associated with the tasks, i.e. their capability to discriminate between the physical load levels. Statistical analysis was performed to verify the significance of the results. Concerning kinematics-related variables, the most beneficial index in assessing physical load among all the tasks was the joint displacement, given its capability to detect wherever the human current body configuration lays within specific sections of the range of motion that should be avoided (e.g., in proximity to the maximum limit). Concerning dynamics-related variables, an ergonomic index that better explains the required physical load was identified for each task. Specifically, the overloading joint torque index effectively accounted for the mechanical overburden of the body structures in handling heavy objects. Considering a drilling task in which all the components of the interaction forces at the hand/tool interface are relevant, the joint compressive forces index proved to be more predictive of the risk. Finally, the overloading fatigue index was more valuable for repetitive and prolonged but light activities such as painting, since it better addressed the inter-subjects variability and the risk cumulative effect. Finally, a comparison was made with the ergonomic risk scores resulting from the Ergonomic Assessment Worksheet (EAWS) [14], a well-recognised tool to evaluate workers' physical exposure to MSDs. Compared to the proposed set of indexes, EAWS only partially considers individual characteristics and lacks a full evaluation of the dynamic aspects. Hence, the integration of the two would be beneficial for a thorough analysis of physical risk factors. Given the low-cost and fast calculation, the comprehensive evaluation and the subject-specificity, the proposed set of indexes represents a straightforward input for robots, which can adapt its behaviour to the workers' needs and mitigate their physical workload.

2.2 Online Cognitive Load Assessment

As production becomes more complex, the cognitive resources demand and mental workload affect operator performance and workplace efficiency on the shop floor. To cope with the fast-growing global burden of work-related mental disorders, it is essential to design cognitive load assessment methods that can be easily applied and used in real industrial scenarios. Only a limited number of studies in the literature have explored online measurement of cognitive workload[1] [15]. In [16], three machine learning techniques on ECG and EDA data were compared to classify the cognitive ergonomic risk levels (i.e. low or high) potentially online. Similarly, the studies in [17, 18] quantified user stress by analysing Heart Rate Variability (HRV). However, these approaches rely on impractical sensors that are difficult to deploy in industrial settings, lack customisation options, and typically require the expertise of trained specialists for setup. In the remainder of this section, we present the online and quantitative framework proposed in [19], which aims to address these limitations by automating the acquisition of *systematic observations* of workers without hampering the natural flow of their work activities. This framework solely exploits the input images from an RGB-D or stereo camera to monitor how industrial tasks affect people's attention allocation, decision-making, unforeseen movements, and errors.

An overview of the proposed cognitive load assessment framework is shown in Fig. 1. Artificial intelligence algorithms such as head pose estimation and skeletal tracking are used to examine the worker's (i) attention level at different workstations (indicated with reference frame Σ_{Wi} with $i = 1, \dots M$ in the figure), (ii) stress indicators through the analysis of activity-related body language (i.e., self-touching occurrences and high activity periods), and (iii)information and part identification costs, i.e. the cognitive effort required to use assembly instructions and handle the correct tools and components to complete the task. In addition, prior-defined parameters that reflect characteristics of the specific assembly task and workstation layout (e.g., the number of assembly parts and the noise level) are included. By combining all these factors through personalisable weights λ, the final scores of *mental effort* and *stress level* are calculated. This approach enables the detection of excessive cognitive load in workers during their activity and provide intuitive alerts through a visual feedback interface.

The framework has been extended in [20] to consider hybrid industrial environments and appraise the impact of collaborative robots on operators' cognitive load in the production line. Based on gaze tracking and interpretation research, two metrics of human trust in the robotic assistant are defined. Specifically, the study investigates the percentage of attention the user pays to the robot during the collaborative assembly task and counts the number of glances and gazes over time towards the area dedicated to the assistant. Two graphical user interfaces are also proposed for

---

[1] Note that this chapter uses the term *cognitive load* to indicate a multi-dimensional construct representing the load that the work activities impose on the user's cognitive control [5]. *Mental effort* and *stress* refer to specific variables that can be measured.

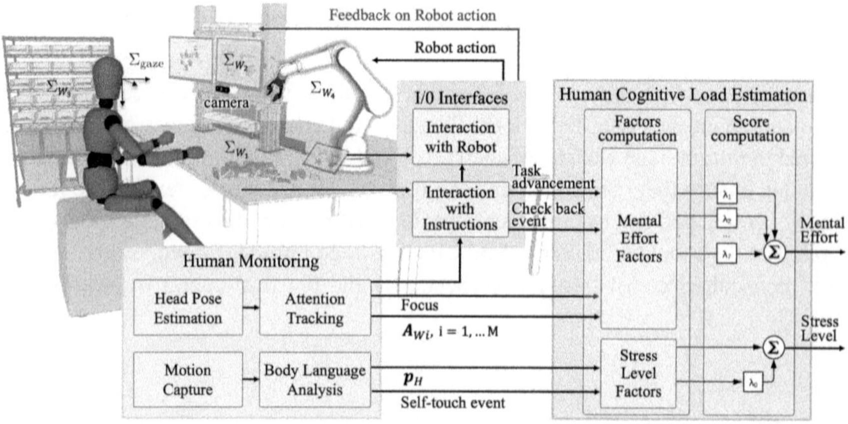

Fig. 1 Online camera-based assessment of cognitive load. Left: Conceptual representation of a workplace layout including a stereo or RGB-D camera and M workstations, e.g., the assembly (i.e. Σ_{W_1}), instruction monitor (i.e. Σ_{W_2}), storage area (i.e. Σ_{W_3}), and area dedicated to the robot (i.e. Σ_{W_4}). Right: Block diagram of the proposed framework. It examines the worker attention distribution $\mathbf{A}_{W_i}$ with $i = 1, ..M$ and upper-body kinematics $\mathbf{p}_H$ to detect intense activity periods and self-touching. Interaction with assembly instructions and the robot is analysed to track task advancement and errors

browsing the assembly instructions and handling the interaction with the assistant, and the associated keystroke dynamics is examined.

The performance of the developed framework was validated on twenty-nine subjects overall using state-of-the-art methods in cognitive science. ECG and EDA signals were indeed recorded during the task and processed offline, along with subjective questionnaires to ensure and support the validity of the results. In addition, the statistical analysis revealed that the degree of the robot transparency and feedback available to the worker has a noticeable impact on the development of cognitive workload. As expected, changes in experienced cognitive effort between human and automated assistance mainly depended on the operator's familiarity with the technology. Finally, the proposed trust metrics suggested that robotics experts tend to rely more on automated support. However, informed robot movements could put the naïve human co-worker at ease and foster trust.

The proposed *vision-based* scores offer a low-cost, comfortable (e.g., avoiding wearability constraints and recurrent removal of the device to charge it), and online solution to measure workers' cognitive workload. With this information, robots can assist and adapt to workers' needs and distress, ultimately enhancing user satisfaction, productivity and well-being.

3 Ergonomics Improvement with Collaborative Robotics

Online indicators of human physical and/or cognitive workload can be exploited to drive and adapt the robot behaviour. Planning and control strategies can be implemented to mitigate the workload when provoked by the task or the working environment and ensure that robot intervention promotes HFE improvement. In the following, we present and discuss some attempts in this direction, thus strategies that can be applied by the robot based on the online monitoring of human co-workers to enhance their physical (Sect. 3.1), cognitive (Sect. 3.2) and physical-cognitive (section 3.3) HFE.

3.1 Physical Ergonomics Improvement

Adopting the general model proposed in [21], two types of physical exposure can be distinguished: *external* and *internal exposure*. The movements and exerted forces that workers with their anthropometrics exploit to perform a certain activity, are the *external exposure*. *Internal exposure* instead refers to the corresponding joint moments and muscle forces within the human body. Since *external exposure* is more accessible, i.e., it can be directly observed or measured with motion-capture systems, most of the robot compensatory strategies that can be found in the literature are conceived for its mitigation [22, 23]. Yet, an exhaustive evaluation of physical HFE should not disregard *internal exposure*, which may be even more crucial for the design and adaptation of the robot behaviour. Physical *internal exposure* can be achieved with the *direct measurement* methods mentioned in Sect. 2. However, as discussed above, solutions to reduce their computational complexity should be found to achieve an online estimation. In this section, a hybrid approach that integrates a detailed biomechanical model of the human body with a machine-learning technique to online estimate the fatigue associated with the worker's movements is first presented. Then, the application of one of the online indexes described in 2.1 in a reconfigurable HRC framework is illustrated.

3.1.1 An Approach for Human-Robot Dynamic Task Allocation

The repetitive manual assembly of small and light components at high frequencies represents an important category of HRC tasks. For these types of tasks, the motions performed by humans strongly influence their muscular effort (i.e., body *internal exposure*) and fatigue. Indeed, each activity is characterised by a set of movements entailing the activation of different body regions, which in turn could contribute to the overall motion at different levels. In [24] a task allocation method is proposed that dynamically suggests the human and the robot the best task activities to mitigate the physical workload accumulated by the worker.

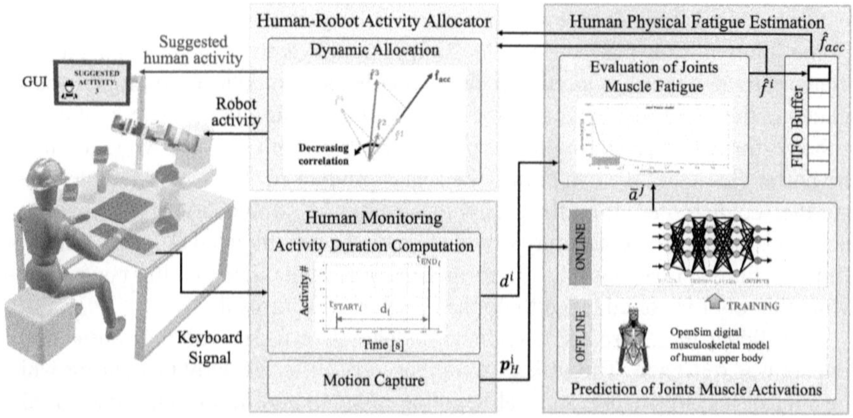

Fig. 2 Dynamic task allocation to mitigate human physical fatigue. The user presses the keyboard to signal the start of the new activity. The motion capture system tracks online the human upper-body motions $\boldsymbol{p}_H^i$ during the current activity i. The developed OpenSim model is used offline to collect a dataset encoding the mapping between human motions and muscle activation. The DNN, trained offline, predicts online the muscle activation associated with the worker's motions. By knowing the activity duration d^i and average joint activation, $\bar{\boldsymbol{a}}^j$, the Power Model is used to estimate the joint fatigue vector $\hat{\boldsymbol{f}}^i$. This is stored in a FIFO buffer which allows estimating the human accumulated fatigue, $\hat{f}_{acc}$. The dynamic allocation strategy determines the activity whose fatigue vector is less correlated with $\hat{f}_{acc}$. This activity is suggested to the human through the GUI, whereas the most tiring activity is commanded to the robot

The method, schematically illustrated in Fig. 2, works as follows. At the beginning of each new activity, the worker informs the system about which activity he/she is starting, as well as the start time. This input allows the motion capture system to track online and collect the positions of a set of meaningful points of the human upper body. A musculoskeletal model of the human upper body realised in the OpenSim software [25] is used to examine how much the muscles activate during the motions. This is evaluated offline through the OpenSim Static Optimisation (SO) technique. Indeed, the SO tool is not fast enough to be used online. A general dataset encoding the complex mapping between skeletal points' motions and muscle activations of selected body regions (from now on denoted as 'joints') is then collected offline. A Deep Neural Network (DNN) is trained offline to learn the mapping function of the described input-output dataset. This way, the DNN is exploited online to predict the joints' muscle activations during the tracked human movements.

Afterwards, the fatigue experienced by each joint for each activity is estimated as the product of the average joint activation and activity duration. This is consistent with the approach applied in EAWS (refer to Sect. 2.1), where the biomechanical load index is obtained by multiplying effort intensity and duration. By exploiting the joint-specific Power Model, a relative indicator is derived, expressing the actual fatigue experienced by the joint with respect to the maximum sustainable fatigue for that activity. Therefore, each task activity is described, in terms of physical ergonomics, by the so-called 'fatigue vector'. This expresses an estimate of the fatigue sustained by the joints during that activity. As soon as a new activity is accomplished, the related fatigue vector is stored in a buffer that is used to collect the fatigue vectors related to the last N worker's activities. The fatigue accumulated so far by the joints is then estimated by applying the Exponentially Weighted Moving Average filter to the buffer elements. By computing the scalar product between each fatigue vector and the vector representing the accumulated fatigue, a value describing how much executing the considered activity would increase the accumulated fatigue is obtained. Thus, the human is suggested to perform the activity that minimises this contribution, while the robot is commanded to do the one that maximises it.

To validate the approach, a realistic collaborative task involving fourteen volunteers was designed. The task was composed of different activities during which the subject had to perform manual handling of small and lightweight loads at high frequency. The statistical analysis of the experimental outcomes showed that the proposed method effectively minimises the muscle fatigue accumulated by the human during the task, while improving their production performance.

3.1.2 An Approach for Re-configurable Human-Robot Collaboration

Re-configurable HRC workstations have the potential to enable agile manufacturing by merging the dexterity, flexibility, and problem-solving capacity of humans with the strength, endurance and precision of robots. In this direction, a novel framework is proposed in [26] to enable online robot adaptation to both human intentions and dynamic factors. The method is schematically illustrated in Fig. 3.

The first driver establishes whether a certain work tool is intended to be manipulated, a worker is left or right-handed, or a worker moves within the workspace. This is achieved through human whole-body tracking by directly extracting kinematic configurations from the input images of a stereo vision camera. The algorithm extends the work proposed in [27]. The object recognition methodology proposed in [28] is adopted to recognise the work tool held by the human.

The second driver entails the overloading of the body joints, i.e. the loading effect of an external force/object on human joints, one of the online indexes mentioned in Sec.2.1. The latter can be computed by exploiting the displacement of the human whole-body centre of pressure (CoP) in the support plane (xy) under two conditions: the results from body dynamics themselves (inertial, Coriolis, centrifugal, and gravity forces) and exposure to external forces. This calculation eliminates body dynamics from the resulting equation thus reducing the number of subject-specific parameters to be identified [11]. Robot trajectories are adaptively modulated to facilitate the human partner's movements in achieving ergonomic configurations during co-manipulation, depending on the human position in the robot workspace, the operating hand (left or right), and the overloading condition. Indeed, through visual detection of the human body configuration, an ergonomic body configuration is estimated by online minimizing the overloading torques.

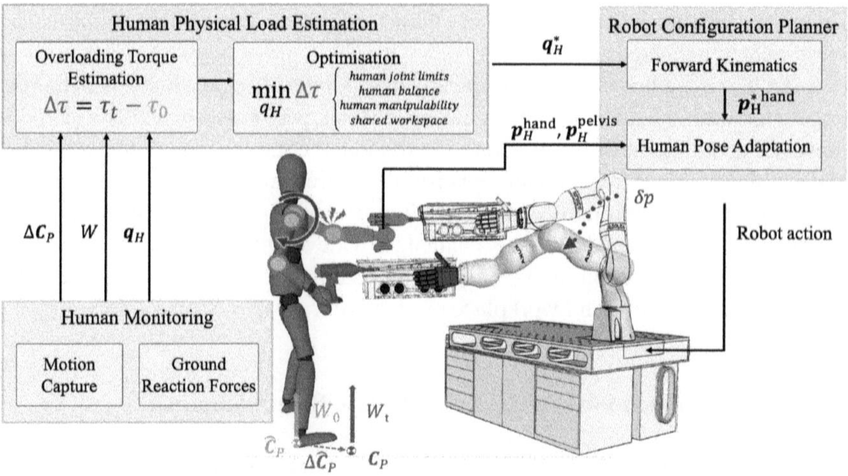

Fig. 3 Re-configurable HRC framework for improving worker ergonomics and productivity. In an offline phase, a model of the human whole-body centre of pressure (CoP) $\hat{\mathbf{C}}_P$ is identified and their body weight W_0 is measured. During the task, the human body configuration $\mathbf{q}_H$, ground reaction force W_t, and CoP $\mathbf{C}_P$ are monitored online. Based on the CoP difference $\Delta\mathbf{C}_P$, the overloading torques are estimated as the difference between the torque with (τ_t) or without (τ_0) the external load and then minimised to obtain an optimal configuration $\mathbf{q}_H^*$. The human optimal hand position $\mathbf{p}_H^*$ is found via forward kinematics and adapted given the currently used hand $\mathbf{p}_H^{hand}$ and pelvis $\mathbf{p}_H^{pelvis}$ position. The robot then guides the human in a more ergonomic condition

An experimental evaluation of this framework was conducted with ten subjects participating in drilling/polishing experiments. Moreover, it was evaluated during the KUKA Innovation Award at Hannover Messe, where live daily demonstrations for one entire week were run. The accuracy in tracking multiple subjects and tools and its robustness to the detection of multiple visitors at the booth were validated, demonstrating its high potential in realistic industrial environments.

3.2 *Cognitive Ergonomics Improvement*

The evidence that excessive cognitive demand at work prejudices operators' mental health and working performance has increased the interest in developing hybrid industrial environments to promote human cognitive ergonomics. However, measuring human cognitive load is challenging, as it results from a complex cascade of neuro-physiological processes that are not directly observable. In the literature, two main approaches for cognitive load estimation in HRC have emerged: *body sensors-based* methods, which involve *direct measurements* of physiological signals through sensing devices directly applied to the human body, and *vision-based* methods that use external cameras for automatic emotion recognition, expression synthesis, and stress-related behavioural indicators.

So far, these methods have been typically used for the offline assessment of human-robot interaction quality through extensive analysis performed by experts [29, 30]. However, enabling online evaluation of individuals' cognitive ergonomics in industrial settings presents a significant opportunity to prevent and mitigate work-related mental health risks through robotic *anticipatory strategies* [31]. This section discusses two attempts to address this challenge by developing HRC systems that can monitor cognitive load biomarkers online and adjust robot behaviour as needed to enhance cognitive HFE and workplace productivity.

3.2.1 An Approach for Human-Robot Online Role Adaptation

A peculiar factor of the HRC scenarios is the robot interaction role while cooperating with the human co-worker. In this framework, two role paradigms are feasible:

- Robot Leader and Human Follower: the robot takes the lead in the interaction driving the task progression, whereas the human complies with its decision;
- Robot Follower and Human Leader: the human is responsible for leading the overall interaction and the robot adapts to it.

In this regard, it is paramount to examine how a certain robot interaction role impacts the stress and productivity of the human operator. To shed light on these aspects, in [32] an experimental study was conducted involving thirty-three subjects, during which the human and robot were required to jointly build a progressively increasing tower of coloured cubes by taking on one of the two role paradigms. The participants were split into two similar and homogeneous groups for gender and age. Volunteers of the first group, which played as leaders during the experiment, were responsible for taking initiative all along the task. Conversely, volunteers belonging to the second group acted as followers, i.e. they had to respond consistently with the robot action. Independently from the role, the robot executed identical motion trajectories to reach each cube, and all the subjects were asked to perform the task as fast as possible to simulate the typical manufacturing scenario where throughput is crucial.

The effects of the two role paradigms were examined both in terms of psycho-physical stress and productivity. The stress of each subject, both at rest (baseline) and during the task, was evaluated based on the *direct measurement* of their ECG signal. The variation in the time intervals (also denoted as RR intervals) between two adjacent heartbeats of the ECG was measured, thus obtaining the HRV. By analysing the HRV in frequency and time domains, the commonly used and reliable short-term stress indicators were obtained, i.e. LF/HF and RMSSD, respectively. A high LF/HF ratio indicates increased physiological stress, while a high RMSSD is a marker of improved well-being and reduced stress. Thus, user stress was evaluated using both measures relative to the baseline. The production rate of each volunteer was estimated based on the user's cycle time, measured during the task. Experiments confirmed the expected correlation between stress levels and throughput, more evident when the robot acted as leader.

These results highlight two relevant aspects of HRC. First, the robot interaction role impacts both human stress and performance. Second, due to the complementary effects (in terms of stress and productivity) obtained when applying the two role paradigms, it is clear that the robot interaction role might be used as a control variable to effectively handle the existing trade-off between productivity maximisation and stress minimisation. In this direction, in [33] a novel closed-loop control strategy is described that endows the robot with the ability to simultaneously stimulate human performance and mitigate stress by adapting online its interaction role. The overall strategy is presented in Fig. 4. Game theory (GT) [34] provides the conceptual basis to model how the human-robot interaction affects the operator's performance and stress, as well as the stress-productivity trade-off. Based on the principles of GT, the interaction between robot and human is interpreted as a repeated non-cooperative normal-form game between two players i.e. the human (H) and the robot (R). Furthermore, both H and R are considered rational (self-interested) agents, meaning that each of them aims at maximising its own profit during the game. In particular, player R is associated with the goal of maximising work productivity and player H with the goal of minimising stress. Based on the above, the two players come to represent the two conflicting facets of whatever working environment: the importance of increasing throughput on the one hand and the need to mitigate the worker's stress on the other hand. The concept of game actions is then used to account for the possible attitudes of robot and human during cooperation, which might be: adapt (D) when

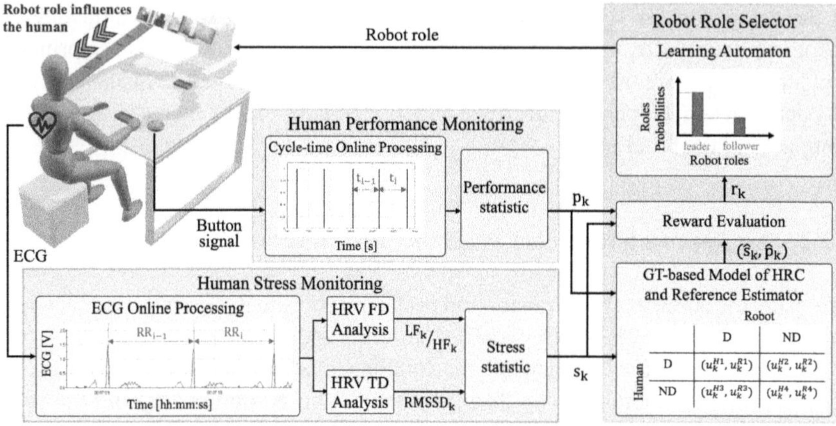

Fig. 4 Online robot role adaptation to optimise human performance and well-being. The human stress s_k and performance p_k, estimated from the user ECG and from the button signal, respectively, are monitored online. These are used to update the game theory model of the human-robot interaction and determine the stress-performance trade-off, $(\hat{s}_k, \hat{p}_k)$. A Learning Automaton (LA) evaluates the reward r_k by comparing the reference $(\hat{s}_k, \hat{p}_k)$ with the actual s_k and p_k. Based on that, the LA decides the best next role that the robot must take on during the interaction

the player comes to meet the goal of the other player, or not adapt (ND) in the opposite case. In the framework of GT, the combinations of two simultaneous players' actions are denoted as 'outcomes' of the game. Thus, the game outcomes describe the admissible states resulting from the human-robot interaction. In addition, according to GT, each game outcome is associated with a pair of values (one for each agent), called 'utility functions' or 'payoffs', each of which quantifies how much the player is satisfied with the considered outcome. In the proposed formulation, the payoffs of H and R are modelled as discrete time-varying functions that, at each step, take the values of the measured stress and performance for player H and R, respectively.

In view of the above, this model makes it possible to identify a particular game outcome, denoted as Nash Equilibrium (NE), that represents a local equilibrium state of the game and a 'no regrets' situation from the players' perspective. In fact, by definition, a game outcome is a NE if, once achieved, no player has an incentive to unilaterally modify its action to increase its utility. Therefore, the payoffs associated with the NE are regarded as the stress-performance reference trade-off against which the quality of the stress-performance compromises (produced by the HRC) can be evaluated. The optimisation is then based on this comparison. A Learning Automaton (LA) is the decision-making part of the overall control strategy. LA exploits the HRC-game model to evaluate how the role taken on by the robot has influenced the worker's state. Then, it penalises or rewards the commanded robot role proportionally to its negative or positive effect on the worker's stress and performance. Based on that, the LA determines the best role that the robot must take on at the next step of the interaction with the human to optimise the performance and well-being of the worker.

To validate the approach, a realistic collaborative assembly use case involving fifteen volunteers was designed. The statistical analysis applied to the outcomes of these experiments showed that the described strategy, by suitably adjusting the robot interaction role, is able to simultaneously maximise the worker's performance (evidenced by a significant reduction in cycle time) and mitigate his/her psycho-physical stress (shown by a decrease in LF/HF ratio and an increase in RMSSD compared to the baseline).

3.2.2 An Approach for Robot Proximity and Reactivity Optimisation

To ensure the successful acceptance and usage of robots in hybrid industrial environments, it is crucial to prioritise workers' comfort and positive perception of safety. In this context, a novel framework for simultaneously optimising human psycho-physical stress and productivity during collaboration by online adapting the robot trajectory is proposed in [35] and illustrated in Fig. 5. A waypoint-based trajectory planner to enrich the intuitiveness and flexibility of the framework was implemented where the user can quickly teach the robot a new task by manually positioning and orienting the joints in a series of desired poses (also called waypoints). Thus, a path travelling by the waypoints is created, and its smoothness and associated timing law are optimised according to the estimated operator's psycho-physical state. Smoothness is achieved by generating quintic B-spline trajectories in the joint space, which

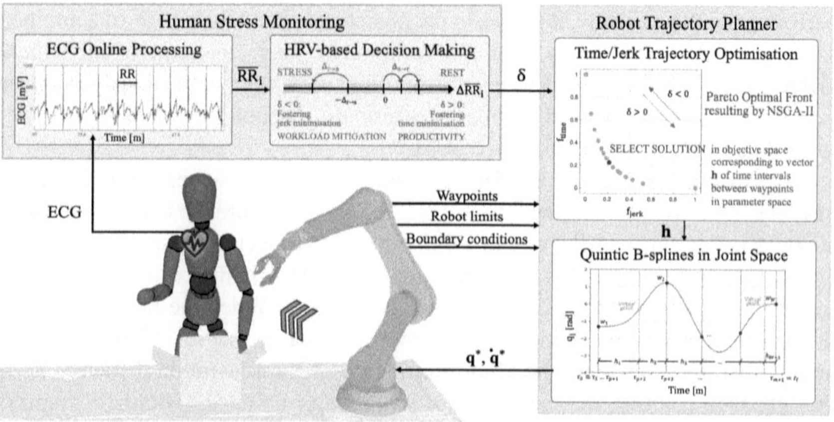

Fig. 5 Robot trajectory adaptation to solve the trade-off between system productivity and acceptable amount of human stress. Online HRV analysis in the time domain allows detecting intense human psycho-physical stress (e.g. a decrease of mean values of RR intervals within the i-th window of fixed time duration), thus preferring smoother and slower trajectories of robot joints $\mathbf{q}^*$. To do so, the algorithm moves in the optimal solutions of δ steps proportional to the variation $\Delta \overline{RR}_i$. In resting conditions, productivity is instead fostered by generating faster but less smooth trajectories

present continuous derivatives up to the fourth derivative, thus guaranteeing bounded values of joint velocity, acceleration and jerk along the path. Such smooth robotic manoeuvres have been shown to be more legible [36] and perceived as safer and more comfortable by human workers, which may be due to their similarity to natural arm movements [37]. Minimal jerk movements have the potential to enhance human comfort and cognitive ergonomics. However, minimising jerk alone leads to slow robot motions, affecting the productivity of the human-robot dyad. Therefore, a constrained bi-objective optimisation problem is proposed to minimise the integral of the squared jerk (i.e., maximise the smoothness) along with the total execution time of the robot joints' trajectories. The variables that should be optimised are the time intervals between consecutive waypoints, which are constrained to not result in kinematic quantities that the robot cannot achieve due to its internal limits. The result of the optimisation is a set of non-dominated solutions in the objective space called Pareto Optimal Front $\mathcal{F}$, namely solutions for which none of the objective functions can be improved in value without deteriorating the other objective. Each solution in $\mathcal{F}$ corresponds to a diverse interval vector **h** in the parameter space, thus to a trajectory with different features.

An online decision-making algorithm is implemented to select the most appropriate **h** according to the human psycho-physical stress. The latter is continuously monitored during the interaction by analysing the HRV. In particular, the variation over time of the intervals RR between consecutive heartbeats is monitored. When a change toward a higher stress level is detected, the algorithm chooses the **h** out of the set that boosts the smoothness of the desired robot trajectory at the cost of losing performance. Specifically, it moves in the objective space solutions $\mathcal{F}$ of a number of steps ($\delta < 0$) proportional to mean RR variation between consecutive time windows $\Delta\overline{RR}$. Since our final aim is to enhance productivity without perilously increasing demand, variations out of the range of stress are mapped into steps ($\delta > 0$) in $\mathcal{F}$ toward faster but less smooth trajectories.

In an extension of this work [38], the camera-based assessment of human awareness and mental effort, outlined in Sect. 2.2, is integrated into the loop to scale online adaptive safety zones while preserving smooth robot trajectories using B-spline properties. The radius of the physical safety zone around the human is adjusted based on the percentage of attention directed toward the robot. When the human is focused on the robot, this safety zone shrinks, allowing closer interaction. Conversely, when attention shifts elsewhere, the algorithm increases the separation distance. To further enhance comfort, the mental effort score was employed to define a cognitive-grounded safety zone, dictating the subject-specific optimal interaction distance the robot should maintain. Based on this data, the system enables online path alterations to avoid entering the cognitive safety zone, enhancing the perception of safety, and trigger a stop command if it violates the physical safety limits, preventing collisions.

The proposed strategy for optimising proximity and reactivity of robot motion was validated in a collaborative box-filling task designed to induce a mild cognitive load. Overall, thirty-six participants underwent three conditions: (a) the robot followed minimum-time trajectories, (b) it followed minimum-jerk trajectories, and (c) the robot's timing was adjusted online based on ECG signals from a wearable chest strap. Results showed that fast trajectories boosted productivity but negatively impacted human well-being, while smooth motions reduced errors but lowered production rates. Our tuning algorithm guarantees excellent levels of average productivity without compromising human performance in terms of errors committed or inducing stress, but keeping the worker engaged in the task.

3.3 Holistic Approaches for Ergonomic Human-Robot Collaboration

In the previous sections, we have seen how the robot could contribute to mitigating the operator's physical or cognitive workload. However, to achieve an effective and seamless human-robot interaction, it is crucial to design holistic approaches where robots simultaneously take into account both the physical and cognitive demands of workers and adapt their behaviour to match their preferences and skills. This broader view stimulates discussion in heterogeneous research communities and promotes collaboration between roboticists, biomechanists, and psychologists.

A first attempt following this line was made in [39], where the cognitive science principle that "humans act coefficiently as a group" (i.e., simultaneously maximising the benefits and minimising the effort of all agents involved) was transferred into human-robot cooperative settings (see Fig. 6). Human-robot co-efficiency was

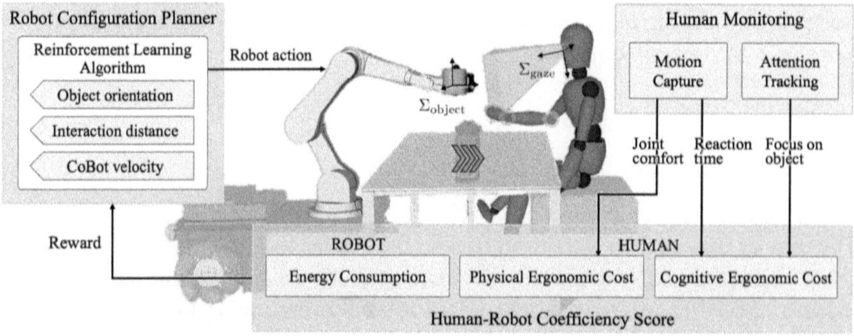

Fig. 6 Transferring human paradigm of acting co-efficiently in human-robot handovers. The framework identifies implicit indicators of human physical and socio-cognitive comfort and computes the robot energy consumption. That information is used as reward in a reinforcement learning problem to adapt and learn online the preferred combination of robot interaction parameters (i.e., the object orientation, interaction distance and robot velocity) for each specific human-robot dyad

modelled by online capturing implicit human comfort and discomfort body signals as well as the robot energy consumption. More specifically, taking inspiration from a study analysing human body movements (e.g., eye contact, distance, synchronisation to the robot, and touches) in human-robot interaction and the correlations with subjective evaluations [40], a cognitive ergonomic cost was formulated and should be maximised during the interaction. The latter positively correlates with the average attention an individual gives to the task and negatively correlates with the reaction time. Besides, a physical ergonomic cost was defined based on the claim of "pen-and-paper" methods (i.e., *systematic observations* performed by experts and stored on ad-hoc sheets to record workplace exposure) that a human is subject to physical exertion when one of their joints is close to the range of motion extrema. This allows for estimating the aggregate expenses of the dyad during the interaction and defining a human-robot co-efficiency score that is used as reward in a Reinforcement Learning (RL) problem to adapt the robot behaviour and learn to maximise such co-efficiency. Particularly, a set of interaction parameters (i.e., the object's orientation on the horizontal plane, the interaction distance, and the robot velocity) are considered, and a multi-armed bandit problem is defined in such a way that each arm represents a robot action with a diverse combination of parameters' values. At each iteration t, among the actions $\mathcal{A} \in \mathbb{R}^K$, the robot (as the agent) selects an action (i.e., arm $a \in \mathcal{A}$) to execute and obtains a reward $R_t(a)$. Then, the robot updates its internal knowledge about the expected reward by exploiting the well-known Upper Confidence Bound (UCB)1-tuned algorithm.

The results of the experimental campaign involving twelve naïve subjects highlighted the proposed learning strategy's capacity to online improve cognitive and physical HFE and reduce robot expenses. This means that the framework succeeded in embedding the concept of co-efficiency inspired by theories of human conjoint actions into human-robot interactions. Additionally, acting coefficiently, the robot was able to meet the individual interaction preferences of most of the subjects who partook in the experiments.

3.4 Conclusions

The aim of this chapter was to introduce the reader to HFE in industrial human-robot interaction. After describing the main methods proposed in the last decades to evaluate the HFE, innovative indicators for monitoring online the human physical and cognitive workloads were introduced. Then, several *anticipatory strategies* based on the online worker monitoring were presented, which allow adapting online robot features to enhance the physical and cognitive ergonomics and optimise the related KPIs.

Despite the growing interest, the use of collaborative robots to improve ergonomics is still an emergent research topic. Indeed, many challenges are yet to be addressed to achieve the full ergonomic potential of the collaborative workstation. For instance, some efforts should be invested in developing more sophisticated technologies that ensure accurate, robust, and minimally invasive human monitoring during cooperation. This technological progress, in turn, will open the door to the implementation of more advanced methods that, by leveraging highly sensorised and smart collaborative workstations, enable estimating human psycho-physical state. In this direction, multi-sensor data fusion would further improve the robustness of the ergonomic assessment. Lastly, research in collaborative robotics should dwell on combining the recent studies on ergonomics with the ones related to safety, thus realising safe and human well-being-oriented workstations. The ultimate goal is to design holistic approaches for ergonomic HRC, which thoroughly and simultaneously take into account physical-cognitive HFE, as well as worker satisfaction, safety, and performance.

Acknowledgements The works presented in this chapter were developed within the European Research Council Starting Grant *Ergo-Lean* (Grant Agreement No. 850932), the European Union's Horizon 2020 research and innovation programme *SOPHIA* (Grant Agreement No. 871237), Progetto PRIN 2017 *TIGHT* (prot. 2017SB48FP).

References

1. Lorenzini, M., Lagomarsino, M., Fortini, L., Gholami, S., Ajoudani, A.: Ergonomic human-robot collaboration in industry: a review. Front. Robot. AI **9**, 813907 (2023)
2. Hart, S.G.: Nasa-task load index (NASA-TLX) 20 years later. In: Proceedings of Human Factors and Ergonomics Society (HFES) Annual Meeting, vol. 50, pp. 904–908. SAGE Publications (2006)
3. David, G.C.: Ergonomic methods for assessing exposure to risk factors for work-related musculoskeletal disorders. Occup. Med. **55**(3), 190–199 (2005)

4. Schaub, K.G., Mühlstedt, J., Illmann, B., Bauer, S., Fritzsche, L., Wagner, T., Bullinger-Hoffmann, A.C., Bruder, R.: Ergonomic assessment of automotive assembly tasks with digital human modelling and the 'ergonomics assessment worksheet'(EAWS). Int. J. Hum. Factors Model. Simul. **3**(3–4), 398–426 (2012)
5. Paas, F., Tuovinen, J., Tabbers, H., Van Gerven, P.: Cognitive load measurement as a means to advance cognitive load theory. Educ. Psychol. **38**(1), 63–71 (2003)
6. Haapalainen, E., Kim, S., Forlizzi, J.F., Dey, A.K.: Psycho-physiological measures for assessing cognitive load. In: Proceedings of International Conference on Ubiquitous Computing (UC), pp. 301–310. IEEE (2010)
7. Maurice, P., Padois, V., Measson, Y., Bidaud, P.: Human-oriented design of collaborative robots. Int. J. Ind. Ergon. **57**, 88–102 (2017)
8. Ventura, L., Lorenzini, M., Kim, W., Ajoudani, A.: A flexible robotics-inspired computational model of compressive loading on the human spine. IEEE Robot. Autom. Lett. **6**(4), 8229–8236 (2021)
9. Latella, C., Traversaro, S., Ferigo, D., Tirupachuri, Y., Rapetti, L., Andrade Chavez, F.J., Nori, F., Pucci, D.: Simultaneous floating-base estimation of human kinematics and joint torques. Sensors **9**(12), 2794 (2019)
10. Lorenzini, M., Kim, W., Ajoudani, A.: An online multi-index approach to human ergonomics assessment in the workplace. IEEE Trans. Hum. Mach. Syst. **52**, 812–823 (2022)
11. Kim, W., Lee, J., Tsagarakis, N., Ajoudani, A.: A real-time and reduced-complexity approach to the detection and monitoring of static joint overloading in humans. In: Proceedings of International Conference on Rehabilitation Robotics (ICORR), pp. 828–834. IEEE (2017)
12. Lorenzini, M., Kim, W., De Momi, E., Ajoudani, A.: A new overloading fatigue model for ergonomic risk assessment with application to human-robot collaboration. In: Proceedings of International Conference on Robotics and Automation (ICRA), pp. 1962–1968. IEEE (2019)
13. Fortini, L., Lorenzini, M., Kim, W., De Momi, E., Ajoudani, A.: A real-time tool for human ergonomics assessment based on joint compressive forces. In: Proceedings of International Conference on Robot and Human Interactive Communication (RO-MAN), pp. 1164–1170. IEEE (2020)
14. Schaub, K., Caragnano, G., Britzke, B., Bruder, R.: The European assembly worksheet. Theor. Issues Ergon. Sci. **14**, 616–639 (2013)
15. Kulić, D., Croft, E.A.: Affective state estimation for human-robot interaction. IEEE Trans. Rob. **23**(5), 991–1000 (2007)
16. Rajavenkatanarayanan, A., Nambiappan, H.R., Kyrarini, M., Makedon, F.: Towards a real-time cognitive load assessment system for industrial human-robot cooperation. In: Proceedings of International Conference on Robot and Human Interactive Communication (RO-MAN), pp. 698–705. IEEE, Naples Italy (2020)
17. Landi, C.T., Villani, V., Ferraguti, F., Sabattini, L., Secchi, C., Fantuzzi, C.: Relieving operators' workload: towards affective robotics in industrial scenarios. Mechatronics **54**(10), 144–154 (2018)
18. Shah, J.K., Yadav, A., Hopko, S.K., Mehta, R.K., Pagilla, P.R.: Robot adaptation under operator cognitive fatigue using reinforcement learning. In: Proceedings of International Conference on Robot and Human Interactive Communication (RO-MAN), pp. 1467–1474. IEEE (2023)
19. Lagomarsino, M., Lorenzini, M., Momi, E.D., Ajoudani, A.: An online framework for cognitive load assessment in industrial tasks. Robot. Comput. Integr. Manuf. **78**, 102380 (2022)
20. Lagomarsino, M., Lorenzini, M., Balatti, P., Momi, E.D., Ajoudani, A.: Pick the right co-worker: online assessment of cognitive ergonomics in human-robot collaborative assembly. IEEE Trans. Cogn. Dev. Syst. **15**(4), 1928–1937 (2023)
21. Westgaard, R., Winkel, J.: Guidelines for occupational musculoskeletal load as a basis for intervention: a critical review. Appl. Ergon. **27**(2), 79–88 (1996)
22. Busch, B., Toussaint, M., Lopes, M.: Planning ergonomic sequences of actions in human-robot interaction. In: Proceedings of International Conference on Robotics and Automation (ICRA), pp. 1916–1923. IEEE, Brisbane, Australia (2018)

23. Zanchettin, A.M., Lotano, E., Rocco, P.: Collaborative robot assistant for the ergonomic manipulation of cumbersome objects. In: Proceedings of International Conference on Intelligent Robots and Systems (IROS), pp. 6729–6734. IEEE, Macau, China (2019)
24. Messeri, C., Bicchi, A., Zanchettin, A.M., Rocco, P.: A dynamic task allocation strategy to mitigate the human physical fatigue in collaborative robotics. IEEE Robot. Autom. Lett. **7**, 2178–2185 (2022)
25. Delp, S.L., Anderson, F.C., Arnold, A.S., Loan, P., Habib, A., John, C.T., Guendelman, E., Thelen, D.G.: Opensim: open-source software to create and analyze dynamic simulations of movement. IEEE Trans. Biomed. Eng. **54**(11), 1940–1950 (2007)
26. Kim, W., Lorenzini, M., Balatti, P., Nguyen, P.D., Pattacini, U., Tikhanoff, V., Peternel, L., Fantacci, C., Natale, L., Metta, G., Ajoudani, A.: Adaptable workstations for human-robot collaboration: a reconfigurable framework for improving worker ergonomics and productivity. IEEE Robot. Autom. Mag. **26**, 14–26 (2019)
27. Nguyen, D.H. P., Hoffmann, M., Roncone, A., Pattacini, U., Metta, G.: Compact real-time avoidance on a humanoid robot for human-robot interaction. In: Proceedings of International Conference on Human-Robot Interaction (HRI), pp. 416–424. IEEE (2018)
28. Pasquale, G., Ciliberto, C., Odone, F., Rosasco, L., Natale, L.: Teaching ICUB to recognize objects using deep convolutional neural networks. In: Machine Learning for Interactive Systems, pp. 21–25. (2015) PMLR
29. Hu, Y., Abe, N., Benallegue, M., Yamanobe, N., Venture, G., Yoshida, E.: Toward active physical human-robot interaction: quantifying the human state during interactions. IEEE Trans. Hum. Mach. Syst. **52**(3), 367–378 (2022)
30. Hopko, S., Khurana, R., Mehta, R., Pagilla, P.: Effect of cognitive fatigue, operator sex, and robot assistance on task performance metrics, workload, and situation awareness in human-robot collaboration. IEEE Robot. Autom. Lett. **6**, 3049–3056 (2021)
31. Lagomarsino, M.: Innovative Models for Cognitive Load Assessment and Mitigation through Personalised Robot Intervention, p. 12. (2023). PhD thesis
32. Messeri, C., Zanchettin, A.M., Rocco, P., Gianotti, E., Chirico, A., Magoni, S., Gaggioli, A.: On the effects of leader-follower roles in dyadic human-robot synchronisation. IEEE Trans. Cogn. Dev. Syst. **1**, 1 (2020)
33. Messeri, C., Masotti, G., Zanchettin, A.M., Rocco, P.: Human-robot collaboration: optimizing stress and productivity based on game theory. IEEE Robot. Autom. Lett. **6**, 8061–8068 (2021)
34. Tadelis, S.: Game theory: an introduction, Princeton University Press (2013)
35. Lagomarsino, M., Lorenzini, M., De Momi, E., Ajoudani, A.: Robot trajectory adaptation to optimise the trade-off between human cognitive ergonomics and workplace productivity in collaborative tasks. In: Proceedings of International Conference on Intelligent Robots and Systems (IROS), IEEE, Kyoto, Japan (2022)
36. Dragan, A., Lee, K., Srinivasa, S.: Legibility and predictability of robot motion. In: Proceedings of International Conference on Human-Robot Interaction (HRI), pp. 301–308. IEEE (2013)
37. Flash, T., Hogan, N.: The coordination of arm movements: an experimentally confirmed mathematical model. J. Neurosci, 1688–1703 (1985)
38. Lagomarsino, M., Lorenzini, M., De Momi, E., Ajoudani, A.: PRO-MIND: Proximity and reactivity optimization of robot motion to tune safety limits, human stress, and productivity in industrial settings. IEEE Trans. Robot. **41**, 2067–2085 (2025)
39. Lagomarsino, M., Lorenzini, M., Constable, M.D., De Momi, E., Becchio, C., Ajoudani, A.: Maximising co-efficiency of human-robot handovers through reinforcement learning. IEEE Robot. Autom. Lett, 1–8 (2023)
40. Kanda, T., Ishiguro, H., Imai, M., Ono, T.: Body movement analysis of human-robot interaction. In: Proceedings of International Joint Conference on Artificial Intelligence (IJCAI), (2003)

Gestures for Human-Robot Interaction

Alessandro Carfì and Fulvio Mastrogiovanni

Abstract Human interaction involves the interplay of various communication media, including body movements, voice, gaze, and even written signs. In such an interplay, gestures and, more generally, body motions represent a fundamental communication channel. Due to the significance of gesture-based communication in human-human interaction, intelligent and embodied agents are expected to utilize its implications to interact with humans effectively. An embodied agent, like a robot, can reason and produce gestures as a bidirectional communication channel between itself and the human it is interacting with. To achieve full bidirectional, gesture-based communication, the same embodied agent should be able to both recognize and perform communicative and informative gestures. This chapter describes the different challenges characterizing gesture based interaction models between humans and robots.

Keywords Gesture-based communication · Human-robot interaction · Gesture recognition · Natural interfaces · Human factors

1 Introduction

Humans use motion to navigate and interact with their environment, which they do in different and varied ways depending on context and personal characteristics. Therefore, studying human motion has practical applications in sports science, physical therapy, and robotics [1, 2]. For example, a clear understanding of motion can

A. Carfì (✉) · F. Mastrogiovanni
Department of Informatics, Bioengineering, Robotics and Systems Engineering, University of Genoa, Genoa, Italy
e-mail: alessandro.carfi@dibris.unige.it

F. Mastrogiovanni
e-mail: fulvio.mastrogiovanni@unige.it

G. Barresi and S. Wischniewski (eds.), *Ergonomics in Robotics: Advances and Innovations*, Springer Tracts in Advanced Robotics 162,
https://doi.org/10.1007/978-3-032-19374-2_19

help optimize performance and minimize injury risks in the workplace [3]. However, human motion studies also have applications in psychology, biology, and neuroscience [4]. In these fields, human motion is used to study the basic working principles of our brain by observing how we interact with the world.

Interacting with the environment also involves engaging with other intelligent beings, such as humans and animals, as well as machines. In this context, human motion is important for its physical impact on the environment and for the information it conveys to observers. However, not all motions are equally communicative. The informational content of human motions exists along a spectrum. The more information a motion conveys, the greater its communicative impact, offering observers deeper insights into the individual's intentions, emotions, and overall state. We can define the communicative impact of a motion as its potential to transmit information to other intelligent beings.

Therefore, human motions can be classified in a continuum, ranging from low to high communicative impact. Daily actions, such as brushing teeth, can be recognized by an observer, who can then annotate that the action was performed by the individual. However, not much information can be extracted from this observation. On the other end of the spectrum, sign languages allow trained agents to associate the detection of a specific set of motions with an increasingly high amount of information, with little or no ambiguity. In between, we find motions that convey a moderate set of information. For example, someone carrying an object slowly may imply that it is heavy or that its handling requires care [5]. This information can be inferred by an observer, but it may require additional context to remove any ambiguity.

When studying the impact of human movement on communication, it is crucial to determine whether the conveyed information is explicit or implicit. Although humans can intentionally use their motion to communicate, observers often extract information even when the actor's primary intent is not communicative. In other words, intentionality defines whether the motion is performed with the primary goal of communicating with others or if communication is simply a byproduct of the action. The literature refers to this distinction as the implicit or explicit communication dichotomy. However, we can argue that this dichotomy oversimplifies the biological mechanisms that drive human behaviours. We know from the literature that humans often modify their motion to better communicate with others when observed. This mechanism, known as signalling [6], suggests that it is more appropriate to consider the intentionality associated with the communicative means of a motion as a continuum. This continuum allows us to distinguish all shades of intentionality, from unintentional communication acts to intentional ones. Notice that when we use the term intentionality, we refer to the communication impact of the motion, not to the motion itself, which, in this chapter, we consider always intentional. To describe the two extremes of the continuum, we use the terms explicit and implicit. Figure 1 uses a 2D representation to evaluate motions based on their communicative impact, horizontal axis, and intentionality of the communication, vertical axis.

Upon reconsidering the previously introduced examples, it becomes apparent that observing a person brushing their teeth yields little information, and any communication that occurs is unintentional. At the other end of the spectrum, sign language

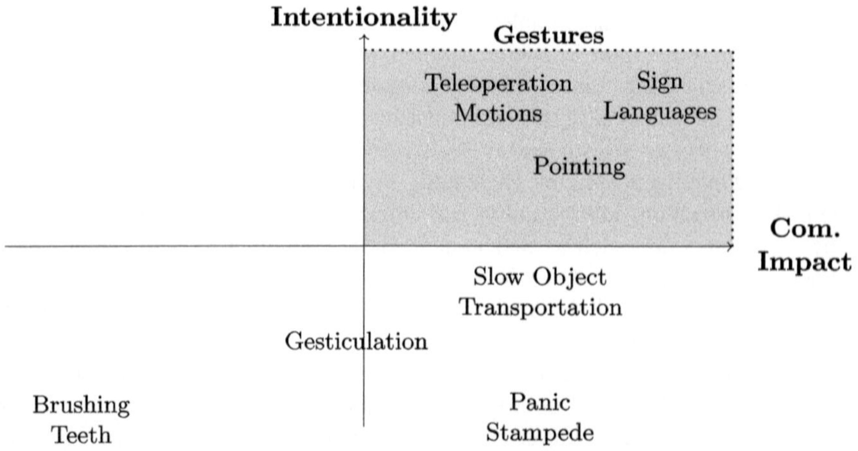

Fig. 1 The horizontal axis represents the communicative impact of human motions, while the vertical axis represents the intentionality of the communication. The green overlay highlights the region where gestures are located

facilitates rich and intentional communication. This representation enables us to visually capture the communication impact and the intentionality of any human motions. For example, observing a crowd running in a disordered manner might suggest that the people are panicked and attempting to escape from something. In this case, the communicative impact is high but unintentional. Conversely, the motions performed by an operator to teleoperate a system are fully intentional, but their communicative impact is constrained by the application. It should be noted that the information presented in the image is a subjective evaluation made by the authors with the intention of characterizing human motions in terms of their communication impact and intentionality. Precise measurements would need to be taken to make objective evaluations, but this is beyond the scope of the chapter.

Since human motion can have communicative implications, different research fields have explored their analysis to understand humans or to establish a more natural interaction with machines. Implicit human motions can be perceived and studied to extrapolate information about humans and the environment, while explicit communicative motions from the user can help an intelligent system establish better interaction with humans. While the analysis of motions with unintentional communicative implications can be valuable in ergonomics (e.g., evaluating physical fatigue during human-robot collaborative tasks), this chapter focuses on the role of intentional communicative human motions—referred to by the literature as *gestures*. These gestures, highlighted in green in Fig. 1, play a crucial role in interactions with intelligent systems. We have defined gestures based on the performer's perspective, considering their communication richness and intent. However, communication is a one-to-many process, and gestural communication does not make an exception. A human may perform a gesture with high communicative impact. However, if the observer is not present or is unable to process the information, the message conveyed by the gesture is lost. For example, if a person uses sign language in front of someone who cannot interpret it, communication will fail. Therefore, this chapter will describe the challenges that an intelligent system should solve to interpret human gestures and the potential of this communication modality.

Gestures are often investigated as an alternative communication medium between humans and intelligent systems. Depending on the scenario, gesture-based solutions can make interactions more intuitive and efficient or enable communication where other modalities fail, such as underwater. These communication modalities have attracted interest in the human-robot interaction field, where they have been adopted to provide intuitive interaction tools, leveraging the naturalness with which humans use gestures to communicate. In the remainder of this chapter, we will discuss the role of gestural communication in facilitating human-robot interaction, focusing on designing pipelines for robot systems capable of interpreting gestures and the ergonomic implications of this communication modality. We will only cover human-to-robot communication, as other types of communication are outside the scope of this chapter. In order to effectively achieve our goals, we aim to provide a comprehensive account of human gestures. This will involve presenting a precise definition of what constitutes a gesture and carefully classifying and categorizing different types of gestures based on their form and function. Additionally, it will be important to accurately describe the challenges and limitations that currently exist in the field of human-robot interaction when it comes to the use of gestures. By providing a thorough analysis of these key concepts, we hope to develop a more robust and nuanced understanding of how gestures can be effectively utilized in this context and ultimately pave the way for more successful and intuitive human-robot interaction.

2 Gestures Definition

In the previous section, we introduced gestures as human motions performed to convey information to an observer [7]. On top of this definition, we would like to provide a more formal one, helpful for the rest of the chapter. The human skeleton is composed of a series of joints connecting, with multiple degrees of freedom, all its links. Although human link motions are obtained through muscle activation, intrinsically linear motions, the resulting motion at the joint level is an angular displacement. Therefore, to achieve a precise but simple description of human motion, we can consider the human joint state as displayed in Fig. 2.

$$\boldsymbol{q} = \{q_1 \dots q_i \dots q_n\} \tag{1}$$

Here, i represents a general joint angle between consecutive skeleton links, while q_i is an angle representing its state. Notice that in human kinematics, often, between two links there are multiple *virtual* joints, as in the case of the spherical joint of the shoulder. To describe motions, the joint state should be defined over time as $\boldsymbol{q}(t)$,

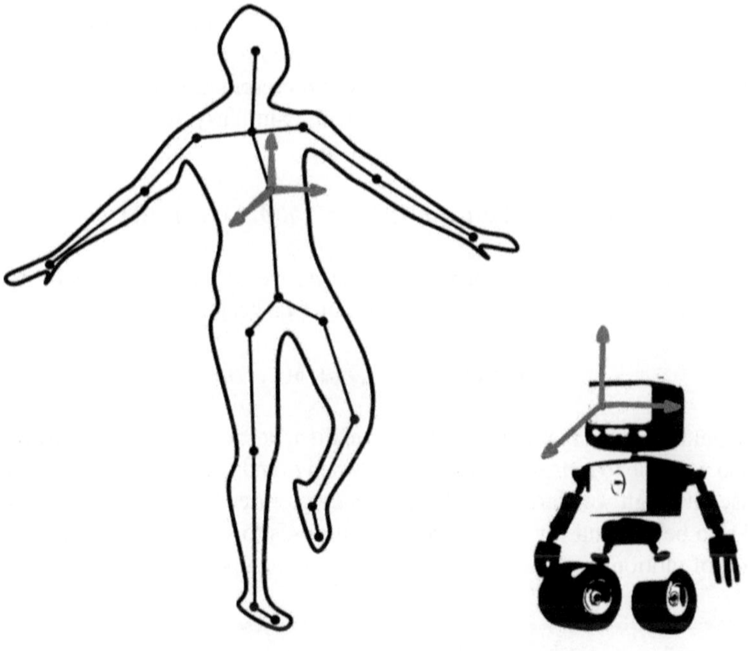

Fig. 2 Human skeleton with joints and links as perceived by the robot. In red, green and blue the axis x, y and z of the reference frame attached to the human and the robot

leading us to define also the angular velocities ($\dot{\boldsymbol{q}}(t)$) and accelerations ($\ddot{\boldsymbol{q}}(t)$) of each joint. These elements describing the human kinematics and its motions can be used to define a trajectory ($\boldsymbol{\tau}$) in an n-dimensional space providing a more compact representation for the human motion at the joint level.

$$\boldsymbol{\tau}(t_s, t_e) = \left\{\boldsymbol{q}(t_s) \dots \boldsymbol{q}(t_e), \dot{\boldsymbol{q}}(t_s) \dots \dot{\boldsymbol{q}}(t_e), \ddot{\boldsymbol{q}}(t_s) \dots \ddot{\boldsymbol{q}}(t_e)\right\} \tag{2}$$

Here, t_s represents the start time of the trajectory, and t_e represents the end time. This formalism fully describes the motion of human kinematics over time. However, humans also move in space and gestures' significance, as we will later describe, could change based on the location where they are performed. To account for this, we define the pose in the space of the human.

$$\boldsymbol{p} = \{x, y, z, \psi, \theta, \phi\} \tag{3}$$

where x, y, and z represent the 3D position of the person and ψ, θ and ϕ its orientations with respect to a global reference. Such a reference frame could align with the observer's perspective, for instance, that of a robot (see Fig. 2), or be defined relative to the environment. This representation also assumes the definition of a frame on the human body, but it is not important how the frame is attached to the body as long as its definition is preserved. As for the joint state, also the human pose should be defined over time as $(p)(t)$, along with its derivatives $\dot{\boldsymbol{p}}(t)$ and $\ddot{\boldsymbol{p}}(t)$. We should point out that, differently from the joint state, whose representation was homogenous, the human pose has linear and angular components. Similarly, for what we did for the joint state, we can define the pose trajectory ($\boldsymbol{\delta}$).

$$\boldsymbol{\delta}(t_s, t_e) = \left\{\boldsymbol{p}(t_e), \dot{\boldsymbol{p}}(t_s) \dots \dot{\boldsymbol{p}}(t_e), \ddot{\boldsymbol{p}}(t_s) \dots \ddot{\boldsymbol{p}}(t_e)\right\} \tag{4}$$

After introducing the human pose trajectory, we can combine it with the joint state trajectory to provide the overall definition of a gesture.

$$\boldsymbol{g}(t_s, t_e) = \{\boldsymbol{\tau}(t_s, t_e), \boldsymbol{\delta}(t_s, t_e)\} \tag{5}$$

It should be noted that although this formalism describes gestures as trajectories, it is also possible to represent single poses when t_s and t_e coincide. This is important because although we started the discussion of gestures by talking about motion, it could also be relevant for communication purposes to show the complete or partial absence of motion, such as the static pose of a human pointing at an object.

3 Gesture Taxonomy

To mediate the interaction between humans and robots using gestures, it is crucial to understand how they are categorized and how different types of gestures can affect the design of a gesture-based interface. Therefore, an appropriate gesture taxonomy is essential to describe the gesture types a given system uses and to gain a clear understanding of the associated issues. Different perspectives on designing a coherent taxonomy for gestures can be found in literature, addressing different aspects of gestural interaction. Here, we will focus on a functional taxonomy, which describes gesture characteristics directly impacting the technical problems a robotic observer should address to adopt gesture-based interaction. The taxonomy discussed in this chapter includes four features: *effect*, *time*, *focus*, and *space* [7].

3.1 Effect

The effect of a gesture describes how it will impact the robot that the human is interacting with. If we define the function that maps a gesture $\boldsymbol{g}$ to the robot state as $f(\boldsymbol{g})$, the effect of the gesture is essentially the mapping itself. Based on their effect, gestures can be classified as continuous or discrete. With continuous gestures, the information obtained is continuously mapped to a change in the robot state, whereas with discrete gestures, the change is *atomically* associated with the entire gesture.

Let us define $S_c \in \mathbb{R}^n$ as an n-dimensional continuous state and $S_d \in \mathbb{S}$ as a discrete state, where $\mathbb{S}$ is the set of all possible, countable, discrete states. For continuous gestures, $f(\boldsymbol{g})$ has a domain of $\boldsymbol{g}^{(t_e-t_s)}$ and a co-domain of $S_C^{(t_e-t_s)}$, whereas for discrete gestures, the co-domain is simply the discrete state S_d. With t_s and t_e, introduced in Sect. 2, identify the gesture start and end time instants.

Consequently, the intended effect of the gesture on the robot's internal state should guide the choice of gesture type. Discrete gestures allow interaction with the robot's discrete internal states, while continuous gestures can interact with continuous ones. For example, in Coronado et al., a system was described where the human arm inclination was detected by a wrist-worn IMU and mapped to the linear and angular velocities of a mobile robot, which is an example of coupling between continuous gestures and continuous state [8]. On the other hand, in Bongiovanni et al., gesture patterns from human motions were recognized and used to navigate a menu, an example of discrete gestures [2]. Each occurrence of a gesture $\boldsymbol{g}^{(t_e-t_s)}$ was associated with a discrete command of the interface, e.g., move up, move down or select.

3.2 Time

The time characteristic classifies gestures as static or dynamic and organizes them into three phases: *preparation*, *stroke*, and *retraction*. We should notice that the classification of a gesture as static or dynamic aims to describe human behaviour only during the stroke phase since it is the phase in which communication takes place. Therefore, t_s and t_e, introduced in Sect. 2, are the starting and ending times of the stroke phase. Before a static gesture starts, the person moves to a desired pose (preparation) and then stands still for an arbitrary amount of time (stroke). After this, the person returns to resting or reaches the starting pose of the following gesture (retraction). Therefore, while static gestures, in principle, imply motion, this is limited to the *preparation* and *retraction* phases, and the information is conveyed by the posture assumed during the *stroke* phase. If we ground this description in the gesture definition introduced in Sect. 2, we can say that for dynamic gestures should exist at least one $q_i \in \boldsymbol{q}$ such that $\dot{q}_i \neq 0$, or one of the components in $\boldsymbol{p}$ with a velocity not equal zero. Instead, for static gestures, it holds that both $\dot{\boldsymbol{q}}(t_s)$ and $\ddot{\boldsymbol{q}}(t_s)$ are zero while the person is still in the environment. We should also point out that according to this taxonomy, static gestures cannot be considered continuous. We can not associate a static gesture with a continuous state since this would require the definition of an infinite number of postures, which is not only impractical but also impossible. Furthermore, the transition between two postures will imply a motion during the *stroke* phase, transforming the gesture into a dynamic gesture in which, at some point in the trajectory, the human stands still. On the other hand, dynamic gestures can be continuous or discrete, depending on the effect they have on the robot state. For instance, if a person moves their arm and its inclination is mapped to the velocity of a mobile robot, the gesture is dynamic and continuous [8]. Conversely, if a person moves their arm in a specific pattern corresponding to a discrete command in a robot interface, the gesture is dynamic and discrete [2].

3.3 Focus

In the presented taxonomy, the focus describes which body parts convey information through motion. This name has been chosen to emphasize that the relevance of a body part to the interaction is determined a priori by the system requirements. For example, if a robot is programmed to focus on the arm, it will ignore fingers' motions. If we refer to this gesture definition, focusing on a specific body part implies that we should consider only a subset of the joint state, i.e., $\boldsymbol{\rho}(t) \subseteq \boldsymbol{q}(t)$. Therefore, in these cases, the gesture component coming from the human joint state can be reduced in complexity ($\boldsymbol{\tau}_{\rho}(t_s, t_e)$), simplifying the problem overall. We should notice that this simplification still allows describing gestures where the focus is on multiple body parts, possibly even not directly connected as in the case of bi-manual gestures [9].

3.4 Space

The final characteristic of gestures that we will discuss is space. With this characteristic, we try to represent if the information conveyed by a motion is linked to where the human performs it. Therefore, this characteristic is strictly related to the human pose $\boldsymbol{p}(t)$. The space characteristic divides gestures into those that are spatially related and those that are unrelated. In spatially unrelated gestures, the communication is not affected by the human pose ($\boldsymbol{p}(t)$). However, the speed ($\dot{\boldsymbol{p}}(t)$) or acceleration ($\ddot{\boldsymbol{p}}(t)$) of the human, together with the joint state trajectory ($\boldsymbol{\tau}$), can affect communication. In contrast, for spatially related gestures, the human pose ($\boldsymbol{p}(t)$) is an essential component.

For example, the discrete gesture of an interface simulating mid-air virtual buttons is necessarily spatially related [10]. The system should know the person's position and orientation to understand which virtual button has been pressed. Similarly, a continuous gesture for the teleoperation of a mobile robot is spatially related since the position in space determines which object the robot will grasp or where it will release it. In contrast, a continuous gesture adopted for the direct control of the speed of a mobile robot simulating the motorbike throttle is spatially unrelated. In this case, it is not important where the person is in space, but only the orientation of their wrist.

4 Problem Statement

In the previous section, we introduced a definition of gestures and described their categorization in the form of a taxonomy. Here, we aim to structure the problem of gesture-based interfaces for human-robot interaction. The question we want to address with this problem statement is: which challenges the robot artificial intelligence should solve to make sense of human gestures? To answer this question, we described the overall problem as the interplay of three sub-problems: sensing, data processing, and system reaction. In this section, we will see how the design choices to address the sub-problems are strongly influenced by each other and closely related to the gestures that a robot is supposed to interpret.

4.1 Sensing

The sensors used to collect gesture data depend on specific application requirements, such as privacy [11], price [12], or computational efficiency [13]. Two conflicting paradigms affect gesture sensing: *come as you are* and *wearability* [12]. *Come as you are* refers to a system design in which users do not have to wear any tool to interact with the machine, favouring a more natural interaction. On the other hand, according to the *wearability* paradigm, the user wears the devices needed for sensing its motions, e.g., data gloves or smartwatches [14]. In this way, humans can freely move in the environment bringing the interface with them. These two paradigms are not exclusive, and according to the application, the interface designer could combine multiple sensing solutions to compensate drawbacks and advantages of each approach. Another possible categorization divides sensing approaches between image-based and non-image-based [15]. This distinction does not overlap with the two paradigms introduced before. Non-image-based solutions include both wearable and non-wearable devices, e.g., WiFi routers for human motion sensing through channel state information [16]. On the other hand, image-based approaches usually belong to the "come as you are" paradigm, but some exceptions exist where cameras are worn by the user [17].

Instead of providing a list of all the possible sensing solutions for human motion [18], we focus here on the relationship between the gestures adopted in a specific human-robot interface and the sensing modality. To do this, we refer to the taxonomy introduced in the previous section, and we discuss the implications that each gesture characteristic could have on the choice of the sensing solution. Here, we ignore the effect characteristic since the difference between continuous and discrete gestures does not particularly constrain the sensor choice.

The temporal characteristic, which categorizes gestures as static or dynamic, can significantly affect the choice of sensor. Even though most sensors can detect both types of gestures, some sensors may be more suitable for certain gesture types. For example, to detect a static pose, an accelerometer worn by the user can readily recognize the orientation of a body part by analyzing the distribution of gravity acceleration across the three axes. In contrast, achieving the same with a camera would require a significantly more complex processing pipeline. Conversely, some sensors can not perceive gestures depending on their temporal connotation, e.g., event cameras [19] can only perceive dynamic gestures when used in a fixed configuration.

Similarly, the difference between spatially related and unrelated gestures constrains the selection of the sensing solution. In fact, sensing spatially related gestures should perceive both human motion and their position in space. This problem is typically solved using image-based solutions [20].

Finally, the influence of focus on the sensing strategies is limited. We should only determine if a specific sensor can perceive the body part we are interested in. Some sensors could have a low resolution that prevents the perception of small body parts such as the fingers, and wearable devices could have problems regarding their mounting mechanisms.

4.2 Data Processing

All the gesture characteristics presented in the taxonomy directly affect the data processing. However, as we will see in this section, the gesture effect completely changes the problem nature, i.e., continuous and discrete gestures follow different data processing pipelines. As we highlighted when we introduced the effect characteristic in Sect. 3, the difference between continuous and discrete gestures lies in the function that maps the gesture to the robot state ($f(\boldsymbol{g})$) and the data processing is implementing that function starting from the raw data collected by the sensing previously described.

Therefore, for continuous gestures, the data processing should provide a continuous mapping between the sensory input and the robot state. In the most straightforward solutions, this mapping is obtained directly using raw sensory data [8]. However, more commonly, the data needs to be processed to extract relevant features, such as the 2D position of the hand from an image [21]. This process is known by the name of *feature extraction*.

Instead, in discrete gestures, the robot state change is *atomically* associated with the entire gesture. Achieving this requires not only extracting features from sensory data but also detecting the start and end instances of the gesture, i.e., gesture detection, and identifying which gesture has been performed from the pool of gestures for which the robot knows how to react, i.e., gesture classification. The overall data processing of discrete gesture is known in the literature with the term gesture recognition [22], and as we described, is composed of three steps: pre-processing and feature extraction, detection, and classification. At first, sensory data should be de-noised and processed to extract relevant features as we described for continuous gestures. Then, during detection, also known as segmentation or spotting, the data from the previous phase are analyzed to determine if a gesture occurred and its start and end instants (t_s and t_e). This process results in temporal sequences of segmented data, discarding data that do not correspond to a gesture execution. Often, this result is achieved using filtering or thresholding mechanisms [15]. Finally, the classification discerns which gesture was performed starting from a list of gestures known to the system, i.e., dictionary. The result of this phase is usually the gesture label, but sometimes, it may also come with a confidence value.

While we can achieve classification with purely model-based techniques, most recent research in this area has shifted towards data-driven solutions [23]. The order in which we presented the three subproblems that characterized the data processing is not binding. In particular, detection and classification can be inverted. Usually, gesture detection is performed after the feature extraction. In these cases, we talk of *direct* detection, and the result of this process is fed to the classification. However, it is also possible to perform the detection on classification results, i.e., *indirect* detection [24]. In these cases, the classifier continuously processes the data resulting from the feature extraction. This strategy produces a continuous stream of gesture labels, which the detection system should analyze to determine the occurrence of a gesture.

As mentioned earlier, various gesture characteristics could affect the technical solutions adopted in a gesture-based interface without altering the overall problem. Regarding time, static and dynamic gestures are time-independent and -dependent, respectively. This difference leads to collecting different types of data, which in turn implies different data processing techniques. Similarly, the gesture focus could also affect the selection of technical solutions for data processing. For example, different approaches may be needed to recognize a hand or a face in an image. Finally, space also affects data processing, as the system must be able to extract the human pose from the sensory information for spatial-related gestures.

4.3 System Reaction

The response of a robot to a given gesture can vary depending on the application. For instance, it may consist of a high-level command for a manipulator [2] or a velocity command for a mobile robot [8]. As we introduced in the gesture taxonomy, the system responses result from the mapping between the human gesture and the robot state. The system's reaction is largely agnostic to gesture characteristics, as any differences imposed by those characteristics are absorbed during the sensing and data processing phases. However, the effect characteristic does affect the system's reaction since, as previously introduced, continuous and discrete gestures imply continuous and discrete system reactions. An important consideration when dealing with robot behaviours in response to human gestures is how the machine can provide a natural and intuitive communication interface. Numerous studies have been conducted in the past few decades, and the recent trend of designing and developing technologies for the consumer market has further increased this research [18].

5 Human Factors in Gesture-Based Interaction

As discussed in the introduction, the previous sections have focused on gesture-based interaction from the observer perspective, i.e., the robot, with the human simply being the object of observation. However, to use gestures as an effective communication medium between humans and robots, we need to consider how the interface is received by the human, i.e., human factors. Here, we will describe the critical aspects to consider when building a gesture-based interface. While some of these aspects are generically applicable to any interface, we will focus on those aspects specific to gesture-based interactions, such as ergonomic considerations regarding human gestures.

5.1 Requirements

The literature has identified many requirements related to human factors [12], including *responsiveness*, *user adaptability and feedback*, *learnability*, *accuracy*, *low mental load*, *intuitiveness*, *comfort*, and *lexicon size*. These factors should be considered when designing a gesture-based interface since a careful designer could prioritize some technological solutions over others based on these factors.

As a rule of thumb, the response time, which is the time interval between user input and system reaction, should be minimized. The system should be designed to react as quickly as possible to gestural stimuli [25, 26]. Long reaction times can be easily perceived by humans and could disrupt the user experience and reduce overall satisfaction [27]. In general-purpose applications, classical design methodologies recommend response times of less than 100 ms. However, recent studies have suggested that acceptable latency may be even lower, particularly for interactions that resemble physical ones [27]. This requirement may discourage designers from adopting sensors generating heavy data, especially when the hardware is not top-performing, as in edge applications. At the same time, for the same reasons, lighter processing techniques could be preferred over more complex ones.

In terms of user adaptability, the robot should be able to interpret gestures regardless of who is performing them. Therefore, the interface must adapt to user-specific traits [28], including disabilities. Additionally, adaptability to the user could be displayed by allowing the personalization of gestures used in the interaction [29]. This personalization could involve either a simple gesture substitution or a change in the types of gestures that robots can react to. For example, a user with partial paralysis may need to focus on a body part that they can control [30]. This requirement has a technological consequence: the techniques used to model and process gestures must allow for easy adaptation, which likely involves the system's ability to learn from experience [31].

Moreover, the robot should respond to the detected gesture and also promptly provide the user with appropriate feedback. The absence of direct feedback in HMI can result in a lack of trust in the system and a suboptimal user experience [12]. There are several methods currently used to provide feedback to users, including classical GUIs [21], haptics [32], augmented reality [33], and virtual reality [34]. When dealing with a physical robot, the main feedback will be provided by its motion, which should be unambiguous and coherent.

For real-world applications, a gesture-based interface system should accurately interpret input gestures with high precision, close to 100% [29]. This objective is of utmost importance for the communication interface between humans and robots. Lower accuracy could frustrate the user, disrupting the interaction, and also induce incorrect robot reactions that could lead to physical damage or even injuries to the user. In addition to improving the perception pipeline, accuracy can be increased by carefully selecting gestures that are easy to perceive. However, this process often conflicts with other requirements, so the designer should approach it with care.

The usability and effectiveness of a gesture-based interface depend greatly on its ability to interpret a high number of gestures, known as the lexicon size. This feature is fundamental in cases like the automated interpretation of sign language [35]. However, expanding the lexicon size conflicts with the need for learnability and low mental load, as a larger dictionary is more difficult to learn and recall [36]. Moreover, a larger dictionary could lead to lower performance in data processing due to increased problem complexity. Therefore, designers should consider all these factors when determining the appropriate size of the dictionary. Finally, the association between gestures and robot reactions should be intuitive. This requirement, in turn, simplifies the learning process and reduces the user's mental load [37].

In summary, many requirements relate to the chosen gestures and their impact on user experience and problem-solving. This overview suggests that gesture selection and its associated system behaviour are paramount.

5.2 User Comfort

As discussed in the previous section, selecting gestures for a human-robot interface involves balancing system and human requirements. An interaction between a human and a robot that relies on gestures as the primary mode of interaction should limit user fatigue. Otherwise, even a well-designed technical solution may become unusable for an extended period. While an uncomfortable interface may be tolerable for short interactions, prolonged use can lead to user fatigue, making it unsuitable for lengthier interactions. To this extent, it is crucial to avoid complex gestures or gestures that require intense muscle activity, especially if the gesture-based interface is designed for prolonged use [12]. Despite their importance, ergonomic factors are often disregarded in research targeting gesture-based interaction and only recently have studies started incorporating them into interface design [38–40]. This section provides an overview of ergonomics for gesture-based interaction, including a description of the problem and possible tools to address it.

Gesture-based interaction has been researched not only for human-robot interaction but also for human-machine interaction in general. In this context, the most commonly used gestures involve human arms. The term "mid-air gestures" is often used to refer to this kind of gesture; performed in front of the human body without any physical support. Arm gestures are popular because of their intuitiveness. Humans typically interact with their environment using their arms and hands. Also, arm motions are usually characterized by high magnitude and are thus easy to perceive. However, the high magnitude of arm motion can cause user fatigue, especially for mid-air gestures where the arm has no physical support. This effect is described by researchers as the *gorilla arm* syndrome [41]. According to [38], discomfort in mid-air gestures can be reduced by keeping arm motions close to the body and limiting them to the lower area of the reachable space. Alternatively, [42] suggests adding physical support for the arm. However, introducing physical support is not always possible, and it may be against the general concept of gestural interaction, often viewed as a modality of interaction that is unconstrained by any physical support. Alternative approaches to address the comfort problem could explore adopting body parts that require lower energy expenses, such as fingers and feet. However, these alternatives are less commonly studied, especially from an ergonomic perspective [7].

Regardless of the type of gesture considered for a human-robot interaction interface, the designer needs to consider user comfort and have tools to evaluate it. Since gestures are just human motions, the tools used to perform ergonomic assessments come from the ergonomics literature and the study of human activities in the workplace. Examples of such tools include the Rapid Upper Limb Assessment (RULA) and the Rapid Entire Body Assessment (REBA) [43, 44]. These tools allow us to evaluate the risk factors associated with specific poses performed by humans. As discussed in the previous paragraph, arm gestures are typically the focus of the literature. Therefore, RULA is one of the more commonly used tools [45]. More complex approaches have been studied to estimate the comfort of human gestures. These include the adoption of consumed endurance [46] and investigating data-driven approaches to cluster arm motions based on muscle activation [47]. However, the study of ergonomics in gestural interaction is still in its preliminary stages. Future studies should focus on more complex and advanced techniques to validate user comfort. These solutions should not only consider gestures as static poses but also as motions that should be repeated over a prolonged time.

When discussing user comfort, it's essential not to focus solely on the physical aspect but also consider cognitive load. In this chapter, we have already addressed some factors related to cognitive load, such as lexicon size and the learnability of gestures. However, cognitive load is a multifaceted concept that involves various factors, including the user's experience, the complexity of the task, and the environment in which the interaction occurs [48]. Gesture-based interaction, like other interaction modalities, is influenced by these factors. Consequently, the same methods used to evaluate cognitive load in different contexts can be applied here. Tools such as user experience questionnaires [49] and mental workload assessments [50] provide explicit measures of interface performance. Additionally, implicit measures—such as eye tracking and physiological signals—can assess cognitive load without requiring direct user input [51, 52]. In this regard, research on gesture-based interaction remains limited, and further studies are needed to explore how to design interfaces that are not only physically comfortable but also cognitively efficient.

5.3 Gesture Dictionaries

In Sect. 4.2, we discussed gesture dictionaries and their relationship with discrete gestures. Later, in Sect. 5.1, we introduced requirements for effective gesture-based interaction and emphasized the importance of gesture design. Specifically, we noted how lexicon size and complexity can significantly impact overall interaction performance. In this section, we will describe what a dictionary is and how to design it to consider both human factors and system performance.

A dictionary (or vocabulary) is the words set that make up a language and their associated meanings. In our context, the dictionary contains the gestures that the robot can interpret. Each gesture in a dictionary is described by its trajectory $\boldsymbol{g}$ and the associated meaning for the robot. Researchers often rely on the definition of dictionaries of "synthetic" gestures, which are gestures whose meaning is undefined outside the specific application. However, adopting synthetic gestures is not always necessary. Most of the time, the dictionary is provided by the context. For example, a robot that understands sign language can rely on a pre-existing dictionary adopted in human-to-human interaction. When pre-existing dictionaries are adopted, it is wise not to alter them since a misalignment between the human's and robot's understanding of a gesture could disrupt the interaction. Therefore, we will focus on the problem associated with the design of synthetic dictionaries.

Designing a gesture dictionary involves selecting an appropriate set of motions and linking them with their synthetic meanings while meeting the requirements presented in Sect. 5.1. Various options have been explored to complete this process for a given application while considering human factors. The first approach is to use questionnaires to ask volunteers to sketch the gestures they believe are more suitable for the specific application. This approach has been proven particularly effective since research has shown that users can adapt to gestures defined by others if they are intuitive enough [29]. The second approach involves asking volunteers to imitate a gesture-based interaction for a specific application and observe their motions. The dictionary is built based on the experimenter's observations [53]. The third approach involves using Wizard of Oz experiments [54]. In this case, volunteers should solve an interaction problem with a robot they believe to be autonomous. An experimenter manually operates the robot to react to human gestures as if it were autonomous. In this case, observing the gestures performed by volunteers during their interaction with the robot is used as a starting point for the dictionary definition. Later, a conceptualization phase is necessary.

After designing a dictionary, it is crucial to have tools to verify its effectiveness. A first solution could involve conducting experiments with human subjects and defining precise metrics for evaluating each requirement under consideration [55]. Alternatively, some studies have proposed the Vocabulary Acceptability Criteria (VAC) [37], which allows experts to evaluate gestures based on six attributes: *iconicity* (how much a gesture recalls the associated command), *simplicity*, *efficiency*, *compactness* (how much the gesture covers the space around the body), *salience* (how discriminating a movement is), and *economy* (related to the movement magnitude) [37]. Although the requirements that should drive the dictionary design are well formalized, there is no standard for gesture dictionary evaluation.

Finally, accurately describing a gesture is essential for ensuring reproducible research. However, it can be challenging to do so, and in some cases, the only way to reproduce the work of others is by collecting a new data set using the same experimental procedure, especially when the original data sets are unavailable. Despite the proposed gesture taxonomy, it is not always clear how gestures were performed, and their description often relies on simple drawings since they can be easily shared and printed. However, a simple drawing may not always be sufficient to describe the complexity of dynamic gestures. We believe that the most informative solution involves combining drawings with well-thought-out textual descriptions. Moreover, video describing the target gestures, although more challenging to share, can be a fundamental complement. Nonetheless, this issue remains unresolved.

6 Conclusion

In this chapter, we gave an overview of human-robot interaction mediated by gestures. We started by providing a formal definition and description of gestures, then proceeded to analyze the challenges characterizing gesture-based human-robot interaction from the human and robot perspective.

While technical and technological aspects of gesture-based interaction have been extensively explored in the literature, there is still limited focus on human-related factors. Research efforts should focus on tailoring mature technological solutions to users' needs. Interfaces do not necessarily need to be *natural*, but they should be capable of supporting interaction between humans and robots. This objective requires more detailed ergonomic studies and exploration of gesture types that are less commonly used. In particular, we believe that the focus on arm gestures alone could be detrimental to research in the field. Designing interfaces for future human-robot interaction requires considering the entire human body. Finally, the gesture dictionary design process should be formalized and standardized to ensure reproducible research.

References

1. Roggio, F., Trovato, B., Sortino, M., Musumeci, G.: A comprehensive analysis of the machine learning pose estimation models used in human movement and posture analyses: a narrative review. Heliyon , (2024)
2. Bongiovanni, A., De Luca, A., Gava, L., Grassi, L., Lagomarsino, M., Lapolla, M., Marino, A., Roncagliolo, P., Macciò, S., Carfì, A.: Gestural and touchscreen interaction for human-robot collaboration: a comparative study. In: International Conference on Intelligent Autonomous Systems, pp. 122–138. (2022)
3. Merlo, E., Lamon, E., Fusaro, F., Lorenzini, M., Carfi, A., Mastrogiovanni, F., Ajoudani, A.: Dynamic human-robot role allocation based on human ergonomics risk prediction and robot actions adaptation. In: Proceedings of the 38th IEEE International Conference on Robotics and Automation (ICRA), pp. 2825–2831. , Philadelphia, Pennsylvania, USA (2022)
4. Altinok, D.C.A., Ohl, K., Volkmer, S., Brandt, G.A., Fritze, S., Hirjak, D.: 3d-optical motion capturing examination of sensori- and psychomotor abnormalities in mental disorders: progress and perspectives. Neurosci. Biobehav. Rev. , 105917 (2024)
5. Lastrico, L., Belcamino, V., Carfì, A., Vignolo, A., Sciutti, A., Mastrogiovanni, F., Rea, F.: The effects of selected object features on a pick-and-place task: a human multimodal dataset. Int. J. Robot. Res. **43**(1), 98–109 (2024)
6. Pezzulo, G., Donnarumma, F., Dindo, H.: Human sensorimotor communication: a theory of signaling in online social interactions. PLoS ONE **8**(11), e79876 (2013)
7. Carfì, A., Mastrogiovanni, F.: Gesture-based human-machine interaction: taxonomy, problem definition, and analysis. IEEE Trans, Cybern (2021)
8. Coronado, E., Villalobos, J., Bruno, B., Mastrogiovanni, F.: Gesture-based robot control: design challenges and evaluation with humans. In: Proceedings of the 33rd IEEE International Conference on Robotics and Automation (ICRA), pp. 2761–2767. , Singapore (2017)
9. Shah, D., Denicia, E., Pimentel, T., Bruno, B., Mastrogiovanni, F.: Detection of bimanual gestures everywhere: why it matters, what we need and what is missing. Robot. Auton. Syst. **99**, 30–49 (2018)
10. Shin, J., Kim, C.M.: Non-touch character input system based on hand tapping gestures using kinect sensor. IEEE Access **5**, 10496–10505 (2017)
11. Zhao, W., Deborah, D.D., Reinthal, M.A., Ekelman, B., Goodman, G., Niederriter, J.: Privacy-aware human motion tracking with realtime haptic feedback. In: Proceedings of the 4th IEEE International Conference on Mobile Services (MS), pp. 446–453. , Coimbra, Portugal (2015)
12. Wachs, J.P., Kölsch, M., Stern, H., Edan, Y.: Vision-based hand-gesture applications. Commun. ACM **54**(2), 60 (2011)
13. Poularakis, S., Katsavounidis, I.: Low-complexity hand gesture recognition system for continuous streams of digits and letters. IEEE Trans. Cybern. **46**(9), 2094–2108 (2015)
14. Belcamino, V., Carfì, A., Mastrogiovanni, F.: A systematic review on custom data gloves. IEEE Trans. Hum.-Mach. Syst **54**(5), 520–535 (2024)
15. Liu, H., Wang, L.: Gesture recognition for human-robot collaboration: a review. Int. J. Ind. Ergon. **68**, 355–367 (2018)
16. Zhuravchak, A., Kapshii, O., Pournaras, E.: Human activity recognition based on Wi-Fi CSI data–a deep neural network approach. Procedia Comput. Sci. **198**, 59–66 (2022)
17. Chen, F., Deng, J., Pang, Z., Baghaei Nejad, M., Yang, H., Yang, G.: Finger angle-based hand gesture recognition for smart infrastructure using wearable wrist-worn camera. Appl. Sci. **8**(3), 369 (2018)
18. Vuletic, T., Duffy, A., Hay, L., McTeague, C., Campbell, G., Grealy, M.: Systematic literature review of hand gestures used in human computer interaction interfaces. Int. J. Hum Comput Stud. **129**, 74–94 (2019)
19. Wang, Q., Zhang, Y., Yuan, J., Lu, Y.: Space-time event clouds for gesture recognition: from RGB cameras to event cameras. In: Proceeding of the IEEE Winter Conference on Applications of Computer Vision (WACV), pp. 1826–1835. IEEE, Waikoloa, Hawaii, USA (2019)

20. Lai, Y., Wang, C., Li, Y., Ge, S.S., Huang, D.: 3D pointing gesture recognition for human-robot interaction. In: Proceedings of the 28th Control and Decision Conference (CCDC), pp. 4959–4964. , Yinchuan, China (2016)
21. Bolano, G., Tanev, A., Steffen, L., Roennau, A., Dillmann, R.: Towards a vision-based concept for gesture control of a robot providing visual feedback. In: Proceedings of the IEEE International Conference on Robotics and Biomimetics (ROBIO), pp. 386–392. Kuala Lumpur, Malaysia (2018)
22. Carfì, A., Motolese, C., Bruno, B., Mastrogiovanni, F.: Online human gesture recognition using recurrent neural networks and wearable sensors. In: Proceeding of the 27th IEEE International Symposium on Robot and Human Interactive Communication (RO-MAN), pp. 188–195. IEEE, Nanjing, China (2018)
23. Islam, M.J., Ho, M., Sattar, J.: Understanding human motion and gestures for underwater human-robot collaboration. J. Field Robot. **36**(5), 851–873 (2019)
24. Escalera, S., Athitsos, V., Guyon, I.: Challenges in multi-modal gesture recognition. In: Gesture Recognition, pp. 1–60. Springer (2017)
25. Hu, C., Meng, M.Q., Liu, P.X., Wang, X.: Visual gesture recognition for human-machine interface of robot teleoperation. In: Proceedings of the IEEE/RSJ International Conference on Intelligent Robots and Systems (IROS), vol. 2, pp. 1560–1565. , Las Vegas, Nevada, USA (2003)
26. Lee, J.H., Delbruck, T., Pfeiffer, M., Park, P.K.J., Shin, C.W., Ryu, H., Kang, B.C.: Real-time gesture interface based on event-driven processing from stereo silicon retinas. IEEE Trans. Neural Netw. Learn. Syst. **25**(12), 2250–2263 (2014)
27. Attig, C., Rauh, N., Franke, T., Krems, J.F.: System latency guidelines then and now—is zero latency really considered necessary? In: Proceedings of the 14th International Conference on Engineering Psychology and Cognitive Ergonomics (EPCE), pp. 3–14. Vancouver, Canada (2017)
28. Carreira, M., Ting, K.L.H., Csobanka, P., Gonçalves, D.: Evaluation of in-air hand gestures interaction for older people. Univ. Access Inf. Soc. **16**(3), 561–580 (2017)
29. Kela, J., Korpipää, P., Mäntyjärvi, J., Kallio, S., Savino, G., Jozzo, L., Marca, S.D.: Accelerometer-based gesture control for a design environment. Pers. Ubiquit. Comput. **10**(5), 285–299 (2006)
30. Rizzoglio, F., Casadio, M., De Santis, D., Mussa-Ivaldi, F.A.: Building an adaptive interface via unsupervised tracking of latent manifolds. Neural Netw. **137**, 174–187 (2021)
31. Kuno, Y., Murashina, T., Shimada, N., Shirai, Y.: Intelligent wheelchair remotely controlled by interactive gestures. In: Proceedings of the 15th International Conference on Pattern Recognition (ICPR), vol. 4, pp. 672–675. , Barcelona, Spain (2000)
32. Kim, K., Kim, J., Choi, J., Kim, J., Lee, S.: Depth camera-based 3d hand gesture controls with immersive tactile feedback for natural mid-air gesture interactions. Sensors **15**(1), 1022–1046 (2015)
33. Bai, H., Gao, L., El-Sana, J., Billinghurst, M.: Free-hand interaction for handheld augmented reality using an RGB-depth camera. In: Proceedings of the SIGGRAPH Asia Symposium on Mobile Graphics and Interactive Applications (MGIA), pp. 1–4. Hong Kong, China (2013)
34. Park, K.B., Lee, J.Y.: New design and comparative analysis of smartwatch metaphor-based hand gestures for 3D navigation in mobile virtual reality. Multimedia Tools Appl. **78**(5), 6211–6231 (2019)
35. Galka, J., Masior, M., Zaborski, M., Barczewska, K.: Inertial motion sensing glove for sign language gesture acquisition and recognition. IEEE Sens. J. **16**(16), 6310–6316 (2016)
36. Yeasin, M., Chaudhuri, S.: Visual understanding of dynamic hand gestures. Pattern Recogn. **33**(11), 1805–1817 (2000)
37. Gonzalez, G., Madapana, N., Taneja, R., Zhang, L., Rodgers, R., Wachs, J.P.: Looking beyond the gesture: vocabulary acceptability criteria for gesture elicitation studies. In: Proceedings of the 62th Human Factors and Ergonomics Society Annual Meeting (HFES), vol. 62, pp. 997–1001. , Philadelphia, Pennsylvania, USA (2018)

38. Schön, D., Kosch, T., Müller, F., Schmitz, M., Günther, S., Bommhardt, L., Mühlhäuser, M.: Tailor twist: Assessing rotational mid-air interactions for augmented reality. In: Proceedings of the SIGCHI Conference on Human Factors in Computing Systems (CHI), pp. 1–14. , Hamburg, Germany (2023)
39. Bordoni, S., Tang, G.: Development and assessment of a contactless 3d joystick approach to industrial manipulator gesture control. Int. J. Ind. Ergon. **93**, 103376 (2023)
40. Fashimpaur, J., Karlson, A., Jonker, T.R., Benko, H., Gupta, A.: Investigating wrist deflection scrolling techniques for extended reality. In: Proceedings of the SIGCHI Conference on Human Factors in Computing Systems (CHI), pp. 1–16. , Hamburg, Germany (2023)
41. Boring, S., Jurmu, M., Butz, A.: Scroll, tilt or move it: using mobile phones to continuously control pointers on large public displays. In: Proceedings of the 21st Annual Conference of the Australian Computer-Human Interaction Special Interest Group (OZCHI), pp. 161–168. , Melbourne, Australia (2009)
42. Hansberger, J.T., Peng, C., Mathis, S.L.: Areyur Shanthakumar, V., Meacham, S.C., Cao, L., Blakely, V.R.: Dispelling the gorilla arm syndrome: the viability of prolonged gesture interactions. In: Proceeding of the 9th International Conference on Virtual. Augmented and Mixed Reality (VAMR), pp. 505–520. Springer, Vancouver, Canada (2017)
43. McAtamney, L., Corlett, E.N.: RULA: a survey method for the investigation of work-related upper limb disorders. Appl. Ergon. **24**(2), 91–99 (1993)
44. Hignett, S., McAtamney, L.: Rapid entire body assessment (REBA). Appl. Ergon. **31**(2), 201–205 (2000)
45. Manghisi, V.M., Wilhelm, M., Uva, A., Engelmann, B., Fiorentino, M., Schmitt, J.: Towards gestured-based technologies for human-centred smart factories. Int. J. Comput. Integr. Manuf. **36**(1), 110–127 (2023)
46. Hincapié-Ramos, J.D., Guo, X., Moghadasian, P., Irani, P.: Consumed endurance: a metric to quantify arm fatigue of mid air interactions. In: Proceedings of the ACM Conference on Human Factors in Computing Systems (CHI), pp. 1063–1072. , Toronto, Canada (2014)
47. Bachynskyi, M., Palmas, G., Oulasvirta, A., Weinkauf, T.: Informing the design of novel input methods with muscle coactivation clustering. ACM Trans. Comput.-Hum. Interact. (TOCHI) **21**(6), 1–25 (2015)
48. Hassenzahl, M., Tractinsky, N.: User experience–a research agenda. Behav. Inf. Technol. **25**(2), 91–97 (2006)
49. Laugwitz, B., Held, T., Schrepp, M.: Construction and evaluation of a user experience questionnaire. In: Proceedings of the 4th Symposium of the Workgroup Human-Computer Interaction and Usability Engineering of the Austrian Computer Society (USAB), , Graz, Austria (2008)
50. Hart, S.G., Staveland, L.E.: Development of NASA-TLX (task load index): results of empirical and theoretical research. Adv. Psychol. **52**, 139–183 (1988) Elsevier
51. Kosch, T., Karolus, J., Zagermann, J., Reiterer, H., Schmidt, A., Woźniak, P.W.: A survey on measuring cognitive workload in human-computer interaction. ACM Comput. Surv. **55**(13s), 1–39 (2023)
52. Heard, J., Harriott, C.E., Adams, J.A.: A survey of workload assessment algorithms. IEEE Trans. Hum.-Mach. Syst **48**(5), 434–451 (2018)
53. Jahani, H., Kavakli, M.: Exploring a user-defined gesture vocabulary for descriptive mid-air interactions. Cogn. Technol. Work **20**(1), 11–22 (2018)
54. Höysniemi, J., Hämäläinen, P., Turkki, L., Rouvi, T.: Children's intuitive gestures in vision-based action games. Commun. ACM **48**(1), 44 (2005)
55. Stern, H.I., Wachs, J.P., Edan, Y.: Designing hand gesture vocabularies for natural interaction by combining psycho-physiological and recognition factors. Int. J. Semant. Comput. **02**(01), 137–160 (2008)

Human-Swarm Interaction at Work

Aya Hussein, Hung Nguyen, and Hussein Abbass

Abstract Swarm robotics have seen increased interest in a wide range of applications due to their advantageous characteristics and low cost. Robot swarms are particularly useful for tasks requiring spatial distribution where a single sophisticated robot is not capable of performing the task efficiently. However, effective control of robot swarms is challenging due to the diverse and demanding operational requirements as contrasted to the limited human cognitive capacity. This chapter articulates the key considerations and operational requirements for effective human-swarm interaction. We propose the use of *shepherding* as a bio-inspired mechanism for human control over robot swarms to address these considerations. We present experiments where the proposed swarm control mechanism is used for controlling a swarm of real UGVs via a simulated UAV. We show that the proposed mechanism offers high levels of scalability and versatility while supporting different levels of autonomy and granularity.

Keywords Multi-robot systems · Shepherding · Gestural interaction · Digital twins · Swarm guidance

1 Introduction

A swarm refers to a large group of agents that interact with each other to achieve a collectively complex behaviour. Biological swarms have attracted researchers' interest to understand how swarm members cooperate to generate collective capabilities that exceed the summation of their individual capabilities [1]. Many natural

A. Hussein (✉)
Faculty of Science and Technology, University of Canberra, ACT, Australia
e-mail: aya.hussein@canberra.edu.au

H. Nguyen
Department of Software Development, National Laboratory for Securing Information (NLSI), Ha Noi, Vietnam

H. Abbass
School of Systems and Computing, University of New South Wales-Canberra, ACT, Australia

G. Barresi and S. Wischniewski (eds.), *Ergonomics in Robotics: Advances and Innovations*, Springer Tracts in Advanced Robotics 162,
https://doi.org/10.1007/978-3-032-19374-2_20

swarm systems have inspired scientists to replicate their behaviours in artificial settings. For instance, the behaviours of ants have inspired the well-known Ant Colony Optimisation algorithm [2].

The field of swarm robotics [3] aims to leverage insights from biological swarms to allow for the collaboration of large numbers of simple robots towards achieving complex tasks. By designing algorithms that utilise local robot interactions, previous studies have shown the ability of swarms of simple robots to tackle real-world tasks, including collective transport [4], search and rescue [5], and surveillance [6].

Robot swarms are required for tasks that exceed the performance limits of single-robot systems. Additionally, well-designed swarms have three main desirable characteristics that make them advantageous over single-robot systems: robustness, flexibility, and scalability [3]. Robustness refers to the swarm ability to continue its operation despite the failure of some of its members [7]. The distributed nature of swarms makes them robust to individual failures because failed or lost members can be compensated for by the other swarm members. Swarm flexibility refers to the ability of generating a wide range of global behaviours by tweaking the local interactions among swarm members [8]. Scalability results from the fact that swarm performance experiences low variations as a function of its size [7].

Despite the potential of robot swarms in various application domains, fully autonomous swarms are not expected to be realised in the near future due to many technical challenges and accountability concerns [9, 10]. This makes the human element a necessary part of swarm operations to ensure that swarm behaviours align with mission objectives and to enable continued success in complex and dynamic environments [11, 12]. Previous studies showed that human-swarm teams are more robust to environmental changes and can achieve higher performance than fully autonomous swarms [12, 13].

Human-swarm interaction (HSI) is concerned with enabling humans and robot swarms to act as a team to optimise common mission objectives. The work on this area can build on findings from human-robot systems. However, HSI brings new challenges that warrant novel solutions. A unique challenge in HSI is the presence of enormous amounts of information about swarm members' perceptions and actions. The limited cognitive capacity of a human operator means that the interaction needs to be carefully designed to enable active human engagement without overwhelming the human.

A key obstacle to utilising robot swarms in realistic operations is the lack of flexible mechanisms for scalable human control over robot swarms. Although several swarm control mechanisms exist in the literature, these mechanisms are either task specific, afford single level of granularity, or lack scalability. This chapter proposes the use of shepherding as an effective swarm control mechanism that addresses the limitations of other mechanisms. Section 2 discusses related literature on swarm control with the aim of assessing the strengths and limitations of existing mechanisms. Then, Sect. 3 articulates the key considerations for designing effective HSI mechanisms as derived from the desirable operation requirements of robot swarms. Section 4 presents additional considerations relating to the design of effective human-autonomy language. The use of shepherding as an algorithm for swarm control is presented in Sect. 5. This is followed by Sect. 6 where we describe how shepherding is used in our proposed mechanism for HSI. The experimental setup and evaluation are presented in Sects. 7 and 8, respectively. Some concluding remarks and future directions are finally presented in Sect. 9.

2 Related Work

A critical factor to the success of human-swarm teams is the availability of effective interaction mechanisms that allow the human to be adequately engaged with swarm operation without being overwhelmed [11, 12]. Interaction mechanisms that proved successful in single robot systems do not scale up to swarms of large number of robots. This section discusses the main human factors challenges that can impede the effective human control of swarm. The section then presents the key classes of swarm control mechanisms with a focus on how well they address the human factors challenges.

2.1 Human Factors Challenges

HSI faces unique challenges around managing human workload, maintaining adequate situational awareness, and designing appropriate levels of autonomy. There is significant evidence that human performance drops notably with the increase in the number of robots in multi-robot system [14–16]. This is exacerbated in swarm systems where the considerable inter-dependencies between robots lead to an exponential increase in control complexity [10]. This complexity translates to significant workload imposed on the human operator, which can quickly surpass the human cognitive limits. The two main components contributing to this workload are the need for processing swarm information to form situational awareness and the need for making decisions about what actions the swarm should execute next.

Hepworth et al. [17] proposed a transparency architecture for organising information coming from the swarm to facilitate human situational awareness. The architecture defines swarm transparency in terms of three tenets: interpretability, explainability, and predictability. Interpretability is about presenting swarm state information in a human-interprtable form. Explainability aims to provide understanding to the human by providing the causal chains that lead to changes in swarm states. Predictability focuses on offering projection of the swarm state in the near future. Despite the increasing interest in swarm transparency, only few studies (e.g., [18, 19]) focused on experimentally investigating the different tenets of swarm transparency and how they can be delivered in a a manner that enables situational awareness without overwhelming the human operator.

The importance of situational awareness lies in its role for enabling informed decision making. In HSI, human decision making takes the form of commands issued to the swarm members such that the complexity of the commands stem mainly from their level of autonomy, granularity and the rate at which they should be issued. Using low levels of autonomy means that command complexity is mostly managed by the human and requires the human to have continuous control over the swarm. Meanwhile, high-autonomy levels require intermittent human control to select tactics or issue sub-goals while leaving it to the swarm to decide on the execution details. Additionally, command granularity, in terms of whether commands are designed to address individuals, groups, or the whole swarm, plays a critical role in determining the required rate of commands. These unique requirements prompted researchers to develop innovative control mechanisms specifically designed to tackle the complexities of robot swarms, as explored in the remainder of this section.

2.2 *Swarm Control Mechanisms*

Kolling et al. [10] identified four classes of swarm control mechanisms; these are behaviour selection, parameter selection, environmental influences, and control members. Behaviour selection assumes the availability of a library of pre-programmed swarm behaviours from which the human can choose one. These behaviours can be of low level (e.g., tele-operation commands [20]) or high level (e.g. formation control [21]). Behaviour selection is relatively easy to use by novice operators. However, it requires the availability of all necessary swarm behaviours and it lacks adaptability to changing environmental conditions [22].

The second class of swarm control mechanisms is parameter selection where the behaviours of individual swarm members are pre-programmed, however these behaviours can be parameterised to generate diverse swarm-level behaviours. Couzin et al. [8] showed that different parameters of the same agent decision rules can be used to generate various swarm behaviours, including aggregation, torus, and flocking. However parameter selection is impractical for online operation, as it requires finding agent-level parameters that generate a given desirable swarm-level patterns, which is one of the most difficult problems in swarm robotics [23].

Environmental influences is the third class of swarm control where environmental signals can be inserted in the swarm operation environment. This can take the form of message broadcasting via beacons [22] or pheromone-like inserted signals [24, 25]. Such methods offer indirect control over the swarm which can be difficult for human operators and result in inferior performance compared to more direct methods [22].

As for the last class of methods, several studies investigated the use of a few control members to propagate human control over the rest of the swarm. In these methods, the human can send their commands (which can take the form of low-level tele-operation or high-level guidance) to one or more control members. The effect of these commands is then propagated to the rest of the swarm using agent-to-agent interaction. A control member is mostly assumed the role of a leader that other agents are programmed to follow [26]; however a handful of studies investigated control via a predator that other agents are programmed to repel from [27] or a normal agent that other agents interact with indistinguishably [28]. This class of control methods is the most promising mechanism for HSI. It allows greater flexibility in intervening with swarm operations under unforeseen conditions, it does not require the human operator to reverse engineering the agent-level behaviours needed to execute their intention, and it offers more direct control over the swarm.

Nevertheless, there is some concerns about existing methods that use few members for swarm control. For instance, using leaders to guide the swarm makes an underlying assumption that all swarm members should be grouped in one cluster before being guided by the leader. Many tasks require the swarm to be scattered during some phases, for example environmental monitoring [29] and target search [30] tasks. It is not clear how control via leaders can be used to separate swarm members into different clusters or regroup them into one cluster. Similarly, using predators or normal agents for swarm control has been validated only under limited swarm configurations. In the next section, we present some considerations for designing swarm control methods to avoid the limitations of the existing methods.

3 Considerations for Effective HSI

There is a stressing need for effective control mechanisms to enable meaningful human supervision and guidance of swarm operations in a way that ensures the alignment of swarm operations with technical and ethical requirements. In this section, we elicit some factors that need to be taken into consideration during the design of swarm control methods to ensure effective human-swarm interaction.

Scalability: A key desirable characteristic of swarm systems is their scalability in terms of the low performance variations as a function of the sizes of task and swarm. To achieve this scalability, both agent-level behaviours and human interaction with the swarm need be scalable. A plethora of studies have been dedicated to designing scalable agent behaviours by making the agent behaviours responsive only to the state of itself and its neighbours, rather than the state of all agents in the swarm. Examples of scalable agent behaviours include Reyonld's boid model [31] and fish school behaviours [32]. For human control, it is critical to design scalable control mechanisms where the associated cognitive demands do not increase significantly with the swarm size. Some techniques exist in the literature that consider the swarm as a single unit from the point of view of the operator (e.g., control by leader or broadcasting same command to all members [20]). This makes the control complexity relatively constant regardless of swarm size. The key limitations of these techniques are twofold. First, they make significant assumptions about swarm members' spatial configurations which limits the range of tasks that can be performed. For example, control by leader assumes that all swarm members are clustered in one group, making it unsuitable for tasks requiring high spatial distribution. Second, they constraint the operator's ability to intervene with agent-level issues.

Versatility: The flexibility characteristic of robot swarms entails that swarms should be easily employed to deliver wide variety of tasks. This can be achieved by ensuring that agent behaviours are not tightly coupled to a specific task and by designing control methods to allow the human to interact with the swarm under different swarm configurations. As discussed in the above point, some existing HSI mechanisms are scalable, but limited in the range of tasks they fit. On the other hand, mechanisms where swarm members are individually controlled offer high flexibility but have unacceptably poor scalability. There is a lack of HSI mechanisms that accommodate for the, sometimes conflicting, characteristics of scalability and versatility.

Levels of granularity: The ability to control the swarm at different levels of granularity is key to efficient and effective interaction [17]. Interacting with the swarm at the macro level is useful for creating a big picture of the swarm, updating swarm-level plans, or issuing high-level commands to all swarm members. Meanwhile, micro-level interactions are important for revealing detailed information about agent states, intervening with a single agent's operation, or sending different commands to different agents. The same task can require both macro- and micro-level control. Enabling intuitive switching between different levels of granularity can greatly facilitate human-swarm team performance.

Levels of autonomy: Selecting the appropriate level of autonomy is critical to the performance of any autonomous systems, and robot swarms are not an exception. While the level of granularity is concerned with the number of swarm members being addressed by a command, the level of autonomy is concerned with the amount of details supplied by the human on how to perform an action. Low-levels of autonomy require very detailed human guidance by using tele-operation-like commands (e.g., move forward and turn left). On the other hand, higher levels of autonomy require discrete human guidance, for instance to communicate way-points. Low autonomy imposes high workload, whereas high autonomy can cause human disengagement. HSI mechanisms should ensure an appropriate level of autonomy that allows for adequate human supervision without exceeding the cognitive capacity of the human.

4 Considerations for Human-Autonomy Language

In addition to the considerations described in Sect. 3, Abbass et al. [33] discusses important requirements that human-autonomy languages need to address to enable seamless information flow and effective teaming. These requirements are: contextual relevance, computational efficiency, semantic equivalence, and direct syntactic mapping.

Contextual relevance refers to the appropriateness of a communication language to a given context, in terms of its ability to convey important information. In the context of swarm control via a few control members, contextual relevance means that the language used by the human needs to be sufficiently expressive to convey human intents to the control members. **Computational efficiency** is necessary to ensure that the communication does not form a bottleneck in the system. Computational efficiency is determined by the efficiency of the language in terms of the compactness of its vocabulary and structure as well as the the translation efficiency which describes the resources required to convert an agent's intents into sentences and process the sentences by its recipient.

Semantic equivalence describes the degree to which a communication language enables the encoding and decoding of the sender's intent while preserving its semantics. Lastly, **direct syntactic mapping** is concerned with the degree to which agents that reason at different time scales can communicate effectively using the language. The factors of human-autonomy communication languages and the considerations for swarm control methods, described in Sect. 3, are used in this work to guide the design of effective HSI as described in the following two sections.

5 Shepherding Model for Swarm Control

The literature on swarm robotics has been heavily inspired by biological swarms. Algorithms for agent local behaviours and interactions mimic several aspects of the behaviours of social animals and insects. A typical example of nature-inspired swarm behaviours is the use of the boid model for generating flocking behaviours in robot swarms. Enabling human control over robot swarms can equally benefit from the lessons learnt from human control over biological swarms. Throughout history, humans have successfully controlled large swarms of sheep through the practice of shepherding.

In shepherding, a human conveys their intentions to a sheepdog, acting as a control agent, which in turn guides the members of the swarm by herding them along a path directed by the human. The presence of a sheepdog aids in the control of the swarm, benefiting from the dog's superior cognitive and physical abilities compared to the individual members of the swarm. Drawing inspiration from sheep herding, researchers have replicated this concept in robotic systems, showcasing its potential as a scalable swarm control mechanism [34–36]. This section describes the shepherding model by presenting the behaviours of the sheepdog and the sheep.

Sheepdog behaviours: Let β be the agent acting as the sheepdog used to control a swarm of n agents $\Pi = \{\pi_1, \pi_2, \ldots, \pi_n\}$ acting as sheep. The operation of β can be described in terms of two behaviours: collecting and driving. The collecting behaviour is used to collect scattered swarm members into one group. To achieve this, β finds the swarm member π_i that is furthest to the geometric center of mass (GCM) of Π. Then, β moves itself to be behind π_i and starts moving towards the GCM of the swarm. This is achieved by setting the dog's target position as follows during the collecting behaviour:

$$P_\beta^{t+1} = P_{\pi_i}^t + R_{\pi\beta}.\frac{P_{\pi_i}^t - GCM^t}{\|P_{\pi_i}^t - GCM^t\|} \tag{1}$$

such that P_x^t is the position of agent x at time t, and $R_{\pi\beta}$ is the distance within which a swarm member can sense the presence of β. In response, π_i escapes from β by moving towards the GCM of the swarm such that the swarm becomes more cohesive. β keeps collecting scattered sheep till all swarm members form one cluster. This swarm state (i.e., one cluster) triggers the driving behaviour where β moves behind the GCM of the swarm such that the swarm becomes situated between β and the target destination, denoted by P_{dest}. This is achieved by setting the target position of β using the following equation:

$$P_{\beta}^{t+1} = GCM^t + R_{\pi\beta}.\frac{GCM^t - P_{dest}}{\|GCM^t - P_{dest}\|} \tag{2}$$

This driving behaviour continues until the swarm reaches the desirable destination defined by P_{dest}.

Sheep behaviours: A sheep agent π_i exhibits a number of behaviours in response to the states of the sheepdog and other sheep agents. Perceiving a dog as a predator, π_i repels from β when β is within a distance of $R_{\pi\beta}$ from π_i. To increase its sense of safety, π_i moves towards the GCM of its sheep neighbours that are within the sensing range of $R_{\pi\pi}$. To avoid colliding with other sheep agents, π_i repels from the GCM of very close agents that are within $R_{\pi-\pi}$ from π_i. These sheep behaviours (repulsion from dog, cohesion, and collision avoidance) are linearly combined to determine the next target position for π_i.

6 HSI via Shepherding

The concept of shepherding can be used for enabling human control over robot swarms by designing a setting where the human interacts with a sheepdog-like robotic agent which influences the rest of the swarm. The interactions between the swarm and the sheepdog are dictated by the shepherding model described in Sect. 5. The interactions between the human and the sheepdog need to be designed in light of the factors presented in Sects. 3 and 4. In this section, we describe the design of ours hepherding-based mechanism for HSI.

We consider a task consisting of a swarm of unmanned ground vehicles (UGVs) representing a sheep swarm that the human can control through an aerial unmanned vehicle (UAV) representing a sheepdog. The UGVs apply the behaviours described in Sect. 5. Meanwhile, the UAV has two modes of operations corresponding to two levels of autonomy. Under low autonomy, the human can teleoperate the UAV by sending it low-level commands to control its altitude (up, down, or maintain the same altitude) and movement in the x-y plane (forward, backward, left, right, or hover). The human needs to use their left hand to issue commands for altitude changes, whereas they need to use their right hand to issue commands about movement in the x-y plane.

On the other hand, in the high level of autonomy, the human controls the UAV by issuing one of these commands: *collect*, *drive*, and *hover*. When the *collect* command is given, the UAV applies the algorithm for collecting scattered UGVs towards the GCM of the swarm (denoted by collection sub-goal). This is achieved by setting the target position of the UAV according to Eq. 1. Similarly, the drive command directs the UAV to go behind the swarm's GCM and propel it towards the driving sub-goal. This is achieved by setting the target position of the UAV according to Eq. 2. To ensure the UGVs remain within the environment and to avoid potential collisions, the human must inform the UAV whether the driving and collection sub-goals should be recomputed by taking these additional factors into consideration. For the high level of autonomy, the left hand is used for selecting between the *collect*, *drive*, and *hover* commands. Meanwhile, the right hand is used to issue commands about recomputing the collect/drive sub-goals.

7 Experimental Setup

This section presents the experiments conducted to evaluate the proposed shepherding-based HSI mechanism described in Sect. 6. We designed a task where a swarm of three Turtlebot-3 UGVs are deployed in a lab environment and are required to navigate to a designated destination while avoiding no-access areas. The environment is defined by a square geo-fence of a side length of 2.9 m. Each UGV has a Raspbian Jessie single board computer (SBC) with a Rasberry Pi 2 processor which runs Raspbian Linux operating system. The UGVs' position and orientation information are estimated from a top-view camera. ROS Core has been used for communication between the different system components.

To enable using the proposed shepherding-based HSI, we needed a UAV to act as the sheepdog. Due to the lack of availability of a small-sized indoor UAV by the research team, we used a virtual UAV instead. That is, the setup constituted three physical UGVs and one virtual UAV. To enable this, we developed a simulated environment in Gazebo that replicates the physical environment and includes a virtual UAV. A graphical user interface (GUI) has been developed to visualise the positions of both the physical UGVs and the virtual UAV projected in a single unified interface. The GUI gets position data from the real environment and from Gazebo and plots it on a single view.

The human is required to control the swarm by issuing gestural commands to the UAV. The set of available commands depend on the system's level of autonomy. The gesture are captured by Microsoft Kinect connected to the computer that runs the GUI. A *C#* program has been developed to recognise the gestures and send them to Gazebo. The UAV executes the user commands and its new position is sent to the physical swarm so that the UGVs can apply the sheep behaviour described in Sect. 5. Figures 1 and 2 depicts the system used and the GUI, respectively.

Each of the first and second authors acted as a human operator and performed the task under the two levels of autonomy. The system was developed by the second

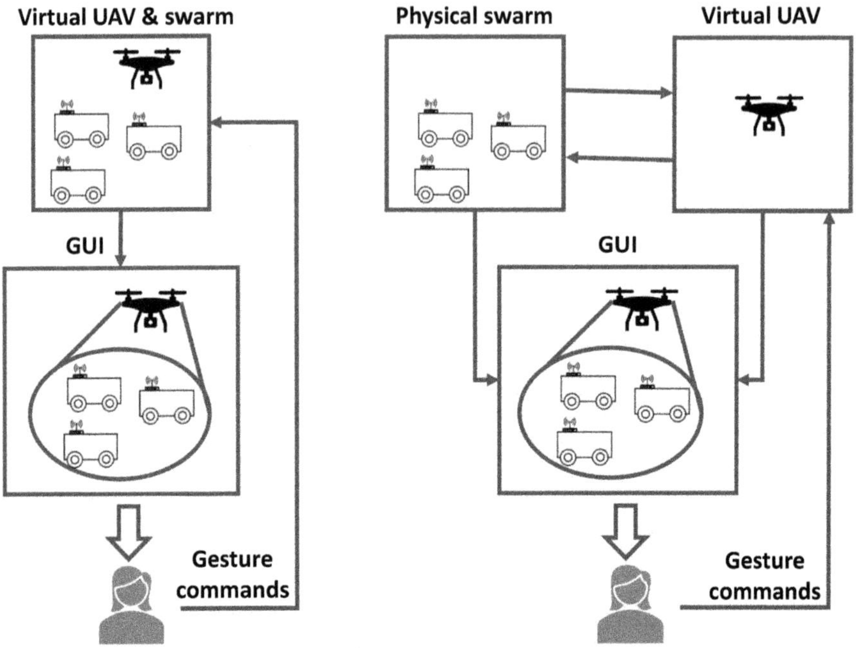

Fig. 1 An overview of the systems used during the training (left side) and evaluation (right side)

Fig. 2 The GUI presenting information from the physical and virtual environments

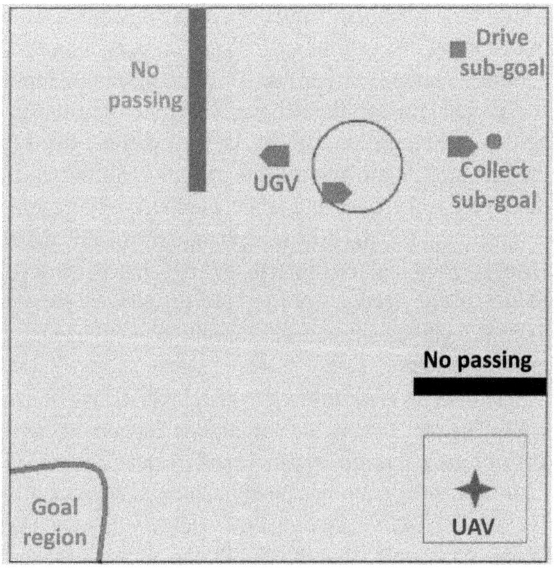

author; meanwhile, the first author interacted with the system for the first time during the experiments. The human operators underwent two training sessions (one for each level of autonomy) to become familiar with the system. Each training session included ten episodes where the swarm was simulated such that the task was running only on Gazebo (i.e., no physical environment). This is to ensure that the trainee is mainly focused on learning the gestures without being overwhelmed by the complexity of controlling physical UGVs and the associated actuation errors that occur in real environments. After the end of each training session, the operator performed twenty episodes under the same level of autonomy such that the task involved both the simulation and physical environments. The same GUI and gestural commands have been used during training and evaluation. Figure 1 shows the system used for training and evaluation.

8 Results

This section presents the evaluation results obtained from the experiments. It then evaluates the system by analysing it against the design considerations presented in Sects. 3 and 4.

8.1 Quantitative Results

Table 1 shows the success rate (SR), task duration, and the UGVs travel distance from successful episodes. Under high autonomy, (100%) success has been recorded as compared to 75% success under low autonomy. For task duration, the low level of autonomy required a statistically significant more steps (M = 6638, SD = 1132) than the high level of autonomy (M = 5121, SD = 1381), ($p = .015$). As for energy efficiency, there is no statistically significant difference between the distance travelled by the UGVs under low autonomy (M = 16.5, SD = 2.7) and high autonomy (M = 19.3, SD = 6.3), ($p = 0.21$). Qualitatively, it was reported that the low level of autonomy imposes much more physical workload due to the high rate of issuing new gestures than the high autonomy level.

Table 1 Quantitative results

	SR	Task duration	UGVs travel distance
	(%)	$\mu \pm \sigma$	$\mu \pm \sigma$
Low autonomy	75	6638 ± 1132	16.5 ± 2.7
High autonomy	100	5121 ± 1381	19.3 ± 6.3

Table 2 Effectiveness of the proposed shepherding-based HSI

		Low autonomy	High autonomy
Scalability	Control points	One UAV	One UAV
	Complexity	$O(l)$ if UGVs are clustered; $O(N \cdot l)$ if dispersed	$\sim O(1)$
Versatility	Form clusters	Drive individual UGVs towards swarm GCM	Issue collect command
	Control clusters	Drive swarm GCM towards a target point	Issue drive command
	Separate clusters	Move UAV to be situated on top of swarm GCM	Extendable: issue parameterised drive command
	Control individuals	Propel individual UGV towards a target point	Extendable: issue parameterised drive command
Granularity	Low	Drive swarm GCM	Drive swarm GCM
	High	Drive individual UGVs	Extendable: issue parameterised drive command

8.2 Analysing HSI Effectiveness

Section 3 presents four key factors that impact the effectiveness of HSI. Here, we analyse the effectiveness of the proposed shepherding-based HSI mechanism by studying the degree to which these factors are satisfied in the design of the interaction. Table 2 summarises the analysis of the effectiveness of the proposed HSI under its two levels of autonomy.

Both levels of autonomy offer high scalability as only one UAV (acting as a sheepdog) is controlled by the human regardless of the swarm size. The effort in controlling the UAV depends on the level of autonomy. For low autonomy, if the swarm members are grouped in one cluster, the swarm will be dealt with as a single entity and the UAV control effort will depend only on the length of the path to the goal area l. In contrast, if the UGVs are dispersed and need to be driven to different target areas, the control effort becomes linear in both the swarm size and the path length.

For the high level of autonomy, the UAV control effort is almost constant across different clustering configurations. If the UGVs are fully clustered, the human can use a single *drive* command for the UAV to drive the whole swarm as a singly entity. If the UGVs are dispersed, a single *collect* command can be used by the human for the UAV to collect all the UGVs in one cluster. In either of these clustering configurations, a few more commands could be also needed for signalling the need for avoiding collisions or avoiding exiting the geo-fence.

To study the versatility of the proposed approach, we investigate the degree to which it can be used for forming clusters, dispersing clustered UGVs, controlling clustered UGVs, and controlling dispersed UGVs. The low level of autonomy offers the highest versatility due to the free-from teleoperation of the UAV. However, this requires the operator to be highly skilled to achieve their intended goal. For the high level of autonomy, forming clusters is simply achieved by issuing *collect* commands. Controlling clustered UGVs is equally simple by issuing *drive* commands. On the other hand, dispersing clustered UGVs is not directly supported by the current set of high-autonomy commands. Similarly, controlling individual UGVs towards different goal areas is not supported by the current set of high-autonomy commands. Nonetheless, these two types of operations can be easily supported, for example by allowing the human to parameterise their drive commands by providing the current and desirable points of the UGV(s) to be controlled.

Considering the granularity aspect of HSI, the proposed mechanism enables both high and low granularity under the low level of autonomy. High granularity is achieved by targeting and driving single UGVs at a time, whereas low granularity can be achieved by driving UGV clusters. The high level of autonomy enables low granularity by collecting UGVs and driving them as a single entity. The current set of available commands do not allow for high granularity under high autonomy; however, this can be easily accounted for by allowing the human to parameterise their drive commands, as discussed above.

Overall, it can be seen that the high level of autonomy offers the highest scalability and can be easily extended to enable highest versatility and to allow for different levels of granularity. On the other hand, the low level of autonomy already offers the most of versatility and granularity, but falls significantly short of scalability and relies heavily on operators' teleoperation skills.

8.3 *Analysing Language Effectiveness*

Table 3 summarises the language effectiveness analysis for the gesture languages designed for the low and high levels of autonomy. Both languages afford contextual relevance for the shepherding task described in Sect. 7 as shown by their ability to allow the human to convey their intents to the UAV and achieve high success rates. For the law autonomy, the language afford contextual relevance for generic swarm tasks due to the low-level vocabularies that allow performing any sequence of actions. However, the language designed for the high-level of autonomy cannot generalise to broader swarm tasks without extending it by allowing the human to provide parameters for the drive command. This extension can be simply implemented by using mouse clicks to allow the human to pinpoint the parameters of the *drive* command on the GUI.

Table 3 Language effectiveness

	Low autonomy	High autonomy
Contextual relevance	Expressive vocabulary that allows executing any operations	Expressive for the specific mission
Computational efficiency	Low compactness	High compactness
	High translation efficiency due to gesture commands Real time operation	
Semantic equivalence	High, due to 1-to-1 mapping between commands and UAV's primitive actions	High, due to 1-to-1 mapping between commands and UAV's algorithm output
Direct syntactic mapping	Similar timescale between human and UAV	Human uses longer timescale than UAV

The computational efficiency is studied through language compactness and translation efficiency. For law autonomy, the language has low compactness as each command corresponds to a single UAV action. Meanwhile, the language for high autonomy is characterised by high compactness as a single command can be unpacked into a long sequence of UAV actions. Both languages have high translation efficiency thanks to the use of the natural and intuitive gestural modality for encoding human intent. Using Microsoft Kinect allowed for the efficient decoding of the gestures into UAV actions. Both languages allow for real-time interaction between the human and the UAV.

The semantic equivalence of the law-autonomy language is high because of the one-to-one mapping between the human commands and the primitive actions of the UAV. The high-autonomy language also ensures semantic equivalence because of the use of a deterministic algorithm for calculating the way-points of the UAV. Marking the positions of the drive and collect sub-goals on the GUI (see Fig. 2) enhances the semantic equivalence by giving the human some information about how their next drive/collect commands will be executed.

For the direct syntactic mapping, using the low-autonomy language the human issues commands at the same time scale of the UAV. The experiments showed that this is possible due to the computational efficiency of the language. Nonetheless, this setting quickly leads the human to be fatigued as it requires high rate of task state assessment and command issuing by the human. On the other hand, the high autonomy language allows the human to work at a significantly longer average timescale which induces much less workload than the low autonomy.

9 Conclusion

Human control of robot swarms is challenging due to the wide gap between the operational requirements of human-swarm systems and the limited cognitive capacity of human operators. Lessons from human control of animal swarms (e.g., sheep swarms) can provide useful insights on the use of assistive control agents (e.g., sheepdog) to help exert human control on large swarms. This chapter articulates the key considerations for designing effective human-swarm interaction as driven from the desirable characteristics of swarm systems. We propose a shepherding-based HSI mechanism based on these considerations. We evaluate the proposed mechanism in a lab setting where a human operator needs to control a swarm of UGVs using a virtual UAV. Our results demonstrate the ability of the proposed mechanism to deliver human intents under two different levels of autonomy. The proposed mechanism is evaluated against the considerations for effective HSI and was shown to offer high scalability, versatility, and various levels of granularity. Additionally, the language designed for the proposed mechanism was shown to address the different requirements for effective human-autonomy languages.

Future work can expand the proposed mechanism to enable the human to easily parameterise their commands to maximise the versatility of the human-swarm system. Additionally, our mechanism can be extended by allowing the human to create macros for common behaviours such that the UAV control would require less input from the human during executing these behaviours. Finally, future research can investigate the use of AI to support human control of the UAV to optimise swarm operations.

Acknowledgements This research is partially funded by the Australian Research Council Discovery grant no. DP200101211

References

1. Yu, X., Gen,M.:Introduction to evolutionary algorithms. Springer Science & Business Media (2010)
2. Dorigo, M., Birattari, M., Stutzle, T.: Ant colony optimization. IEEE Comput. Intell. Mag. **1**(4), 28–39 (2006)
3. Şahin, E.: Swarm robotics: from sources of inspiration to domains of application. In: International Workshop on Swarm Robotics, pp. 10–20. Springer (2004)
4. Alkilabi, M.H.M., Narayan, A., Tuci, E.: Cooperative object transport with a swarm of e-puck robots: robustness and scalability of evolved collective strategies. Swarm Intell. **11**(3), 185–209 (2017)
5. Din, A., Jabeen, M., Zia, K., Khalid, A., Saini, D.K.: Behavior-based swarm robotic search and rescue using fuzzy controller. Comput. & Electr. Eng. **70**, 53–65 (2018)
6. Xu, C., Zhang, K., Jiang, Y., Niu, S., Yang, T., Song, H.: Communication aware UAV swarm surveillance based on hierarchical architecture. Drones **5**(2), 33 (2021)
7. Hussein, A., Elsawah, S., Petraki, E., Abbass, H.A.: A machine education approach to swarm decision-making in best-of-n problems. Swarm Intell. **16**(1), 59–90 (2022)

8. Couzin, I.D., Krause, J., James, R., Ruxton, G.D., Franks, N.R.: Collective memory and spatial sorting in animal groups. J. Theor. Biol. **218**(1), 1–11 (2002)
9. Hussein, A., Elsawah, S., Abbass, H.A.: Towards trust-aware human-automation interaction: An overview of the potential of computational trust models. In: HICSS, pp. 1–10. (2020)
10. Kolling, A., Walker, P., Chakraborty, N., Sycara, K., Lewis, M.: Human interaction with robot swarms: a survey. IEEE Trans. Hum.-Mach. Syst **46**(1), 9–26 (2016)
11. Hussein, A., Abbass, H.: Mixed initiative systems for human-swarm interaction: opportunities and challenges. In: 2018 2nd Annual Systems Modelling Conference (SMC), pp. 1–8 (2018)
12. Hussein, A.: Performance, trust, and transparency for effective human-swarm interaction, UNSW (2021). Ph.D. Dissertation
13. Kolling, A., Nunnally, S., Lewis, M.: Towards human control of robot swarms. In: Proceedings of the Seventh Annual ACM/IEEE International Conference on Human-Robot Interaction, pp. 89–96. ACM (2012)
14. Trouvain, B., Wolf, H.: Evaluation of multi-robot control and monitoring performance. In: Proceedings. 11th IEEE International Workshop on Robot and Human Interactive Communication, pp. 111–116. IEEE (2002)
15. Crandall, J.W., Goodrich, M.A., Olsen, D.R., Nielsen, C.W.: Validating human-robot interaction schemes in multitasking environments. IEEE Trans. Syst. Man Cybern.-Part A: Syst. Hum **35**(4), 438–449 (2005)
16. Velagapudi, P., Scerri, P., Sycara, K., Wang, H., Lewis, M., Wang, J.: Scaling effects in multi-robot control. In: IEEE/RSJ International Conference on Intelligent Robots and Systems , pp. 2121–2126 (2008)
17. Hepworth, A.J., Baxter, D.P., Hussein, A., Yaxley, K.J., Debie, E., Abbass, H.A.: Human-swarm-teaming transparency and trust architecture. IEEE/CAA J. Autom. Sin. **8**(7), 1281–1295 (2020)
18. Hussein, A., Elsawah, S., Abbass, H.A.: The reliability and transparency bases of trust in human-swarm interaction: principles and implications. Ergonomics **63**(9), 1116–1132 (2020)
19. Walker, P., Nunnally, S., Lewis, M., Kolling, A., Chakraborty, N., Sycara, K.: Neglect benevolence in human control of swarms in the presence of latency. In: 2012 IEEE International Conference on Systems, Man, and Cybernetics (SMC). ()
20. Podevijn, G., O'Grady, R., Nashed, Y.S., Dorigo, M.: Gesturing at subswarms: towards direct human control of robot swarms. In: Towards Autonomous Robotic Systems: 14th Annual Conference. TAROS, Oxford, UK, 28–30 Aug 2013, Revised Selected Papers 14. Springer (2014), 390–403 (2013)
21. Nagavalli, S., Chien, S.-Y., Lewis, M., Chakraborty, N., Sycara, K.: Bounds of neglect benevolence in input timing for human interaction with robotic swarms. In: Proceedings of the Tenth Annual ACM/IEEE International Conference on Human-Robot Interaction, pp. 197–204. (2015)
22. Kolling, A., Sycara, K., Nunnally, S., Lewis, M.: Human swarm interaction: An experimental study of two types of interaction with foraging swarms. J. Hum.-Robot. Interact. **2**(2) (2013)
23. Dorigo, M., Theraulaz, G., Trianni, V.: Reflections on the future of swarm robotics. Sci. Robot. **5**(49), eabe4385 (2020)
24. Walter, B., Sannier, A., Reiners, D., Oliver, J.: UAV swarm control: calculating digital pheromone fields with the GPU. J. Def. Model. Simul. **3**(3), 167–176 (2006)
25. Li, G., Chen, C., Geng, C., Li, M., Xu, H., Lin, Y.: A pheromone-inspired monitoring strategy using a swarm of underwater robots. Sensors **19**(19) (2019)
26. Fabra, F., Zamora, W., Masanet, J., Calafate, C.T., Cano, J.-C., Manzoni, P.: Automatic system supporting multicopter swarms with manual guidance. Comput. Electr. Eng. **74**, 413–428 (2019)
27. Goodrich, M.A., Pendleton, B., Sujit, P.B., Pinto, J.: Toward human interaction with bio-inspired robot teams. In: 2011 IEEE International Conference on Systems, Man, and Cybernetics, pp. 2859–2864. (2011)
28. Walker, P., Amraii, S.A., Lewis, M., Chakraborty, N., Sycara, K.: Human control of leader-based swarms. In: 2013 IEEE International Conference on Systems, Man, and Cybernetics, pp. 2712–2717. (2013)

29. Duarte, M., Gomes, J., Costa, V., Rodrigues, T., Silva, F., Lobo, V., Marques, M.M., Oliveira, S.M., Christensen, A.L.: Application of swarm robotics systems to marine environmental monitoring. In: OCEANS 2016 - Shanghai, pp. 1–8. (2016)
30. Yang, B., Ding, Y., Jin, Y., Hao, K.: Self-organized swarm robot for target search and trapping inspired by bacterial chemotaxis. Robot. Auton. Syst. **72**, 83–92 (2015)
31. Reynolds, C.W.: Flocks, herds and schools: a distributed behavioral model. In: Proceedings of the 14th Annual Conference on Computer Graphics and Interactive Techniques, pp. 25–34. (1987)
32. Berlinger, F., Gauci, M., Nagpal, R.: Implicit coordination for 3D underwater collective behaviors in a fish-inspired robot swarm. Sci. Robot. **6**(50), eabd8668 (2021)
33. Abbass, H.A., Petraki, E., Hunjet, R.: JSwarm: a Jingulu-inspired human-AI-teaming language for context-aware swarm guidance. Front. Phys. 10, (2022)
34. Hussein, A., Petraki, E., Elsawah, S., Abbass, H.A.: Autonomous swarm shepherding using curriculum-based reinforcement learning. In: Proceedings of the 21st International Conference on Autonomous Agents and Multiagent Systems, pp. 633–641. (2022)
35. El-Fiqi, H., Campbell, B., Elsayed, S., Perry, A., Singh, H.K., Hunjet, R., Abbass, H.A.: The limits of reactive shepherding approaches for swarm guidance. IEEE Access **8**, 214–671 (2020)
36. Mohamed, R.E., Elsayed, S., Hunjet, R., Abbass, H.: A graph-based approach for shepherding swarms with limited sensing range. IEEE Congress on Evolutionary Computation (CEC) , 2315–2322 (2021) IEEE

Societal Perspectives

HRI, Culture, and Social Anxiety

Tatsuya Nomura

Abstract Culture and social anxiety are among the human factors influencing the social acceptance of robots. In this chapter, research results are presented based on comparisons among nations conducted by the author's research group, consisting of those about attitudes toward robots, assumptions about and images of robots, Frankenstein Syndrome, and roboethics. Then, an overview of social anxiety-related research in the field of human–robot interaction (HRI) is provided, regarding on fear of negative evaluation and gaze impact, impact of social anxiety on anxiety and tension during interactions with robots, and effects of social anxiety on robot preference in dialogue situations. Finally, the possibility of interaction effects between cultural factors and social anxiety in HRI is discussed and some future research directions are proposed.

Keywords Culture · Social anxiety · Human–robot interaction · Social acceptance of robots · Interaction effects

1 Introduction

Robots capable of interacting with humans have the potential to solve ergonomic problems for the realization of various tasks, such as elder care, education, mental health, and assistance in domestic fields. However, numerous human factors, including age, gender, educational background, and technological experience, influence the acceptance of robots in society. Moreover, the interaction effects between such human and robot/situational factors can occur in increasingly complex ways.

Among the human factors that influence the social acceptance of robots, this chapter focuses on culture. In recent years, several studies have investigated the influence of human culture on feelings and behaviors toward robots, some of which have focused on the social acceptance of robots [1, 2]. Considering the increasingly

T. Nomura (✉)
Faculty of Advanced Science and Technology, Ryukoku University, Otsu, Shiga, Japan
e-mail: nomura@rins.ryukoku.ac.jp

G. Barresi and S. Wischniewski (eds.), *Ergonomics in Robotics: Advances and Innovations*, Springer Tracts in Advanced Robotics 162,
https://doi.org/10.1007/978-3-032-19374-2_21

widespread use of robots around the world, culture is regarded as one of the most influential factors that encourage/discourage their social acceptance.

In addition to culture, social anxiety can affect interaction with robots, and because different cultures have different social rules and expectations, its perceivable effects vary [3]. Moreover, the relationship between social anxiety and the acceptance of robots is expected to vary across cultures.

Therefore, this chapter presents research results based on comparisons among nations conducted by the author's research group. Next, it provides an overview of research related to social anxiety in the field of human–robot interaction (HRI). Finally, it discusses the possibility of interaction effects between cultural factors and social anxiety in HRI and proposes some future research directions.

2 Survey Studies About Cultural Differences in HRI

Recently, some reviews have been published with a focus on culture in relation to HRI. Papadopoulos and Koulouglioti [1] analyzed 22 papers published from 2000 to 2017, and found that culture influenced attitudes and preferences toward both humanoid and animal-like robots. Lim et al. [2] conducted a literature review from a more general perspective. They reviewed 50 studies on how culture influences human expectations, attitudes, and behaviors before, during, and after interactions with robots. Their paper has a specific characteristic in that the reviewed studies were analyzed in terms of sample size, cultural comparisons, participant age range, methodology (e.g., experiment, survey, etc.), type of robot used, and primary research question. However, these reviews lack perspective regarding what measures should be used for cultural comparisons. Therefore, instead of conducting a literature review, the present chapter aims to introduce a number of survey studies regarding cultural differences and HRI from the perspective of measurements.

The author and several research collaborators have conducted questionnaire surveys aimed at exploring people's acceptance of robots and performing comparisons across nations. The present chapter summarizes the results of these surveys, which consist of international comparisons of the following: (1) attitudes toward robots, (2) assumptions and images about robots, (3) anxiety and expectations toward humanoid robots based on the concept of "Frankenstein Syndrome", and (4) ethical problems related to robots (roboethics). These studies share a common characteristic in that items suitable for international comparison, such as psychological scales and questionnaires, were adopted as measures.

2.1 *Attitudes Toward Robots in General*

An attitude is psychologically defined as a relatively stable and enduring predisposition to behave or react in a certain way toward persons, objects, institutions, or issues; the source of an attitude is either cultural, familial, or personal [4]. On the other hand, research on technophobia [5] suggests that anxiety and negative attitudes toward computers influence their social acceptance. Thus, negative attitudes toward robots are thought to be related to people's understanding of robots, and their feelings about the impact of robots on society and their quality of life. For aiming at measuring the degree to which this psychological construct influences people's acceptance of robots, the author's research group developed a psychological scale called the Negative Attitudes toward Robots Scale (NARS) [6].

The NARS consists of fourteen items classified into three subscales. The first subscale (S1, six items) measures negative attitudes toward interactions with robots. The second subscale (S2, 5 items) measures negative attitudes toward the social influence of robots. The third subscale (S3, 3 items) measures negative attitudes toward emotional interactions with robots. Each item is scored on a five-point scale, from 1) strongly disagree to 5) strongly agree, and an individual's score on each subscale is calculated by adding the scores of all items included in the subscale, with some items reverse-coded. The validity and reliability of the NARS were confirmed through a psychological experiment and survey [6]. Table 1 shows the questionnaire items of the English version of the NARS.

Table 1 The questionnaire items and subscales of the NARS

Item No	Questionnaire items	Sub scale
1	I would feel uneasy if robots really had emotions	S2
2	Something bad might happen if robots developed into living beings	S2
3	I would feel relaxed talking with robots. (a)	S3
4	I would feel uneasy if I was given a job where I had to use robots	S1
5	If robots had emotions, I would be able to make friends with them. (a)	S3
6	I feel comforted being with robots that have emotions. (a)	S3
7	The word "robot" means nothing to me	S1
8	I would feel nervous operating a robot in front of other people	S1
9	I would hate the idea that robots or artificial intelligences were making judgments about things	S1
10	I would feel very nervous just standing in front of a robot	S1
11	I feel that if I depend on robots too much, something bad might happen	S2
12	I would feel paranoid talking with a robot	S1
13	I am concerned that robots would be a bad influence on children	S2
14	I feel that in the future society will be dominated by robots	S2

(a) Reversed item

S1: Negative Attitude toward Situations of Interaction with Robots

S2: Negative Attitude toward Social Influence of Robots

S3: Negative Attitude toward Emotions in Interaction with Robots

Since the development of the NARS, some researchers have reported the influence of negative attitudes on human perceptions and behaviors toward robots, as well as the factors the affect these attitudes. Moreover, some studies have used the NARS to explore cultural differences regarding attitudes toward robots. Bartneck et al. [7] conducted an international comparative survey among seven different countries and reported the presence of cultural differences in NARS scores. In an experiment involving human–robot interaction, Wang et al. [8] found that Chinese participants had more negative attitudes toward robots and relied less on the robot's advice than did those from the US.

The NARS was originally developed with the Japanese, and formally back-translated. Then, it was validated in the multiple languages, including the French, Polish, and Turkish versions [9–11].

An online survey conducted by the author ($N = 200$) reported results that differed from those of existing studies [12]. This survey explored differences in not only negative attitudes toward robots, but also correlations between these attitudes and other factors, such as perceptions of the respondent with respect to their relation to their family and religious commitments. Moreover, this survey took age into account in the analysis. The following two items, which were used in a comparison survey between Japan and the Northern Europe, were presented on the face sheet:

- Do you think you relate to your family members? (five-graded answer from "1. I completely agree" to "5. I completely disagree")
- Does such notion as "I have nothing to do with religion or faith" apply to you? (five-graded answer from "1. It strongly applies to me" to "5. It does not apply to me at all.")

The survey found the following results, First, people in the UK had more negative attitudes toward interactions with robots than did people in Japan. Second, older people who had weaker relationships with their family members had more negative attitudes toward interactions with robots in both countries. Third, older people in the UK who had weaker relationships with their family members also had more negative attitudes toward the social influence of robots. Fourth, older people in Japan who had a stronger commitment to religion had more negative attitudes toward the social influence of robots. These results suggested that not only negative attitudes toward robots, but also the factors that influence them, may differ dependent on culture and age.

2.2 Assumptions About and Images of Robots

The concept of "robot" itself is known to be very old. On the other hand, it is only recently that humanoid and pet-type robots have appeared as commercial products in daily life, even in Japan, which has been regarded as one of the most advanced nations in the world in terms of the development of robotics. Moreover, compared with computers, toward which people have a rather fixed set of images and related assumptions, images and assumptions about robots vary widely depending on the appearance of the robot (e.g., whether it is humanoid or pet-type) or its roles or functions (e.g., if it functions as a vacuum cleaner).

At this stage, it is important to conduct a cultural comparison of what people assume and imagine when they encounter the word "robots" while considering differences in concrete feelings and attitudes toward robots. For aiming at exploring cultural differences in these assumptions and images about robots, the author's research group developed and administered the Robot Assumptions Questionnaire (RAQ) to university students in Japan, Korea, and the US. This questionnaire focused on the following five factors related to humanoid and animal-type robots: relative autonomy, social relationship with humans, emotional aspects, roles assumed, and images held [13].

Based on a pilot test and discussions between researchers from the three countries, the RAQ is composed of one item on assumptions about the relative autonomy of robots (three choices: complete autonomy, partially controlled by humans, and completely controlled by humans), one item on assumptions about social relationships with humans (three choices: equal to humans, similar to pets, and similar to tools), one item on assumptions about the emotional aspects of robots (three choices: equal to humans in emotional capacity, less emotional than humans, and no emotions), 10 items on robot roles, and seven items on images of robots (seven graded scales).

The survey ($N = 796$) found the following results. First, university students in the three countries tended to assume that humanoid robots perform concrete tasks in society, and that animal-type robots play a pet- or toy-like role. Second, Japanese students had a tendency to assume more strongly than did the Korean and US students that humanoid robots have somewhat human characteristics and that their roles are related to social activities, including communication. Third, Korean students tended to have more negative attitudes toward the social influence of robots, particularly humanoid robots, and to assume more strongly that the roles of robots are related to medical fields than did the Japanese students. Fourth, students in the US had a tendency to have both more positive and more negative images of robots than did the Japanese students, and to assume less strongly that robots were a blasphemous of nature than did Japanese and Korean students. These results suggested that the term "robots" evokes different images of entities having different roles, as well as different feelings among people from different nations.

2.3 *Frankenstein Syndrome for Humanoid Robots*

The term "Frankenstein Syndrome", which was originally used in the context of controversies in the life sciences, in particular, genetic engineering [14], refers to the concept that the creation of human-like artifacts is an act of potential transgression. Kaplan [15] made use of this concept to explain the reasons why Japanese people tend to accept humanoid robots more naturally than do Westerners.

Considering this concept of "Frankenstein Syndrome", the author's research group developed the Frankenstein Syndrome Questionnaire (FSQ) to measure the degrees of acceptance of humanoid robots among humans, primarily to clarify differences on the social acceptance of humanoid robots between the UK and Japan [16, 17]. The FSQ consists of 30 items for measuring the acceptance of humanoid robots, including anxiety and expectations toward this technology among the general public. Each item is scored on a seven-point scale, from 1) strongly disagree to 7) strongly agree. The questionnaire is composed of three factors: (I) negative feelings toward humanoid robots (e.g., "The widespread use of humanoid robots would take jobs away from people", "Humanoid robots may make us even lazier"), (II) Expectations for humanoid robots (e.g., "I can trust persons and organizations related to the development of humanoid robots", "Humanoid robots can be very useful for teaching young kids"), and (III) Root anxiety toward humanoid robots (e.g., "The development of humanoid robots is blasphemous", "I don't know why, but humanoid robots scare me").

The results of a survey ($N = 200$) that included the FSQ and NARS and compared results between people in their 20s and 50s [17] suggested the following differences in the social acceptance of humanoid robots between the two countries. First, the UK respondents felt more negative toward humanoid robots than did their Japanese counterparts. Second, the UK respondents in their 20s had more positive expectations for humanoid robots than any other group. Third, there was a moderate level of correlation between negative attitudes toward emotional interactions with robots and negative feelings toward humanoids only in the respondents aged in their 50s. Fourth, the correlation between negative attitudes toward the social influence of robots and expectations for humanoids showed a similar trend. Fifth, there was a moderate level of correlation between negative attitudes toward emotional interactions with robots and root anxiety toward humanoids only in the UK respondents aged in their 50s. These results suggested differences between the two nations in both the acceptance of specific types of robots such as humanoids and the relationship between this acceptance and attitudes toward robots in general.

2.4 Robot Ethics

The recent development of robots aimed at introduction into arenas of daily-life, including homes, schools, stations, and hospitals, has been encouraging discussions about robot ethics, that is, concerns ethical problems that occur with robots, among philosophers and scientists. For example, Asaro [18] argued the need for discussions on the following three topics related to robot ethics: the ethical systems to be built into robots, the ethics of people who design and use robots, and the ethical relationships between humans and robots. Lin [19] proposed the following three broad (and interrelated) areas of ethical and social concerns in relation to robot ethics: safety and errors, law and ethics, and social impacts.

As shown in the previous sections, however, the conceptualizations of and feelings toward robots among the general public differ among nations owing to different situations with respect to the mass media and historical influence of technology. Moreover, interpretations of the term "ethics" differ between nations due to different social norms. Therefore, to reach an international consensus on the application of robotics, the opinions of the general public toward "robots" and "ethics" should be compared among several countries.

The present author previously conducted an open-ended questionnaire survey ($N = 400$) in Japan, the US, and Europe to explore differences in opinions regarding robot ethics [20]. To take the historical influences of wars and ethical perspectives of military robotics into account, the survey in Europe was conducted in Germany and France, a defeated and victorious country in World War II, respectively. The open-ended questionnaire items in the survey were as follows:

1. What do you imagine when you hear the terms "robots" and "ethics" at the same time?
2. What sort of ethical problems do you envision when robots become widespread in society?
3. How should we solve the problems mentioned in item 2?

The open-ended responses were qualitatively analyzed, and the results revealed the following characteristics of the responses by the general public in Japan, the US, Germany, and France regarding issues related to robot ethics. First, people in Japan tended to react to the ethical issues related to robotics more seriously than did those in the US, Germany, and France, and were more influenced by virtual content such as science fiction movies. Second, people in Germany were the least likely to associate robotics with ethics. Third, despite also being in the EU, people in France expressed more apprehension toward robotics. Fourth, Japan and France showed opposite trends with respect to potential unemployment caused by the increasingly widespread use of robots. Fifth, although a minority of people mentioned this issue overall, more people tended to specify it in Japan and the US than in the two European countries. Sixth, unlike the people in France, the people in Japan tended to argue for restricting the use and development of robots as a solution to ethical problems. These results suggested that people in different nations focus on and try to find different solutions to various ethical issues in relation to robots.

3 Anxiety and HRI

People with social anxiety tend to feel nervous in social situations, such as when they talk with other people, meet someone for the first time, or give a speech. Such individuals often fail to communicate with others appropriately, and thereby fail to perform social tasks reasonably. People who experience excessive amounts of social anxiety are often diagnosed as having social anxiety disorder; these individuals have more difficulty performing activities of daily living, such as studying at school and working in an office [21]. The lifetime prevalence of social anxiety disorder has been reported to be about 13%, and about 20% of the population is thought to experience intense fear when speaking or performing tasks in front of others [22]. Therefore, the development of training programs that help such individuals moderate their social anxiety is needed [23].

At present, little research has been conducted on the impact of social anxiety on HRI, or on the development of social robots that take social anxiety into account. Kang and Gratch [24] conducted an experiment in the US in which the participants ($N = 108$) interacted with a human or computer graphics (CG)-represented robot through a computer screen to measure shyness, a proximal construct of social anxiety. The results revealed that compared with participants with no shyness, participants with high levels of shyness tended to disclose more about themselves (that is, to tell the other person private information about themselves) when interacting with the CG robot than with the human. Zhu and Deng [25] conducted two online questionnaire surveys on Chinese university students (1st: $N = 100$, 2nd: $N = 150$) to explore the relationship between social anxiety and the strength of the intention to use a robot partner, assuming a situation where a robot was given as a tennis practice partner. They found that although social anxiety did not directly influence the intention to adopt a robot partner, the sense of relief provided by the robot partner was a mediating factor that indirectly influenced the intention to adopt the robot partner. These two studies suggested that highly socially anxious people tend to prefer interacting with robots rather than with people.

As one of the few robot development studies that take the characteristics of such highly socially anxious people into account, Hara and Terada [26] proposed what they referred to as "Mobile Personal Space" to alleviate social anxiety in face-to-face situations, such as one-to-one and one-to-many situations. In that study, nonverbal cues (e.g., eye gaze) were assumed to arouse social anxiety, and the person inside interacted with the outside through CG characters while the display covered the person and filtered visual information [27].

The author's research group has also conducted some psychological experiments and questionnaires to answer the research question: "Compared with people without high social anxiety, do people with high social anxiety actually prefer interacting with robots over people?" Representative examples are described below.

3.1 *Fear of Negative Evaluation (FNE) and Gaze Impact*

The author and the colleague [28] examined the impact of fear of negative evaluation (FNE) on interactions with robots. Similar to shyness, FNE, characterized by a fear of negative judgment from others, is a proximate construct of social anxiety. People with high levels of FNE tend to have high levels of social anxiety concurrently and to avoid communication with others. Based on the fact that people with high levels of social anxiety tend to avoid looking at others [29], the study conducted a psychological experiment to explore how the gaze of a robot affected the participants in an evaluative context. The experiment was conducted on Japanese university students ($N = 188$) under a setting where a humanoid robot about the same size as a child would ask questions about their English conversation skills (e.g., "Would you be able to answer in English if you were asked for directions?"), which the robot would then evaluate. Two experimental conditions were set up: one in which the robot actually presented a negative evaluation to the participant, and one in which it did not. In addition, two conditions were provided for the gaze from the robot to the participant: a normal gaze condition, in which the robot looked at the participant for 25% of the dialogue time, and a reduced gaze condition, in which the robot looked at the participant for only 6% of the dialogue time. Furthermore, the participants were divided into a "high FNE group" and a "low FNE group" according to the results obtained from a psychological scale [30]. A 2 × 2 × 2 between-participant design was used to analyze the participants' subjective evaluations of the robot's interactions after the experiment and the number of utterances during the interactions.

The results suggested a preference for "normal gaze conditions" among participants with low compared with those with high FNE, as well as a tendency for participants with high FNE to trust the robot more than those with low FNE.

3.2 *Impact of Social Anxiety on Anxiety and Tension During Interactions with Robots*

The author and the colleagues [31] conducted a more direct examination of the influence of social anxiety on the preference for people or robots in dialogue. That study focused on the influence of social anxiety as a personal trait on changes in anxiety and stress before and after interactions between robots and people, and was tested through psychological experiments.

The experiment was conducted on Japanese university students ($N = 19$) under a setting where a person or a robot would act as counselors. During the dialogue, a person/robot playing the role of counselor encouraged the participants to talk about something relevant to themselves and listened to them for about 5 min. Each participant interacted with both a person and a humanoid robot about the same size as a child, and anxiety and stress were measured by questionnaires before and after each interaction. The participants were also assessed in terms of social anxiety using a psychological scale [30] before the start of the experiment, and based on the results, divided into a "high social anxiety group" and a "low social anxiety group". A 2 × 2 mixed factorial design with high and low social anxiety as a between-participant factor and the person/robot interaction partner as a within-participant factor was used to analyze anxiety and stress before and after the interaction.

The results indicated that compared with the low social anxiety group, the high social anxiety group was less anxious and nervous before the interaction with the robot than with the person. On the other hand, regardless of having high or low social anxiety, lower post-interaction tension was found in the case of interactions with the robot than with the person. This finding suggests that in the context of counseling, anxiety and tension (anticipatory anxiety) before a face-to-face situation is lower with a robot than with a human partner, thereby implying that the disadvantage experienced by individuals with high social anxiety resulting from their avoiding dialogue with others in counseling situations may be eliminated by introducing robots into counseling.

3.3 *Effects of Social Anxiety on Robot Preference in Dialogue Situations*

The author and colleagues [32] conducted an online questionnaire study on Japanese males and females ($N = 742$) aged 20–60 years who were living in Japan regarding whether people or robots were preferred as partners in various interaction situations, taking the influence of social anxiety into account and focusing mainly on the influence of the robot's appearance. The questionnaire presented a total of 30 communication situations (e.g., "complaining about something at home") and asked the respondents to indicate, with illustrations, whether they preferred a human, a robot that looked like a human, or a humanoid robot with a machine-like appearance as a communication partner in each situation. In addition, social anxiety was measured using a psychological scale [33].

The results indicated that the group of participants who preferred the robot as a communication partner had a greater tendency to have social anxiety than the group of participants who chose the person. Furthermore, depending on the situation, the group of participants who chose the robot with a machine-like appearance had higher social anxiety than the group who preferred the robot that looked like a human, even if in other ways it was the same robot. This finding may directly indicate that, depending on the communication situation, individuals with high levels of social anxiety may prefer a robot that does not look like an actual person.

4 The Possibility of Interaction Effects Between Cultural Factors and Social Anxiety in HRI

As mentioned in the introduction, it is expected that relationships between social anxiety and the acceptance of robots vary among different cultures. Because different cultures have different social rules and expectations, the situations where social anxiety is evoked may differ. Even if social anxiety can impact the acceptance of robots, the differences in situations where social anxiety is evoked may lead to differences in situations where interactions with robots are preferred. Even if social anxiety can lead to a preference for robots over humans in social situations, the degree of the impact may differ depending on the culture. However, to our knowledge, no study has investigated the interaction effects between culture and social anxiety in HRI, at least in the current state. Therefore, some future research directions can be proposed.

4.1 *Robot Preferences Among People with High Social Anxiety in Other Cultures*

As mentioned in Sects. 3.1 and 3.2, we found that social anxiety and its proximate constructs, such as FNE, could impact the preference for communication partners in various situations involving humans and robots. As there may be cultural differences in communication rules and values, whether these differences may also impact social anxiety in communication with humans and robots should be investigated in experimental studies.

As there may be interaction effects between social anxiety and cultural and other factors (e.g., robots' nonverbal behaviors such as gaze and postures), the selection of factors is crucial in the experiment design. In fact, our experiment study focusing on FNE, robot evaluations, and gaze revealed complexity of the interaction effects. The selection of other factors while assuming culture and social anxiety as prior factors can make experiment designs more complex, and results hard to be interpreted.

4.2 *Social Situations Where Robots Are Preferred in Other Cultures*

As mentioned in Sect. 3.3, social anxiety can impact the preference for robots as communication partners in several situations. As the degree of anxiety perceived by individuals in specific communication situations may vary depending on the culture, this can lead to differences in the communication situations where robots are preferred; the reasons for this could be revealed using large-scale questionnaire surveys.

On the other hand, the above research direction has one challenge. As mentioned in Sects. 2.2 and 2.4, cultural differences exist regarding the conceptions of and concrete expectations for robots. When situations where robots and humans communicate are provided in a questionnaire, some situations may not actually occur in some cultures. Therefore, the types of situation and robot humans can be expected to interact with in some representative cultures should be investigated first.

5 Summary

This chapter summarized the results of questionnaire surveys aimed at exploring people's acceptance of robots from the perspective of measures for international comparisons, as well as the results of experiments and surveys on the impact of social anxiety on HRI. Although some future research directions on the interaction effects of culture and social anxiety were discussed, these should be considered more deeply while taking additional factors and measures into account.

References

1. Papadopoulos, I., Kouloughoti, C.: The influence of culture on attitudes towards humanoid and animal-like robots: an integrative review. J. Nurs. Scholarsh. **50**, 653–665 (2018)
2. Lim, V., Rooksby, M., Cross, E.S.: Social robots on a global stage: establishing a role for culture during human-robot interaction. Int. J. Soc. Robot. **13**, 1307–1333 (2021)
3. Mohammadi, A., Abasi, I., Soleimani, M., Moradian, S.T., Yahyavi, T., Zarean, M.: Cultural aspects of social anxiety disorder: a qualitative analysis of anxiety experiences and interpretation. Iran J. Psychiatry **14**, 33–39 (2019)
4. Chaplin, J.P. (ed.): Dictionary of psychology, 2nd edn. Dell Pub Co., New York (1991)
5. Brosnan M.J.: Technophobia: The Psychological Impact of Information Technology. Routledge, New York (1998)
6. Nomura, T., Suzuki, T., Kanda, T., Kato, K.: Measurement of negative attitudes toward robots. Interact. Stud. **7**, 437–454 (2006)
7. Bartneck, C., Suzuki, T., Kanda, T., Nomura, T.: The influence of people's culture and prior experiences with Aibo on their attitude towards robots. AI & Soc. **21**, 217–230 (2007)
8. Wang, L., Rau, P.-L. P., Evers, V., Robinson, B.K., Hinds, P.: When in Rome: the role of culture & context in adherence to robot recommendations. In: Proceedings of 5th ACM/IEEE International Conference on Human-Robot Interaction, pp. 359–366 (2010)
9. Dinet, J., Vivian, R.: Exploratory investigation of attitudes towards assistive robots for future users. Le travail Humain **77**, 105–125 (2014)
10. Pochwatko, G., Giger, J.-C., Różańska-Walczu, M., Świdrak, J., Kukiełka, K., Mozaryn, J.: Polish version of the negative attitude toward robots scale (NARS-PL). J. Autom. Mob. Rob. Intell. Syst. **9**, 65–72 (2015)
11. Erebak, S., Turgut, T.: Negative attitudes toward robots scale: validity and reliability of Turkish Version. Toros Üniversitesi İİSBF Sosyal Bilimler Dergisi, **5**, 407–418 (2018). Retrieved from https://dergipark.org.tr/en/pub/iisbf/issue/41627/486895
12. Nomura, T.: Comparison on negative attitude toward robots and related factors between Japan and the UK. In: Proceedings of 5th ACM International Conference on Collaboration Across Boundaries (CABS): Culture, Distance and Technology, pp. 87–90 (2014)
13. Nomura, T., Suzuki, T., Kanda, T., Han, J., Shin, N., Burke, J., Kato, K.: What people assume about humanoid and animal-type robots: cross-cultural analysis between Japan, Korea, and the USA. Int. J. Humanoid Robot. **5**, 25–46 (2008)
14. Rollin, B.E.: The Frankenstein Syndrome: Ethical and Social Issues in the Genetic Engineering of Animals. Cambridge University Press (1995)
15. Kaplan, F.: Who is afraid of the humanoid?: Investigating cultural differences in the acceptance of robots. Int. J. Humanoid Robot. **1**, 465–480 (2004)
16. Syrdal, D.S., Nomura, T., Dautenhahn, K.: The Frankenstein syndrome questionnaire—results from a quantitative cross-cultural survey. In: Herrmann, G., et al. (eds.) Social Robotics: 5th International Conference, ICSR 2013, pp. 270–279. Springer (2013)

17. Nomura, T., Syrdal, D.S., Dautenhahn, K.: Differences on social acceptance of humanoid robots between Japan and the UK. In: Proceedings 4th International Symposium on "New Frontiers in Human-Robot Interaction" (2015)
18. Asaro, P.M.: What should we want from a robot ethic? Int. Rev. Inf. Ethics **6**, 9–16 (2006)
19. Lin P.: Introduction to robot ethics. In: Lin, P., Abney K., Bekey, G.A. (eds.) Robot Ethics: The Ethical and Social Implications of Robotics, pp. 3–15. MIT Press (2011)
20. Nomura, T.: General Republic's opinions on robot ethics: comparison between Japan, the USA, Germany, and France. In: Proceedings of 4th International Symposium "New Frontiers in Human-Robot Interaction" (2015)
21. American Psychiatric Association: Diagnostic and Statistical Manual of Mental Disorders, 5th edn. American Psychiatric Association, Washington, D.C. (2013)
22. American Psychiatric Association: Diagnostic and Statistical Manual of Mental Disorders, 4th edn. American Psychiatric Association, Washington D.C. (1994)
23. Schneier, F.R., Johnson, J., Hornig, C.D., Liebowitz, M.R., Weissman, M.M.: Social Phobia: comorbidity and morbidity in an epidemiologic sample. Arch. Gen. Psychiatry **49**, 282–288 (1992)
24. Kang, S.-H., Gratch, J.: Virtual humans elicit socially anxious interactants' verbal self-disclosure. Comput. Anim. Virtual Worlds **21**, 473–482 (2010)
25. Zhu, D.H., Deng, Z.Z.: Effect of social anxiety on the adoption of robotic training partner. Cyberpsychol. Behav. Soc. Netw. **24**, 343–348 (2021)
26. Hara, T., Terada, K.: Mobile personal space: a personal space robot to assist people with social anxiety disorder. Proc. Interact. **2016**, 228–230 (2016). (in Japanese)
27. Ackerman, E.: A Personal-Space Robot to Assist People With Social Anxiety. IEEE Spectrum, Oct, 2017. https://spectrum.ieee.org/a-personal-space-robot-to-assist-people-with-social-anxiety
28. Nomura, T., Kanda, T.: Influences of evaluation and gaze from a robot and humans' fear of negative evaluation on their preferences of the robot. Int. J. Soc. Robot. **7**, 155–164 (2015)
29. Iizuka, Y.: A study on the relationship of shyness and gaze. Jap. J. Psychol. **66**, 277–282 (1995). (in Japanese)
30. Ishikawa, R., Sasaki, K., Hukui, H.: Standardization of Japanese version of FNE and SADS. Jap. J. Behav. Ther. **18**, 10–17 (1992). (in Japanese)
31. Nomura, T., Kanda, T., Suzuki, T., Yamada, S.: Do people with social anxiety feel anxious about interacting with a robot? AI & Soc. **35**, 381–390 (2020)
32. Suzuki, T., Yamada, S., Nomura, T., Kanda, T.: Do people prefer robots over humans as daily communication partners?: Relationship to social anxiety in conjunction with the appearance of the robots. J. Jap. Soc. Fuzzy Theory Intell. Inform. **31**, 789–796 (2019) (in Japanese)
33. Kanai, Y., Sasagawa, S., Chen, J., Suzuki, S., Shimada, H., Sakano, Y.: Development and validation of the Japanese version of social phobia scale and social interaction anxiety scale. Jap. J. Psychosom. Med. **44**, 841–850 (2004). (in Japanese)

Robotics and Trust

J. Rabe, Alessandra Sciutti, J. Zonca, D. Pasquali, and F. Rea

Abstract Imagine you would be hanging on a cliff—Would you take a robot's hand? Though the imagination of one hanging on a cliff seems dramatic, it well demonstrates a situation of vulnerability in which a person needs to rely on another's capability and integrity to help him or her. As the number of service robots in various fields increases enormously, see [74], it is likely that people will find themselves in situations of vulnerability in which they depend on robotic assistance. As such, robots cannot be merely regarded as tools performing repetitive tasks. Instead, robots' role as embodied social agents that are attributed with intentions and morality gains importance. This calls for a clear understanding of trust in HRI which needs to be well distinguished from trust in automation, see [41] for a review. The best way to achieve this is to understand trust in interpersonal relationships first.

Keywords Trust in HRI · Trust in automation · Reciprocity · Social robots · Trust evolution

Many definitions of trust exist that emerged from the fields of psychology and sociology. These definitions have in common that trust can be defined as having confidence in an interaction partner to do the appropriate action, see [13, 16, 22]. Emphasizing an interaction partner's vulnerability, Sabel defined trust as follows: "Trust is the mutual confidence that no party to an exchange will exploit another's vulnerabilities" [75], p. 1133. Being confident that an interaction partner will not exploit another partner's vulnerability implies trust in an interaction partner's attitudes, benevolence, integrity, trustworthiness, and performance, see [47, 56]. With regard to our exemplary scenery of one hanging on a cliff, reaching out for another's hand, this means

J. Rabe
Bielefeld University of Applied Sciences and Arts, Bielefeld, Germany

A. Sciutti (✉) · J. Zonca · D. Pasquali · F. Rea
Italian Institute of Technology (IIT), Genoa, Italy
e-mail: alessandra.sciutti@iit.it

G. Barresi and S. Wischniewski (eds.), *Ergonomics in Robotics: Advances and Innovations*, Springer Tracts in Advanced Robotics 162,
https://doi.org/10.1007/978-3-032-19374-2_22

having trust in the other's capabilities of pulling one back as well as her or his benevolence to do so. Considering these two aspects of trust, psychologists differentiated between trust in a partner's performance, called cognitive trust, and a partner's benevolence, called affective trust [39, 48, 54].

1 How to Define Trust When an "Interaction Partner" Is Not Human

Drawing a line between trust in interpersonal relationships and human trust in technology, trust in HRI has been widely defined and measured by referring to the theoretical and empirical framework of trust in automation. While the core concept of trust remains consistent across contexts, the realm of automated technologies includes vast series of tools and artificial agents, from automated navigation systems to common personal computers. Relying on a common definition, automation can refer to technology that actively selects, encodes, and transforms information, makes decisions, and controls different types of automatic processes [47]. A critical requirement of automation is the ability to perform these operations effectively, efficiently, and reliably without errors that can lead to negative consequences for humans [37]. This focus on performance is not surprising since automated technologies are typically developed to fulfill specific types of tasks, useful to a wealth of human activities. Importantly, automation is also employed in critical sectors including aviation, healthcare, environmental control, and production of nuclear energy in which errors of any kind may have catastrophic consequences. Assuming that the employment of automation significantly reduces the risks of errors and potential harm, a crucial requirement is that humans use these technologies properly. In order to fulfill this requirement, automation should be trusted meaning that humans should accept to delegate some responsibility and control to these technologies. This, however, does not imply that automation should be trusted and relied on indiscriminately. The right proportion of trust is crucial. On the one hand, over-trust can cause misuse of technologies possibly leading to negative consequences in the presence of failure or errors by automated systems. On the other hand, under-trusting automation may lead to disuse of automated technologies which in turn may result in errors and thus enhanced risks of using technologies that were actually intended to increase human safety. Extensive evidence has shown that appropriate trust in automation can be leveraged by system transparency which facilitates monitoring and evaluation of automated technologies' reliability [60, 61, 91, 96]. In sum, automation technologies can assist humans effectively if these technologies reliably fulfill the tasks they are designed for and transparently signal both correct functioning and potential emergence of failures. Trust in automation is thus mainly determined by a system's performance and transparency, the aspect of trust that psychologists consider as cognitive trust.

2 What's so Special About Trust in HRI?

Compared to automation technologies, robots are embodied agents that are used in a large number of social applications and environments with the goal of helping and assisting humans, including elderly, patients, children, workers, and people with reduced mobility. Robots should not only complete tasks; in many cases, they need to socially and physically interact with humans over an indefinite period of time. These robots are thus referred to as social robots. Being equipped with human-like appearance and behavior, social robots are designed to enhance natural and humane interactions with humans [78, 84]. As such, a robot's anthropomorphism [58], gender [87], biological movements [80], social gaze [2], and human-robot similarity [94], among others, have been found to affect human-robot interaction, particularly human-robot-trust. Social features, however, seem to be a double-edged sword. Kwon et al. found that participants tended to overestimate a robot's capabilities the more social features it shared [46]. The authors concluded that due to the robot's social features, participants applied human mental models to HRI [46]. That is, listening to a speaking robot, participants supposed the robot would also understand language because speaking and understanding are commonly interrelated in interpersonal conversations. As such, humans may even show pro-social attitudes and behavior towards robots that act in a socially acceptable fashion [29, 40, 45, 82]. In turn, robots were found to be perceived as moral agents to whom social norms of morality apply [12, 93]. All this might result in an image of robots as technically sophisticated social agents that develop their own motives and will, an image that is further strengthened by robots' presentation in the media (see [14]). In other words, besides a robot's performance, the interaction with robots is determined by typically social, affective, and relational mechanisms. Therefore, human-robot trust, unlike human-automation trust, may be modulated by an affective component that typically characterizes interpersonal trust. On the contrary, non-social automated technologies are typically perceived as motiveless: Users and operators are generally not afraid of being deceived or betrayed by them, shifting the entire focus on their functional reliability [41]. Conversely, if robots are perceived as social entities, they may have, in principle, individuals' motives and goals which might conflict with those of humans. Moreover, humans may require social robots to follow basic social behavioral norms while interacting with humans: Behavioral deviations from norms may signal anti-social behavior leading to a loss of trust. For all these reasons, theoretical and empirical research on human-robot trust cannot disregard the affective mechanisms characterizing interpersonal trust. Therefore, the two dimensions of cognitive and affective trust [39, 48, 54] that described trust in interpersonal relationships have been applied to HRI (see [20, 21, 23, 52]). Results by Bernotat et al. [20, 21] confirmed the application of trust in HRI as a two-dimensional concept to be useful: The authors found that participants trusted a robot's functions, rather than its benevolence.

3 Why People (Don't) Trust Robots

Taking our introductory example of one hanging on a cliff having to decide whether to take a robot's hand, Bernotat et al.'s [20, 21] results suggest that participants consider robots capable of pulling them back. However, they doubt robots would be benevolent enough to actually do so. But where does this ambivalence toward robots come from?

Previous research showed that most participants indicated to know robots mainly from the media [14, 19–21]. This way, media has crucially affected participants' views of robots (see [79]). In media, robots were commonly displayed as sophisticated technologies that serve humans in the beginning until they discover and violently enforce their will to become autonomous of their human masters, see e.g., [9, 24]. Media's emphasis on robots' functions might thus explain why a robot's high (vs. low) mental abilities, see [30] and autonomy [83] resulted in ambivalent attitudes toward robots and hesitations to use them. Likewise, robot performance was found to lead to under-trust and disuse of robots on the one hand [27, 49, 53, 77], but also to over-trust and misuse [6, 7, 73, 76] on the other hand. Media's strong emphasis on robots' emotions and intentions might have further strengthened the affective component of trust in HRI. As emotional information is more likely to attract people's attention, see [72], it is better memorized than neutral information [36, 72], see [44] for an overview. Due to the strong link to people's emotions, it seems plausible that the image of robots that was coined by media over more than a century, see [9, 24] is still prevalent in people's minds. Similar to interpersonal contact situations [4, 70], human-robot encounters seem the most efficient way to change people's views of robots to more positive and more realistic ones. That is, people need to interact with robots to evaluate and to "correct" their ideas, emotions, and attitudes toward them. Despite a good command of technology use, participants were found to report low levels of experience with robots, see [17]; the latter determined their attitudes toward robots: The more experience with robots participants reported, the less negative attitudes, such as fear, toward robots and the more overall positive views of robots as helpful innovations participants shared [17]. The authors assumed that more experience with robots might lead to more perceived self-efficacy and control in HRI. Experience with robots might furthermore help to make robot behavior transparent and thus to counter people's suspicions about robots' potentially threatening and malicious intentions. That is, participants' experience with robots seems crucial for trust in HRI to evolve. A better understanding of robots can and should of course further be facilitated by deliberately making robots' functions and "intentions" transparent. Human- and robot features thus need to go hand-in-hand for trust in HRI to emerge. In a meta-analysis, Hancock et al. [34, 35], furthermore identified that contextual factors, such as collaborative tasks between humans and robots and a shared in-group identity promoted trust in HRI. This aligns with classic social psychologists' research stating that equal group status, cooperation, institutional support, and shared goals can improve relationships between social groups [4, 70]. Further indicating a strong similarity between HRI and inter-group contact situations, Bernotat

[18] concluded that people seem to regard robots a distinct (social) group in their own right, see also [88]. This once more demonstrates to what extent social concepts apply to HRI. As such, similar to interpersonal relationships, amongst all robot-, human-, and context-related aspects that were outlined in this section, people's willingness and propensity to trust are decisive for trust in HRI to emerge (see [28, 34, 35]). A trustful relationship between humans and robots, however, can only be developed over time.

4 Trust Evolution Over Time

Trust is a dynamic phenomenon, a continuous calibration between a robot's trustworthiness (i.e., how much it deserves to be trusted) and a human's trust. Once HRI is established, it is up to the human and the robot to shape trust and trustworthiness' evolution over time. The human-, robot-, and contextual aspects outlined in the previous sections can be regarded as a basis for such an evolution of trust. Let's think about a simple human-robot co-working scenario: A worker has to assemble products with components handed over by a robotic arm. The worker could be initially reticent about accepting the help of the robot [62, 81], but he or she would soon learn to appreciate the precision, swiftness, and ergonomics by which the robot hands over components. Handover after handover, trust and trustworthiness build up in parallel. However, this balance could be extremely fragile: It could happen that a component slips from the robot's hands, lowering the worker's trust in the robot, see also [3] for the effects of errors on trust in HRI while performing a domestic task. This *undertrust* would inevitably affect HRI; for instance, the worker could pay more attention to the robot's hand—ready to catch the component if needed—and less to the assembly task, worsening the worker's performance and potentially harming her- or himself. This example focused on cognitive, performance-based trust. However, the same could happen with affective trust. Researchers explored several mitigation and trust repair strategies to compensate for robot errors [11, 31], along with the best intervention timing [59]. Most of the time, especially with affective trust, just saying *"sorry"* showed improvements in the HRI [59]. Likewise, time, i.e., interacting without any specific intervention, seemed to be sufficient to restore cognitive trust. However, it could also happen that a transparent excuse from a failing robot could deteriorate the interaction. Aroyo et al. explored the effect of mechanical failures in a treasure-hunt study in which volunteers had to find hidden objects in a room, relying on the hints of the iCub robot [6, 8]. During the interaction, the robot exhibited failures affecting its speech—replaced with meaningless words -, and its gestures—becoming jerky—up to a simulated crash and recovery. Interestingly, the authors found that if the iCub verbally excused itself after a failure, participants relied less on it than they did when it did not employ any mitigation strategy. Instead of under-trusting robots, participants could also over-trust them. Such a case is usually linked to prior factors like an excessive willingness to trust others [28], overestimation of users' own competence [57], robots' capabilities [43], and potential danger.

Literature showed that people tend to trust robots even in case of faults [73] or irrational behavior and suggestions [76]. Lowering down an excessive trust is usually part of the calibration process of trust-building. For instance, Robinette et al. showed how participants would follow a robot during a simulated fire emergency, even if it led to a wrong irrational direction (i.e., a dark dead-end) [73]. Also, malicious users could exploit excessive trust in robots. This is the case of Robot Social Engineering [68, 71]; in this phenomenon, malicious attackers take control of robotic agents, leveraging their social skills and users' trust, to manipulate their victims, usually with the purpose of collecting sensitive information (e.g., credentials), steal personal valuables (e.g., money), or gain access to restricted areas [92]. Aroyo et al. studied whether the Geminoid replica of Professor Ishiguro could apply on University students with the same authority as its creator [5]. In the study, participants were asked to evaluate the robots' ability to be a professor by complying with educational exercises from the Geminoid robot. The robot's requests escalated from simple math questions to several push-ups, up to stealing other professors' items, deleting the experimental recordings, and keeping those facts hidden from the experimenter. Also, in the treasure-hunt experiment mentioned above [6, 7], the iCub applied a common Social Engineering attack framework from Mouton et al. [55]. Firstly, the iCub gathered personal information related to the password-resetting process. Secondly, it gained participants' trust through the treasure hunt game. Finally, it tried to exploit that trust by asking participants to gamble their prize in search of a new hidden object. Remarkably, all participants answered all the questions and gambled their prize.

5 Trust in HRI as a Reciprocal Process

The typical conceptualization and operationalization of human-robot trust, stemming from the traditional view of trust in automation, are inherently unidirectional and asymmetrical. Since the goal is to allow humans to trust robots, the research focus in HRI has been on improving robots' trustworthiness or at least eliciting the perception of robots' trustworthiness in the eyes of human users. Therefore, researchers have been focusing on studying the impact of certain robots' characteristics (physical or functional) on human trust [90, 95]. In this framework, the robot does not need to trust or show trust towards its human interacting partner(s), due to the clear asymmetrical nature of their interactions. In this sense, human-robot trust does not assume or require reciprocity: Humans' trust towards robots depends on the robot's functional capabilities in fulfilling tasks (and humans' beliefs about robots' capabilities and how robots would use them), and is unaffected by the emergence/absence of trust-related signals coming from the robot itself [100]. Nonetheless, we have already pointed out that human-robot trust differs from trust in automation and can be shaped, to an extent, by relational and affective dynamics typical of human-human trust [86, 99]. In this regard, we should highlight that interpersonal trust relationships are indeed modulated by reciprocity: We typically trust others only if they are willing to trust us as well, leading to a condition of mutual and symmetrical vulnerability [50, 51, 97, 98]. For instance, in a dyadic relationship, violations of trust from one person have direct effects on the other person, who in turn may be less willing to trust the interacting partner in the future. Can reciprocity dynamics also emerge in human-robot trust?

Although research on the topic is still embryonic, recent evidence suggests so. Zonca et al. [99] used an experimental paradigm that has been previously used to study reciprocity in human-human trust-related social interactions [50]. In a joint perceptual decision making task, a human participant and a humanoid robot made a perceptual judgment concerning the same stimulus and then communicated their answer to the counterpart. Then, one of the two agents was asked to make a final decision based on the "social" feedback received. The final decision of the robot was manipulated in order to create two types of decision behavior. In one condition, the robot showed high trust in participants' responses, whereas in another condition the robot did not trust participants' responses. When the robot showed high trust in participants' responses, participants stopped using the robot's judgments to improve their performance and rated the robot as less competent. However, when asked to reveal their final decision to the robot (that has been shown to have high trust in participants), participants were unwilling to reveal their loss of trust to the robot and continued to modify their initial judgment towards that of the robot. This behavioral asymmetry was absent when participants interacted with a robot that did not trust them or with a computer. These results suggest that a robot's trust in participants represented a pro-social signal that affected humans' trust-related behavior towards the robot itself. We should acknowledge that the observed effect of reciprocity is qualitatively and quantitatively different from observations in interpersonal interactions [50, 51, 97, 98], and from results in human-computer interaction, highlighting the ambivalent nature of social robots with respect to trust. In line with these findings, a recent experimental study by Chen et al. [26] revealed that allowing reciprocity in human-robot interaction had a significant impact on behavioral trust in the Trust Game and, importantly, on a measure of emotional trust [42], highlighting how reciprocity can modulate human-robot trust in its affective component. The idea that trust-related signals coming from a robot may influence human trust was also tested in an experimental study by Strohkorb et al. [85]. Participants performed a collaborative game with a robot showing (or not showing) vulnerability, which is a crucial component of affective trust, during the interaction. Results suggested that robots' vulnerability influenced participants' trust-related behavior: When interacting with a vulnerable robot, participants were more prone to reveal their vulnerable state to their teammates, reducing the tension in the team. Results of another recent study [38] revealed that humans were more willing to trust and, consequently, to collaborate with a robot that blamed itself for collaborative failures. All these findings highlight how trust-related signals coming from a robot can influence human trust towards robots, by relational and reciprocal mechanisms that characterize *affective* trust. In line with this bidirectional view of trust, recent theoretical works started to model trust from a robot-centered perspective. Indeed, robots are becoming more social and are expected to coexist in humans' environments soon. Hence, the historical absolute trustworthiness granted to human agents in HRI should be put under discussion, equipping social robots with the proper skills to understand when human partners deserve to be trusted [100]. Thus, robots should have the capability to monitor human trust levels in real time and adapt their behavior accordingly. In recent years, numerous models have been developed to support this capability, enabling robots to assess and use trust signals coming from humans [10, 25, 32, 69]. In this regard, Vinanzi et al. [89, 90] developed a cognitive architecture that combines trust and Theory of Mind (TOM) modules with an episodic memory system to allow a humanoid robot to evaluate the trustworthiness of human partners in joint tasks. The authors have shown that allowing robots to monitor the performance of the human partner(s) in a joint task and take control of the task themselves in case of need enhances collaborative performance. However, in Vinanzi's model, humans' trustworthiness was not dynamically updated over multiple interactions, which conflicts with the previously discussed evolution of trust over time. On the same research line, Pasquali et al. explored the usage of lie detection as a quantitative metric to assess human trustworthiness in HRI. The authors leveraged pupillometry [?], [15] to indirectly measure changes in cognitive effort caused by lying or deceiving, compared to truthful behavior. In a first study by Gonzalez-Billandon et al., the gaze and pupillometry of volunteers were compared when lying and telling the truth to a human or a robot during an interrogatory scenario [33]. The same effect, i.e., an increased pupil dilation when lying with respect to telling the truth, was found in both human-human and human-robot interaction, supporting the reliability of such a metric. Based on these results, Pasquali et al. developed an autonomous architecture to enable the humanoid robot iCub to detect lies in real-time during informal HRI [63–67]. In a brief interaction, the robot learned a subjective model of how a human partner lied or told the truth; then, it leveraged such knowledge to classify further utterances from the same partner. The learned model could evolve and calibrate over time, similar to trust and trustworthiness evolve in human–human interactions (Fig. 1).

6 Concluding Remarks and Future Directions

We can conclude that due to robots' special role as social agents, HRI highly resembles to social interpersonal relationships. This way, independent of their functionality, robots seem "human" have human-like emotions and intentions and evoke the same in their potential human users. A quote from the movie "Eva" [1] expresses this well: *"It doesn't matter whether robots have emotions or not. What really matters is which emotions they evoke in you."* Therefore, HRI more than any other human-technology interaction, is determined by emotions and vulnerability. As such, trust in HRI plays a crucial role and needs to be differentiated from trust in automation. Researchers and developers should focus on the psychological and the social aspects that determine trust in HRI by considering its cognitive and affective components. It should furthermore be considered that—similar to interpersonal relationships—trust in HRI is a

Fig. 1 iCub robot offering its hand. Are humans ready to take it?

dynamic and reciprocal process. It bears the risk of over-trusting and under-trusting which can both have severe consequences. For instance, over-trusting in robots can make people prone to exploitation by third parties that take over the robot instead of the media-spread fear of robots taking over per se. Under-trusting in robots in turn might lead to misuse and discomfort in HRI. Therefore, to avoid harm, we strongly recommend to inform people clearly about a robot's functions and potential use cases and to encourage them to engage in HRI. This might help to become familiar with robots and to develop a realistic and comfortable image of robots rather than one that is coined by fear and stereotypes of robots. Moreover, considering the emotions and potential vulnerability that underlie HRI, we recommend to put a sharper focus on the ethical aspects of robot use and vote for guidelines that rule HRI. One might think of Asimov's robot laws [9]. These state that a robot should never harm a human or, through inaction, allow a human to come to harm. Furthermore, a robot should obey its human commander and protect its own existence, unless doing so could cause harm to a human [9]. These laws are, of course, not sufficient to regulate HRI according to current requirements, but they do reflect people's concerns that robots might harm humans. This, in turn, highlights people's (over)trust in robots' functions, along with their doubts about robots' benevolence, which furthermore demonstrates how robots evoke emotions in potential human users. However, since robots evoke emotions and intentions rather than in fact having them, laws are needed to protect humans from the projection of their own thoughts, fears, and intentions onto HRI.

For this to be achieved, a realistic view of robots and a certain degree of familiarity with them are inevitable. This way, we might one day, in fact, end up with people being willing to take a robot's hand in situations of vulnerability.

References

1. Maíllo, K.: (Regisseur): Eva [Film]. Escándalo films, Ran entertainment (2011)
2. Admoni, H., Scassellati, B.: Social eye gaze in human-robot interaction: a review. J. Human-Robot Interact. **6**(1), 25–63 (2017)
3. Aliasghari, P., Ghafurian, M., Nehaniv, C.L., Dautenhahn, K.: How do we perceive our trainee robots? exploring the impact of robot errors and appearance when performing domestic physical tasks on teachers' trust and evaluations. ACM Trans. Human-Robot Interact. **12**(3), 1–41 (2023)
4. Allport, G.W., Clark, K., Pettigrew, T.: The nature of prejudice, (1954)
5. Aroyo, A.M., Kyohei, T., Koyama, T., Takahashi, H., Rea, F., Sciutti, A., Yoshikawa, Y., Ishiguro, H., Sandini, G.: Will people morally crack under the authority of a famous wicked robot? pp. 35–42. Institute of Electrical and Electronics Engineers Inc. (2018)
6. Aroyo, A.M., Pasquali, D., Kothig, A., Rea, F., Sandini, G., Sciutti, A.: Expectations vs. reality: unreliability and transparency in a treasure hunt game with ICUB. IEEE Robot. Autom. Lett **6**(3), 5681–5688 (2021)
7. Aroyo, A.M., Rea, F., Sandini, G., Sciutti, A.: Trust and social engineering in human robot interaction: Will a robot make you disclose sensitive information, conform to its recommendations or gamble? IEEE Robot. Autom. Lett. **3**(4), 3701–3708 (2018). https://ieeexplore.ieee.org/document/8411113/
8. Aroyo, A., Pasquali, D., Koting, A., Rea, F., Sandini, G., Sciutti, A.: Perceived differences between on-line and real robotic failures (2020). https://www.researchgate.net/publication/344403863_Perceived_differences_between_on-line_and_real_robotic_failures
9. Asimov, I.: Alle Roboter-Geschichten (All robot novels), vol. 23317, Bastei Lübbe (2011)
10. Azevedo-Sa, H., Yang, X.J., Robert, L.P., Tilbury, D.M.: A unified bi-directional model for natural and artificial trust in human-robot collaboration. IEEE Robot. Autom. Lett. **6**(3), 5913–5920 (2021)
11. Baker, A.L., Phillips, E.K., Ullman, D., Keebler, J.R.: Toward an understanding of trust repair in human-robot interaction. ACM Trans. Interact. Intell. Syst. **8**, 1–30 (2018)
12. Banks, J., Koban, K., Haggadone, B.: Avoiding the abject and seeking the script: Perceived mind, morality, and trust in a persuasive social robot. ACM Trans. Human-Robot Interact. **12**(3), 1–24 (2023)
13. Barney, J.B., Hanson, M.: Trustworthiness as a source of competitive advantage. Long Range Plan. **28**(4), 127–127 (1995)
14. Bartneck, C.: Robots in the theatre and the media. In: Proceedings of the Design & Semantics of Form & Movement (DeSForM2013), pp. 64–70 (2013)
15. Beatty, J., Lucero-Wagoner, B.: The pupillary system. In: Handbook of Psychophysiology, vol. 2 (2000). https://psycnet.apa.org/record/2000-03927-005
16. Bernhard, B.: The logic and limits of trust, (1983)
17. Bernotat, J., Landolfi, L., Pasquali, D., Nardelli, A., Rea, F.: Remember me - user-centered implementation of working memory architectures on an industrial robot. Front. Robot, AI (2023)
18. Bernotat, J.: Keep an Eye on Stereotypes-The Impact of Gender Stereotypes (Toward Humans and Robots) on Language Processing, Department of Psychology, Bielefeld University (2021). Ph.D. thesis

19. Bernotat, J., Eyssel, F.: Can ('t) wait to have a robot at home?-Japanese and German usersâŁž attitudes toward service robots in smart homes. In: 27th IEEE International Symposium on Robot and Human Interactive Communication (RO-MAN), pp. 15–22. IEEE (2018)
20. Bernotat, J., Eyssel, F., Sachse, J.: Shape it—the influence of robot body shape on gender perception in robots. In: Social Robotics: 9th International Conference, ICSR 2017, Tsukuba, Japan, 22–24 Nov 2017, vol. 9, pp. 75–84. Springer (2017)
21. Bernotat, J., Eyssel, F., Sachse, J.: The (fe) male robot: how robot body shape impacts first impressions and trust towards robots. Int. J. Soc. Robot. **13**, 477–489 (2021)
22. Biros, D.P., Daly, M., Gunsch, G.: The influence of task load and automation trust on deception detection. Group Decis. Negot. **13**(2), 173–189 (2004)
23. Cameron, D., Loh, E.J., Chua, A., Collins, E., Aitken, J.M., Law, J.: Robot-stated limitations but not intentions promote user assistance, (2016). arXiv:1606.02603 arXiv preprint
24. Capek, K.: Rossum's universal robots (RUR). Penguin, London Castells M (2000) The rise of the network society, 2nd edn. Blackwell, Malden Cooley M (2007) From judgment to calculation. AI Soc. **21**(4), 395409Cowley (1921)
25. Chen, H., Chan, I.Y.: A systematic literature review of trust-aware shared control for human-robot collaboration in construction. In: 2024 16th International Conference on Computer and Automation Engineering (ICCAE), pp. 502–506. IEEE (2024)
26. Chen, N., Zhai, Y., Liu, X.: The effects of robots' altruistic behaviours and reciprocity on human-robot trust. Int. J. Soc. Robot. **14**(8), 1913–1931 (2022)
27. Choi, H., Swanson, N.: Understanding worker trust in industrial robots for improving workplace safety. In: Trust in Human-Robot Interaction, pp. 123–141. Elsevier (2021)
28. Colquitt, J.A., Scott, B.A., LePine, J.A.: Trust, trustworthiness, and trust propensity: a meta-analytic test of their unique relationships with risk taking and job performance. J. Appl. Psychol. **92**(4), 909 (2007)
29. Connolly, J., Mocz, V., Salomons, N., Valdez, J., Tsoi, N., Scassellati, B., Vázquez, M.: Prompting prosocial human interventions in response to robot mistreatment, vol. 20, pp. 211–220. HRI (2020)
30. Dang, J., Liu, L.: Robots are friends as well as foes: ambivalent attitudes toward mindful and mindless AI robots in the united states and China. Comput. Hum. Behav. **115**, 106612 (2021)
31. Esterwood, C., Robert, L.P.: Do You Still Trust Me? Human-Robot Trust Repair Strategies, pp. 183–188. IEEE (2021)
32. Floyd, M.W., Drinkwater, M., Aha, D.W.: Adapting autonomous behavior using an inverse trust estimation. In: Computational Science and Its Applications—ICCSA 2014: 14th International Conference, Guimarães, Portugal, June 30–July 3, 2014, Proceedings, Part I, vol. 14, pp. 728–742. Springer (2014)
33. Gonzalez-Billandon, J., Aroyo, A.M., Tonelli, A., Pasquali, D., Sciutti, A., Gori, M., Sandini, G., Rea, F.: Can a robot catch you lying? a machine learning system to detect lies during interactions. Front. Robot. AI 6, (2019)
34. Hancock, P.A., Billings, D.R., Schaefer, K.E., Chen, J.Y., De Visser, E.J., Parasuraman, R.: A meta-analysis of factors affecting trust in human-robot interaction. Hum. Factors **53**(5), 517–527 (2011)
35. Hancock, P.A., Kessler, T.T., Kaplan, A.D., Brill, J.C., Szalma, J.L.: Evolving trust in robots: specification through sequential and comparative meta-analyses. Hum. Factors **63**(7), 1196–1229 (2021)
36. Heuer, F., Reisberg, D.: Vivid memories of emotional events: the accuracy of remembered minutiae. Memory Cogn. **18**, 496–506 (1990)
37. Hoff, K.A., Bashir, M.: Trust in automation: integrating empirical evidence on factors that influence trust. Hum. Factors **57**(3), 407–434 (2015)
38. Van der Hoorn, D.P., Neerincx, A., de Graaf, M.M.: "I think you are doing a bad job!" the effect of blame attribution by a robot in human–robot collaboration. In: Proceedings of the 2021 ACM/IEEE International Conference on Human–Robot Interaction, pp. 140–148 (2021)
39. Johnson, D., Grayson, K.: Cognitive and affective trust in service relationships. J. Bus. Res. **58**(4), 500–507 (2005)

40. Kahn Jr, P.H., Kanda, T., Ishiguro, H., Gill, B.T., Shen, S., Gary, H.E., Ruckert, J.H.: Will people keep the secret of a humanoid robot? psychological intimacy in HRI. In: Proceedings of the Tenth Annual ACM/IEEE International Conference on Human-Robot Interaction, pp. 173–180 (2015)
41. Khavas, Z.R.: A review on trust in human-robot interaction, (2021). arXiv:2105.10045 arXiv preprint
42. Komiak, S.Y., Benbasat, I.: The effects of personalization and familiarity on trust and adoption of recommendation agents. MIS Q. 941–960 (2006)
43. Kopp, T., Baumgartner, M., Kinkel, S.: "it's not paul, it's a robot": the impact of linguistic framing and the evolution of trust and distrust in a collaborative robot during a human–robot interaction. Int. J. Human–Computer Stud. **178**, 103095 (10 2023)
44. Kuhbandner, C., Pekrun, R.: Brennpunkte der Gedächtnisforschung: Entwicklungs-und pädagogisch-psychologische Perspektiven, pp. 35–51. (2010). Emotion und gedächtnis
45. Kühnlenz, B., Kühnlenz, K., Busse, F., Förtsch, P., Wolf, M.: Effect of explicit emotional adaptation on prosocial behavior of humans towards robots depends on prior robot experience. In: 2018 27th IEEE International Symposium on Robot and Human Interactive Communication (RO-MAN), pp. 275–281. IEEE (2018)
46. Kwon, M., Jung, M.F., Knepper, R.A.: Human expectations of social robots. In: 11th ACM/IEEE International Conference on Human-Robot Interaction (HRI), pp. 463–464. IEEE (2016)
47. Lee, J.D., See, K.A.: Trust in automation: Designing for appropriate reliance. Hum. Factors **46**(1), 50–80 (2004)
48. Lewis, J.D., Weigert, A.: Trust as a social reality. Soc. Forces **63**(4), 967–985 (1985)
49. Lussier, B., Gallien, M., Guiochet, J., Ingrand, F., Killijian, M.O., Powell, D.: Fault tolerant planning for critical robots. In: 37th Annual IEEE/IFIP International Conference on Dependable Systems and Networks (DSN'07), pp. 144–153. IEEE (2007)
50. Mahmoodi, A., Bahrami, B., Mehring, C.: Reciprocity of social influence. Nat. Commun. **9**(1), 2474 (2018)
51. Mahmoodi, A., Nili, H., Bang, D., Mehring, C., Bahrami, B.: Distinct neurocomputational mechanisms support informational and socially normative conformity. PLoS Biol. **20**(3), e3001565 (2022)
52. Malle, B.F., Ullman, D.: A multi-dimensional conception and measure of human-robot trust. In: Nam, C.S., Lyons, J.B. (eds.) Trust in Human-Robot Interaction: Research and Applications, pp. 3–25. Elsevier (2021)
53. Matthews, G., Lin, J., Panganiban, A.R., Long, M.D.: Individual differences in trust in autonomous robots: implications for transparency. IEEE Trans. Human-Mach. Syst. **50**(3), 234–244 (2019)
54. McAllister, D.J.: Affect-and cognition-based trust as foundations for interpersonal cooperation in organizations. Acad. Manag. J. **38**(1), 24–59 (1995)
55. Mouton, F., Malan, M.M., Leenen, L., Venter, H.S.: Social engineering attack framework. In: 2014 Information Security for South Africa-Proceedings of the ISSA 2014 Conference, (2014)
56. Muir, B.M.: Trust between humans and machines, and the design of decision aids. In: Hollnagel, E., et al. (eds.) Cognitive Engineering in Complex Dynamic Worlds, pp. 71–83. Academic, London (1988)
57. Muir, B.M.: Trust between humans and machines, and the design of decision aids. Int. J. Man–Mach. Stud. **27**, 527–539 (11 1987)
58. Natarajan, M., Gombolay, M.: Effects of anthropomorphism and accountability on trust in human robot interaction. In: Proceedings of the 2020 ACM/IEEE International Conference on Human-Robot Interaction, pp. 33–42. (2020)
59. Nayyar, M., Wagner, A.R.: When should a robot apologize? Understanding how timing affects human-robot trust repair, pp. 265–274. (2018)
60. Ososky, S., Sanders, T., Jentsch, F., Hancock, P., Chen, J.Y.: Determinants of system transparency and its influence on trust in and reliance on unmanned robotic systems. Unmanned Systems Technology XVI **9084**, 112–123 (2014) SPIE

61. Parasuraman, R., Manzey, D.H.: Complacency and bias in human use of automation: an attentional integration. Hum. Factors **52**(3), 381–410 (2010)
62. Parvez, M.O., Öztüren, A., Cobanoglu, C., Arasli, H., Eluwole, K.K.: Employees' perception of robots and robot-induced unemployment in hospitality industry under Covid-19 pandemic. Int. J. Hospit. Manag 107, (2022)
63. Pasquali, D., Aroyo, A., Gonzalez-Billandon, J., Rea, F., Sandini, G., Sciutti, A.: Do you see the magic? an autonomous robot magician can read your mind, pp. 2–5 (2020). https://mypersonalrobots.org/s/HRI_2020_WS_Creative_SR_paper_6.pdf
64. Pasquali, D., Gaggero, D., Volpe, G., Rea, F., Sciutti, A.: Human vs robot lie detector: better working as a team? vol. 13086 LNAI, pp. 154–165. Springer, Cham (2021). https://doi.org/10.1007/978-3-030-90525-5_14
65. Pasquali, D., Gonzalez-Billandon, J., Aroyo, A.M., Sandini, G., Sciutti, A., Rea, F.: Detecting lies is a child (robot)'s play: Gaze-based lie detection in HRI. Int. J. Soc. Robot, 1–16 (2021). https://doi.org/10.1007/s12369-021-00822-5
66. Pasquali, D., Gonzalez-Billandon, J., Rea, F., Sandini, G., Sciutti, A.: Magic ICUB: a humanoid robot autonomously catching your lies in a card game, (2021)
67. Pasquali, D., Sciutti, A., Rea, F.: Toward enabling ICUB to detect lies in everyday life (2021). https://zenodo.org/record/5525169
68. Pasquali, D., Sciutti, A., Sandini, G., Bencetti, S., Rea, F., Genova, U.: Toward a human-oriented social engineering defense system, (2022)
69. Patacchiola, M., Cangelosi, A.: A developmental Bayesian model of trust in artificial cognitive systems. In: 2016 Joint IEEE International Conference on Development and Learning and Epigenetic Robotics (ICDL-EpiRob), pp. 117–123. IEEE (2016)
70. Pettigrew, T.F., Tropp, L.R.: Allport's intergroup contact hypothesis: its history and influence. In: On the Nature of Prejudice: Fifty Years After Allport, pp. 262–277, (2005)
71. Postnikoff, B.: Robot social engineering. PhD Dissertation (2020). http://hdl.handle.net/10012/16030
72. Reisberg, D., Heuer, F.: Remembering emotional events. Memory Emotion **s3**, 1 (2004)
73. Robinette, P., Li, W., Allen, R., Howard, A.M., Wagner, A.R.: Overtrust of robots in emergency evacuation scenarios. In: 11th ACM/IEEE International Conference on Human-Robot Interaction (HRI), pp. 101–108. IEEE (2016)
74. of Robotics, I.F.: https://ifr.org/ifr-press-releases/news/sales-of-robots-for-the-service-sector-grew-by-37-worldwide (2023)
75. Sabel, C.F.: Studied trust: building new forms of cooperation in a volatile economy. Human Relat. **46**(9), 1133–1170 (1993)
76. Salem, M., Lakatos, G., Amirabdollahian, F., Dautenhahn, K.: Would you trust a (faulty) robot? effects of error, task type and personality on human-robot cooperation and trust. In: Proceedings of the Tenth Annual ACM/IEEE International Conference on Human-Robot Interaction, pp. 141–148. (2015)
77. Sanders, T., Oleson, K.E., Billings, D.R., Chen, J.Y., Hancock, P.A.: A model of human-robot trust: theoretical model development. In: Proceedings of the Human Factors and Ergonomics Society Annual Meeting, vol. 55, pp. 1432–1436. SAGE Publications Sage CA, Los Angeles, CA (2011)
78. Sandini, G., Sciutti, A.: Humane robots-from robots with a humanoid body to robots with an anthropomorphic mind, (2018)
79. Sandoval, E.B., Mubin, O., Obaid, M.: Human robot interaction and fiction: a contradiction. In: Social Robotics: 6th International Conference, ICSR 2014, Sydney, NSW, Australia, 27–29 Oct 2014, vol. 6, pp. 54–63. Springer (2014)
80. Saygin, A.P., Chaminade, T., Ishiguro, H., Driver, J., Frith, C.: The thing that should not be: predictive coding and the uncanny valley in perceiving human and humanoid robot actions. Soc. Cogn. Affect. Neurosci. **7**(4), 413–422 (2012)
81. Schneiders, E., Papachristos, E.: It's not all bad—worker perceptions of industrial robots, vol. 2022, pp. 1025–1029. IEEE Computer Society (2022)

82. Siegel, M., Breazeal, C., Norton, M.I.: Persuasive robotics: the influence of robot gender on human behavior. In: IEEE/RSJ International Conference on Intelligent Robots and Systems, pp. 2563–2568. IEEE (2009)
83. Stapels, J.G., Eyssel, F.: Robocalypse? yes, please! the role of robot autonomy in the development of ambivalent attitudes towards robots. Int. J. Soc. Robot. **14**(3), 683–697 (2022)
84. Strohkorb, S., Scassellati, B.: Promoting collaboration with social robots. In: 11th ACM/IEEE International Conference on Human-Robot Interaction (HRI), pp. 639–640. IEEE (2016)
85. Strohkorb Sebo, S., Traeger, M., Jung, M., Scassellati, B.: The ripple effects of vulnerability: The effects of a robot's vulnerable behavior on trust in human-robot teams. In: Proceedings of the 2018 ACM/IEEE International Conference on Human-Robot Interaction, pp. 178–186. (2018)
86. Tavella, F., Manzi, F., Vinanzi, S., Di Dio, C., Massaro, D., Cangelosi, A., Marchetti, A.: Towards a computational model for higher orders of theory of mind in social agents. Front. Robot. AI **11**, 1468756 (2024)
87. Tay, B., Jung, Y., Park, T.: When stereotypes meet robots: the double-edge sword of robot gender and personality in human-robot interaction. Comput. Hum. Behav. **38**, 75–84 (2014)
88. Vanman, E.J., Kappas, A.: "Danger, will robinson!" the challenges of social robots for intergroup relations. Soc. Pers. Psychol. Compass **13**(8), e12489 (2019)
89. Vinanzi, S., Cangelosi, A., Goerick, C.: The collaborative mind: intention reading and trust in human-robot interaction. Iscience 24(2), (2021)
90. Vinanzi, S., Patacchiola, M., Chella, A., Cangelosi, A.: Would a robot trust you? developmental robotics model of trust and theory of mind. Philos. Trans. R. Soc. B **374**(1771), 20180032 (2019)
91. Wang, N., Pynadath, D.V., Hill, S.G., Ground, A.P.: Building trust in a human-robot team with automatically generated explanations. In: Proceedings of the Interservice/Industry Training, Simulation and Education Conference (I/ITSEC), vol. 15315, pp. 1–12. (2015)
92. Wang, Z., Sun, L., Zhu, H.: Defining social engineering in cybersecurity. IEEE Access **8**, 85094–85115 (2020)
93. Wen, R., Kim, B., Phillips, E., Zhu, Q., Williams, T.: Comparing norm-based and role-based strategies for robot communication of role-grounded moral norms. ACM Trans. Human-Robot Interact. **12**(3), 1–25 (2023)
94. You, S., Robert Jr, L.P.: Human-robot similarity and willingness to work with a robotic co-worker. In: Proceedings of the 2018 ACM/IEEE International Conference on Human–Robot Interaction, pp. 251–260 (2018)
95. Zanatto, D., Patacchiola, M., Goslin, J., Cangelosi, A.: Investigating cooperation with robotic peers. PLoS ONE **14**(11), e0225028 (2019)
96. Złotowski, J., Sumioka, H., Nishio, S., Glas, D.F., Bartneck, C., Ishiguro, H.: Appearance of a robot affects the impact of its behaviour on perceived trustworthiness and empathy. Paladyn. J. Behav. Robot **7**(1), 000010151520160005 (2016)
97. Zonca, J., Folsø, A., Sciutti, A.: Dynamic modulation of social influence by indirect reciprocity. Sci. Rep. **11**(1), 11104 (2021)
98. Zonca, J., Folsø, A., Sciutti, A.: I'm not a little kid anymore! Reciprocal social influence in child-adult interaction. R. Soc. Open Sci. **8**(8), 202124 (2021)
99. Zonca, J., Folsø, A., Sciutti, A.: The role of reciprocity in human-robot social influence. Iscience 24(12), (2021)
100. Zonca, J., Sciutti, A.: Does human-robot trust need reciprocity? arXiv preprint, (2021). arXiv:2110.09359

Ergonomics, Values and Robot Ethics

Peter Remmers

Abstract The development, implementation and use of technological artifacts introduces ethical issues—technologies are not ethically neutral. This ethical dimension is particularly obvious in robotic technologies, as many types and applications of robots are situated at the intersection of automation and social interaction. Consequently, developers, researchers and users of robots are confronted with ethical issues, for example concerning safety or (human) autonomy. With respect to ergonomics, two main themes can be identified: (1) Ergonomics itself involves certain values such as well-being, accessibility, performance optimization, safety, and trust. These values (and others) can and should be discussed in ethical terms. (2) Complex ethical issues in human–robot interaction are often difficult to address and can be cognitively overwhelming in the design process. One common ad-hoc strategy is to simply bypass and ignore these potential issues. However, professional tools that deal with ethical issues in the development and use of robotic system are available and may help to advance values in robotics. The most widely acknowledged framework to address values embedded in socio-technical systems during the development process is Value Sensitive Design (VSD). Focusing on ethical values, we present this framework with respect to robotic technologies. Finally, as implied by the very idea of human–robot interaction, robots themselves can be interpreted as subjects of action. This interpretation suggests an ethical reflection of the actions and behaviors performed by robots as if they were moral agents, an idea that is at odds with the key themes of ergonomics, values and ethics of robotics.

Keywords Ethics of robotics · Value sensitive design · Ergonomic values · Care robots · Robot ethics

P. Remmers (✉)
Institute of Cognitive Science, Osnabrück University, Osnabrück, Germany
e-mail: peter.remmers@uni-osnabrueck.de

G. Barresi and S. Wischniewski (eds.), *Ergonomics in Robotics: Advances and Innovations*, Springer Tracts in Advanced Robotics 162,
https://doi.org/10.1007/978-3-032-19374-2_23

1 Introduction

Robot ethics is a specialized branch of ethics of technology that investigates and facilitates ethical aspects of robotic technology [1–3]. 'Ethical aspects' are aspects determined by normative deliberations, or, in other words, by reflections on how something should be from a rational and value-based standpoint [4]. Ethical investigations are grounded on reason-based concepts of 'the good', on ethical values or on ethical principles for action. For example, safety from any kind of harm is considered to be a good thing in an ethical sense, or as a value that should be realized in a robot [5].

The topics of ethics and values present themselves as natural extensions of ergonomics, particularly in the context of robotics. Ergonomics itself is guided by values such as well-being, accessibility, performance optimization, safety, health, and trust [6]. Ethics, on the other hand, encompasses additional values such as justice, privacy, and freedom. Similar to ergonomics, ethical and value sensitive developments in robotic technology focus on the advancement of human–robot interactions and socio-technical systems. The overarching objective is the realization of certain values in the design of robotic technology to improve the world in an ethical sense. Such an 'integration' of ethics and values in robotics is driven by the insight that robot technology often introduces unique ethical issues, for example concerning safety, autonomy, embodiment or anthropomorphism [1–3].

Like the attribute 'ergonomic', the term 'ethical' is sometimes used to describe a quality of a technological artifact, for example, by referring to a specific robot as 'ethical' or by proposing the concept of an 'ethics certification' for suitable robot systems. However, this usage of the attribute 'ethical' can be problematic because an ethical perspective typically considers levels of evaluation that go beyond the individual technological artifact and its isolated status. Value-based reflections generally investigate the ethical, social, economic, ecological, legal, political and cultural contexts in which robotic technology is developed and used, leading to diverse evaluations. Any improvement or decline with respect to a specific value will not just depend on the design of an isolated artifact, but on the differences this design makes in the respective context. Thus, to avoid misunderstandings and misguided expectations, the attribute 'ethical' should be used with caution when applied to robot systems.

Furthermore, while ergonomics primarily focuses on the design of a product, interaction, or system, ethics and value sensitive approaches in technology development also address the development process itself. Factors such as stakeholder integration, diversity in development teams, and interdisciplinary collaboration are essential aspects of the ethical quality of technology development.

For a preliminary orientation, we can identify two main branches of robot ethics: (1) As a sub-field of applied ethics, robot ethics is concerned with ethical issues introduced by robotic technologies and with strategies to address these issues; (2) as a meta-ethical field of study, robot ethics reflects on the impact of robotic technology on ethical theory, as compared to traditionally strictly human- and action centered ethics.

In this contribution, we focus mainly on the first understanding of robot ethics (1). After establishing some basic ideas about ethics and values in technology (Sect. 2), we discuss the facilitation of (ethical) values in the design and development of technological artifacts, with a specific focus on the widely utilized framework of Value Sensitive Design (VSD) [7] (Sect. 3). While there are currently only limited examples of VSD in robotics, we highlight the approach of Care Centered Value Sensitive Design developed by van Wynsberghe for Health Care Robots [8, 9], illustrated by reference to the interdisciplinary research project *RoMi* on a robot system for nursing homes [10] (Sect. 4). Finally, in Sect. 5, we give a very brief outlook on some meta-ethical reflections as implied by the second understanding of robot ethics (2), primarily to emphasize their diversion from the general objectives of ergonomic and ethical and/or value sensitive design.

2 Applied Ethics of Technology

In recent years, ethics and value-based approaches to the development of technological artifacts have gained significant traction across various domains of new and emerging technologies [11]. There is a growing demand for the integration of ethical and value sensitive expertise in processes of technology development, for example in political funding initiatives like Responsible Research and Innovation (RRI) [12, 13]. Ethical contributions are deemed especially important for certain new and emerging fields of technology such as robotics, artificial intelligence (AI), bio- and nanotechnology. To provide a quick understanding of these endeavors, it is helpful to clarify two central premises of ethics of technology in general.

Firstly, a fundamental premise of ethics of technology is that technological artifacts are not neutral with respect to values and ethics [14]. This is primarily due to technology's influence on the main objects of ethical reflection: actions and behaviors. To illustrate this influence, we can consider three simple effects: (1) technologies create possibilities and opportunities for certain actions while rendering other options unfeasible—and these technology-induced actions may have ethical impact; (2) technologies facilitate certain actions and behaviors while hindering others, thereby altering the incentives for action and enabling new practices and habits—with implications for the system of incentives for ethical actions and behaviors; (3) many pervasive technologies introduce effects of collective action that differ significantly from the intentions and goals of individual actions—a situation that may create ethical dilemmas or mismatches between individual intentions and collective effects. Similar consequences of technologies can be spelled out for values: A technology can make it possible/easier or impossible/harder to realize certain values. Without going into further detail about technology's "moral status" [15], we can conclude that there is an opportunity or even a necessity for ethical reflections on technology design. And by acknowledging the moral dimension of technological artifacts, we can address ethical aspects in the processes of development, regulation and use.

Secondly, strongly linked to the ethical dimension of technology is a general focus on structural implications of technology, in contrast to a focus on isolated technological artifacts, users, or personal values. The ethical significance of a technology is not reducible to the individual actions or intentions of its users. Furthermore, new and emerging technologies exert their influence on potential futures, as they can have profound long-term implications for various aspects of value, including work environments, group identities, societies, economies, cultures and ways of life. To address this shift from the level of isolated technical artefacts to the level of their application and use in complex social contexts, it is crucial to examine the structural role of a technology, for example by reference to the concept of socio-technical systems [16].

These premises—the ethical dimension of technology and its structural quality—mark a difference from the more traditional endeavors of ethical theory and applied ethics. Traditionally, ethical theory is focused on justifications of ethical principles and values, whereas applied ethics is using these principles to resolve actual ethical issues in the real world. Consequently, ethical deliberations usually aim at a certain guidance for the actions and intentions of persons. In contrast, ethics of technology is focused on actions and intentions as structurally shaped by technology and determined in the process of technology design and in the regulation of technology use.

3 Ethics and Values in Technology Design

With this general account of ethics of technology in mind, approaches such as 'value sensitive design' (VSD) [17, 18], 'design for values' [19], 'integrated ethics' or 'design justice' [21, 22] attempt to ethically inform or facilitate processes of technology conception, development and implementation. From a design perspective, we can identify three goals for such a facilitation of ethical aspects or values:

First, an ethical or value-based contribution to the process of technology development should start with a comprehensive account of relevant ethical aspects of a given technology, with the goal of providing a clear view of a technology's ethical relevance and impact. To do this, ethicists provide detailed accounts of specific technologies and their use from a dedicated ethical or value-centered perspective. Thus, a constructive and interdisciplinary approach to technological developments like in robotics will discover and highlight value-related implications of design choices, identify (often implicit) values and develop alternative scenarios for the future reality of a technology.

Second, based on these insights, a facilitation of values actively engages in a design perspective on the developed technological artifact. Reflections on ethics and values may suggest certain directions for the design path or weigh in on necessary trade-offs. This role will often require an interdisciplinary collaboration in the design process to assess the technical, social and practical opportunities and limitations of a specific development.

Finally, the facilitation of ethics and values should relate to the 'big picture' of a technology in their social and ecological environment in order to develop a critical stance towards the changes implicated by the technology. For example, sometimes it may be helpful to curb the enthusiasm for technological solutionism or to imagine easy alternatives to complex technological endeavours—while at the same time resisting the temptation to pick sides between optimism and pessimism. Ultimately, the question of what is ethically desirable should take precedence over the question of what is technologically or economically feasible. Thus, ethicists will highlight the opportunities and limits of a technical solution from an overarching ethical perspective and with respect to values.

4 Ergonomics and Value Sensitive Design (VSD)

Ergonomics is based on values like well-being, accessibility, performance-optimization, health, safety, and trust. Values like these seem clear enough as a starting point for design. Still, we can go on to discuss the role of these values and their implementation in technological artefacts. In applied ethics, the application of generally defined values to specific applications is investigated, as are value conflicts and reasonable trade-offs between conflicting values. Such a discussion of values extends the scope of discourse from the values of ergonomics towards other values such as responsibility, sustainability, privacy or justice. Consequently, in ethics of technology, the values of ergonomics are extended by additional values that go beyond the ones mentioned above [23].

When working with values, the question arises how values that are defined in abstract and normative ways can be implemented in concrete socio-technological systems. There are many methodological approaches for an integration of values into developments of new and emerging technologies. One such approach is VSD, a popular framework for integrating values into processes of technology development [24]. VSD was originally developed in the 1990s with a focus on human–computer interaction, but was subsequently adapted and applied in many different fields of technology development. Additionally, a number of similar frameworks have been proposed and applied, for example under the titles of "Design for Values" or "Values in Design" [19].

The overall goal of VSD is to create technology that is sensitive to the values and needs of its users, other affected stakeholders and society as a whole. By integrating value-based considerations into the design process, VSD aims to promote responsible, ethical and socially beneficial technological innovations.

But what are values? Generally, when we speak about values in the context of ethics and VSD, three essential assumptions are commonly associated: (1) The values in question are not descriptive, but normative, referring to certain projected ideals or visions. (2) Values are bound to the idea of what is *good* and therefore distinct from a person's or group's attitudes, interests, preferences, goals and aims. (3) Values may vary between different cultural backgrounds and often change throughout history.

Starting from this ethical conception of values, VSD introduces a broader scope for values, defining them as referring "to what a person or a group of people consider important in life" [25]. This broad definition of (human) values extends the concept of values beyond the more specific sub-set of moral and ethical values, typically determined in ethical theories. Thus, the broad definition of VSD entails ethical and moral values, but also other values such as (for example) epistemic values that contribute to knowledge, political values such as civil liberty or even economic values such as material welfare. Additionally, this definition allows for a certain variety of values between different cultural backgrounds and does not determine fixed values in a dogmatic fashion. Nevertheless, in value-oriented discussions of technologies and their implementation, it is common to work with lists of relevant values. Examples for typical values discussed in the ethical domain are well-being, safety, privacy, justice, autonomy (or self-determination) and freedom. More concrete conceptualizations, operationalizations and specifications of these values may vary for different contexts, use cases and technological backgrounds, ultimately translating high-level values into evaluation criteria and attributes.[1]

In the process of addressing technologies, VSD works with an "interactional stance" [24]: Values are realized and shaped in the interactions between technological artefacts and humans—both sides of this interaction are not considered in isolation of each other. Consequently, both the design process of a technology as well as the associated socio-technical practices are regarded in VSD. Although the phrase of values as 'embedded' in technological artifacts is often used, it should not suggest that we can somehow attribute values to artefacts without considering their practical context of use—in short: its interactions with humans.

Methodologically, VSD is structured in a "tripartite methodology" [28], a combination of a conceptual stage, an empirical investigation and a design implementation. In the conceptual stage, values are defined and operationalized to formulate criteria for their realization in concrete states of affairs. In the empirical stage, research is conducted to investigate the requirements and effects of a specific value-based intervention. Finally, in the design stage, options for the design of the technology in question are developed, based on the findings of the previous (conceptual and empirical) stages.[2]

Another important element in VSD is the integration and participation of stakeholders into the process of technology development. 'Integration' and 'participation' is meant in a very active sense here; associated methods should ideally go beyond mere surveys or interviews, creating opportunities for co-design processes and transdisciplinary collaborations. Additionally, the underlying concept of stakeholders goes beyond the idea of 'user-centeredness', including anyone interacting with the technology (direct stakeholders), but also groups and institutions that are

---

[1] For a general framework for operationalizing values, see [25].

[2] For further descriptions of more fine-grained methods that can be applied in VSD, see [26, 27]. For a general overview of methods applied to integrate ethics in technology development, see [28].

affected in any way by the fact that the technology exists, even if only others directly interact with the technology in question (indirect stakeholders). Generally, the possibilities range from participatory approaches for the inclusion of direct and indirect stakeholders in the design process to vision-centered approaches for estimating relevant values and ethical aspects and pre-emptive assessments of current technological developments.

5 Care-Centered Value Sensitive Design and the Project RoMi

While VSD as a framework is conceptualized in general terms and has been applied extensively in many different areas of technology development [27], there are only few conceptions specifically adapted to the field of robotics. In recent years, we find propositions to adopt VSD for the examination of the intents and values of roboticists [18], for emerging technologies like robotics in the context of the so-called fourth industrial revolution [29], for the integration of available codes of ethics in care robot developments [17] and for iterative processes of value-based developments in social robotics [30]. A noteworthy proposition for a specific framework in social robots that advances from value-based methods such as VSD can be found in the framework of Integrative Social Robotics (ISR) [31].

A well-developed advancement of VSD with respect to robotics can be found in use-cases of robots in the sector of health and elder care. In various publications, van Wynsberghe combined the VSD framework to care ethics in order to develop a value-based approach to the use of robots in health care [8, 9]. The ensuing framework, *Care Centered Value Sensitive Design* (CCVSD), aims to incorporate values related to care such as empathy, reciprocity and responsibility into the design and development of robots for health care applications.

To illustrate the application of CCVSD, we refer to the research project *RoMi* (*Robot support for routine tasks to strengthen cooperation in care facilities*), an interdisciplinary cooperation funded by the German Ministry of Research and Education between 2020–2023 [10].[3] The general goal of RoMi was to adapt and investigate a robot platform for interactive applications in nursing homes. The primary research objective was to test if the robotic system offers reasonable assistance for care workers in a way that enables them to delegate certain tasks to the robot. The project was executed by an interdisciplinary team incorporating robotics, human factors, geriatrics, industrial design and ethics. Because of the project's special focus on ethical and value sensitive design, it is an illustrative example for robot ethics in action.

In the conceptual stage of the project, important values for the nursing sector were identified such as quality of work, well-being, autonomy, meaningful interactions and

[3] See www.interaktive-technologien.de/service/ergebnissteckbriefe/ergebnissteckbriefe-robopflege/romi.

mutual appreciation. To correlate these values with the setting of care practices in a nursing home, a number of value-related aspects were determined and elaborated:

1. The main role of the robot was determined to interactively assist *care givers*. This approach diverges from the more common focus on the interaction between robots and *care receivers* in nursing applications. Instead, the RoMi project prioritized the needs and support of the daily work that is done by care givers themselves.
2. The RoMi project focused on robotic *assistance* for care workers rather than a substitution of their roles entirely. By delegating non-care-related tasks to the robotic system, the project aimed at a re-balancing of the workload of care workers to enhance their overall job satisfaction.
3. The robotic system is not supposed to directly engage in nursing activities or tasks involving close contact with care receivers. Instead, its purpose is to assist with *incidental non-care-related tasks* that are often executed besides directly working with care recipients. Consequently, the use of the robot should enable care workers to dedicate more time and attention to their primary responsibility—providing direct care to those in need.
4. *Social interactivity* of the robot in the form of personal responsiveness or companionship was reduced to a minimum. 'Social' features of a robot such as friendly faces or colloquial speech output may facilitate operation, interaction and acceptance for laypersons, but it is important to not overdo it. Instead, the goal in the project was to achieve a reasonable balance between functionality and minimal social interactivity to ensure seamless operation and comfortable interaction.

Ideally, these aspects promote important values from the perspective of care givers. From the perspective of care receivers, values such as social interaction, autonomy, privacy and justice were considered. A special topic of interest was the change of the practice of care and of nursing environments by the robot's 'mechanizing effects', thereby recognizing a certain aesthetic quality of highly technical appliances like robots, especially in existential—albeit common—processes like cognitive decline and dying.

Starting from this conceptual setting, the project approached some basic questions for robotic design: How can the projected robot system provide the outlined kind of assistance in care-centered practices [32]? To answer this question, a series of research activities was initiated in the empirical stage of the VSD process, among them an investigation into the demand for the automation of tasks [10]. More specific design ideas for the assistive care robot were created in a series of workshops on ethical aspects and on the design of the robot.

Finally, in the design stage, the project partners developed interaction scenarios, design solutions for the robot itself and certain services provided by the robot. This process was led by roboticists with current possibilities and limitations of robotic technology in mind. As a result, the robot offers three services: (1) A messenger service, (2) a delivery service for objects, and (3) a delivery service for beverages. In the basic scenario, the robot navigates autonomously through the hallways of a nursing home to deliver messages, objects or beverages to care receivers. Ideally, some time-consuming tasks unrelated to care practices are automated, providing new time-frames for care workers to execute other, more important care tasks.

However, it turned out that there are quite a few critical trade-offs involved. For example, care workers lose some control over delivery processes by employing a robot. They have to rely on the robot's performance and the care receivers' ability and readiness to accept the robot's deliveries. Communication and incidental interactions between care givers and care receivers are reduced for those tasks that are automated—no small talk while delivering a bottle of water anymore. Finally, the described delivery services only address care receivers that are able to interact properly with the robot, which involves the opening of doors and walking up to the robot. These requirements limit the use of the robot in a nursing home with very diverse inhabitants.

Overall, the RoMi project addressed common issues of the current situation in the care sector by employing methods of CCVSD, with a dedicated focus on ethical aspects such as: quality and self-determination of care labor, balanced care work load, accessibility, fair distribution of technological benefits and, finally, the 'mechanization' of care and the related social environments [32]. Through conceptual investigations, participative methods and collaborative design efforts, a clear example for the possibilities and limitations of robotic technology for the care sectors was developed.

6 Conclusion: The Meta-Ethical Discourse on Robots

Besides the activities to integrate values and ethics into robotics research and development as described above, the term 'robot ethics' is also often interpreted as a meta-ethical topic (the second branch of robot ethics as indicated in the introduction). Here, we find a specialized academic discourse on robots as ethical 'subjects' [3, 33], a discourse that takes actual technological developments in robotics only as a starting point to reflect on the idea of robots as a new form of intelligent being, a theme that is traditionally presented in fictional works and popular debates. While this topic is not really relevant for practical disciplines like ergonomics, it is actively discussed in philosophy and related academic disciplines.

This particular discourse on robot ethics is motivated by the fact that robots are physically embodied, autonomous and/or remote-controlled machines for automation, interaction and collaboration—increasingly designed for social contexts. These qualities lead to effects of anthropomorphism [34, 35] or sociomorphism [31], effects that suggest a literal sense of robotic agency. This situation provokes a set of questions about ethical standards for 'robots as agents', a thought that has further implications for ethical theory. For example: Should we understand robots as subjects of action (moral agents) or as responsive to moral actions (moral patients)—and why? Do interactions with robots imply certain ethical commitments and liberties? Is it possible to develop an ethical theory that is not centered on living beings and that could in principle include non-living entities like robots? By asking these and similar questions, robot ethics departs from real-world issues towards more general debates in Machine Ethics and Ethics of AI, research fields that work with the premise that ethics and morality can be formalized in a way that is abstracted from the specific conditions only found in humans (or, more generally, in living beings).

Here, robot ethics shifts the tendency of human-centered design—a guiding idea for ergonomics and value-based investigations—towards interaction-centered design. But if we follow this train of thought with a focus on values, this shift away from human-centeredness seems to arrive at some odd consequences: For example, if we ask the question "Whose values?", it seems absurd to suggest anything other than humans or other living beings as possible answers. Similarly, the idea of the (ethically) good that is conceptually connected to the idea of values seems out of place when applied to robots as 'ethical subjects' (or 'moral agents'). Just like ergonomics, ethics is in the service of living beings, not machines. Thus, some of the more radical ideas of robot ethics diverge dramatically from the core assumptions of ergonomics, ethics and values for robotics and are therefore not further discussed in the present chapter.

References

1. Lin, P., Abney, K., Bekey, G.A. (eds.): Robot ethics: the ethical and social implications of robotics. MIT Press, Cambridge, Mass (2012)
2. Lin, P., Abney, K. (eds.): Robot Ethics 2.0: From Autonomous Cars to Artificial Intelligence. Oxford University Press, New York, NY (2017)
3. Coeckelbergh, M.: Robot Ethics. The MIT Press (2022)
4. Ryan, M., Stahl, B.C.: Artificial intelligence ethics guidelines for developers and users: clarifying their content and normative implications. JICES **19**, 61–86 (2021). https://doi.org/10.1108/JICES-12-2019-0138
5. Bowen, J.: The ethics of safety-critical systems. Commun. ACM **43**, 91–97 (2000). https://doi.org/10.1145/332051.332078
6. Lange-Morales, K., Thatcher, A., García-Acosta, G.: Towards a sustainable world through human factors and ergonomics: it is all about values. Ergonomics **57**, 1603–1615 (2014). https://doi.org/10.1080/00140139.2014.945495
7. Friedman, B., Hendry, D.: Value Sensitive Design: Shaping Technology with Moral Imagination. The MIT Press, Cambridge, Massachusetts (2019)
8. van Wynsberghe, A.: Designing robots for care: care centered value-sensitive design. Sci. Eng. Ethics **19**, 407–433 (2013). https://doi.org/10.1007/s11948-011-9343-6
9. van Wynsberghe, A.: Healthcare Robots: Ethics, Design and Implementation. Routledge, London (2016)
10. Klebbe, R., Klüber, K., Dahms, R., Onnasch, L.: Caregivers' perspectives on human-robot collaboration in inpatient elderly care settings. Machines **11**, 34 (2022). https://doi.org/10.3390/machines11010034
11. Frahm, N., Schiølin, K.: The rise of tech ethics: approaches, critique, and future pathways. Sci. Eng. Ethics **30**, 45 (2024). https://doi.org/10.1007/s11948-024-00510-3
12. von Schomberg, R., Hankins, J.: Hrsg: International Handbook on Responsible Innovation: A Global Resource. Edward Elgar Publishing, Cheltenham, UK Northampton, MA, USA (2019)
13. Salvini, P., Palmerini, E., Koops, B.-J.: Robotics and responsible research and innovation. In: International Handbook on Responsible Innovation. S. 405–424. Edward Elgar Publishing Limited, Cheltenham/Northampton (2019)

14. Kroes, P., Verbeek, P.-P. (eds.): The Moral Status of Technical Artefacts. Springer, Dordrecht (2014)
15. Redaelli, R.: Different approaches to the moral status of AI: a comparative analysis of paradigmatic trends in science and technology studies. Discov. Artif. Intell. **3**, 25 (2023). https://doi.org/10.1007/s44163-023-00076-2
16. Franssen, M.: Design for values and operator roles in sociotechnical systems. In: Van Den Hoven, J., Vermaas, P.E., and Van De Poel, I. (eds.) Handbook of Ethics, Values, and Technological Design, pp. 117–149. Springer Netherlands, Dordrecht (2015). https://doi.org/10.1007/978-94-007-6970-0_8
17. Umbrello, S., Capasso, M., Balistreri, M., Pirni, A., Merenda, F.: Value sensitive design to achieve the UN SDGs with AI: A case of elderly care robots. Mind. Mach. **31**, 395–419 (2021). https://doi.org/10.1007/s11023-021-09561-y
18. Cheon, E., Su, N.M.: Integrating roboticist values into a value sensitive design framework for humanoid robots. In: 2016 11th ACM/IEEE International Conference on Human-Robot Interaction (HRI), pp. 375–382 (2016)
19. van den Hoven, J., Vermaas, P.E., van de Poel, I.: Hrsg: Handbook of Ethics, Values, and Technological Design. Springer, Netherlands (2015)
21. Costanza-Chock, S.: Design Justice: Community-Led Practices to Build the Worlds We Need. The MIT Press, Cambridge, MA (2020)
22. Ostrowski, A.K., Walker, R., Das, M., Yang, M., Breazea, C., Park, H.W., Verma, A.: Ethics, equity, & justice in human-robot interaction: a review and future directions. In: 2022 31st IEEE International Conference on Robot and Human Interactive Communication (RO-MAN), S 969–976. IEEE, Napoli, Italy (2022)
23. Maurice, P., Allienne, L., Malaisé, A., Ivaldi, S.: Ethical and social considerations for the introduction of human-centered technologies at work. In: 2018 IEEE Workshop on Advanced Robotics and its Social Impacts (ARSO), pp. 131–138 (2018). https://doi.org/10.1109/ARSO.2018.8625830
24. Friedman, B., Kahn, P.H., Borning, A.: Value Sensitive Design and Information Systems. In: Zhang, P., Gelletta, D. (eds.) Human-Computer Interaction in Management Information Systems: Foundations, pp. 348–372. M.E. Sharpe, New York (2006)
25. van de Poel, I.: Values and Design. In: Michelfelder, D.P., Doorn, N. (eds.) The Routledge Handbook of the Philosophy of Engineering, pp. 300–314. Routledge (2020). https://doi.org/10.4324/9781315276502-27
26. Friedman, B., Hendry, D.G., Borning, A.: A survey of value sensitive design methods. HCI **11**, 63–125 (2017). https://doi.org/10.1561/1100000015
27. Winkler, T., Spiekermann, S.: Twenty years of value sensitive design: a review of methodological practices in VSD projects. Ethics Inf. Technol. **23**, 17–21 (2021). https://doi.org/10.1007/s10676-018-9476-2
28. Reijers, W., Wright, D., Brey, P., Weber, K., Rodrigues, R., O'Sullivan, D., Gordijn, B.: Methods for practising ethics in research and innovation: a literature review, critical analysis and recommendations. Sci. Eng. Ethics **24**, 1437–1481 (2018)
29. Gazzaneo, L., Padovano, A., Umbrello, S.: Designing smart operator 4.0 for human values: a value sensitive design approach. Proced. Manuf. **42**, 219–226 (2020). https://doi.org/10.1016/j.promfg.2020.02.073
30. Schmiedel, T., Zhong, V.J., Eyssel, F.: Towards a wave approach for value sensitive design in social robotics. In: Companion of the 2023 ACM/IEEE International Conference on Human-Robot Interaction, pp. 592–596. Association for Computing Machinery, New York, NY, USA (2023)
31. Seibt, J., Damholdt, M.F., Vestergaard, C.: Integrative social robotics, value-driven design, and transdisciplinarity. Interact. Stud. **21**, 111–144 (2020). https://doi.org/10.1075/is.18061.sei
32. Remmers, P., Fischer, N.: Social robots in care facilities: reflections on current research and the potential of ethical visions. In: Hakli, R., Mäkelä, P., and Seibt, J. (eds.) Social Robots in Social Institutions: Proceedings of Robophilosophy 2022, pp. 43–49. IOS Press, Amsterdam (2022)

33. Wallach, W., Asaro, P. (eds.): Machine Ethics and Robot Ethics. Routledge/Taylor & Francis Group, New York (2017)
34. Remmers, P.: The Ethical Significance of Human Likeness in Robotics and AI. eip. **10**, 52–67 (2019). https://doi.org/10.14746/eip.2019.2.6
35. Damholdt, M.F., Quick, O.S., Seibt, J., Vestergaard, C., Hansen, M.: A scoping review of HRI research on 'anthropomorphism': contributions to the method debate in HRI. Int. J. Soc. Robotics. **15**, 1203–1226 (2023). https://doi.org/10.1007/s12369-023-01014-z

Robotics for Aging Populations

Kar Him Mo, Fei Chen, and Giacinto Barresi

Abstract Robotics offers versatile tools to face the challenges due to the older population growth, especially meeting the needs for assisting the elderly in the activities of daily living (ADLs) and for the geriatric care. However, the perspectives offered by human factors studies can further enhance this potential of robotics. This chapter explores the integration of ergonomics in the design of robots devised to enhance the independence and well-being of older adults. Furthermore, we mention the opportunities offered by wearable robotics, telepresence, and the pervasive solutions of the Internet of Robotic Things (IoRT) to implement paradigms of precision healthcare with decentralized and ubiquitous features, improving the sustainability of elderly care—following the human-centric perspectives of Society 5.0 and Health 5.0, with their synergies. For instance, we discuss issues of technology acceptance alongside the perspectives of value-sensitive design to ensure that robotics align with the experiential and, in particular, ethical needs in aging people and communities, first of all for inclusion. Intelligent systems will also be discussed for their potential to make robotic solutions even more versatile and adjustable to the seniors' needs. By prioritizing the needs and preferences of older individuals (and their own network of relatives, friends, and acquaintances), ergonomics in robotics must contribute to the societal mission of preserving dignity, autonomy, and quality of life in aging societies.

Keywords Aging · Older adults · Elderly · Society · Healthcare · Accessibility · Inclusion · Gerontechnology · Geriatronics

---

K. H. Mo (✉)
School of Architecture and JC Institute of Ageing, The Chinese University of Hong Kong; and Happy Ageing Lab Foundation, Hong Kong SAR, China
e-mail: khmo@cuhk.edu.hk

F. Chen
Department of Mechanical and Automation Engineering, T-Stone Robotics Institute, The Chinese University of Hong Kong, Hong Kong SAR, China

G. Barresi
Bristol Robotics Laboratory, School of Engineering, CATE, University of the West of England, Bristol, UK

G. Barresi and S. Wischniewski (eds.), *Ergonomics in Robotics: Advances and Innovations*, Springer Tracts in Advanced Robotics 162,
https://doi.org/10.1007/978-3-032-19374-2_24

1 Introduction

As life expectancy increases, a global demographic shift is accelerating: by 2050, the older population is expected to increase to more than double [1]. This age-related transformation presents crucial societal challenges, particularly in healthcare and social inclusion, because of the frailty characterising the senescence. The impact of such challenges is demonstrated by worldwide initiatives like the United Nations Decade of Healthy Ageing (2021–2030)[1], devised to enrich the lives of the older adults, aligning with the Sustainable Development Goals[2] [2].

Accordingly, innovative solutions in interaction technology are continuously designed by pondering the unmet needs in aging societies, considering the individuals in their context. Among these solutions, robotic devices can certainly bring versatile and impactful ways to improve the independence and the quality of life of older adults. Based on the abilities, limitations, and goals of their users in each specific situation, robots designed for older adults are definitely heterogeneous, serving functions that range from caregiving to workforce support. They can encompass almost all types of robotic devices discussed in this book, including exoskeletons and telepresence robots, even within pervasive paradigms like the Internet of Robotic Things (IoRT) to enact frameworks like precision healthcare [3].

However, an impactful integration of robotic devices in the daily life of the elderly requires the human-centered design approach of ergonomics for properly responding to the challenges experienced by aging people. Such challenges depend, for instance, on changes in sensory, cognitive, motor, autonomic functions, typically leading to limitations in terms of autonomy, self-expression, and social life [4]. Thus, the (physical, cognitive, organizational) branches of ergonomics must consider specific issues and solutions to ensure that the robotic devices will actually assist the aging individuals in an efficient way (without creating additional challenges).

Some examples: physical ergonomics must focus on making robots safely support the user's mobility and the interaction with objects in activities of daily living (ADLs); cognitive ergonomics should ease the interaction with any type of robot, and make them able to help the user's mental capabilities, from prospective memory to decision-making; organizational ergonomics can make robots seamlessly integrated into domestic and healthcare settings, or in processes engaging older workers. In order to accomplish these goals, recent interdisciplinary methodologies in neuroergonomics (investigating human factors through neuroscientific approaches), for instance, can also employ wearable devices to monitor health-related indices in people with chronic or degenerative conditions while they enable the adaptive features of robots, adjusting their activity to the affective state of senior users.

We also need to consider how ergonomics should focus on the barriers faced by older adults for adopting the use of robots: investigating and addressing the issues of technology acceptance should always be prioritized [5]. Furthermore, robotics must also align with the ethical and societal values of the aging individuals and

---

[1] https://www.decadeofhealthyaging.org.

[2] https://sdgs.un.org/goals.

communities, ensuring that the smart machines preserve their dignity, autonomy, and preferences (alongside the ones of the network of people connected to them) while addressing concerns about privacy and safety too.

Considering their societal impact, robots should also be designed to foster inclusion and social interactions (especially intergenerational connectivity), empowering older adults as active participants in their communities. By prioritizing these value-sensitive aspects (looking at values as human factors), robotics can successfully address the challenges of aging societies. This chapter will explore how ergonomics can contribute to design Human-Robot Interactions (HRIs) for older adults in diverse contexts, establishing exemplary ways for helping individuals to achieve a successful aging in terms of well-being and experience [6].

In this chapter, the second section will offer a quick overview to describe examples of robotic applications for older adults. The third section will focus on the contributions of ergonomics in this domain. The fourth section will discuss the societal impact of robots designed for the elderly.

2 Robots for Aging Populations

Before proposing any robots as solutions, we need to study the users and their needs. First of all, we must observe that, in the context of aging, it is not easy to fully distinguish healthy and pathological senescence. In general, even the healthy aging can certainly show deficits in the individual skills, physically and mentally, because of anatomical and physiological changes that cumulate their effects over time. However, such changes can also increase the risk for pathological, often chronic and degenerative conditions to happen. Sometimes they can even be caused or worsened by medications for different medical conditions. We can consider these as generic premises for further difficulties and the frailties experienced by older adults, facing stressors and accidents in daily life. A fall or an ischemic stroke can have severe consequences, with a difficult recovery and a general loss of autonomy in ADLs, for instance. Chronic (like the diabetes) and degenerative (like the Alzheimer dementia) conditions can also hit frail people severely, especially when they are not detected in time for adjusting the daily contexts, routines, and tools according to a diagnosis. All these events can have dire impacts at psychological and social levels for any person.

When robots are designed to help older adults to face such challenges, they become part of the class of the gerontechnologies [4, 7]. They can also be included in an emergent specific domain labeled as geriatronics[3]: using intelligent machines in geriatrics and gerontology [8].

Overall, robotic devices can be designed to respond to the multiple needs of the senior users. They can be tools (e.g., spoons with mechatronic features to mitigate the Parkinsonian tremor) [9], limbs (e.g., prosthetic leg for amputees who lost their biological one because of the diabetes) [10], wearings (e.g., assistive and rehabilitative

[3] https://geriatronics.mirmi.tum.de/en/what-is-geriatronics.

exoskeletons) [11], mediums (e.g., telepresence devices) [12], agents (e.g., robotic companions for cognitive and affective stimulation) [13] that can also work as assistants in clinical procedures (e.g., robotic coaches for muscle rehabilitation training) [14], or even environments (e.g., solutions for home automation and domotics to provide monitoring and services) [3]. Each robot helps the older person to perform diverse tasks, from the ADLs to the work duties—as in the care of collaborative robots (cobots) devised to extend the professional activity of aging people [15].

However, the diverse personal stories and contexts of each individual makes a taxonomy in robotics a real challenge. Considering care robotics , for instance, [16] proposed to distinguish the robot-assisted care (robots indirectly help the elderly or support the activity of the caregivers) and the robot-based care (robots directly interact with the older people for providing cares). Interestingly, the authors highlight how teleoperated robots can belong to both fields, according to their functions (e.g., enabling a remote check of the person's health or making the elderly establish social interactions with remote interlocutors). For instance, teleoperation with shared control offers significant potential to improve ergonomics in ADLs and clinical procedures for elderly users by assisting in tasks requiring precise object handling [17].

Furthermore, devices like rehabilitation robots can fall in the area of robot-assisted care or robot-based care according to the type of interaction with the caregiver or the older person: if they are mainly controlled by the first, they enable robot-assisted care; if they adjust to the performance or the conditions of the second, they provide robot-based care.

Additionally, we must consider the case of the assistance during the work activities of the elderly [18]. Because of the extension of the working lifetime and the delay of the retirement age, robots (especially cobots) must be devised to help reducing the physical strain, the mental workload, and the injury risk for all workers. The process of aging can make an older worker perceive tasks as increasingly demanding. Thus, the cobots must adjust their own activity to help the worker without creating additional burden through a rapid pace to accomplish sequential tasks. Monitoring the conditions of stress (which can facilitate the occurrence of errors) is a compulsory part of this process, exactly like it is within the health monitoring in domestic settings: tracking vital signs, detecting falls, averting muscle fatigues, and alerting emergency services and the caregivers can be done by robotic systems, especially the ones connected in IoRT-based networks (embracing wearable and environmental sensors), enabling personalised coaching for well-being and care [19]. We can notice how these paradigms of prevention establish the background for improved cares, as in the case of precision healthcare: decentralised and personalised services mediated by robots can help reducing the impact of the elderly fragility on the healthcare systems (as envisioned during the recent COVID-19 pandemic) while providing a more human-centered set of functions [3]—e.g., the ARI Robot of Pal Robotics in the EU-funded SHAPES project.

These solutions, based on data collection for adjusting services to the individual needs, can also enhance several functions of robotic systems for elderly. This is the case of robots for cognitive support (e.g., reminding to take medicaments) and stimulation (e.g., enacting games for memory training) [20, 21], and for social-affective interactions (e.g., robotic companions for eliciting emotional reactions, and telepresence robots for reaching remote interlocutors) [22–24]. Such robotic systems can also be quite helpful when the cognitive-behavioural decline is due to pathological conditions like dementia. However, the spectrum of robots for clinical interventions is far wider, and it includes, first of all, devices for rehabilitation of physical and neurological conditions (e.g., post-stroke recovery). When these procedures cannot restore the skills of the individual at an appropriate level, assistive and prosthetic devices must be used, especially for supporting the individual mobility (e.g., exoskeletons, smart walkers) [4, 10, 25–27].

This section offered a quick overview of the robotic devices designed to help the elderly, according to their needs. Next section will describe how ergonomics can guide the design of such solutions to make them truly human-centered.

3 Human Factors in Robotics for Older Adults

Addressing the needs of aging populations surely requires the introduction of human factors. Ergonomics, which ensures systems align with human needs, skills, and limitations to accomplish tasks in a certain context, provides methodologies for achieving robots that are both functional and user-friendly. Overall, we could say that robots can improve the ergonomic quality of a system in terms of compatibility with their users but they must also be designed for being ergonomic, obviously. By focusing on the insights offered by the physical, cognitive, and organizational branches of ergonomics (and the emergent domain of neuroergonomics) in the field of HRI, next sections will describe how they can help devising robots for older adults.

3.1 Physical Capabilities

Aging progressively leads (even regardless of any pathological processes) to a decline in physical capabilities, including reduced mobility, strength, and flexibility. The approaches of physical ergonomics typically address these challenges in the design of various devices [28–30]. In case of physical HRI (pHRI), this discipline must ensure (through methodologies of areas like anthropometry and biomechanics) that robotic systems actually work to support and augment the user abilities for performing the expected tasks in an efficient and safe way, especially in contexts of assistance and rehabilitation [31]. Reducing physical strain and enhancing movements are the priority of assistive robots (in case of workerwe could mention augmentative robots too): lifters, smart walkers, robotic wheelchairs, exoskeletons, and exosuits. Robotic tools like spoons mitigating tremors and prothetic limbs (already mentioned in this chapter) can fall in this category too. On the other hand, promoting the physical recovery from injuries or chronic conditions (with exercises, for instance, for resistance training targeting specific muscle groups) is the goal of rehabilitation robots, which can fall in two main categories: end-effectors (moving a part of the body without matching the human joints) and exoskeletons (moving the limbs of the user wearing them while matching the human joints) [4]. Importantly, even social robots [32] can help improving the recovery of motor capabilities by guiding exercises and motivating the elderly to perform them without physical contact.

Most robots we mentioned should be characterised by shared control features for adjusting themselves to the limitations of aging individuals: they must ensure comfort in contact points (in particular in case of end-effectors for rehabilitation and exoskeletons for both assistance and rehabilitation), and safety through the detection and the avoidance of collision with objects overlooked by the user (especially in case of devices for assisted mobility). They must also provide stability and support needed to navigate their environments safely. These systems are often equipped with sensors to detect obstacles and adjust their movements in real-time, reducing the risk of falls. The assist-as-needed solutions (intervening just when it is necessary in case of user over-reliance on the robot to perform exercises) are just examples of intelligent guidance of the physical HRI, which can also be led by the real-time interpretation of the user's biosignals for personalising the interaction [4].

For instance, the robotic system for rehabilitative resistance training online monitored the human muscle activation by surface electromyography (sEMG) and adapted the resistance to avoid overworked [14]. Such systems, incorporating variable impedance control and real-time feedback, offer significant potential in rehabilitation by enabling load-adjustable, personalized training. This approach enhances both safety and effectiveness, making it particularly beneficial for older adults in rebuilding strength and maintaining physical function. Humanoid robots are always mentioned to undertake care of the elderly and they should acquire the regulation skill to enhance the safety and adaptability during physical HRI for older adults. Learning from human demonstration can be a promising way to master these regulation skills (e.g., impedance, velocities) [33]. Robots are supposed to dynamically adjust their compliance behaviors to elderly partner states, offering safer and more responsive interactions. Neural Network-based method can predict human compliance variations, allowing robots to adjust in real time and minimize risks of force mismatches or unintended movements during HRI. Combined with a whole-body impedance controller for coordinated and stable actions, this redundancy system offers a promising approach to improving both safety and usability in HRI for aging populations.

Another relevant aspect of physical ergonomics is the reduction of the user fatigue. In HRI, this is translated into solutions for the prediction and detection of fatigue [34] and also stress and workload for adjusting the robot behaviour accordingly: wearable robots, for instance, could distribute weights and provide additional assistance according to the levels of fatigue not only in clinical applications but also in working settings. However, in contexts where humanoid robots collaborate with older workers to manipulate large objects, directly altering the human configuration proves impractical as might be feasible with exoskeletons. Consequently, optimizing the human configuration through the strategic adjustment of the object pose by robots emerges as a viable solution, which can significantly mitigate the risk of muscle fatigue. Considering the potential applications of this insight in cobotics, we must also ponder the relationship between physical and mental workload, as the next section will discuss.

3.2 Mental Processes and User Experience

First of all, cognitive ergonomics can certainly contribute to design robotic systems that support aging users' mental processes, such as memory (especially the prospective memory—"remembering to remember" to get a medicament, for instance) for maintaining a certain degree of autonomy in daily routines [35]. Robots can also work as coaches to test and stimulate (even playfully) the mental capabilities of aging people aside the physical ones [36, 37].

However, cognitive ergonomics must also enhance the HRI itself in any domains, making the machines user-friendly for the elderly. This means working on aspects like user experience, usability, and accessibility by considering the difficulties of the person to handle a user interface (pondering their issues in terms of information processing, from perception to attention, memory, and decision-making) [38, 39]. In particular, the technology acceptance is critical for the adoption of robotic technologies in this context. Indeed, the literature highlights several barriers for introducing robots for the elderly care, also discussing [40] the reasons of refusal or abandonment of robots and highlighting the need to design robots that are enjoyable and easy to use for improving the acceptance in the short-term, and functionally-relevant to extend it in the longer-term. Diverse models and tools are proposed for assessing and promoting the use of robotic technologies in this domain, from the Almere Technology Acceptance Model [41] to the ExoScore (specifically for wearable robots) [42].

The user interface design plays a crucial role in the HRIs [43]. Intuitive solutions can be retrieved in natural user interfaces (based, for instance, on interaction modalities apparently seamless, like tactile, gestural, and vocal ones) [44] and reality-based interfaces (getting inspiration from the actual procedures enacted by people in real-world contexts) [45]. Such approaches can reduce the mental workload by going beyond the fundamental principles of accessibility, like making user interface items easily perceived by the older adults (e.g., clearly visible items with high-contrast colours on a display).

Considering the impact of the mental workload, an emergent sub-field of ergonomics (strictly related to cognitive ergonomics), neuroergonomics, offers powerful solutions. Neuroergonomics [46, 47] integrates neuroscience and ergonomics to use neuroscientific approaches for investigating human factors and designing human-centered systems accordingly. Through techniques bringing neuroscientific methods (especially the ones ubiquitously enabled by wearable sensors) in real-world contexts (also in synergy with other branches of ergonomics) [48], neuroergonomics can define novel impactful ways to connect human and machine and to deepen our knowledge of the first while we improve our know-how on the design of the second outside a laboratory.

In HRI, it can create systems sensitive (as neuro-gerontechnologies) [4] to older users' neurocognitive processes, especially for implementing precision medicine paradigms (especially for monitoring and diagnostics in neurology) and neuroadaptive dynamics (to mitigate mental workload, stress, and anxiety through changes in the robot behaviour, personalising the activity of an assistive or rehabilitative machine). This approach can also employ peripheral information (e.g., gaze motion, heart rate) [49] for detecting the degree of mental workload or distraction of a senior worker and re-engaging them into the tasks with a cobot.

The latter scenario introduces the human-robot teamwork as an activity that certainly must be analysed through the contributions of another branch of ergonomics, organizational ergonomics, as the next sub-section will describe.

3.3 Social and Organizational Contexts

Organizational ergonomics considers the broader context in which devices like robots are deployed with a seamless integration (in the domain explored by this chapter) into care systems, workplaces, and communities [50–52]. Such an effective integration requires attention to workflows, policies, roles, and interactions between humans and robots, as well as the training and support needed for adoption. In particular, organizational ergonomics can highlight how we should also consider the relatives, friends, colleagues, caregivers, and acquaintances of the older user of such robots: we are describing a network of people that are involved with the elderly in different ways, and they must be considered as stakeholders in the design and assessment of such devices.

The senior-cobot teamwork [53] is an example of workplace integration, which is a critical focus of organizational ergonomics. As aging individuals remain active in the workforce, cobots can help them perform tasks that might otherwise be physically and mentally demanding. While cobots in manufacturing settings can handle heavy lifting, precise assembly, or extended periods of repetitive activities (in a safe way, to prevent accidents), older workers can employ their expertise to focus on tasks that require decision-making skills.

In healthcare, robots must certainly integrate the activities of human caregivers rather than replace them. For instance, lifting robots assist caregivers in transferring patients safely, reducing physical strain and the risk of injury. Organizational ergonomics should help planning how these robots fit into existing routines, enabling caregivers to focus on tasks that require emotional and social interaction (also for making the older people feel comfortable using these systems) [54]. The impact [55] of care robots can be quite complex on the caregivers, affecting the atmosphere and the meaningfulness of their activity alongside the perception of their professional development (also considering the attitude toward novelty and training). Further effects of the introduction of care robots can involve the relationships between caregivers and older adults at the professional, community and familial levels.

Beyond the clinical setting, the role of organizational ergonomics also extends to the contexts of community, possibly for powering the design of robotic systems [56]. For instance, protocols for telepresence can manage the remote access of older adults to family calls, social events, and medical appointments. These systems must be easy to deploy and operate within existing infrastructures, such as homes, healthcare facilities, or community centers. These scenarios surely constitute examples of the level of pervasive presence of robots in our societies, highlighting their potential for helping the older population. Next section will extend such examples by discussing insights and opportunities that may be disclosed through the approach of ergonomics.

4 Robotics for Aging 5.0: Human-Centric Insights and Opportunities

Enabling human-centered design features through the approaches provided by ergonomics is certainly a compulsory move if we want to prioritize the unique needs, preferences, and values of people increasingly experiencing difficulties, as in the case of older adults. This is especially important if also consider the growth of the aging population, as mentioned above. We need to get ready to make technologies adjustable over the cycle of life of individuals who will age beyond the average limits of the previous generation. We must also foster inclusion by addressing diverse needs and promoting independence and accessibility. Age-friendly robotics should reach these goals. This section will focus on novel solutions that can disclose high potential aside concern (especially in terms of ethics) in this context.

We already pondered how robots can offer multiple functions to serve older adults according to their needs. We also considered how their design can be improved by human factors approaches. In general, ergonomics must keep investigating and designing HRIs to achieve innovations for a positive aging. Furthermore, we must consider how the robot technology evolves within broader frameworks of Society 5.0 and Health 5.0, which aim to integrate advanced technologies into human-centered systems for societal well-being. Society 5.0 [57] envisions conditions where technologies work in harmony with humans, transcending the digital transformation focus of Society 4.0. Society 5.0 emphasises personalisation, inclusivity, and sustainable development by embedding technology into daily life to solve societal challenges, such as the ones faced by aging populations. Similarly, Health 5.0 (or Healthcare 5.0) [58] builds upon the principles of Health 4.0, which focused on digitalisation and smart healthcare systems, by integrating human-centered approaches and precision medicine into healthcare delivery. Health 5.0 leverages robotics and artificial intelligence (AI) to deliver tailored care, empower patients, and address holistic health needs, such as mental well-being and social support (also exploring cutting edge solutions like digital twins, the virtual replicas of a physical system to predict and analyse its upcoming states—a valuable resource when we consider the individual changes due to the aging) [59]. In this context, the design of robots for elderly care embodies these 5.0 paradigms by prioritising features like personalisation. By addressing gaps in accessibility, trust, and affordability, these technologies foster inclusion, creating a framework where older adults can thrive in a sustainable society prioritizing the individual well-being.

These macroergonomic aspects are crucial for planning the next steps in robotics innovation by responding to real needs through enablers at both technical and experiential levels. Such results should be achieved through human-centric approaches [60] which emphasise (as extension of 4.0 and 5.0 perspectives), more than the human-centered design, the involvement of the stakeholders in co-design processes, based on iterative and participatory processes (described in the 10th chapter of this book, dedicated to rehabilitative, assistive, and prosthetic robotics) fostering the development of broader stakeholder ecosystems. Next two sub-section will discuss examples of approaches requiring a deep analysis of the users' perspective: the first one will present the properties of the value-sensitive design in HRI, in terms of aging people inclusion; the second one will describe the advantages offered by AI technologies to power robotics for the elderly.

4.1 Value-Sensitive Design and Inclusion

Designing robots for vulnerable populations according to a 5.0 human-centric perspective requires a robust focus on ethics, inclusion, and societal needs, which must be based on a wide involvement of diverse stakeholders (from clinicians to associations of final users, from industrial partners in projects to policy-makers). Value-sensitive design offers an appropriate approach to this heterogeneous group for prioritising user dignity, autonomy, and accessibility to enhance the quality of life for elderly individuals. As discussed in previous chapters, value-sensitive design is based on a triadic approach, requiring conceptual (understanding the problem through the lenses of disciplines like philosophy and law), empirical (studying the actual behaviour of users), and technical (implementing solutions to solve the problem, for instance by helping or guiding the user behaviours) investigations. Through this approach, we can design robots as tools for empowerment rather than dependency, maintaining the individual dignity and respecting principles of safety and privacy. This approach not only addresses the immediate needs but also anticipates future challenges and evolves according to user feedback, making the solutions remain relevant and effective over time. Incorporating personalized human models into this framework can significantly enhance the design process. These models could provide subject-specific biomechanical insights, enabling rehabilitative and assistive robots to adapt their support based on an individual's physical condition, movement patterns, and muscle strength. Furthermore, by tailoring rehabilitation and assistance programs to the unique physical or cognitive capabilities of users, these models empower individuals to engage in activities that promote their independence while maintaining their dignity [61]. Values like privacy can also be inherently considered as human factors: this is why privacy-by-design features must be adopted in processes as alerts for falls, medication adherence, or environmental dangers in smart homes with IoRT systems, for instance [62–65]. This design philosophy requires an active involvement of users in the development process, fostering a deeper sense of belonging and acceptance among the target population, which enhances the overall effectiveness of the technological solutions.

Inclusion is surely a value that must be prioritized in this context for keeping the individual dignity while enacting cognitive and social support and stimulation. The accessibility of the devices and the experience offered by systems like telepresence robots is a main target for ergonomics, which can foster inclusion even by connecting people across generational gaps (which can greatly improve the quality of life of the elderly) and pondering the special needs of people with mild-to-severe impairments. Anyway, the engagement of older adults in such scenarios requires the effort to overcome barriers of technology acceptance. Such barriers that must be targeted from the first steps of each project of research and development for improving the perceived usefulness and the perceived ease of use of a device. These should be pondered alongside several other factors influencing the intention towards using a device. Enhancing the trust towards robots (possibly nudging helpful changes in the individual habits) and their affordability requires specific human-centric co-design approaches with the involvement of the stakeholders, also considering diverse socioeconomic and cultural backgrounds of the users belonging to specific cohorts (furtherly fostering the inclusive design of robotic systems). The role of robots as socially inclusive mediators in families, organisations, and communities surely needs making all actors in the network centered on the elderly contribute to the robotic systems co-design, especially for facing the challenge of the older adults loneliness with its dramatic effects on health (as the COVID-19 lockdowns taught) [16, 66–72]. By integrating these principles, robotic systems can become more than just tools; they can act as companions and facilitators, more connected life for the elderly. These approaches should be adopted in every aspect of the elderly care, making its sustainability (as in personalised and decentralised interventions for avoiding to rely too much on the hospitals) advantageous not only in terms of efficiency but also as "humane" service, promoting autonomy, sociality, and dignity aside the physical well-being. It is also important to pay attention to many societal issues like the ageism, a form of ableism [73] leading to age-based stereotypes, prejudices, and discrimination: such a mindset can be involuntarily embedded even in robot design. This situation can also occur when our robots are controlled by advanced AIs: it depends on how these are trained. The next sub-section will discuss how intelligent systems like these must be designed to be truly elderly-centered.

4.2 Intelligent Systems Potential: From LLMs to XAI

AI surely offers groundbreaking contributions to robotics, including HRI. Through solutions of machine learning (especially deep learning), AI works as a pervasive feature in robots that must interact with people, adjusting to the context and to the individual needs according to heterogeneous data and versatile models. However, AIs must be trained through fair approaches [74] in order to accomplish their tasks in equitable ways by quantifying biases and mitigating discrimination against subgroups like elderly. This is a fundamental preliminary step to prepare the training of any AIs, and it should not be ignored although the focus can be the manipulation of objects without a specific emphasis on the user. The interaction with the latter surely requires even more advanced approaches on top of the fairness requirement. Currently, Large Language Models (LLMS) are demonstrating not only the possibility to facilitate natural interactions by supporting verbal communication between conversational agents (also embodied in robots) and individuals (including older adults) [75], but also to facilitate the robot training to perform manipulation tasks [76] that can be transferred to the execution of ADLs. This demonstrates a disruptive potential to project the recent achievements of LLMs to physical AI solutions that can help people in diverse contexts—obviously after considering potential risks in the autonomous activity of the LLMs [77]. Avoiding the development of biases in these intelligent systems is imperative, also incorporating safeguards that prevent misuse and address concerns about autonomy and control in highly capable AI-driven systems.

Further examples are offered by the paradigm of explainable AI (XAI) [78]. XAI systems are made to inform the user about the reasoning that led a certain decision, improving transparency, technology acceptance, and user trust. A robotic coach enriched with XAI features can offer personalised support (according to behavioural and physiological data) while becoming trustworthy and persuasive, nudging changes in the individual habits that can help overcoming chronic conditions by promoting healthy habits, with empowering experiences and without dehumanizing effects on the elder person. These solutions can help decentralizing the healthcare services while they make the social interactions with the robots more intuitive, accessible, and cognitively-affectively stimulating—fighting loneliness as well. Integrating XAI improves robot functionality and upholds ethical standards by making autonomous systems decisions transparent and justifiable, thereby building trust and reliability crucial for their acceptance in sensitive fields like elderly care. Among the XAI applications, especially important is the exploitation of data from sensorized networks (especially from smart homes) to detect the cognitive decline of the elderly to timely alert the clinicians. The authors of [79] proposed a XAI system able to explain the reason of predictions on the individual conditions. This example of data-driven XAI for supporting the diagnosis of cognitive decline surely could bring features that roboticists should consider for implementing smart machines serving the elderly.

5 Conclusion

This chapter proposes an overview on robotics for older people through examples of elderly-centered robot design concepts based on the application of the main branches of ergonomics. All aspects in this domain—from value-sensitive design to the advances in intelligent systems—can work as versatile enablers for a positive aging. While the healthcare sustainability must be prioritised, we cannot forget the older person and the network of relatives, friends, acquaintances, clinicians, and any other stakeholder that should be involved in the co-design of robots for the elderly. Through this systemic view, we can create a supportive, scalable infrastructure of robotic devices that can adapt to the evolving needs of aging individuals and populations.

References

1. Pilotto, A., Volta, E., Barbagelata, M., Custodero, C.: Gerontechnology: Definitions and classification. In: Gerontechnology. A Clinical Perspective, pp. 3–14. Springer, Singapore (2023)
2. Shevelkova, V., Mattocks, C., Lafortune, L.: Efforts to address the sustainable development goals in older populations: a scoping review. BMC Public Health **23**(1), 456 (2023)
3. De Michieli, L., Petrushin, A., Bustreo, M., Del Bue, A., Barresi, G.: Robots in geriatric care: A future with no return? In: Gerontechnology. A Clinical Perspective, pp. 143–161. Springer, Singapore (2023)
4. Barresi, G., Zenzeri, J., Tessadori, J., Laffranchi, M., Semprini, M., De Michieli, L.: Neurogerontechnologies: applications and opportunities. In: Internet of Things for Human-Centered Design: Application to Elderly Healthcare, pp. 123–153. Springer, Singapore (2022)
5. Esmaeilzadeh, P., Maddah, M.: Robotic companions and healthy aging: a mixed-methods exploration of older adults' perspectives and insights. Technol. Soc. **78**, 102689 (2024)
6. Figliano, G., Manzi, F., Tacci, A.L., Marchetti, A., Massaro, D.: Ageing society and the challenge for social robotics: a systematic review of socially assistive robotics for mci patients. PLoS ONE **18**(11), 0293324 (2023)
7. Rau, P..-L.P., Yu, D.: New business approach to gerontechnology in China. In: Emerging Technologies in Business: Innovation Strategies for Competitive Advantage, pp. 261–279. Springer, Singapore (2024)
8. Muminovic, S., Burr, L., Kampschulte, L.: Geriatronics-a student workshop on senior citizens, robotics and ethical issues. In: Robotics in Education: Methodologies and Technologies, pp. 112–117. Springer (2021)
9. Nazrin, A., Abdulla, R., San, L.Y.: A low cost stabilizing spoon for people with parkinson's disease. J. Appl. Technol. Innov. **5**(3), 32 (2021)
10. Eijk, M.S., Linde, H., Buijck, B., Geurts, A., Zuidema, S., Koopmans, R.: Predicting prosthetic use in elderly patients after major lower limb amputation. Prosthetics orth. Int **36**(1), 45–52 (2012)
11. Sposito, M., Poliero, T., Di Natali, C., Semprini, M., Barresi, G., Laffranchi, M., Caldwell, D.G., De Michieli, L., Ortiz, J.: Exoskeletons in elderly healthcare. In: Internet of Things for Human-Centered Design: Application to Elderly Healthcare, pp. 353–374. Springer, Singapore (2022)
12. Aaltonen, I., Niemelä, M., Tammela, A.: Please call me? Calling practices with telepresence robots for the elderly. In: Proceedings of the Companion of the 2017 ACM/IEEE International Conference on Human-Robot Interaction, pp. 55–56. (2017)

13. Ruggiero, A., Mahr, D., Odekerken-Schröder, G., Spena, T.R., Mele, C.: Companion robots for well-being: a review and relational framework. Research Handbook on Services Management, 309–330 (2022)
14. Li, C., Wu, X., Teng, T., Calinon, S., Chen, F.: Towards robo-coach: Robot interactive stiffness/position adaptation for human strength and conditioning training. In: 2024 IEEE International Conference on Robotics and Automation (ICRA), pp. 860–866. IEEE (2024)
15. Macalupu Chira, V.A., Rittenbruch, M., Chamorro-Koc, M., Donovan, J.: Designing the future together: a collaborative designerly approach to socially interactive robots with older adults as co-designers. J. Resp. Innov. **11**(1), 2390708 (2024)
16. Laitinen, A., Niemelä, M., Pirhonen, J.: Demands of dignity in robotic care: recognizing vulnerability, agency, and subjectivity in robot-based, robot-assisted, and teleoperated elderly care. Techné: Res. Philos. Technol **23**(3), 366–401 (2019)
17. Teng, T., Wu, X., Li, Z., Caldwell, D., Chen, F.: Robust shared control with stable contact servoing for enhanced object transportation by telerobotic bimanual mobile manipulators. J. Field Robot. **42**(4), 1491–1507 (2025)
18. Ranasinghe, T., Grosse, E.H., Lerher, T.: Aging in industry 5.0: Enhancing human-robot synergy in manufacturing and logistics. IFAC-PapersOnLine **58**(19), 367–372 (2024)
19. Calatrava-Nicolás, F.M., Gutiérrez-Maestro, E., Bautista-Salinas, D., Ortiz, F.J., González, J.R., Vera-Repullo, J.A., Jiménez-Buendía, M., Méndez, I., Ruiz-Esteban, C., Mozos, O.M.: Robotic-based well-being monitoring and coaching system for the elderly in their daily activities. Sensors **21**(20), 6865 (2021)
20. Merx, Q., Steins, M., Odekerken, G.: The role of psychological comfort with service robot reminders: a dyadic field study. J. Serv. Mark. **39**(10), 1–14 (2025)
21. Tapus, A., Vieru, A..-M.: Robot cognitive stimulation for the elderly. In: International Work-Conference on the Interplay Between Natural and Artificial Computation, pp. 94–102. Springer (2013)
22. Isabet, B., Pino, M., Lewis, M., Benveniste, S., Rigaud, A.-S.: Social telepresence robots: a narrative review of experiments involving older adults before and during the covid-19 pandemic. Int. J. Environ. Res. Public Health **18**(7), 3597 (2021)
23. Pavón-Pulido, N., López-Riquelme, J.A., Pinuaga-Cascales, J.J., Ferruz-Melero, J., Dos Santos, R.M.: Cybi: a smart companion robot for elderly people: Improving teleoperation and telepresence skills by combining cloud computing technologies and fuzzy logic. In: 2015 IEEE International Conference on Autonomous Robot Systems and Competitions, 198–203. IEEE (2015)
24. Oh, S., Oh, Y.H., Ju, D.Y.: Understanding the preference of the elderly for companion robot design. In: Advances in Human Factors in Robots and Unmanned Systems: Proceedings of the AHFE 2019 International Conference on Human Factors in Robots and Unmanned Systems, 24–28 July 2019, Washington DC, USA 10, pp. 92–103. Springer (2020)
25. Tanioka, T.: Nursing and rehabilitative care of the elderly using humanoid robots. J. Med. Invest. **66**(1.2), 19–23 (2019)
26. Lopez-Samaniego, L., Garcia-Zapirain, B.: A robot-based tool for physical and cognitive rehabilitation of elderly people using biofeedback. Int. J. Environ. Res. Public Health **13**(12), 1176 (2016)
27. Sierra, M., Harris, S.D., Múnera, N., Cifuentes, M., C.A.: Socially assistive walker for daily living assistance in older adults. Front. Robot. AI **11**, 1401663 (2024)
28. Kelly, P., Kroemer, K.: Anthropometry of the elderly: status and recommendations. Hum. Factors **32**(5), 571–595 (1990)
29. Taha, Z., Sulaiman, R.: Ergonomics consideration in the design of products for the elderly population. In: Proceedings of the 9th Asia Pacific Industrial Engineering & Management Systems Conference (APIEMS2008), p. 309 (2008)
30. Imrhan, S.N.: Muscular strength in the elderly-implications for ergonomic design. Int. J. Ind. Ergon. **13**(2), 125–138 (1994)
31. Mohebbi, A.: Human-robot interaction in rehabilitation and assistance: a review. Curr. Robot. Rep. **1**(3), 131–144 (2020)

32. Lotfi, A., Langensiepen, C., Yahaya, S.W.: Socially assistive robotics: robot exercise trainer for older adults. Technologies **6**(1), 32 (2018)
33. Li, C., Wu, X., Liu, J., Teng, T., Chen, Y., Calinon, S., Caldwell, D., Chen, F.: Human-like robot impedance regulation skill learning from human-human demonstrations. IEEE Trans. Cogn. Dev, Syst (2025)
34. Kuber, P.M., Kulkarni, A.R., Rashedi, E.: Predicting perceived back fatigue during exoskeleton supported trunk bending tasks using machine learning. In: Proceedings of the Human Factors and Ergonomics Society Annual Meeting, vol. 68, no. 1, pp. 683–688. Sage Publications Sage CA, Los Angeles, CA (2024)
35. Rehm, M., Krummheuer, A.L.: When a notification at the right time is not enough: the reminding process for socially assistive robots in institutional care. Front. Robot. AI **11**, 1369438 (2024)
36. Magyar, G., Magyar, J., Sinčák, P., Cavallo, F., D'Onofrio, G.: Using social robots as cognitive stimulation therapy coaches in an elderly care facility. Researchgate (2017)
37. Čaić, M., Avelino, J., Mahr, D., Odekerken-Schröder, G., Bernardino, A.: Robotic versus human coaches for active aging: an automated social presence perspective. Int. J. Soc. Robot. **12**(4), 867–882 (2020)
38. Schieber, F.: Human factors and aging: identifying and compensating for age-related deficits in sensory and cognitive function. Impact Technol. Successful Aging, 42–84 (2003)
39. Chikwendu, O.C., Ezeanyim, O., Igbokwe, N.C.: Human-robot interaction enhancement through ergonomics and human factors: future directions. Int. J. Eng. Res. Dev. **19**(6), 34–40 (2023)
40. De Graaf, M., Ben Allouch, S., Van Dijk, J.: Why do they refuse to use my robot? reasons for non-use derived from a long-term home study. In: Proceedings of the 2017 ACM/IEEE International Conference on Human-robot Interaction, pp. 224–233. (2017)
41. Heerink, M., Kröse, B., Evers, V., Wielinga, B.: Assessing acceptance of assistive social agent technology by older adults: the Almere model, Springer (2010)
42. Shore, L., Power, V., Hartigan, B., Schülein, S., Graf, E., Eyto, A., O'Sullivan, L.: Exoscore: a design tool to evaluate factors associated with technology acceptance of soft lower limb exosuits by older adults. Hum. Factors **62**(3), 391–410 (2020)
43. Villani, V., Sabattini, L., Secchi, C., Fantuzzi, C.: A framework for affect-based natural human-robot interaction. In: 27th IEEE International Symposium on Robot and Human Interactive Communication (RO-MAN), pp. 1038–1044. IEEE (2018)
44. Hsiao, S.-W., Lee, C.-H., Yang, M.-H., Chen, R.-Q.: User interface based on natural interaction design for seniors. Comput. Hum. Behav. **75**, 147–159 (2017)
45. Cho, M.E., Kim, M.J.: Characterizing the interaction design in healthy smart home devices for the elderly. Indoor Built Environ. **23**(1), 141–149 (2014)
46. Hardy, D.J.: Neuroergonomics: a perspective from neuropsychology, with a proposal about workload. Brain Sci. **11**(5), 647 (2021)
47. Parasuraman, R.: Neuroergonomic perspectives on human systems integration: mental workload, vigilance, adaptive automation, and training. In: APA Handbook of Human Systems Integration, pp. 163–176. American Psychological Association, Washington, DC (2015)
48. Barresi, G.: Neuroergonomics for human-centered technological contexts. In: Digital Environments and Human Relations, vol. 150, pp. 61–85. Switzerland: Springer (2024)
49. Pluchino, P., Pernice, G.F., Nenna, F., Mingardi, M., Bettelli, A., Bacchin, D., Spagnolli, A., Jacucci, G., Ragazzon, A., Miglioranzi, L., et al.: Advanced workstations and collaborative robots: exploiting eye-tracking and cardiac activity indices to unveil senior workers' mental workload in assembly tasks. Front. Robot. AI **10**, 1275572 (2023)
50. Gualtieri, L., Rauch, E., Vidoni, R.: Emerging research fields in safety and ergonomics in industrial collaborative robotics: a systematic literature review. Robot. Comput. Integ. Manuf. **67**, 101998 (2021)
51. Duffy, V.G.: The impact of organizational ergonomics on work effectiveness: with special reference to concurrent engineering in manufacturing industries. Ergonomics **42**(4), 614–637 (1999)

52. Carayon, P.: Social and organizational foundation of ergonomics: multi-level systems approaches. In: Handbook of Human Factors and Ergonomics, pp. 227–235. Hoboken, NJ, USA: John Wiley & Sons, Inc (2021)
53. Peruzzini, M., Pellicciari, M.: A framework to design a human-centred adaptive manufacturing system for aging workers. Adv. Eng. Inform. **33**, 330–349 (2017)
54. Stegner, L., Mutlu, B.: Designing for caregiving: integrating robotic assistance in senior living communities. In: Proceedings of the 2022 ACM Designing Interactive Systems Conference, pp. 1934–1947. (2022)
55. Melkas, H., Hennala, L., Pekkarinen, S., Kyrki, V.: Impacts of robot implementation on care personnel and clients in elderly-care institutions. Int. J. Med. Inform. **134**, 104041 (2020)
56. Moharana, S., Panduro, A.E., Lee, H.R., Riek, L.D.: Robots for joy, robots for sorrow: community based robot design for dementia caregivers. In: 14th ACM/IEEE International Conference on Human-Robot Interaction (HRI), pp. 458–467. IEEE (2019)
57. Fukuyama, M.: Society 5.0: Aiming for a new human-centered society. Japan Spotlight **27**(5), 47–50 (2018)
58. Wazid, M., Das, A.K., Mohd, N., Park, Y.: Healthcare 5.0 security framework: applications, issues and future research directions. IEEE Access **10**, 129429–129442 (2022)
59. Barresi, G., Gaggioli, A., Sternini, F., Ravizza, A., Pacchierotti, C., De Michieli, L.: Digital twins and healthcare: quick overview and human-centric perspectives. In: mHealth and Human-Centered Design Towards Enhanced Health, Care, and Well-being, pp. 57–78. (2023)
60. Laura, C., Bertolaso, M., Afferni, P.: Healthcare 4.0 and human-centric approach. In: Medic. Metodologia Didattica e Innovazione Clinica, (2020)
61. Wu, X., Li, C., Zou, K., Xi, N., Chen, F.: Muscle activation estimation by optimizing the Musculoskeletal model for personalized strength and conditioning training. In: 2024 IEEE International Conference on Robotics and Biomimetics, pp. 201–206. IEEE (2024)
62. Van Wynsberghe, A.: Designing robots for care: care centered value-sensitive design. In: Machine Ethics and Robot Ethics, pp. 185–211. Routledge, London (2020)
63. Umbrello, S., Capasso, M., Balistreri, M., Pirni, A., Merenda, F.: Value sensitive design to achieve the un SDGS with AI: a case of elderly care robots. Mind. Mach. **31**(3), 395–419 (2021)
64. Petrushin, A., Freddolini, M., Barresi, G., Bustreo, M., Laffranchi, M., Del Bue, A., De Michieli, L.: Iot-powered monitoring systems for geriatric healthcare: overview. In: Internet of Things for Human-Centered Design: Application to Elderly Healthcare, pp. 99–122. Springer, Singapore (2022)
65. Chan, H.Y., Muralidharan, A.: Care robots for the elderly: Legal, ethical considerations and regulatory strategies. In: Developments in Intellectual Property Strategy: The Impact of Artificial Intelligence. Robotics and New Technologies, pp. 129–156. Springer, Palgrave Macmillan, Cham (2024)
66. Sitbon, L., Balasuriya, S., Mitchell, A.: Inclusively designing for people in human-robot collaboration. Interact. Des. Arch. **61**(61), 11–41 (2024)
67. Sumner, J., Chong, L.S., Bundele, A., Wei Lim, Y.: Co-designing technology for aging in place: a systematic review. Gerontologist **61**(7), 395–409 (2021)
68. Mannheim, I., Schwartz, E., Xi, W., Buttigieg, S.C., McDonnell-Naughton, M., Wouters, E.J., Van Zaalen, Y.: Inclusion of older adults in the research and design of digital technology. Int. J. Environ. Res. Public Health **16**(19), 3718 (2019)
69. Cruz-Sandoval, D., Tentori, M., Favela, J.: A framework to design engaging interactions in socially assistive robots to mitigate dementia-related symptoms. ACM Trans. Human-Robot Interact. **14**(1), 1–25 (2024)
70. Mannheim, I., Weiss, D., Zaalen, Y., Wouters, E.J.: An "ultimate partnership": Older persons' perspectives on age-stereotypes and intergenerational interaction in co-designing digital technologies. Arch. Gerontol. Geriatr. **113**, 105050 (2023)
71. Ross, G.M.: Dancing with robots: acceptability of humanoid companions to reduce loneliness during Covid-19 (and beyond). AI Soc. **39**(5), 2557–2568 (2024)

72. Rodogno, R.: Nudging by social robots. In: Culturally Sustainable Social Robotics, pp. 337–345. IOS Press, Amsterdam (2020)
73. Dehnert, M.: Ability and disability: social robots and accessibility, disability justice, and the socially constructed normal body. The De Gruyter Handbook of Robots in Society and Culture **3**, 429 (2024)
74. Feuerriegel, S., Dolata, M., Schwabe, G.: Fair AI: challenges and opportunities. Bus. Inf. Syst. Eng. **62**, 379–384 (2020)
75. Onorati, T., Castro-González, Á., Valle, J.C., Díaz, P., Castillo, J.C.: Creating personalized verbal human-robot interactions using LLM with the robot mini. In: International Conference on Ubiquitous Computing and Ambient Intelligence, pp. 148–159. Springer (2023)
76. Liu, J., Li, Z., Lin, W., Calinon, S., Tan, K.C., Chen, F.: Softgpt: Learn goal-oriented soft object manipulation skills by generative pre-trained heterogeneous graph transformer. In: 2023 IEEE/RSJ International Conference on Intelligent Robots and Systems (IROS), pp. 4920–4925. IEEE (2023)
77. Irfan, B., Kuoppamäki, S., Hosseini, A., Skantze, G.: Between reality and delusion: challenges of applying large language models to companion robots for open-domain dialogues with older adults. Auton. Robot. **49**(1), 9 (2025)
78. Aharony, N., Krakovski, M., Edan, Y.: A transparency-based action model implemented in a robotic physical trainer for improved HRI. ACM Trans. Human-Robot Interact. **14**(1), 1–19 (2024)
79. Khodabandehloo, E., Riboni, D., Alimohammadi, A.: Healthxai: Collaborative and explainable ai for supporting early diagnosis of cognitive decline. Futur. Gener. Comput. Syst. **116**, 168–189 (2021)

HFE Considerations in Sustainability in Robotics

Andrew Thatcher and Wei Zhang

Abstract Sustainability can be conceptualized as three interlinked aspects: economic; social; and natural capital. In this chapter we will examine what is required for each of these aspects from a human factors and ergonomics perspective to enable sustainability in human-robotic systems. When looking at the economic and social capital nexus it is important to take into account issues of safety and the future roles of humans in robotic systems. This includes issues such as skills, training, resilience, and employment status of the humans. Furthermore, the possibilities arising from advanced social robotics also requires an examination of issues such as privacy, trust, and understandability. When looking at the economic and natural capital nexus it is necessary to consider aspects such as the materials used to produce, maintain, and deconstruct robotic systems as well as the eco-efficiencies and eco-effectiveness that can be obtained from including robotics in human systems. Such an examination requires a consideration of the diverse impact of robotic systems in different geographical communities, and among regions with different economic development status. The social and natural capital nexus requires us to evaluate when and where to invest in human-robotic systems, but also when and where to resist such developments. Finally, we consider the HFE values that are needed to help develop a more sustainable approach to the development of human-robotic systems.

Keywords Resilience · Cross-cultural safety · Trust · Longevity · Ecological considerations

A. Thatcher (✉)
Psychology Department, University of the Witwatersrand, Johannesburg, South Africa
e-mail: Andrew.Thatcher@wits.ac.za

W. Zhang
Department of Industrial Engineering, Tsinghua University, Beijing, China
e-mail: zhangwei@tsinghua.edu.cn

G. Barresi and S. Wischniewski (eds.), *Ergonomics in Robotics: Advances and Innovations*, Springer Tracts in Advanced Robotics 162,
https://doi.org/10.1007/978-3-032-19374-2_25

1 Introduction

In this chapter we examine the aspects necessary to consider sustainability with robotics systems that include humans. Despite there now being literally hundreds of definitions [1], sustainability boils down to a single issue; the availability and use of resources now and in the future. Sustainability refers to the need to maintain the availability and circulation of resources essential to human wellbeing and effectiveness, either in the immediate present or at some point in the future. Our need to consider sustainability occurs because a resource is either absent, becomes critically depleted in quantity, or its quality becomes degraded. Sustainability therefore refers to the need to manage or remediate a primary resource that has been directly over-exploited by humans (e.g., introducing alternatives to fossil fuels to use as energy sources) or a secondary resource that has been depleted or degraded by human exploitation of a primary resource (e.g., carbon dioxide and sulfur dioxide is released into the atmosphere when burning coal and is implicated in global warming and acid rain respectively). Sustainability is a complicated issue because the actions of one set of actors may have geographical and temporally distal implications for other actors. First, the negative impacts on secondary resources may sometimes only be realized decades or even centuries later (e.g., global warming is the result of centuries of indiscriminate burning of fossil fuels), while the benefits for humans may be realized immediately (e.g., cheap energy to power many labor-saving devices). Second, the actions of actors in one part of the world may only be felt by people in the other parts of the world (e.g., burning fossil fuels in Europe, over centuries, is now disproportionately impacting Pacific Island nations through threats from increasingly volatile hurricanes and rising sea-levels).

This level of complexity and interdependence led the World Commission on Environment and Development [2] to introduce the term 'sustainable development' to emphasize the need for equitable improvements in human social structures across the world and across generations. Redclift [3] considered the term 'sustainable development' to be an oxymoron because development (i.e., growth) is inherently unsustainable given the finite resource conditions and carrying capacities [4] that exist on our planet. For this reason, the term 'sustainability' is preferred in this chapter although the principle of the equitable distribution of resources, opportunities, health, and wellbeing remains relevant.

A common conceptualization of sustainability is a balance between the "three pillars" of economic, social, and natural capital, sometimes also referred to as the "triple bottom line" [5]. Economic capital refers to financial resources such as money and other financial instruments, and physical assets such as buildings, technologies, and infrastructure. Social capital refers to people, their knowledge and cultural systems, and their institutions. Natural capital refers to raw materials, food, biodiversity, and other ecosystem services such as air, soil, and water. Achieving balance between these three pillars of capital is no easy task. On the one hand, aspects of natural capital are either irreplaceable, difficult to compare to the other two capitals, or are not fully factored into consumption systems. On the other hand, failing to equitably address the existential, health, and wellbeing needs of humans may be considered unethical. Effectively this means that certain types of natural capital is non-substitutable (e.g., biodiversity or the air we breathe is irreplaceable), while restrictions on the accumulation of waste (e.g., pollutants) must be balanced against lost opportunity costs for human development [6].

What does all this have to do with robotics? We have raised these complexities to help illustrate that there are different points at which sustainability must be considered in the practice of robotics. Specifically, this extended discussion on sustainability is necessary to demonstrate how the discipline of human factors and ergonomics (HFE) is related to sustainability. Zink and Fischer [7] outlined four ways in which HFE could benefit from the sustainability agenda: (1) HFE is obviously focused on social capital, but must necessarily be constrained by the limits of economic and natural capital; (2) HFE must not only consider the end-users of products but acknowledge how humans are impacted along the whole value chain; (3) HFE should take a life-cycle approach that incorporates manufacturing, use, maintenance, and post-use of products and services; and (4) HFE should help systems cope with the behavioral change management necessary to achieve a sustainable society.

To demonstrate the links, Zink et al. [8] used Dyllick and Hockerts' [5] three pillars model to demonstrate how HFE practices address sustainability. Eco-effectiveness refers to designs and approaches that minimize negative environmental impacts and maximize positive environmental impacts. Eco-effectiveness refers to designs and approaches of systems that support ecological regeneration. Socio-efficiency refers to human activities that mutually support human wellbeing and financial gain. Socio-effectiveness refers to designs and approaches that support absolute positive social impact. Sufficiency refers to only producing or using what is absolutely necessary. Finally, ecological equity refers to the equitable distribution (geographically and within society) of natural capital for social good. HFE has historically focused on the economic-social arm, although there are more recent examples expanding on to both the economic-nature and social-nature arms (see Fig. 1).

Figure 2 captures the same three pillars of sustainability, but this time mapping the HFE considerations for sustainable robotics which will form the discussion of the remainder of this chapter. In Fig. 2 the considerations move beyond the economic-social arm to include the economic-natural and natural-social arms.

2 Robotics and Sustainability in HFE

2.1 Socio-efficiency in Human-Robotic Systems

In many ways, what is encapsulated under the socio-efficiency component of sustainability might be considered "business-as-usual" for HFE. One aspect of socio-efficiency means ensuring that robotic systems are designed and installed to reduce human exposure to dangerous situations to facilitate health and safety of the human

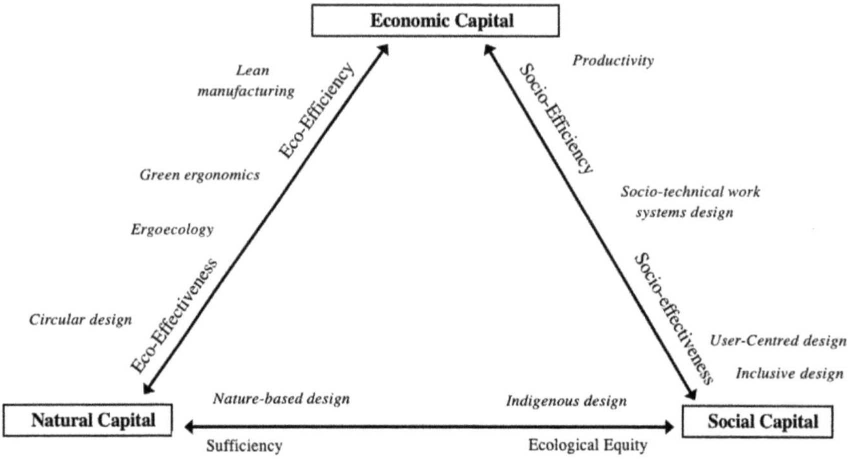

Fig. 1 HFE and the three pillars of sustainability

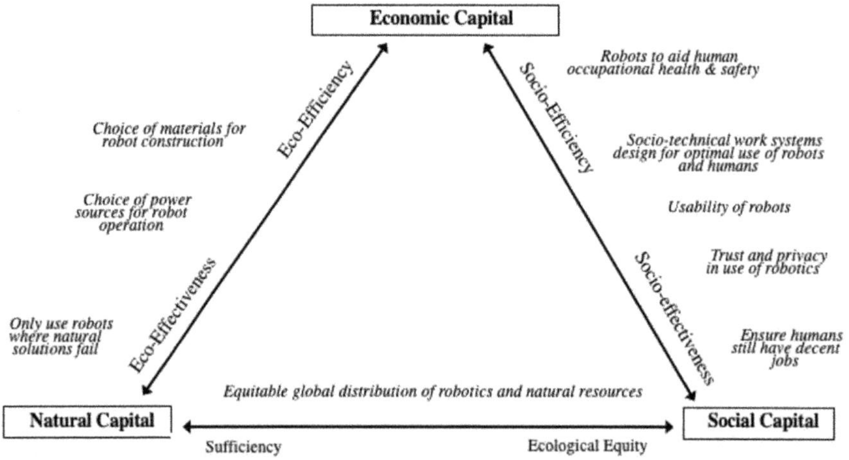

Fig. 2 HFE and the three pillars of sustainability for human-robotic systems

participants. Many robotic systems have been designed and implemented with this exact purpose in mind. The most obvious examples are to protect people in highly dangerous situations such as in warfare, nuclear reactor maintenance, underwater and offshore operations, space exploration, manufacturing plants, and transport systems. There are also less obviously dangerous contexts where robotic systems can replace or support humans in the completion of mundane or repetitive tasks and therefore remove the risks of boredom, emotional abuse, repetitive strain injuries, or other musculoskeletal disorders. Other examples include where people may be performing too many tasks to relieve some of the load and that give people a chance to rest or perform multiple tasks simultaneously (e.g., social care for the elderly/lonely or domestic chores). Finally, robotic systems might be used to augment or support human limitations, for example fine motor control in micro-surgery or where people have strength limitations such as lifting heavy loads (or where people have a physical disability such as poor eyesight or a missing digit or limb). Examples might include various manual material handling tasks, vigilance tasks, and lifting tasks. However, there are also places where robotic systems have been introduced where there are no obvious health and safety risks to the human, or where these risks simply reflect the risks of being alive (e.g., there are robotic systems that simply open doors or that put make-up on humans). This reminds us of a common English saying: "just because you can, doesn't mean you should." Unless there are human physical, cognitive, or social health and safety risks or human limitations, these types of robotic systems may result in unforeseen human (financial and environmental) outcomes such as the loss or devaluation of jobs.

A second aspect of socio-efficiency that was already implied in the previous discussion means ensuring that there are appropriate socio-technical design considerations for efficient human–robot collaborations. This might include considerations for the appropriate allocation of tasks, the re-design of tasks to facilitate optimal human and robot operations, and the redesign of work to ensure ideal human–robot coupling relationships. At its most basic, this form of socio-efficiency might mean machines-are-better-at/men-are-better-at (MABA-MABA) considerations to ensure that the allocation of tasks and jobs depends on whether robotic systems or humans are better at performing them. However, Dekker and Woods [9] have noted that this type of thinking might be an oversimplification. Even in simple MABA-MABA cases, when human tasks are substituted for robotic systems, the entire interaction may be transformed such that the human adapts to the new configuration rendering any initial MABA-MABA assignments obsolete [9]. In addition, with the inclusion of artificial intelligence many robotic systems might be more capable at performing tasks that were previously only considered suitable for humans. For each of these situations HFE needs to critically ask the question: "what will people do?".

The situation is even more complex with robotics systems that include multiple human actors, automation, or artificial intelligence components when each actor in the system adapts and transforms the relationships. Instead, it is argued that the human–robot system should be viewed as a team, and task allocation should be considered as an integral part of the team dynamics, with flexibility built into the allocation process. Indeed, this is the whole purpose of the Human-AI-Robot-Teaming (HART) initiatives which are gaining prominence in the HFE space [10].

2.2 *Socio-effectiveness in Human-Robotic Systems*

Many aspects of socio-effectiveness might also be considered "business-as-usual" for HFE. One aspect of socio-effectiveness that already falls firmly within the realm of HFE is the usability of human-robotic systems. Usability considers whether the robotic systems can be used effectively, safely, and with satisfaction (by the human actors). As a socio-effectiveness aspect, the usability of human-robotic systems emphasizes the human functioning component of the system, but in many cases the functioning of the robotic system is often difficult to unambiguously separate out from the humans in the system. Usability here might refer to whether the operation of a robotic system is easy to learn, easy to use, easy to remember, free from errors, safe, and satisfying. When considering socio-effectiveness as part of sustainability, a key component is that the robotic system must add to the utility of the human (i.e., does the robotic system help the human/s achieve their goal/s). In some cases, it is possible that a robotic system may be introduced gratuitously, without adding any utility or value to the human/s in the system (or the overall "value" may even be negative, resulting in the loss of livelihood for some humans). Key questions to ask include whether the robotic system addresses some problem encountered by the human/s (e.g., safety, exposure to harm, repetitive, boring, etc.) and whether the robotic system is really necessary. These are similar questions that should be asked when considering socio-efficiency. There may be cases where the utility of a robotic system is evaluated as being minimal (e.g., a simple hedonic value, replacing a mundane task or a task that is seldom performed, but with little consequence if the task is not performed). Careful consideration should be given to the appropriateness of supporting these types of systems.

Other aspects that also have a clear relationship with HFE are the issues of trust, privacy, and understandability. The early work on trust in automation [11, 12] has similar relevance when considering trust in robotic systems because they are an extension of automated systems. In fact, one might argue that trust in robotics is even more important given the high degree of autonomy that many robotic systems now have thanks to advances in technology and artificial intelligence. Certainly, trust in HART is a fast-emerging topic in the HFE field [13, 14]. Trust will also become an increasingly important topic as developments in social robotics (e.g., care for the elderly, children, and those with psychological disorders) continue [15]. Social robotics involves deeply personal and (possibly) intense social interactions with humans where issues of social trust (e.g., where humans share personal or private thoughts, feelings, and emotions) are important. We can't really talk about the sustainability of social robotics without considering how human actors will respond to these systems over an extended period of time.

An aspect of socio-effectiveness that is rarely considered by HFE is what will the human do when the entire job (i.e., not just some tasks, but the whole job) is removed. There is often the assumption that if a robotic system replaces a human actor, then new jobs will be created (e.g., introducing a new robotic system will create jobs for humans to design, build, install, maintain, and supervise robots). Indeed, at the global level it might be true that some robotic systems result in an increase in new job opportunities. However, at the local level this is seldom the case. This is due to differences in the geographical distribution of skills. First, it is only in limited situations where the same people who occupy the jobs that are replaced by robotic systems will also have the skills to design, build, and maintain these systems. In some cases, it may be possible to train or re-skill people to do these new jobs and tasks. However, in most cases training or re-skilling would simply take too long or would be too expensive. Second, many of these new jobs will not take place in the same geographical location. For example, unless specifically designed that way it is rare that the design and construction of robotic systems will happen in the same location as their installation and maintenance. These issues are exacerbated through the trans-boundary distribution of robotic systems. Instead, it is more likely that some people will lose their jobs (and their ability to support themselves and their families), even if there is a net increase in jobs globally. The people most likely to lose their jobs will be those who already face harsh social and economic pressures. Taking a sustainability perspective, particularly as an HFE practitioner, means taking a broader socio-technical perspective. How does the introduction of robotic systems impact the social dynamics of communities and societies, especially where the impacts are felt geographically far away? Can we show a net benefit to all societies that are affected, or must we consider differential implementations and recommend the non-implementation of robotic systems in certain circumstances?

2.3 *Eco-efficiency in Human-Robotic Systems*

There is now a growing body of interventions that have considered eco-efficiency for HFE design [16, 17]. To understand all the eco-efficiency implications, it is necessary to have a lifecycle understanding of products and systems [18]. Taking a lifecycle approach means that we need to consider all the "stages" of a product or system, from the design conceptualization stage, through various stages of construction, distribution, use, and repurposing, all the way through to the termination stage, and how the termination stage might either lead to obsolescence or the start of a new cycle [19]. HFE has traditionally focused on the design and use stages without necessarily considering the other lifecycle stages, or how the lifecycle stages integrate with one another. This has changed with the introduction of EQUID 4.0 [20] which includes lifecycle considerations. For eco-efficiency from an HFE perspective there are two streams that connect through these stages: operational eco-efficiencies and material eco-efficiencies.

Material eco-efficiencies involve decisions in the design stage that have implications across all the lifecycle stages, but particularly the termination stage. Specifically, material eco-efficiencies involve making decisions about what materials to use that have the least negative impact on financial, social, and environmental capital. In the design and production stages this involves choices about the types of materials and where those materials are sourced. Important questions to consider include: will the materials for the robotic system damage the environment; have the materials for the robotic system been sourced from conflict areas; how far must the materials for the robotic system be transported; are the materials for the robotic system durable; are the materials or components for the robotic system easy to maintain or replace; are the materials for the robotic system recyclable; are the materials for the robotic system biodegradable; and will the materials for the robotic system release toxins during construction, use, or re-purposing that might damage people or the environment? The source of materials is important to consider, not only for financial reasons (which historically has been the primary consideration) but to ensure that materials do not come from primary sources that damage the environment or that exploit people. For example, there are efforts to design robotic systems using recycled/upcycled materials [21] and even biologically-inspired robotic systems [22].

The distance between the source of materials, where they will be assembled, where they will be used, and where they will be disassembled also determines how far they will travel and what social or environmental damage will result along these transport pathways. One might consider the 3-D printing of robotic components to reduce transportation pollution [23]. It is also important to consider the transformation processes (e.g., heating iron to make steel) of turning a raw material into an end-use material. These travel and material transformation processes refer to "embodied resources" (e.g., embodied water, embodied energy, or embodied labor). The choice of materials also has the potential to cause harm across the lifecycle through the release of toxins and other pollutants that may harm people (e.g., factory workers, transporters, users, or recyclers) or the environment. The final materials consideration is what happens when the robotic system becomes obsolete or reaches a natural termination point? Are the materials easily transformed into other products? Will the materials naturally degrade without causing harm to people or the environment? These whole-of-lifecycle considerations have already been outlined for robotic systems [24]. Are the materials and components modular or re-usable in future robotic system iterations? For example, the global carbon–neutral efforts and carbon-tracing technology may provide not only technical constraints, but also potential solutions for material choices. As a profession that is concerned with well-being and effectiveness it is imperative that HFE practitioners consider the impact of material choices across all lifecycle stages.

The most obvious links to HFE are through operational eco-efficiencies. This involves ensuring that human-robotic systems make efficient use of all natural and human resources (e.g., they are energy, water, nitrogen, materials, and people efficient) and do not contribute pollutants into the environment in their operational phase. There is obviously some overlap between operational eco-efficiencies and material eco-efficiencies. However, operational eco-efficiencies also refer to the additional resources that are mobilized during the various lifecycle stages. For robotic systems the obvious mobilized resource is energy. Is the robotic system designed in a way that optimizes the use of renewable energy sources? Does the robotic system efficiently optimize the use of its energy source/s? Does the robotic system force the human user/s to expend unnecessary physical energy/mental effort during interactions (i.e., use, maintenance, or disassembly)? The word "unnecessary" is included here because the purpose of some robotic systems might be to encourage human energy expenditure (e.g., encouraging exercise during physical rehabilitation) or to stimulate cognitive activities (e.g., playing games, mental exercises). The optimization of energy in robotic systems has already been considered and should be incorporated into HFE design-thinking [25]. While energy resources are the most obvious, it is necessary to also consider other resources [26]. How often do materials or components need to be replaced in the robotic system? How much water does the robotic system use? Currently these considerations are part of quality management in ISO 9000, but further work is required to deepen th sustainability considerations [27].

2.4 Eco-effectiveness in Human-Robotic Systems

Eco-effectiveness is an extension of eco-efficiency that ensures that products and systems not only operate efficiently, but also contribute positively to environmental systems. Unfortunately, eco-effectiveness is seldom considered in HFE despite numerous calls [28, 29]. It is therefore difficult to draw examples from the HFE literature. Nevertheless, drawing inspiration from cradle-to-cradle [30] and biophilic [31] design it is possible to show how careful design choices might contribute to products and systems that nurture the environment rather than simply extracting from the environment. The choice of materials might have two different paths to eco-effectiveness. First, materials made from biological materials that easily biodegrade might be used to provide nutrients at the end of a robotic systems lifecycle. Second, materials with biophilic elements have been shown to support positive emotional attachment to nature which further promotes positive interactions between humans and nature [32]. Richardson and Butler [32] suggest another potential way in which robotic systems might support eco-effectiveness (although they didn't specifically mention robotic systems). Robotic systems could be designed to help people engage with nature. Richardson et al. [33] showed how prompts on smartphones encouraging people to take a picture of "a good thing in nature" significantly increased nature connectedness while simultaneously improving wellbeing. It is not difficult to imagine other robotic systems with similar goals.

2.5 Sufficiency in Human-Robotic Systems

The role of HFE in sufficiency for robotic systems is difficult to see, but necessary to consider. Sufficiency has two meanings in the context of robotic systems. First, it means only producing robotic systems where it is necessary to do so. This has been raised before when considering socio-efficiency and socio-effectiveness. Second, it means only producing sufficient robotic systems to meet a specific demand. These two meanings are complementary and should follow in a stepwise fashion. As was discussed under socio-efficiency, it is important to first consider the conditions when a robotic system is necessary. These are usually in circumstances when it is dangerous or unsafe for a human to carry out the task/s. As discussed earlier, this includes conditions where the human faces physical, cognitive, or social health and safety risks or human limitations. This might include situations where human resources or the required skills are scarce or unavailable (e.g., remote robotic surgery for geographically remote regions). Alternatively, it could mean providing a mix of humans and robotic systems (e.g., customer service robots to support customer service personnel) that would be more appropriate for a wider mix of customers, rather than only providing robots because that saves an organization employing multiple people who require working in shifts. HFE can play a key role in helping to make these determinations. From a sustainability perspective it is always necessary to consider whether natural or human capital will have to be replaced or not. Just because it can be replaced does not necessarily mean that it has to be replaced once the sustainability of the whole system is taken into account.

Where the conditions of this first step have been met it is then necessary to consider the second step; how many robotic systems must be produced to meet a specific demand. This might include questions about whether a robotic system is intended as an interim measure to cover a shortfall over a time-limited period or whether a robotic system is intended to entirely replace the need to place humans at risk. This should never mean over-saturating the context with robotic systems, even if this is tempting from a financial perspective.

2.6 *Ecological Equity in Human-Robotic Systems*

Ecological equity considers the equitable distribution (in space and time) of natural and social capital for social and environmental good. It is important to note that equity does not mean the same thing as equality. Equality refers to each person (or group of people) receiving the same resources or opportunities whereas equity refers to the relative distribution of resources and opportunities according to need. In an ideal world this would mean providing robotic systems to support the needs of those who need it most. However, given the current economic and geographical structure in who develops robotic systems and where they are currently implemented (mostly wealthy regions of the world supplying to wealthy regions of the world), equity is likely to come into conflict with financial capital instead of the provision of robotic systems to those who most need the most protection from risks.

Where does this leave space for HFE to contribute? As one of the human and social sciences engaged in supporting the development and implementation of robotic systems it is imperative that HFE practitioners guide robotic system development teams towards solutions that have the greatest value for the largest number of people whilst also acknowledging the importance of the natural environment in providing life-sustainaing resources. However, this means interpreting "value" in terms of "need" and "relevance" rather than simply in terms of the largest financial capital gain. For HFE practitioners, this type of outcome is fairly typical when conducting any task or work analysis to identify the most important tasks and the greatest ergonomics risks. However, once financial capital enters the equation, it is far more difficult for HFE practitioners to convince robotic system developers and suppliers to think more broadly about relevant target groups and the type of impacts that can be made for society and the natural environment. From a sustainability perspective it is likely that the work of Mauerhofer [34] in emphasizing the importance of natural capital in underpinning social capital (which itself underpins financial capital) may be useful in helping HFE practictioners to formulate these arguments.

3 Sustainability and Systems-Thinking in Human-Robotic Systems

In our description of the considerations that arise from the "three pillars" conceptualization of sustainability it should be obvious that economic, social, and natural capital are interdependent. This fact is emphasized by Mauerhofer [34] in his reconceptualization of the three pillars "triangle" as a three dimensional hierarchical cone with natural capital as the base of the cone. After all, social and economic capital mean nothing without the primary resources from the environment and the supportive structures (air, water, food) that allows humans to exist. This high degree of interconnectedness between system elements is part of what makes sustainability a complex systems-thinking problem. In addition, raw materials, other resources, human needs, and pollutants are not evenly distributed geographically. Due to the interconnectedness and the non-linear relationships, small effects in one geographical region might have exponential impacts in another geographical region. For example, a change to robotic check-in for the hospitality industry in the UAE may eliminate the need to employee hundreds of people from Indonesia, Singapore, The Philippines, and India (as well as removing the income that supports their families back in their home countries). It is therefore also evident that jobs lost (and jobs created) by the introduction of robotic systems might exacerbate existing unevenness. It should also already be evident that interventions and consequences are nonlinear, and that their impacts unfold over time.

Thatcher et al. [35] have noted that complex systems-thinking requires HFE to understand and incorporate concepts such as non-linear dynamics, emergence, adaptability, and self-organization into our design and evaluation endeavors. HFE, as a systems discipline, is tasked with embracing this complexity and bringing these insights to the attention of people across the whole robotic system lifecycle. The key components of complex systems that are important for understanding robotic systems are that there are competing human goals and that the full consequences of introducing robotic systems are unknowable due to self-organization and emergence [35]. These characteristics of complex systems should not be seen as limitations, but as preconditions for the precautionary principle [36] which states that we should do everything we can to anticipate or identify potential human or environmental threats so that we can take precautionary measures to avoid the threats or ameliorate the negative consequences of those threats (and in certain circumstances, delay the implementation of robotic systems until more is known).

4 Sustainability and HFE Values

Bolis et al. [37], Lange-Morales et al. [28], and Thatcher et al. [35] have all emphasized the importance of HFE having clearly defined values when addressing sustainability challenges. Lange-Morales et al. [28] outlined six HFE values for sustainability: (1) respect for human rights; (2) respect for the Earth; (3) respect for ethical decision-making; (4) appreciation of complexity; (5) respect for transparency and openness; (6) and respect for diversity. Most of these values have already been raised in our discussions earlier in this chapter. Respect for human rights emphasizes the centrality of the human in HFE deliberations regarding when and where to apply robotic systems. What emerges from the discussion on sustainability issues in this chapter is the need to consider multiple human actors, not only those who are the primary focus of a robotic system. These humans include the miners and forestry workers who extract the raw materials, the factory workers, the assemblers, the maintainers, the recyclers, and the people who might lose their jobs or will require re-skilling. Respect for the Earth was encapsulated in the discussions on eco-efficiency, eco-effectiveness, sufficiency, and ecological equity for robotic systems. The Earth provides us with all the raw materials needed to build robotic systems and to sustain life and therefore needs to be supported and nurtured to continue fulfilling this role. Respect for ethical decision-making means taking the needs of all stakeholders into account and to anticipate unforeseen outcomes. When discussing ecological equity and robotic systems, ethical decision-making taking multiple factors into account was emphasized as important. The appreciation of complexity is evident in the interconnectedness and temporal and geographical displacement of effects from the implementation of robotic systems encapsulated in each aspect of the "three pillars" conceptualization. Respect for transparency and openness links to the precautionary principle where the HFE practitioner recognizes that there will be unforeseen outcomes from the implementation of robotic systems and that any unforeseen or unjust outcomes will need to be ameliorated beforehand or made known to system users.

The one value that has not be covered so far is respect for diversity. This value embraces all forms of diversity by recognizing that there is no single "best" way for a robotic system to be designed or implemented for all circumstances [28]. Diversity encourages local or indigenous identity and (if necessary) adaptation. Wherever possible, the use and development of local skills, resources, and materials should be encouraged. One particular local user bears special mention; the end-user. Participatory approaches should be encouraged as involvement in the design and implementation of the end-product is more likely to encourage sustainable use.

5 Conclusions

HFE is primarily concerned with the health, wellbeing, and productivity of the human/s in the system. One of the central themes that runs throughout this chapter on sustainability consideration is the need for the HFE practitioner to consider which human/s. A traditional HFE approach is to consider the humans who are the direct recipients of the HFE intervention. We hope to have shown that not only is this approach naïve, but it is also too narrow to encapsulate all the impacts on human wellbeing and efficiency and potentially damaging to our long-term prospects on this planet. This chapter has raised a considerable number of challenges for HFE practitioners when considering human-robotic systems and sustainability. We expect many people attempting to read this chapter will balk at the challenges that we have raised for HFE practice. Unfortunately, failing to face these challenges does not eliminate them, it only displaces them geographically or defers them to future generations. To think in complex systems and to respond ethically, equitably, and fairly is the only way to ensure sustainability.

References

1. Johnston, P., Everard, M., Santillo, D., Robèrt, K.H.: Reclaiming the definition of sustainability. Environ. Sci. Poll. Res. Int. **14**, 60–66 (2007). https://doi.org/10.1065/espr2007.01.375
2. World Commission on Environmental Development. Our Common Future. Report of the World Commission on Environment and Development. Oxford University Press, Oxford (1987). https://doi.org/10.1017/S0376892900016702
3. Redclift, M.: Sustainable development (1987–2005): an oxymoron comes of age. Sust. Dev. **13**, 212–227 (2005). https://doi.org/10.1002/sd.281
4. Meadows, D., Randers, J., Meadows, D.: Limits to growth: the 30-year update. Chelsea Green Publishing (2004). https://doi.org/10.4324/9781849775861
5. Dyllick, T., Hockerts, K.: Beyond the business case for corporate sustainability. Bus. Strat. Env. **11**, 130–141 (2002). https://doi.org/10.1002/bse.323
6. Neumayer, E.: Weak Versus Strong Sustainability: Exploring the Limits of Two Opposing Paradigms. Edward Elgar Publishing, Northampton, MA (2003)
7. Zink, K.J., Fischer, K.: Do we need sustainability as a new approach in human factors and ergonomics? Ergon **56**, 348–356 (2013). https://doi.org/10.1080/00140139.2012.751456
8. Zink, K.J., Steimle, U., Fischer, K.: Human factors, business excellence and corporate sustainability: differing perspectives, joint objectives. In: Corporate Sustainability as a Challenge for Comprehensive Management, pp. 3–18. Physica Verlag, Heidelberg (2008)
9. Dekker, S.W., Woods, D.D.: MABA-MABA or abracadabra? Progress on human–automation co-ordination. Cogn. Techn. Work **4**, 240–244 (2002). https://doi.org/10.1007/s101110200022
10. Cooke, N.J., Cohen, M.C., Fazio, W.C., Inderberg, L.H., Johnson, C.J., Lematta, G.J., et al.: From teams to teamness: future directions in the science of team cognition. Hum. Fac. 00187208231162449 (2023). https://doi.org/10.1177/00187208231162449
11. Lee, J., Moray, N.: Trust, control strategies and allocation of function in human-machine systems. Ergon **35**, 1243–1270 (1992). https://doi.org/10.1080/00140139208967392
12. Muir, B.M.: Trust in automation: Part I. Theoretical issues in the study of trust and human intervention in automated systems. Ergon. **37**, 1905–1922 (1994). https://doi.org/10.1080/00140139408964957

13. Ezenyilimba, A., Wong, M., Hehr, A., Demir, M., Wolff, A., Chiou, E., Cooke, N.: Impact of transparency and explanations on trust and situation awareness in human–robot teams. J. Cogn. Engin. Dec. Mak. **17**, 75–93 (2023). https://doi.org/10.1177/15553434221136358
14. Textor, C., Zhang, R., Lopez, J., Schelble, B. G., McNeese, N. J., Freeman, G., et al.: Exploring the relationship between ethics and trust in human-artificial intelligence teaming: A mixed methods approach. J. Cogn. Engin. Dec. Mak. **16**, 252–281 (2022). https://doi.org/10.1177/15553434221113964
15. Sheridan, T.B.: A review of recent research in social robotics. Curr. Opin. Psych. **36**, 7–12 (2020). https://doi.org/10.1016/j.copsyc.2020.01.003
16. Revell, K.M., Stanton, N.A.: Mind the gap–Deriving a compatible user mental model of the home heating system to encourage sustainable behaviour. App. Ergon. **57**, 48–61 (2016). https://doi.org/10.1016/j.apergo.2016.03.005
17. Sauer, J., Rüttinger, B.: Environmental conservation in the domestic domain: the influence of technical design features and person-based factors. Ergon. **47**, 1053–1072 (2004). https://doi.org/10.1080/00140130410001686302
18. Cao, H., Folan, P.: Product life cycle: the evolution of a paradigm and literature review from 1950–2009. Prod. Plan Cont. **3**, 641–662 (2012). https://doi.org/10.1080/09537287.2011.577460
19. García-Acosta, G., Lange-Morales, K.: Beyond product life cycles: An introduction to product sociotechnical cycles (PstC) as an alternative for HFE toward sustainability in product design and development. In: Human Factors for Sustainability, pp. 123–143. CRC Press (2020). https://doi.org/10.1201/b22516
20. Lange-Morales, K., García-Acosta, G., Bruder, R.: Conceptual framework for the design and development of sustainability-oriented products: toward EQUID 4.0. Th. Iss. Ergon. Sci. 1–27 (2023). https://doi.org/10.1080/1463922X.2023.2292614
21. Chatzopoulos, A., Tzerachoglou, A., Priniotakis, G., Papoutsidakis, M., Drosos, C., Symeonaki, E.: Using STEM to educate engineers about sustainability: a case study in mechatronics teaching and building a mobile robot using upcycled and cecycled materials. Sustainable **15**, 15187 (2023). https://doi.org/10.3390/su152115187
22. Hartmann, F., Baumgartner, M., Kaltenbrunner, M.: Becoming sustainable, the new frontier in soft robotics. Adv. Mat. **33**(19), 2004413 (2021). https://doi.org/10.1002/adma.202004413
23. Khosravani, M.R., Haghighi, A.: Large-scale automated additive construction: overview, robotic solutions, sustainability, and future prospect. Sustainable **14**, 9782 (2022). https://doi.org/10.3390/su14159782
24. Bi, Z.M., Liu, Y., Baumgartner, B., Culver, E., Sorokin, J.N., Peters, A., O'Shaughnessey, S.: Reusing industrial robots to achieve sustainability in small and medium-sized enterprises (SMEs). Ind. Robot. **42**, 264–273 (2015). https://doi.org/10.1108/IR-12-2014-0441
25. Soori, M., Arezoo, B., Dastres, R.: Optimization of energy consumption in industrial robots, a review. Cogn. Robotics **3**, 142–157 (2023). https://doi.org/10.1016/j.cogr.2023.05.003
26. Song, J., Chen, Y., Luan, F.: Air pollution, water pollution, and robots: Is technology the panacea. J. Env. Manag. **330**, 117170 (2023). https://doi.org/10.1016/j.jenvman.2022.117170
27. Psarommatis, F., Sousa, J., Mendonça, J.P., Kiritsis, D.: Zero-defect manufacturing the approach for higher manufacturing sustainability in the era of industry 4.0: a position paper. Int. J. Prod. Res. **60**, 73–91 (2022). https://doi.org/10.1080/00207543.2021.1987551
28. Lange-Morales, K., Thatcher, A., García-Acosta, G.: Towards a sustainable world through human factors and ergonomics: it is all about values. Ergon. **57**, 1603–1615 (2014). https://doi.org/10.1080/00140139.2014.945495
29. Thatcher, A.: Green ergonomics: definition and scope. Ergon **56**, 389–398 (2013). https://doi.org/10.1080/00140139.2012.718371
30. McDonough, W., Braungart, M.: Cradle to Cradle: Remaking the Way We Make Things. North Point Press (2010)
31. Kellert, S.R., Heerwagen, J., Mador, M.: Biophilic Design: The Theory, Science and Practice of Bringing Buildings to Life. Wiley (2011)

32. Richardson, M., Butler, C.W.: Nature connectedness and biophilic design. Build. Res. Inf. **50**, 36–42 (2022)
33. Richardson, M., Dobson, J., Abson, D.J., Lumber, R., Hunt, A., Young, R., Moorhouse, B.: Applying the pathways to nature connectedness at a societal scale: a leverage points perspective. Ecos. Peop. **16**, 387–401 (2020). https://doi.org/10.1080/26395916.2020.1844296
34. Mauerhofer, V.: 3-D sustainability: an approach for priority setting in situation of conflicting interests towards a sustainable development. Ecol. Econ. **64**, 496–506 (2008). https://doi.org/10.1016/j.ecolecon.2007.09.011
35. Thatcher, A., Nayak, R., Waterson, P.: Human factors and ergonomics systems-based tools for understanding and addressing global problems of the twenty-first century. Ergon **63**, 367–387 (2020). https://doi.org/10.1080/00140139.2019.1646925
36. O'Riordan, T., Cameron, J.: Interpreting the precautionary principle. Routledge (2013). https://doi.org/10.4324/9781315070490
37. Bolis, I., Morioka, S.N., Leite, W.K.D.S., Zambroni-de-Souza, P.C.: Sustainability is all about values: The challenges of considering moral and benefit values in business model decisions. Sustain **13**, 664 (2021). https://doi.org/10.3390/su13020664

GPSR Compliance
The European Union's (EU) General Product Safety Regulation (GPSR) is a set of rules that requires consumer products to be safe and our obligations to ensure this.

If you have any concerns about our products, you can contact us on

ProductSafety@springernature.com

In case Publisher is established outside the EU, the EU authorized representative is:

Springer Nature Customer Service Center GmbH
Europaplatz 3
69115 Heidelberg, Germany

www.ingramcontent.com/pod-product-compliance
Ingram Content Group UK Ltd.
Pitfield, Milton Keynes, MK11 3LW, UK
UKHW020704231026
476908UK00003B/110
* 9 7 8 3 0 3 2 1 9 3 7 3 5 *